AF478439

Morphogenetic Hormones of Arthropods

Morphogenetic Hormones of Arthropods
Volume 1

2 Embryonic and Postembryonic Sources

edited by A. P. GUPTA

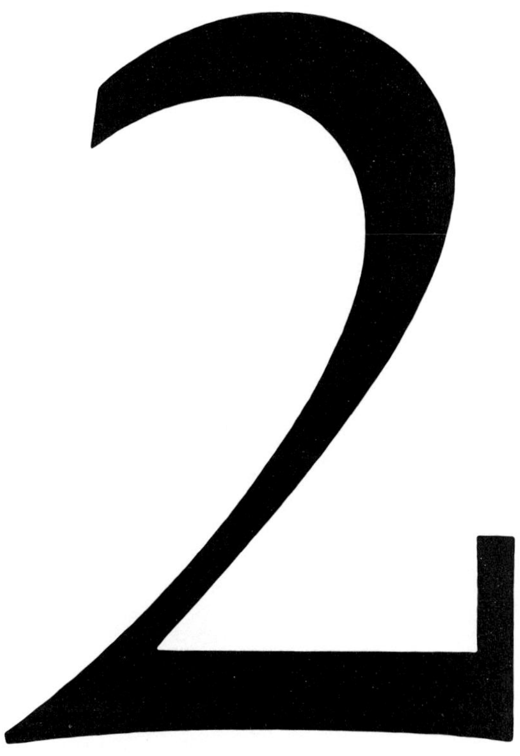

RUTGERS UNIVERSITY PRESS
New Brunswick and London

Library of Congress Cataloging-in-Publication Data

Morphogenetic hormones of arthropods / edited by A. P. Gupta.
    p.   cm.—(Recent advances in comparative arthropod
morphology, physiology, and development)
    Includes bibliographies and indexes.
    Contents: pt. 1. Discoveries, syntheses, metabolism, evolution,
modes of action, and techniques—pt. 2. Embryonic and
postembryonic sources.—pt. 3. Roles in histogenesis, organogenesis, and
morphogenesis
    ISBN 0–8135–1414–2 (pt. 1) ISBN 0–8135–1415–0 (pt. 2)
    1. Arthropoda—Morphogenesis. 2. Juvenile hormones. 3. Hormones.
I. Gupta, A. P., 1928– . II. Series.
    [DNLM: 1. Arthropods. 2. Hormones. 3. Morphogenesis.
    QL 434.7 M871]
QL434.72.M67   1989
595.2'043—dc19
DNLM/DLC
for Library of Congress   89–6029
                                                                 CIP

British Cataloging-in-Publication information available

CONTENTS

PREFACE

This is the second of three parts of Volume 1 on the morphogenetic hormones of arthropods, published as part of an international series entitled *Recent Advances in Comparative Arthropod Morphology, Physiology, and Development*. The first part, subtitled *Discoveries, Syntheses, Metabolism, Evolution, Modes of Action, and Techniques* was published recently and has been well received. As I pointed out in the preface to Part 1, the emphasis in these three parts will be on the morphogenetic hormones in Aquatic and Terrestrial Chelicerata, Crustacea, and Myriapoda. Chapters on Insecta have been included to complement the coverage on other arthropod groups. Although ecdysteroids have been reported from all five major taxa, their glands are largely unknown in many of the subtaxa. The same in true of the glands producing JH-like compounds. Embryonic sources of most of morphogenetic hormones, except in some insects, are unknown. The most conspicuous gaps occur in Myriapoda, Aquatic and Terrestrial Chelicerata, and entomostracan Crustacea. In addition, the roles of androgenic, maxillary, and mandibular glands in some morphogenetic processes are known in only very few species. Clearly, much remains to be discovered, and I hope that this volume will provide impetus and leads for much-needed research in the aforementioned areas.

In any multiauthored book, some overlaps are inevitable, and the series as a whole and Part 2 in particular are no exception. However, overlaps have been kept to a minimum and retained only where necessary. Wherever relevant, such overlaps and divergences of opinions in various chapters have been cross-referenced. Because the subject of taxonomic rankings of major arthropod groups is controversial, each contributor has used his/her preferred taxonomic categories. Furthermore, each contributor has had complete freedom to develop, interpret, and present his/her views. Each chapter presents an in-depth review of the topic it covers. In those chapters by contributors whose native languages are other than English, the editing has retained as much as possible of the original style and flavor, confining itself to essential emendations for the sake of clarity.

The endeavor of organizing such a series could not have been accomplished without the cooperation and assistance of all the contributors who responded to my invitation to participate in this project. To them, I am indebted. To numerous authors, journal publishers, and professional societies who generously and unhesitatingly allowed reproduction of published or unpublished materials in the original or modified forms, I am grateful. And I greatly appreciate the help and cooperation of Karen Reeds, Marilyn Campbell, Barbara Kopel, and Dina Bednarczyk at

Rutgers University Press and the meticulous copyediting by Bert N. Zelman of Publishers Workshop Inc., Brooklyn, New York. As always in the past, the ungrudging support of my wife and children during the preparation of this book has been enormous and has allowed me to devote to this project much time that rightfully belonged to them. To them, I am ever so grateful.

Ayodhya P. Gupta
New Brunswick, New Jersey

CONTRIBUTORS

Jacques Beaulaton
Department of Biology & Animal
    Physiology
University of Antilles & Guyane
P.O. Box 592
97167 Pointe-á-Pitre Cedex
France

J. C. Bonaric
U. S. T. L.
Laboratory of Zoology
Eugéne Bataillon Place
34060 Montpellier
France

Jeannine Caplet
Laboratory of Animal Biology
University of Picardie
Amiens
France

Pierre Cassier
Cytophysiology of Arthropods
University of Pierre and Marie Curie
105 Boulevard Raspail
75006 Paris
France

Michel Descamps
Laboratory of Invertebrate
    Endocrinology
Animal Biology
University of Science & Technology of
    Lille
59655 Villeneuve d'Ascq Cedex
France

August Dorn
Division of Biology
Institute for Zoology
Johannes Gutenberg University
Saar Street 21
6500 Mainz
West Germany

Gertrude W. Hinsch
Department of Biology
University of South Florida
Tampa, FL 33629

Catherine Jamault-Navarro
Laboratory of Animal Biology
University of Picardie
Amiens
France

Christian Juberthie
Laboratoire Souterrain du CNRS
Moulis
09200 Saint Girons
France

Geneviève G. Payen
Laboratory of Reproductive
    Physiology
Division of Crustacean Endocrinology
University of Pierre and Marie Curie
4 Jussieu Place
75252 Paris Cedex 05
France

François Sahli
Laboratory of General & Animal
    Biology
University of Dijon
Dijon
France

Gerhard Seifert
Institute for Special Zoology
Stephan Street 24
Justus Liebig University
D-6300 Giessen
West Germany

Eugene Spaziani
Department of Biology
University of Iowa
Iowa City, Iowa 52242

CONTRIBUTORS

Jacques Beaulaton
Department of Biology & Animal
    Physiology
University of Antilles & Guyane
P.O. Box 592
97167 Pointe-á-Pitre Cedex
France

J. C. Bonaric
U. S. T. L.
Laboratory of Zoology
Eugéne Bataillon Place
34060 Montpellier
France

Jeannine Caplet
Laboratory of Animal Biology
University of Picardie
Amiens
France

Pierre Cassier
Cytophysiology of Arthropods
University of Pierre and Marie Curie
105 Boulevard Raspail
75006 Paris
France

Michel Descamps
Laboratory of Invertebrate
    Endocrinology
Animal Biology
University of Science & Technology of
    Lille
59655 Villeneuve d'Ascq Cedex
France

August Dorn
Division of Biology
Institute for Zoology
Johannes Gutenberg University
Saar Street 21
6500 Mainz
West Germany

Gertrude W. Hinsch
Department of Biology
University of South Florida
Tampa, FL 33629

Catherine Jamault-Navarro
Laboratory of Animal Biology
University of Picardie
Amiens
France

Christian Juberthie
Laboratoire Souterrain du CNRS
Moulis
09200 Saint Girons
France

Geneviève G. Payen
Laboratory of Reproductive
    Physiology
Division of Crustacean Endocrinology
University of Pierre and Marie Curie
4 Jussieu Place
75252 Paris Cedex 05
France

François Sahli
Laboratory of General & Animal
    Biology
University of Dijon
Dijon
France

Gerhard Seifert
Institute for Special Zoology
Stephan Street 24
Justus Liebig University
D-6300 Giessen
West Germany

Eugene Spaziani
Department of Biology
University of Iowa
Iowa City, Iowa 52242

Morphogenetic Hormones of Arthropods

EMBRYONIC SOURCES

Embryonic Sources of Morphogenetic Hormones in Arthropods

1

AUGUST DORN

1.1. Introduction

Since the pioneering work of Kopec (1922), who proposed that the insect brain might have a function similar to that of the vertebrate thyroid in promoting metamorphosis, the hormonal system controlling morphogenetic events in insects has been thoroughly studied and greatly elucidated. The source of hormones, their chemical structure (excepting morphogenetic neurohormones), and their mode of action (concerning ecdysteroids, down to the molecular level) have been clarified at least in insects; considerable information is available also for crustaceans, specifically regarding postembryonic development. Although the most important morphogenetic processes take place during embryogenesis, it was generally believed for a long time that hormones play no role in these events (see Doane, 1973). First reports that ecdysteroids and juvenile hormone are present in insect eggs (see Gilbert and Schneiderman, 1961; Kaplanis et al., 1973, 1975) did not seem to excite any curiosity as to whether such findings had any bearings on the generally accepted concept. Only slowly was a critical mass of information on hormones in the insect embryo compiled (Hoffmann et al., 1980; Dorn, 1972, 1983), and then suddenly interest in the field was aroused, which is still continuing. Unfortunately, the excitement has hardly spread to other arthropod groups; for example, our knowledge of the embryonic hormones in crustaceans and chelicerates is limited (only ticks have aroused interest), and it is virtually absent in myriapods. The following discussion will present a brief overview of the morphogenetic hormones in arthropods and a detailed account of their embryonic origins in various major groups.

1.2. Morphogenetic Hormones and Their Production Sites in Arthropods: An Overview

The insects exhibit three systems of morphogenetic hormones that interact in directing the molting process: neurohormones, ecdysteroids, and juvenile hormones (JHs) (Fig. 1.1). Neurohormones have a superior role. One factor, PTTH (prothoracicotropic hormone), stimulates the prothoracic glands (PTGs) to synthesize and release ecdysone. Probably two brain factors, ATH (allatotropin) and allatostatin respectively activate and inhibit corpus allatum function, which is responsible for the production and release of the JH(s). Other morphogenetic neurohormones are the eclosion hormone, which dictates the time of eclosion, and the diapause hormone of *Bombyx mori*, which determines embryonic diapause. Additional morphogenetic neurohormones may well be detected in the future.

The questions arise as to whether all arthropods operate with the same hormonal inventory and whether they all have homologous endocrine glands.

The presence of ecdysteroids has been demonstrated in all arthropod groups including Pantopoda (Bückmann et al., 1986) and other taxa as well. For example, ecdysteroids were identified in Mollusca (Takemoto et al., 1967; Romer, 1979; Whitehead and Sellheyer, 1982), Annelida (Sauber et al., 1983; Welter et al., 1986), Nemathelminthes (Horn et al., 1974; Dennis, 1977; Mendis et al., 1983; Nirde et al., 1983), Plathelminthes (Mendis et al., 1984), and Coelenterata (Sturaro et al., 1982; Guerriero and Pietra, 1985). It is not clear, however, whether the synthesizing tissues are homologous in all arthropod groups; in fact, the ecdysone source is not established beyond doubt in Chelicerata and Myriapoda.

JH was believed to be exclusive to insects, but at least some crustaceans (Borst et al., 1987; Laufer et al., 1987a) and probably chelicerates have it too (Pound and Oliver, Jr., 1979). In myriapods, apparently little attempt has been made to identify JH or to examine an influence of juvenoids on morphogenesis.

The uncertainty concerning the identity of endocrine glands outside insects and crustaceans makes it very difficult to search for tropic or static neurohormones in chelicerates and myriapods. In some instances, ligation experiments have been carried out, but with contradictory results.

The neuroendocrine systems controlling postembryonic development in the different arthropod groups—in some instances still putative—are summarized in Figs. 1.1–1.4. [*Note:* All figures are grouped at the end of the text, before the reference list, in this chapter.]

1.3. Morphogenetic Hormones and Their Embryonic Sources in Arthropods

1.3.1. *Insecta*

1.3.1.1. ECDYSTEROIDS: DIFFERENTIATION OF PROTHORACIC GLANDS

Titers and Metabolism

Mueller (1962, 1963) and Boohar and Bucklin (1963) were the first to demonstrate molting hormone activity in the eggs of *Melanoplus differentialis*. Mueller tested whole grasshopper embryos by the spermatocyst assay (Schmidt and Williams, 1953), whereas Boohar and Bucklin used the *Calliphora* assay to probe the embryo and the yolk separately. They found that ecdysone activity was present primarily in the yolk and that this activity remained relatively constant regardless of the stage of development. Yolk-free embryos showed a low but detectable level of activity, which was not affected by the termination of diapause. Mueller (1962) proposed the following: "At the blastoderm stage, ecdysone is produced by each cell in amounts adequate for its needs. As development progresses certain tissues become dependent on the hormone, whereas each epidermal cell retains the ability to produce hormone in excess of its needs and releases the oversupply." *M. differentialis* undergoes embryonic molts. Mueller (1962) suggested that the timing of the molts and the nature of the new cuticle are determined by the state of differentiation of the epidermal cells. In fine experimental work, including separation of metathoracic legs from neuroendocrine centers and culturing them *in vitro*, she demonstrated that neither the brain nor the prothoracic glands (PTGs) are needed for cuticle synthesis and apolysis by the explants. These results, and others achieved afterward (Takami, 1963; Sbrenna Micciarelli and Sbrenna, 1972; Cavallin and Fournier, 1981), conflicted with the existing hypothesis of Jones (1953), the father of embryonic insect endocrinology. His studies (1953, 1956a,b) had established the concept that the direction of embryonic molts involves the activity of neurosecretory centers, which stimulate the PTGs. But the synthesis of ecdysteroids by the glands was not demonstrated.

Subsequently, a large number of insect eggs was checked for ecdysteroid activity. First, crude extracts and *Calliphora* or *Musca* assays were applied (see Dorn and Romer, 1976); later, advanced separation methods [thin-layer chromatography (TLC), high-pressure liquid chromatography (HPLC), or countercurrent distribution] were combined with bioassay, radioimmunoassay (RIA), mass spectrometry (MS), mass fragmentometry, or nuclear magnetic resonance (NMR) to establish the identity and quantity of the ecdysteroid(s) present. Practically all eggs examined contained ecdysteroids (see Hoffmann and Lagueux, 1985). Certainly, most astonishing was the fact that already newly deposited eggs contained considerable amounts of ecdysteroids, indicating a maternal supply. Since PTGs, the "classical" source of ecdysone, degenerate during metamorphosis, the question arose: What is the site of synthesis in the reproducing female? Hagedorn and coworkers (1975) found the ovary in *Aedes aegypti* to be the most probable site of origin. This was then experimentally proved in a large number of insect species (in *Locusta migratoria* by Lagueux et al., 1977; Charlet et al., 1979; in *Galleria mellonella* by Bollenbacher et al., 1978; in *Aedes aegypti* by Hagedorn et al, 1979; in *Aedes atropalpus* by Masler et al., 1980; in *Drosophila melanogaster* by Rubenstein et al., 1982; in *Nauphoeta cinerea* by Zhu et al., 1983). In *L. migratoria* and *N. cinerea*, the hormone is synthesized by the follicle cells of mature eggs and, at least in *Locusta*, is almost quantitatively incorporated into the oocyte (Goltzené et al., 1978; Glass et al., 1978). According to Lagueux et al. (1981), the ecdysteroids of the newly deposited eggs are bound to vitellogenin in *L. migratoria*.

As a rule, the newly deposited eggs contain the ecdysteroids primarily in a biologically inactive form, namely, as conjugates (Table 1.1). In larvae and imagoes, ecdysteroid conjugates are known as excretion forms; in

TABLE 1.1. Ecdysteroid Conjugates Isolated
from Arthropod Eggs (Incomplete)[a]

Insecta
  Ecdysone
    [N^6](Isopentenyl)adenosinemonophosphate [C-22]
      (1)
    Palmitate (2)
    Oleate (2)
    Stearate (2)
    Linoleate (2)
    Acyl esters [C-22] (3)
  26-Hydroxyecdysone
    Phosphate [C-26] (4)
    β-D-Glucopyranoside (4)
  2-Deoxyecdysone
    Adenosinemonophosphate [C-22] (1)
    Phosphate (1)

Chelicerata (ticks)
  Ecdysone
    Stearate [C-22] (5)
    Oleate [C-22] (5)
    Linoleate [C-22] (5)
    Palmitate [C-22] (6)
  20-Hydroxyecdysone
    Palmitate [C-22] (5)
    Stearate [C-22] (5)
    Oleate [C-22] (5)
    Linoleate [C-22] (5)

[a]In brackets: site of conjugation (when known). In parentheses: reference(s)—(1) Hoffmann and Lagueux, 1985; (2) Slinger et al., 1986; (3) Whiting and Dinan, 1987; (4) Warren et al., 1986; (5) Connat et al., 1987b; (6) Crosby et al., 1987.

eggs, they apparently represent "storage" forms. Figure 1.5 and Table 1.1 show the great variety of ecdysteroids isolated from eggs of different insect species. Note that putative precursors of ecdysone and 20-hydroxyecdysone also may be present, as well as other metabolites. In some species (e.g., *Oncopeltus fasciatus* and *Manduca sexta*), ecdysone and 20-hydroxyecdysone are absent or close to detection level (Kaplanis et al., 1973, 1975; Warren et al., 1986). In *O. fasciatus*, makisterone A probably has a biological function (Dorn, 1983); in *M. sexta*, conceivably 26-hydroxyecdysone and 20,26-dihydroxyecdysone do (Warren et al., 1986).

Hydrolyzing enzymes that cleave the maternal ecdysteroid conjugates appear early in embryonic development, generating free and supposedly biologically active hormone or precursors, which may be transformed into the active principle. Figure 1.6 shows a hypothetical pathway of synthesis and metabolism of ecdysone and 20-hydroxyecdysone. It is constructed on the basis of observations in eggs from different species. Quite different situations can be met in some species, where the predominant ecdysteroid is divergent from ecdysone or 20-hydroxyecdysone, as pointed out above. Figure 1.7 reveals the ecdysteroids present in 48-h embryos of *M. sexta* identified and quantified by IS-rpHPLC/RIA (infrared spectroscopy–reversed-phase high-pressure liquid chromatography/ radioimmunoassay).

In *M. sexta* (Warren et al., 1986), as in some other insect eggs, titer fluctuations of RIA-active ecdysteroids during embryogenesis could be explained by deconjugation and conjugation of maternal "storage" hormone and maternal precursor molecules. The necessity of a *de novo* synthesis by the embryo, therefore, does not seem to exist. However, there are reports that the amount of maternal ecdysteroids (inclusive ecdysteroid conjugates) are low in eggs of some species: *Nauphoeta cinerea* (Imboden and Lanzrein, 1982), *Leucophaea maderae* (Matz, 1980), and *Clitumnus extradentatus* (Cavallin and Fournier, 1981). Ecdysteroid titers rise in these cases during late embryogenesis. If these findings are correct, a *de novo* ecdysteroid synthesis must take place. The most likely source, of course, would be the PTGs or their analogous glands (ventral glands, ring glands). But other tissues must be considered too. The following questions must be answered in this context: When do the PTGs differentiate? Are there histological indications of a secretory activity? Is there experimental proof of ecdysteroid synthesis by embryonic PTGs? Is there evidence of embryonic ecdysteroid synthesis outside the PTGs?

Site(s) of Synthesis

The PTGs are the source (but probably not the only one) of ecdysteroids during larval development and therefore constitute the primary candidate for the origin of embryonic ecdysteroids. The embryonic origin of the gland is pinpointed to allow an estimation as to when hormone synthesis can be expected. Although there are some conflicting data on the location and mode of PTG differentiation in the older literature (see Dorn, 1972), there is much more agreement in recently published papers (see Rempel et al., 1977). The example presented here to illustrate PTG ontogenesis, *O. fasciatus*, reflects the most common mode of differentiation.

Prior to katatrepsis and dorsal closure, ectodermal cells of the labial segment invaginate from the lateral body wall (Figs. 1.11 and 1.12). These anlagen of the PTGs have, for a short time, a tubular organization. Then the lumen disappears and the growing anlagen come in contact with the rudiments of the salivary gland, which also originate from the labial segment but represent medial invaginations. A major branch of the prothoracic tracheal ingrowth heads craniad and comes alongside the PTG, which adopt a string-like organization in *O. fasciatus* (Fig. 1.16). Differentiation and early growth processes are virtually identical in *M. sexta* (Fig. 1.17) and the beetle *Lytta viridana* (Rempel et al., 1977). After blastokinesis, the PTGs have acquired their definite location and shape (Fig. 1.16). Whether or not they start ecdysone production is controversial.

In early papers (see Dorn, 1972), cyclic cytological changes, correlated with embryonic molts, were interpreted as activity cycles. Experimental work, however, made clear that embryonic molting processes take place also in the absence of PTGs (see above, this section). This, of course, does not rule out an ecdysteroid synthesis of the embryonic PTGs anyway. In three species, the embryonic PTGs were analyzed by electron microscopy in order to obtain signs of activity. In *O. fasciatus*, the PTGs were judged inactive throughout embryonic development on grounds of fine structural analysis (Dorn and Romer, 1976) (Fig. 1.18A). The glands convey a nonspecialized impression with poorly developed organelles, presumably involved in synthetic processes. Sbrenna et al. (1983) published an abstract on the fine structure of PTGs in the embryo of *Schistocerca gregaria*. They concluded that PTGs "at 27–28th stage of embryonic development display a well-marked pattern of endocrine activity, which correlates with ecdysteroids detected by Scalia and Morgan (1982) and embryonic cuticulogenesis" (Sbrenna, 1974; Scalia et al., 1987). Haget et al. (1982) most thoroughly studied *Carausius morosus* and postulated two activity cycles: the first phase supposedly occurs shortly after ventral (= ecdysial) gland appearance and during deposition of the second embryonic cuticle, when dorsal closure takes place (Fig. 1.18B); the second phase is correlated with the third embryonic (= larval) cuticle formation. Certain signs of ecdysteroid synthesis were pointed out: increase in rough ER (endoplasmic reticulum) with some concentric profiles and enlarged tubular cisternae with slightly electron-dense content; increasing numbers of enlarged mitochondria with clear matrices; and distinct cell surface enlargement by invaginations. During the first suspected phase of activity, the volume of the cytoplasm is small in relation to the nucleus and includes lipid droplets, fields of glycogen, and "more or less degranulated canals of the rough ER" (Fig. 1.18B). These characteristics change with maturation of the glands, and lipid droplets and glycogen disappear gradually (Haget et al., 1982).

It is clear that cytological as well as histological characteristics can never serve as a reliable parameter of glandular activity. They may suggest physiological processes, such as hormone production, and are in this respect legitimate and valuable, but must eventually be supported by experimental physicochemical data. From the view of a comparative embryologist, of course, formation and differentiation of the endocrine glands is a fascinating problem of itself.

Clearly, there is no experimental proof for ecdysteroid synthesis by embryonic PTGs in any species. It has been suggested that other tissues, i.e., oenocytes, may produce ecdysteroids during embryogenesis (Dorn and Romer, 1976). Although there is some evidence of an ecdysone synthesis by these cells in postembryonic development (Romer et al., 1974), there are no experimental data concerning embryos. But it should be mentioned that oenocytes differentiate early in embryonic development, become polyploid, and exhibit fine structural features typical of larval organization. Novak and Zambre (1974) suggested the pleuropodia as source of embryonic ecdysteroids. Pleuropodia are ectodermal glands developing from appendages of the first abdominal segment (Dorn and Hoffmann, 1983). According to Slifer (1937), they release a "hatching enzyme" that dissolves embryonic membranes and facilitates hatching. The gland degenerates shortly before or after hatching. The idea that pleuropodia produce molting hormone was not supported by other studies. It has still not been ruled out that the epidermis itself might manufacture ecdysteroids, as was suggested by Mueller (1962).

1.3.1.2.   MORPHOGENETIC NEUROHORMONES: DIFFERENTIATION OF NSCS AND NEUROHEMAL ORGANS

Presence of Neurohormones (PTTH, Allatotropin, Allatostatin, Eclosion Hormone, Diapause Hormone)

PTTH.   PTTH is the best studied morphogenetic neurohormone; nonetheless, first reports on its occurrence in the egg surfaced only in 1986 and 1987. It was found in *Bombyx mori* (Chen et al., 1986) and *M. sexta* (Dorn et al., 1978a). Apparently, there exist two forms with different molecular weights in both lepidopterans. The two PTTHs in *B. mori* also differ in their biological activity: the larger one (20–30 kD) evokes adult molt in

A. Dorn

debrained *B. mori* pupae exclusively (this is the used bioassay), whereas the smaller one (4–5 kD) is only effective in debrained *Samia cynthia ricini* pupae. The larger PTTH was, therefore, termed PTTH-B; the smaller one, PTTH-S (Ishizaki et al., 1983; Fugo et al., 1987). To determine their occurrence in developing embryos, eggs of different stages were extracted according to the protocol shown (Fig. 1.19); a 300-fold purification was achieved for PTTH-B, and 75-fold for PTTH-S. The extracts were subjected to the *Bombyx* and *Samia* bioassays, and the hormone content expressed in PTTH-B and PTTH-S units, respectively (1 unit is designated as the minimal amount, which elicits adult development in more than 50% of the brainless assay pupae).

The search for PTTH-B was unsuccessful in newly deposited and diapausing eggs, and it first became detectable in embryos that had developed almost to the stage of differentiation of their neuroendocrine system, i.e., after blastokinesis and differentiation of the central nervous system. The highest levels were determined immediately before hatching (Chen et al., 1987). The authors consider the possibility that PTTH-B activates the embryonic PTGs to synthesize the ecdysteroids found in later stages of embryogenesis. The source of embryonic PTTH-B is not known.

PTTH-S, in contrast to PTTH-B, was found in newly laid and diapausing eggs of *B. mori* and even in developing ovaries 1 day before adult molt (Fugo et al., 1987). The hormone titer remained about constant from day 0 to day 7 postoviposition, which is about 2 days after differentiation of endocrine glands, and rises sharply until hatching (between day 8 and day 9). Although the authors suggest that PTTH-S in the late development might be synthesized in the embryo's neuroendocrine system, the actual source remains obscure.

PTTH-S obtained from adult heads of *B. mori* has a significant similarity to insulin and insulin-like growth factors in terms of amino acid sequence (Nagasawa et al., 1984). Fugo and coworkers (1987), therefore, speculate that PTTH-S present during early embryogenesis may act as a growth factor. Whether or not PTTH-B activates the embryonic PTG is unclear. Experimental support, such as challenging the inactive PTGs *in vitro*, is missing.

Precisely this approach has been chosen to test the presence of PTTH in the embryo of *M. sexta* (Dorn et al., 1987a). Hormone extracts were made from isolated embryonic brains and head segments of early stages (germ band stage). The head extracts were bioassayed for PTTH activity using the *Manduca* PTG *in vitro* regimen (Bollenbacher et al., 1979). This *in vitro* assay is based on a dose–response correlation of PTG activation by PTTH, and the data yield an activation ratio (A_r). The A_r is the ratio of the amount of ecdysone synthesized and secreted by a PTG in the presence of PTTH to that secreted by the absence of PTTH, i.e., the basal level. The ED_{50} of an extract is the effective dose in brain equivalents necessary to achieve a response equal to 50% of the maximum A_r. The amount of PTTH in an extract can be determined from ED_{50} (Fig. 1.20).

The *in vitro* assay was for 2 h at 25°C, and at the end of the incubation period the ecdysteroid content of the culture medium was determined by RIA as described previously (Warren et al., 1984). Larval (gate II, day 3, fifth instar) and pupal (day 0) PTGs were utilized in the assay. Since ecdysteroids, which may have been present in the extracts, would disturb the assay, care was taken to determine their level and exclude any side effects.

Brain extracts from embryos just prior to hatching yielded a sigmoidal dose–response curve when assayed on larval and pupal PTGs that is representative of PTTH action (Bollenbacher et al., 1979) (Fig. 1.20). The larval glands had a 10-fold greater sensitivity to the embryonic brain extract than the pupal glands. The differential effect of embryonic PTTH activity on larval and pupal PTGs has also been observed for the PTTHs of pupal brains (Bollenbacher et al., 1984). The basis of this difference in sensitivity to PTTH could be functional property of the glands or may be due to different ratios of big and small PTTH in brains at particular developmental stages (O'Brien et al., 1986). PTTH activity was found in all five developmental stages studied: 24, 45, 72–75, 96–99, and 117 h after oviposition (Table 1.2) (Dorn et al., 1987a). From 72 h till hatching (around 117 h), the titer rises drastically. Surprisingly, less developed brains and head segments (45 and 24 h, respectively) also revealed PTTH activity, comparable to the presence of PTTH-S in early stages of *Bombyx* development (Fugo et al., 1987).

The factor with PTTH activity in the brain of *M. sexta* embryos appears to be similar or identical to the PTTH present in the postembryonic stages (see Bollenbacher and Granger, 1985), but proof of this will require chemical verification. Support for the presence of big PTTH in *Manduca* embryos comes from recent immunocytochemical studies using antibodies against this PTTH, which showed that the neurohormone, or an immunochemically similar molecule, is present in 24- and 48-h embryos (Westabrook and Bollenbacher, unpublished information). Indeed, at 48 h, the major axon tracts and dendritic arbors in the protocerebral neuropil are evident. Taken together, the data strongly suggest the presence of PTTH in the embryonic head and brain of *Manduca*. These findings are consistent with the appearance of NSCs in the embryonic brain.

Allatotropin and allatostatin.   Brain factors respectively stimulating (allatotropin) and inhibiting (allatostatin) the corpora allata (CA) have been revealed by bioassays in a number of insects (Tobe and Stay, 1980; Bhaskaran et al., 1980; see De Kort and Granger, 1981, and Raabe, 1982,

TABLE 1.2.   Prothoracicotropic Hormone Activity in Extracts of Brains or Head Portions of Embryos of *Manduca sexta*

| | PTTH activity | | | | |
| | *Larval prothoracic glands* | | | *Pupal prothoracic glands* | |
Embryonic stage (h)	ED_{50}[a] (HEAD OR BRAIN EQUIVALENTS)	RELATIVE CONTENT[b]		ED_{50} (HEAD OR BRAIN EQUIVALENT)	RELATIVE CONTENT
24	>15.75 (2)	<0.17		—	—
45	12.2 (2)	0.24		—	—
72–75	>22.5 (2)	<0.12		118 (2)	0.27
96–99	9.8 (2)	0.29		70 (2)	0.33
117	2.9 (3)	1.0		32.5 (2)	1.0

SOURCE: From Dorn et al. (1987a).

[a] An ED_{50} is derived from a dose–response of prothoracic gland activation and is the number of tissue equivalents necessary to elicit half maximal stimulation.

[b] The relative content was obtained by arbitrarily setting the highest reciprocal of the ED_{50} at 1.0 (117-h value) and converting all other reciprocals of ED_{50} values relative to the 117-h values.

for reviews). These factors are presently being characterized chemically. It is not yet clear whether the factors reach their target organ, the CA, via hemolymph or via direct innervation as suggested by histological observations (Sedlak, 1985). The micrographs on embryonic CA of *Carausius morosus* (Fig. 1.25A,B) prove a direct neurosecretory innervation by two types of neurosecretory nerve fibers. The functional significance of these nerve fibers remains obscure, however, since direct experimental evidence is lacking.

Eclosion hormone.   Eclosion hormone is a neurosecretory peptide that was first found in Lepidoptera. It triggers adult ecdysis by acting on the cerebral nervous system to elicit the ecdysial motor programs and also causes other physiological changes associated with ecdysis (Truman and Riddiford, 1970). Recently, Truman and coworkers (1981) showed that eclosion hormone may control all ecdyses in insects, including embryonic molts; eclosion hormone was demonstrated in the *Hyalophora cecropia* embryo by bioassay. Preliminary experiments indicate that eclosion hormone is present mainly in the head region and appears on day 5 of embryonic development. The titer remains high for the next 2 days but then drops sharply between days 7 and 8, and remains low until the time of hatching at day 10. This drop in stored hormone coincides with the ecdysis of an embryonic cuticle by the developing first-instar larva within the egg (Truman et al., 1981).

Eclosion hormone was also demonstrated in developing embryos of *B. mori* by a bioassay (Fugo et al., 1985). The extraction procedure is shown in Fig. 1.19. A 2100-fold purification was achieved, and the molecular weight of the hormone was estimated to be 7–9 kD by gel filtration on Sephadex G-50. The assay was performed by a whole-insect bioassay: male pharate adults of *Bombyx*, which had already completed adult development and awaited the eclosion gate (controlled by a circadian oscillator) in 9 h, were injected with the extract. In case of eclosion hormone activity, adult ecdysis was precociously elicited 2.5 h after injection. No eclosion hormone activity was detected during early embryogenesis, but first became detectable at the stage of the differentiation of the neuroendocrine system, i.e., at day 6 (hatching occurs at day 10). Then the titer rose dramatically, reaching a maximum 1 day before hatching, whereupon it decreased. In contrast to Truman et al. (1981), Fugo and coworkers (1985) believe that eclosion hormone does not induce embryonic ecdyses; they suggest that it is required for hatching.

The site of eclosion hormone synthesis in the embryo is supposedly the brain, since the hormone has only been detected in head extracts of *Hyalophora* (Truman et al., 1981); postembryonic stages exhibit higher contents in the ventral nerve cord (Taghert et al., 1980). Fugo et al. (1985) extracted whole embryos. Thus, no conclusion as to the source of the hormone can be drawn.

Embryonic diapause and neurosecretion: diapause hormone. Embryonic diapause may occur during early or late embryonic development. In the latter case, endocrine glands are differentiated and a diapause regulation similar to that in pupae seems possible [pupal diapause is initiated in the absence of ecdysteroids; PTGs remain inactive since PTTH although stored in neurosecretory cells (NSCs) is not released into the hemolymph]. Thus in *Aulocara elliotti*, the obligatory diapause is thought to be the consequence of hormonal deficiency, which probably results from lack of neurosecretory stimulus from the brain–corpus cardiacum complex (Neumann-Visscher, 1976). In *Lymantria dispar*, brain NSCs become increasingly stainable with PAF (paraldehyde fuchsin) when diapausing embryos are chilled (Loeb and Hayes, 1980). After 90 days of chilling, NSCs appear packed with neurosecretions, which are apparently transported along the axons. Return of the chilled eggs to 24°C after 120 days at

4°C induces increased transport and consequent loss of PAF-positive substances; so just prior to hatching, little PAF-positive material remains in the brain (Loeb and Hayes, 1980). The authors point out that similar neurosecretory processes take place in lepidopterans, which diapause at more mature stages. It remains to be seen, however, if the neurosecretions accumulated during embryonic diapause indeed represent PTTH, as is implied.

Diapause in *B. mori* occurs at an early embryonic stage and is regulated by the mother's endocrine system (for review see Yamashita and Hasegawa, 1985). The subesophageal ganglion produces a diapause hormone that programs the maturing eggs. Although eggs are programmed at oviposition, it was recently demonstrated that diapause hormone also enters the egg (Kai, 1977). An extraction procedure including a pretreatment with Triton X-100 produced positive results, inasmuch as hormone activity was detected in diapause ovaries and eggs. The extracts induced the deposition of diapause eggs when injected into pupae of polyvoltine race or into pupae with the subesophageal ganglion removed (Kai and Kawai, 1981). Kai and Kawai suggest that diapause hormone might not only act on ovary during the pupal stage to start diapause but could regulate the onset of diapause in laid eggs and also the maintenance of diapause. There is no indication that diapause hormone might be produced during embryogenesis of *B. mori*, in contrast to hormones which are probably involved in direction of diapause during late embryonic stages (see above.)

Differentiation of NSCs and Neurohemal Organs

Neurosecretion in the brain. When we were screening the embryonic brains of *O. fasciatus* and *Carausius morosus* for NSCs (Dorn, 1975a,b, 1985), we could only detect granule-bearing perikarya in the latter species and at rather late stages, i.e., during synthesis of larval (= third embryonic) cuticle. Six to eight perikarya in the pars intercerebralis region contained scattered secretion products (Figs. 1.21A,B). The granules are formed by the Golgi complex and travel via nervi corporis cardiaci to the retrocerebral complex. Morphological features and pathways of the granules prove them to be neurosecretory products, presumably peptidergic in nature. All neurosecretory perikarya of the pars intercerebralis, in which granules accumulate, have the same morphological features, and in particular all granules have the same general appearance: they are round, membrane bounded, with an electron-dense content and a maximal diameter of about 180 nm. Axons projecting from nervi corporis cardiaci into the corpora cardiaca (CC) and CA (Figs. 1.23A and 1.25A) comprise granules of the same type. The neuropil of the pars intercerebralis region and the extrinsic parts of the CC include axons with still another type of neurosecretions (Figs. 1.21B and 1.23A). The granules of these axons are also round, but much smaller than the others, up to 90 nm in diameter. Their content is somewhat withdrawn from the limiting membrane. They are reminiscent of aminergic granules. The perikarya of these axons could not be detected but are presumably located within the brain. This type of neurosecretory axons makes many synaptoid contacts with conventional nerve fibers. They are also found in the CA (Fig. 1.25B).

No signs of granule release could be detected in extrinsic nerve fibers in the CC. But the release processes are presumably very rapid and nerve endings rather rare in the embryonic neurohemal organs, which may account for the failure to detect exocytosis figures or synaptoids (Scharrer, 1983).

In *O. fasciatus*, no NSCs were detected in the brain at any time of embryogenesis. However, a few nerve fibers in the neuropil of the brain contained neurosecretory granules. These nerve fibers left the brain in the nervi corporis cardiaci and branched in the CC and the aortal wall, the neurohemal organ of the hemipterans. Although the origin of these axons could not be traced, it seems highly probable that the appertaining perikarya are located in the brain. All profiles of neurosecretory nerve fibers in the brain and retrocerebral complex contained granules of the same type, electron-dense, with a diameter up to 130 nm. The morphological appearance is different from all types observed in the imago (Unnithan et al., 1971; Unnithan and Nair, 1977; Dorn, 1978a). Neurosecretory fibers were first observed 96 h after oviposition, which is approximately at onset of secretion of larval (= third embryonic) cuticle, in nerve endings of nervi corporis cardiaci in the aortal wall. Neurosecretion seemed to increase toward hatching time. Signs of release were never observed.

Neurosecretion in the CC.   The CC constitute another neurosecretory center. They are believed to represent transformed visceral ganglia, differentiating before katatrepsis (Figs. 1.9, 1.10, 1.13, 1.14, and 1.22A). In *Oncopeltus* three cell clusters evaginate from the dorsal roof of the stomodaeum. The unpaired median cells represent the anlagen of the hypocerebral ganglion, whereas the cell clumps at both sides of it represent the anlagen of the CC (Figs. 1.9, 1.10, and 1.22A). During dorsal closure, the anlagen migrate to their definitive location (Figs. 1.13, 1.14, and 1.15).

In the imago of *O. fasciatus*, the CC include three different NSC types, in *Carausius morosus* at least two (Dorn, 1975b). In the embryo of *Oncopeltus*, first tangible signs of neurosecretion were observed 110 h after oviposition, i.e., after the second embryonic molt, and 13 h before hatching (Fig. 1.22B).

The neurosecretory granules are formed by the Golgi complex and accumulate shortly before hatching. Immediately before hatching, omega-shaped indentations of the cell membrane indicate release of the granules. All secretory cells are of the same type; the granules are round in profile and electron-dense. At the beginning of synthesis in the 110-h embryo of *Oncopeltus*, they are smaller (diameter ~70 nm) than in older embryos (diameter up to 150 nm at hatching time). The granule type does not correspond with any of those found in imaginal CC.

Quite similar observations have been made in the CC of *C. morosus* shortly before hatching. The intrinsic cells, besides the glial cells, produce large granules (up to 250 nm in diameter) (Fig. 1.23A) that are pinched off from the Golgi complex. The secretions are apparently synthesized in cysternae of rough ER, as in the CC of *Oncopeltus*. The nuclei of glandular cells are large and cytoplasm is scant, in contrast with the situation in imagoes (Dorn, 1978a). Cell projections, typical for CC cells in adult insects, are not yet formed. The intrinsic cells surround extensions of the nervus corporis cardiacum (extrinsic part of CC), which form a compact mass of three types of nerve fibers: (1) conventional axons; (2) axons with neurosecretory granules presumably originating in the pars intercerebralis; and (3) axons carrying small granules (up to 90 nm in diameter), intermingled with vesicles of the size of synaptic vesicles (Fig. 1.23A). The origin of the third type of axons is obscure. Similar nerve fibers are seen in the neuropil of the brain (Fig. 1.21B).

Neurosecretion in the CA. In many species, the CA are innervated by neurosecretory nerve fibers, branches of the nervus corporis cardiacum. In *M. sexta*, they even represent a neurohemal organ (Agui et al., 1980). In the imago of *Oncopeltus*, different types of neurosecretory nerve endings are in close contact with the gland cells and form synaptoid structures (Dorn, 1973). The role of this innervation is not yet clear. An allatotropic and/or allatostatic function of one or several of the neurosecretions is probable. The CA in embryos of *Carausius* contain neurosecretory axons during deposition of the larval cuticle (Fig. 1.25A,B). Two types of axons can be seen, identical with the two types described in the nervus corporis cardiacum. The fine structure of the CA cells indicates a glandular activity (Haget et al., 1981), which is also expected from embryonic CA in *Oncopeltus* (Dorn, 1975c). Synthetic activity of embryonic CA have previously been proved by biochemical methods in the case of *Nauphoeta cinerea* (Lanzrein et al., 1984).

Neurosecretion in the stomatogastric nervous system. The occurrence of NSCs in the stomatogastric nervous system is reported in Lepidoptera (Borg et al., 1973; Yin and Chippendale, 1975) and Diptera (Tombes and

Malone, 1977). In other orders, there are apparently no NSCs, but the neuropil of the ganglia contains numerous neurosecretory axons (Dorn, 1978b; Ude et al., 1978) whose origin is largely unknown. In the case of the frontal ganglion of *Carausius*, such axons may come from the brain via the frontal connectives (Dorn, 1978b). Three types of neurosecretory axons can be distinguished in the imago. The embryonic frontal ganglion in this species also shows many neurosecretory fibers (Fig. 1.23B). These, however, are all of the same type. The fine structural characteristics of the granules are identical with those observed in the neuropil of the brain and in the nervus corporis cardiacum. Most conspicuous are the numerous synaptoid structures facing intercellular spaces, glial elements, nerve fibers of the conventional type, or other neurosecretory fibers. It cannot yet be decided whether the neurosecretions should be viewed as neurohormones, neurotransmitters, or neuromodulators.

1.3.1.3. JUVENILE HORMONES: DIFFERENTIATION OF CORPORA ALLATA

Titers, Metabolism and JH Synthesis

Gilbert and Schneiderman (1961) were the first to report the presence of JH in eggs of *Hyalophora*. They found low amounts with a bioassay in newly deposited eggs and an increase during late development. When the CA were extirpated in females, JH was no longer detectable in newly deposited eggs, but embryogenesis was not disturbed, and the titer rose to normal values in late development when CA were differentiated. The authors concluded that JH enters the oocyte by accident with yolk lipids (and has no biological function); later it is supposedly synthesized by the embryonic CA (and may play a role). Implantation experiments with *Oncopeltus* embryos of different ages, carried out by Novak (1951) also hinted a CA activity during the last third of embryogenesis.

A series of papers appeared in the late 1960s and the 1970s dealing with the effects of JH and JH analogues applied to insect eggs (Slama and Williams, 1966; Riddiford and Williams, 1967; Novak, 1969; Riddiford, 1969, 1970; Hunt and Shapiro, 1973; Rohdendorf and Sehnal, 1973; Enslee and Riddiford, 1977; Injeyan et al., 1979). The aim of these experiments was either to discover JH-sensitive morphogenetic processes (which could be governed by endogenous JH) or to pinpoint ovicidal activity (of interest in the context of pest management). Although some interesting results and hypotheses concerning CA function surfaced from this work, proof of the presence of JH during embryogenesis could not be provided by this approach.

JH titer measurements with bioassays came slowly. Using the *Tenebrio* assay, Dorn (1975c) obtained positive results in newly deposited eggs and extremely high values during cuticle deposition in late development of *Oncopeltus*. Since then, Schooley and coworkers have tried to isolate and identify JH in eggs and postembryonic stages of *Oncopeltus*, but failed to determine any of the known JHs and related compounds at any stage of the life cycle, although highly sophisticated physicochemical methods were applied, including gas chromatography/mass spectrometry (GC/MS) with selected ion monitoring (Bergot et al., 1981b; Baker et al., 1988). The discrepancy between the findings of Dorn (1975c), Novak (1951) (see above), Rankin and Riddiford (1978) (the latter found high JH activity in adult *Oncopeltus* using JH deficient *M. sexta* for bioassay), and those of Baker et al. (1988) cannot be explained at present. Great differences between JH activities determined with bioassays and physicochemical methods are also known from other species, e.g., *Macrotermes subhyalinus* (see Lanzrein et al., 1984), *L. migratoria* (Roussel and Schneider, 1979; Roussel and Aubry, 1981; Temin et al., 1986), or *Hyalophora cecropia* (Gilbert and Schneiderman, 1961; Bergot et al., 1981a). There are several possibilities for these discrepancies: unreliability of the used bioassay; presence of compounds that mimic JH in the bioassay; presence of JH forms that are active in the bioassay but are lost in the physicochemical preparation or were not monitored because they are unknown.

Also, relatively high amounts of JH were observed in the embryo of *Nauphoeta cinerea* with the *Galleria* wax test (Imboden et al., 1978). Whereas JH was not demonstrable before dorsal closure, it rose steeply thereafter and exhibited two peaks before hatching. Later on, Lanzrein and coworkers (Lanzrein et al., 1984; Brüning et al., 1985; Bürgin and Lanzrein, 1988) presented conclusive evidence that the embryonic CA of *N. cinerea* synthesize *de novo* JH III and, surprisingly, even larger quantities of methyl farnesoate during late embryogenesis. In these experiments, brain–CC-Ca complexes were isolated from 29-day-old embryos (hatching takes place after about 40 days) and incubated in a medium consisting of larval hemolymph, culture medium M-199, and [*methyl-*^{14}C]methionine. The complexes released labeled JH III and methyl farnesoate, the latter about 10 times higher in quantity. Brains alone did not convert labeled methionine, indicating that JH III and methyl farnesoate were produced by the retrocerebral complex, most probably by the CA. Large quantities of labeled methyl farnesoate were also obtained when labeled methionine was injected into 28-day-old egg cases, which means that synthesis of this compound was no artifact produced *in vitro*. Extracts of 27-day-old egg cases analyzed by HPLC and GC/MS chemical ionization (CI) showed the presence of methyl farnesoate also in normal developing eggs. Between days 22 and 34 of embryogenesis, the ratio JH III : methyl farnesoate is about 1 : 3 but about 4 : 1 between days 35 and 39, synthesized by heads *in vitro* (Lanzrein et al., 1984). When whole egg cases were subjected to

HPLC and GC analyses and titer curves constructed for JH III and methyl farnesoate (Fig. 1.26), both compounds took about the same course, except that shortly before hatching methyl farnesoate decreased rapidly (Brüning et al., 1985). The relative increase of JH III seems to coincide with the breaking up of the chorion and is possibly correlated with changes in consumption and use of oxygen. Methyl farnesoate is a precursor of JH III that lacks epoxidation. Brüning et al. (1985) conclude that the embryonic CA of *Nauphoeta* produce only low levels of methyl farnesoate epoxidase activity and/or that the supply of oxygen is not sufficient for effective conversion into JH III, since this last step has been shown to be O_2 dependent in *Blaberus giganteus* (Hammock, 1975). The biological function of methyl farnesoate is not clear. Fehr and B. Lanzrein (cited in Brüning et al., 1985) found that the compound exerted JH-like effects when injected into larvae, and induced vitellogenin synthesis and oocyte growth when injected into decapitated adult females. That JH III has an important role in embryonic morphogenesis was evidenced by treatment with a precocene analogue of *N. cinerea* eggs (Brüning et al., 1985). JH III synthesis was largely suppressed by this procedure, but not so methyl farnesoate production. Severe developmental disturbances were observed, pointing to the necessity of JH III. This experiment also strengthened the hypothesis that there is no epoxidase function in embryonic CA, since the allatocidal effect of precocene depends on its metabolization to reactive epoxides that destroy the glands (Pratt et al., 1980). The metabolization is carried out by the epoxidases of the CA. Since the embryonic CA are not affected by precocene, it may be deduced that epoxidases are absent.

Recently, Bürgin and Lanzrein (1988) explored the degradation of JH III during the last 15 days of the embryonic development of *N. cinerea*. In an *in vitro* system, these investigators showed that the egg-case envelope metabolized JH at all investigated stages whereas embryo homogenates degraded JH III only shortly before hatching after breaking of the chorion (Fig. 1.27). The major metabolite was JH III acid. Esterases were mainly located in the fat body and/or integument and to some extent in gut and hemolymph. From these findings, the authors concluded that the very high titers of JH III during late embryogenesis of *Nauphoeta* is probably the result of the very low rates of JH III degradation and not so much the result of an enormous production of JH III. Thus, the titer seems to be only loosely regulated in the embryo.

Roe et al. (1987a,b) followed JH metabolism during embryonic development of *Acheta domesticus*. In preovipositional eggs and during early embryogenesis (up to dorsal closure), they found elevated levels of nonspecific JH esterase activity. This is probably needed for degradation of maternally derived JH, which may enter the oocyte passively. (High JH titers are often present in reproducing females.) It has been shown in numerous species that application of JH and juvenoids interrupt early embryonic development, thus indicating that the absence of JH may be required for this developmental stage. JH-specific esterase activity appears in *Acheta* around blastokinesis and is maintained at relatively low levels through hatching (Roe et al., 1987a). During this developmental stage, the JH titer is high.

Whereas embryonic JH degradation in *A. domesticus* and *N. cinerea* is mainly carried out by esterase activity, this is not true for all embryos investigated. R. M. Roe and C. L. Crawford (cited in Roe et al., 1987a) found two routes of JH metabolism—ester hydrolysis and epoxide hydration—in newly laid eggs of *M. sexta*, *Heliothis virescens*, *H. zea*, and *Trichoplusia ni*.

JH titer measurements as well as isolation of JHs and their metabolites have been carried out throughout embryonic development or in selected stages of a number of insects. Bergot et al. (1980, 1981a,b) determined four different JHs (JH 0, iso-JH 0, JH I, and JH II) from butterfly eggs (Fig. 1.8). The occurrence of these JHs and their titer fluctuations during embryonic development is illustrated in Table 1.3. Outside Lepidoptera, only JH III seems to occur, e.g., in *N. cinerea* (Lanzrein et al., 1984), *L. migratoria* (Temin et al., 1986), and *A. domesticus* (Roe et al., 1987a,b). Similar to *Acheta*, but unlike *Nauphoeta*, presumably maternal JH was observed during early embryogenesis of *L. migratoria* (Temin et al., 1986). The levels declined through blastokinesis, increased during late embryogenesis, and declined again as hatching approached. All these observations are supportive of a biological role for JH in embryonic development.

The aforementioned works of Lanzrein and coworkers presented the first proof that embryonic CA synthesize JH. Maternal JH, if present, is apparently degraded. "Storage" JH, comparable to maternal ecdysteroid conjugates, does not apparently exist, but transfer of JH precursors from mother to egg must be considered.

Differentiation of Corpora Allata

The idea that embryonic CA might synthesize JH was put forward long before its experimental proof. Studies of ontogenetic and cytophysiological differentiation of the glands provided the first clues (see Dorn, 1972, 1975c; Haget et al., 1981).

The anlagen of the CA differentiate, according to most recent reports (see Dorn, 1972, 1975c; Rempel et al., 1977; Dorn et al., 1987b), from the maxillary segment (Figs. 1.12 and 1.17). (Divergent modes of development are presumably erroneously described in the older literature—see Dorn, 1972.) As a rule, shortly before katatrepsis and dorsal closure, ectodermal cells located laterally and cranially of the maxillary appendages invaginate. They form either vesicles (e.g., *O. fasciatus*; see Fig. 1.24A) or solid cell masses. The anlagen of the glands move dorsally and come in

TABLE 1.3. JH Titers in Insect Embryos of Various Species and Ages: Titers Are Expressed as Nanograms per Gram Whole Eggs; Limits of Detection Were 0.01–0.02 ng/g, Except Where Noted Otherwise

Species	Age (days)	JH III	II	I	0	Iso-0
O. fasciatus*	3	0.09	—	—	—	—
	4	0.06	—	—	—	—
H. virescens	1	—	0.05	0.1	0.03	ND
	2	—	—	0.03	0.06	ND
M. sexta	0	—	—	—	—	—
(wild-type)	1	—	—	—	—	—
	2	—	0.003	1.47	2.72	1.20
	3	—	0.013	0.59	1.56	1.95
	Hatchling	—	—	0.05	—	—
M. sexta	2	—	0.009	(3.3)[a]	2.1	(3.3)[a]
(black mutant)	3	—	—	0.49	1.6	1.64
H. cecropia	3.5	—	—	0.02	—	—
	5.5	—	0.02	0.52	—	—
	(6.5)[b]	—	—	0.37	—	—
	(6.5)[c]	—	—	0.16	0.13	—
	(8.5)[d]	—	0.05	0.25	0.02	—
	9.5	—	—	0.05	—	—
	Hatchling	—	—	0.05	—	—

SOURCE: Bergot et al. (1981a).
NOTE: * = no JH was found in a recent reinvestigation of the same research group (Baker, personal communication; — = undetectable; ND = not determined.
[a]Peaks not resolved.
[b]Early premolt.
[c]Late premolt.
[d]Postmolt.

contact with the anlagen of the CC (Figs. 1.13, 1.14, and 1.15) or nerves projecting from them, as in *M. sexta* (Dorn, 1987b). The axis brain (NSCs)– CC–CA with its nervous connections is usually established well before the synthesis of the larval cuticle. However, the onset of hormone production and release is difficult to establish. Even in the experiments of Bürgin and Lanzrein (1988)—conversion of JH precursors by embryonic CA *in vitro*—the earliest stages of glandular activity remained open.

Dorn (1975c, 1978a) and Haget et al. (1981) investigated the fine structural differentiation of the CA in the embryo of *O. fasciatus* and *C. morosus*, respectively. It was deduced on grounds of organelle expression that the cells become active in terms of secretion during late embryogenesis. Signs of activity were interpreted (Dorn, 1975c, 1978a; Haget et al., 1981) as follows: increase in the cytoplasm–nucleus ratio, alterations of the nucleoli and appearance of accessory nucleoli, formation of rough ER often arranged in concentric whorls, and change of mitochondrion matrix density (Figs. 1.24B and 1.25). Neurosecretory axons devoid of glial elements can be found projecting into the CA (Fig. 1.25). Haget et al. (1981) suggested two cycles of CA activity in the embryo of *C. morosus*: the first coincides with the completion of dorsal closure; the second with the darkening of the mandibular base. Besides the aforementioned cellular differentiations, these cycles are characterized by the appearance of dilated intercellular spaces. Exocytotic processes directed against the lumen of the vesicular glands are observed during late embryogenesis (Fig. 1.25B). Their significance is not understood.

The data obtained by cytological studies of the CA are quite compatible with the *in vitro* experiments, showing JH (and methyl farnesoate) synthesis of these glands. What remains to be explored is the exact onset of JH production in the CA.

Recently, Hartmann et al. (1987) reported the synthesis of JH III by the serosa of *L. migratoria* prior to JH III production of the CA (i.e., embryonic heads *in vitro*): "The earliest detectable JH III production by CA in isolated heads is on day 6. On the other hand, serosal epithelia of 4- to 5-day-old embryos produce JH III in organ cultures, as has been shown by methylation of [10-^{3}H]JH III acid to [10-^{3}H]JH III, and by incorporation of tritiated CH_3 from L-[*methyl*-^{3}H]methionine in JH III." The separation methods used included TLC and HPLC. Interestingly, JH-binding protein was visualized by immunohistological methods in the serosa and differentiating tissues of ectodermal origin (epidermis, fore- and hindgut, brain, nerve cord, and stomatogastric nervous system) (Hartmann et al., 1987).

1.3.2. *Crustacea*

1.3.2.1. ECDYSTEROIDS: DIFFERENTIATION
OF Y-ORGANS

The presence of ecdysteroids has been demonstrated in eggs of several crustacean species: *Carcinus maenas* (Lachaise and Hoffmann, 1977, 1982;

Lachaise et al., 1986; Goudeau and Lachaise, 1983), *Orchestia gammarellus* (Blanchet et al., 1979), *Callinectes sapidus* (McCarthy, 1979; McCarthy and Skinner, 1979), *Acanthonyx lunulatus* (Chaix and De Reggi, 1982), *Artemia salina* (Spindler et al., 1984), and *Palaemon serratus* (Spindler et al., 1987). Chemical analyses and titer measurements were carried out only in a few cases. In the shore crab *Carcinus maenas*, Lachaise and Hoffmann (1982) detected primarily ponasterone A (Fig. 1.5), only small amounts of 20-hydroxyecdysone (3–10 times lower than ponasterone A), and very little ecdysone by physicochemical methods. The titer of ponasterone A fluctuates especially strongly during early embryogenesis but also shows elevated levels at selected stages of late development (Fig. 1.28). Interestingly, the first high peak occurs around 40 h after egg deposition, i.e., when the first cleavage divisions take place. Not only does the content of free ponasterone A rise from low levels, but also extensive amounts of ponasterone A conjugates amass from low levels; these conjugates are found in newly deposited eggs (Fig. 1.28). This raises two questions: What is the biological role of these ecdysteroids? And where is the hormone synthesized? Although some may be of maternal origin, Fig. 1.28 shows clearly that this cannot account for the titer increase. Lachaise and Hoffmann (1982) speculate: (1) "The very young embryo has the enzymatic equipment for *de novo* synthesis of ecdysone" (i.e., ponasterone A) or (2) "the maternal ecdysteroid (or close precursor) is transferred under a form which is not extracted by our procedure or which is not immunoreactive even after the classical enzymatic hydrolysis." Also not resolved is the origin of the ecdysteroids present during late embryogenesis (Fig. 1.28). The Y-organ is most probably differentiated at that time and may be secretorily active.

A somewhat different situation is met in another crab, *Acanthonyx lunulatus* (Chaix and De Reggi, 1982). The predominant free ecdysteroids are ecdysone and 20-hydroxyecdysone. They were present at four stages of embryonic development (Fig. 1.29): (1) in the newly deposited egg until the beginning of naupliar development (stage I till stage II); (2) at the end of metamerization during the metanauplius stage III, when a "metamerization molt" occurs; (3) during secretion of the "first exoskeleton" (stage IX); (4) at the end of the development of the zoea, shortly before hatching (stage XII). The presence of considerable amounts of high polar ecdysteroids (presumably conjugates of ecdysone and 20-hydroxyecdysone) in newly deposited eggs (Fig. 1.29) strongly parallels the situation in many insect eggs (see Section 1.3.1.1, above). These biologically inactive "storage hormones" are obviously passed on by the mother and may be hydrolyzed during early embryogenesis. The source of ecdysteroids during late embryonic development is unknown. Since the levels of high and low polar ecdysteroids were also low between the peaks of free hormone (Fig. 1.29), a *de novo* synthesis can be expected. The Y-organ would be the most probable site of synthesis, but experimental proof is lacking and, as in insects, other sources cannot be excluded. The role of ecdysteroids in embryonated eggs can only be inferred from correlative considerations. High levels accompany cuticle secretion, but also occur independently of embryonic molts (Chaix and De Reggi, 1982).

Still different circumstances were observed in eggs of the shrimp *Palaemon serratus* (Spindler et al., 1987). As in *Acanthonyx*, but in contrast to *Carcinus*, ecdysone and 20-hydroxyecdysone are the main ecdysteroids, and not ponasterone A. Newly deposited eggs of *Palaemon* contain relatively few ecdysteroids. They consist of ecdysone, 20-hydroxyecdysone, and high and low polar ecdysteroids. The titer of RIA-active ecdysteroids rises during stage C. This is when the blastoderm extends and separates into two parts. The first limb buds develop, the red chromatophores become visible, and the Y-organ appears (see Spindler et al., 1987). The levels of ecdysteroids, primarily 20-hydroxyecdysone, increase further (Fig. 1.30). At an incubation temperature of 11°C, the curve expresses two flat peaks, which become very striking when embryonic development takes place at 19°C (Fig. 1.30). Spindler et al. (1987) ascribe the steep ecdysteroid increase between stages C and D to the synthetic activity of the Y-organ, although the gland reaches its complete differentiation only at stage D. The data on organogenesis were adopted from other studies (Bellon-Humbert, 1983; Le Roux, 1983). Considerable variations of the developmental status may exist within a stage, especially if one considers that embryogenesis lasts about 110 days at 11°C. Thus correlations have to be carefully evaluated. In *Palaemon*, as in *Carcinus*, apparently low amounts of ecdysteroids pass from the mother to the egg.

Owing to the fact that the ecdysteroids were studied in the embryos of only a few crustacean species, no general line can be drawn. In respect to the source of embryonic ecdysteroids, three possibilities exist, as in insects: (1) all embryonic ecdysteroids are derived from the mother, and titer fluctuations of free, biologically active forms are generated by metabolization (including deconjugation and conjugation, synthesis from precursors, and degradation of active forms); (2) all embryonic ecdysteroids are produced in the embryo, either by the Y-organ or other tissues; (3) ecdysteroids present during the early phase of embryogenesis stem from the mother, whereas in later stages they are (in addition) synthesized *de novo* from embryonic tissue (Y-organ or other tissues). So far, neither the capability of ecdysone synthesis by the embryonic Y-organs nor other tissues has been challenged *in vitro*.

As mentioned above, the Y-organ differentiates early in embryonic development. The anlagen of the glands develop from the second maxilla. Y-organs have been interpreted as descendants of head nephridia and as homologous to the insect PTGs, based on the common site of origin (second maxilla and labium, respectively) and innervation from the subesophageal ganglion (Gabe, 1956; Pflugfelder, 1958; Scheer, 1960). Cytological or fine structural studies on the embryonic Y-organ have not yet been carried out, to our knowledge.

1.3.2.2. NEUROHORMONES: DIFFERENTIATION OF NSCS AND NEUROHEMAL ORGANS

Not yet clear is whether or not the Y-organ (or other tissues) produce ecdysteroids during embryonic development; hence, we do not know if MIH (molt-inhibiting hormone) (Fig. 1.2) would be needed to inhibit the embryonic gland should production be suppressed.

It must be considered that morphogenetic hormones may also have special functions in embryogenesis, different from those observed in postembryonic development. There are strong indications that this is true for PTTH in insects (see Section 1.3.1.2, above). Therefore, it might be worthwhile to search for embryonic MIH in any case.

Ontogenesis of the neurosecretory system including neurohemal organs is poorly studied in crustaceans (see Mori and Ando, 1983). Several authors have described the evolution of the central nervous system but left out NSCs. Brain cells differentiate rather early. Neurites and dendrites become distinguishable concomitantly with the expression of eye pigments in the ostracod *Cyprides litoralis* (Weygoldt, 1960). The early differentiation of the central nervous system has also been emphasized by other authors in a variety of species (Scholl, 1963; Weygoldt, 1964, 1975; Zilch, 1974). Thus, the possibility exists that NSCs also evolve early and become secretorily active.

1.3.2.3. JUVENILE HORMONE(S): DIFFERENTIATION OF MANDIBULAR ORGANS

Almost 30 years elapsed between the first observation of JH activity in extracts of *Uca pugilator* and eyestalks of *Homarus americanus* monitored by the *Polyphemus* wax test (Schneiderman and Gilbert, 1958) and the recent physicochemical identification of a JH-like compound (methyl farnesoate) in crustaceans (Laufer et al., 1987a). Indications that JH or a JH-like compound may also be present in crustacean eggs (newly deposited or embryonated) have not yet come to my attention. According to Payen and Costlow (1977), Greer (original data not available) found that egg development is blocked in the isopod *Armadillidium vulgare* after topical application of JH I and JH analogues. But this demonstrates only a sensitivity to JH, not necessarily that the hormone occurs naturally during embryogenesis or that it is needed.

Only recently, Laufer et al. (1987b) determined that the mandibular organ of *Libinia emarginata* was the (probably only) site of "JH" (methyl farnesoate) synthesis. Mandibular organs probably occur in all decapods. Le Roux (1968) found them in 20 species of 10 families. He also studied the organogenesis of mandibular organs in *Inachus phalangium*. The anlagen were visible at the stage of Zoea II and thereafter. They develop from the ectoderm of the mandibular segment and migrate during the megalopa stage to their definite location: between the mandibular and the maxillary segment, they border the posterior base of the apodema of the posterior mandibular adductor above the muscles of the first maxilla (Le Roux, 1968). Chaudonneret (1956) apparently observed the differentiation of this gland before Le Roux and described its innervation by the first maxilla but did not exclude other nervous connections. Although Chaudonneret erroneously considered the mandibular organ as the Y-organ (Le Roux, 1968), he mentioned that the gland could be homologous to the CA of insects, based on the embryological data. It would be highly interesting to examine the capability of JH synthesis of embryonic mandibular organs *in vitro* and to compare it with the results obtained by Lanzrein et al. (1984) in insects (see Section 1.3.1.3, above).

1.3.3. *Myriapoda*

1.3.3.1. ECDYSTEROIDS: DIFFERENTIATION OF "ECDYSIAL GLANDS"

There is only one report, to my knowledge, indicating the presence of ecdysteroids in mature ovaries and ripe eggs of a myriapod, *Lithobius forficatus* (Leubert et al., 1982). Using TLC for separation and RIA for quantification, Leubert et al. demonstrated ecdysone/20-hydroxyecdysone-related compounds (both fractions were eluted together) as well as low and high polarity substances. *In vitro* incubations of mature ovaries signaled the biosynthesis of ecdysone and 20-hydroxyecdysone analyzed by HPLC. The compounds were partly released into the medium (84%), partly retained in the ovary (16%). Biosynthesis was not evidenced by conversion of labeled precursors but by measuring the ecdysteroid contents before and after a 48-h incubation period. The ecdysteroid increase was about five-fold during this time (Leubert et al., 1982).

Table 1.4 shows the ecdysteroid content of ripe (newly deposited) eggs (besides that of oocytes and nonreproductive ovaries) (Leubert et al., 1982). Eggs and immature oocytes contain free ecdysone/20-hydroxyecdysone, but even more high polar ecdysteroids. The high polar ecdysteroids were eluted after TLC and directly subjected to RIA analysis;

TABLE 1.4.   Ecdysteroid Levels in Oocytes, Mature Eggs, and Ovaries (After Reproductive Period) in a Myriapod, *Lithobius forficatus*

| | Ecdysteroids in picogram ecdysone equivalents | | | |
| | *Starting zone* | | *Zone of ecdysone + 20-hydroxyecdysone* | |
	TOTAL	PER MG	TOTAL	PER MG
Oocytes from 72 ovaries (310 mg)	4000	12.9	2500	8.1
Mature eggs (284 eggs = 102 mg)	7500	73.5	5160	50.6
22 ovaries (after reproductive period)	0	0	0	0

SOURCE: From Leubert et al. (1982).

i.e., they were not enzymatically cleaved and then quantified, as is usually done. Since the available antibodies have normally a weaker cross-reactivity to conjugates (which the high polar products presumably represent) than to free ecdysone and 20-hydroxyecdysone, it may be expected that the "real" content of high polar ecdysteroids is still higher (Leubert et al., 1982). In this respect, the *Lithobius* egg exhibits interesting parallels to eggs of a great many insect species (see Section 1.3.1.1, above).

The fate and biological significance of maternal ecdysteroids are obscure. Titer measurements throughout embryogenesis were not carried out. Consequently, the question of an embryonic ecdysteroid source remains open.

The source of ecdysteroids during postembryonic development is not well established (Scheffel, 1965, 1969). Seifert and his coworkers (El-Hifnawi and Seifert, 1972; Rosenberg, 1973; Seifert and Rosenberg, 1974) described strings or clusters of nephrocyte/podocyte-like cells (lymph glands) in the head or anterior body segments of chilopods and diplopods; these cells were posited to represent ecdysial glands. Some experimental evidence supporting this premise exists only in *Lithobius forficatus* (Scheffel, 1969). The lymph glands, as descendants of "common

nephrocytes" (Seifert, 1979), are derivatives of the mesoblast, whereas
PTGs and Y-organs differentiate from ectoderm. Seifert (1979) explains
that the "original ecdysial organs of arthropods [namely, the lymph
glands of myriapods] were, or are, specialized nephrocytes." According
to this theory, the PTGs of insects and Y-organs of decapods would have
to be looked upon as relatively recent phylogenetic acquisitions.

Data on the ontogenetic differentiation of lymph glands were not available to me. In insects, paired strings of cephalic nephrocytes (pericardial
cells) are among the first tissues that reveal their typical cellular features
on both light- and electron-microscopic levels (Dorn, 1972, 1977).
Whether or not the lymph glands of myriapods also differentiate early
and are a source of embryonic ecdysteroids remains to be seen.

1.3.3.2.   NEUROHORMONES: DIFFERENTIATION
OF NSCS AND NEUROHEMAL ORGANS

There is some experimental evidence of a molt-inhibiting brain factor in
myriapods, e.g., *Lithobius forficatus* (Joly, 1966b, 1971; Scheffel, 1969) (see
Fig. 1.4). The presence of such a factor in myriapod embryos is uncertain.
The evolution of the neurosecretory system has not been studied. Consequently a basis for a discussion of neurosecretory processes in the embryo
is lacking.

1.3.3.3.   JUVENILE HORMONE

JH has never been reported to occur in Myriapoda, and it has probably
never been searched for in this arthropod group.

1.3.4.   *Chelicerata*

1.3.4.1.   ECDYSTEROIDS: DIFFERENTIATION
OF "ENDOCRINE TISSUE"

The presence of ecdysteroids in chelicerate eggs has been reported from
ticks only, but from at least seven species (Connat et al., 1987b). Maternal
ecdysteroids are attributed to the egg either as free ecdysone and 20-
hydroxyecdysone, e.g., in *Amblyomma hebraeum* and *A. variegatus* (Connat
et al., 1985, 1987a), or as apolar ecdysteroid conjugates, e.g., in
*Ornithodorus moubata, Boophilus microplus, Rhipicephalus appendiculatus,
Hyalomma dromedarii,* and *Ixodes ricinus* (Connat et al., 1984; Wigglesworth
et al., 1985; Connat et al., 1987b). The labeled apolar ecdysteroid conjugates extracted from eggs after injection of [^{3}H]ecdysone during
vitellogenesis had the same retention times as the long-chain fatty acid
esters conjugated at the C-22 of ecdysone and 20-hydroxyecdysone (Table
1.1) (Connat et al., 1984; Connat and Diehl, 1986). Free ecdysone and 20-

hydroxyecdysone may occur besides the apolar conjugates (Wigglesworth et al., 1985; Whitehead et al., 1986). Polar ecdysteroid conjugates apparently play a minor role. A carrier protein for ecdysteroids has been demonstrated in the egg of *Rhipicephalus appendiculatus* (Whitehead et al., 1986).

Concerning the maternally derived ecdysteroids, Connat et al. (1987b) speculate: "We suggest that either these hormones are artifactually incorporated into the oocytes during vitellogenesis, due to a high spontaneous binding with vitellogenins, or that they could play other roles than the control of embryonic molts." The large quantities of ecdysteroids in tick eggs might act as a feeding deterrent (Connat et al., 1987a).

The production site of the maternally derived ecdysteroids in the egg is not clearly determined. Since the ecdysteroids (in *Amblyomma hebraeum*) appear in the body of the female several days prior to their appearance in the ovary, ecdysteroid synthesis outside the ovary is favored (Connat et al., 1985). Molting hormone-synthesizing glands are not known in ticks. Large, immobile cells that are located in lateral body parts and that may form complexes were thought by Streble (1966) to represent the equivalent of the aranean "endocrine tissue" and to produce ecdysone (see Fig. 1.4). Ticks cease molting before reproduction, but formation of endocuticle still takes place. Whether or not the larval source of ecdysteroids disintegrates after final molt is not known.

Ecdysteroid measurements in relation to embryonic development have yet to be reported. Therefore, correlations of ecdysteroid metabolization with developmental events are impossible, and clues indicating an embryonic source of ecdysteroids are not available.

The genesis of the "endocrine tissue" has been studied in several species of Araneae (Millot, 1930; Legendre, 1959; Streble, 1966). In *Dolomedes fimbricatus* (Legendre, 1959), the first traces of this tissue were noticeable at the stage of neuromere formation, when the coelomic sacs were still complete. Certain cells of the splanchnopleura incorporate yolk. They persist after disaggregation of the coelomic sacs and display a metameric organization, being situated lateral of the ganglion anlagen. They represent the precursors of the "endocrine cells." During concentration of the ganglia in the prosoma, the "endocrine cells" also are dislocated and finally form the posterior part of the "endocrine organ."

Streble (1966) studied the evolution of the "endocrine organ" in *Pardosa amentata*. Like Legendre (1959), he described them as descendants of the mesoderm. After disintegration, the coelomic sacs give rise to slightly differentiated mesenchymal cells. Among them, two cellular elements are distinguishable: the prospective "endocrine cells" with their large nuclei, and the binucleated nephrocytes. The "typical" histological features of the "endocrine cells" appear at the end of the first cocoon stage (Streble,

1966). But Millot (1930) and Legendre (1959) stated in *Dolomedes* a secretory activity of the "endocrine cells" before the first embryonic molt within the chorion. Future work should decide whether or not this secretory activity indeed means synthesis of ecdysteroids.

1.3.4.2. NEUROHORMONES: DIFFERENTIATION OF NSCS AND NEUROHEMAL ORGANS

There are strong indications that the brain, i.e., oral NSCs, synthesizes a molt-stimulating factor (Joly, 1961, 1962, 1966a,b, 1971; Scheffel, 1969) (see Fig. 1.4). This factor may activate the "endocrine tissue" to produce ecdysone. No investigation has been found in the literature concerning the presence of this factor during embryogenesis of chelicerates. In fact, Eckert (1967) found no PAF-stainable cells until the third larval stage of *Coelotes terrestris*. However, failure to demonstrate PAF reactivity is no proof of absence of neurosecretory processes (see Section 1.3.1.2, above); it just means that the number of granules stored in the perikarya is insufficient. Factors may be synthesized and immediately transported to the release site.

Several publications have dealt with the differentiation of the chelicerate nervous system [Kästner, 1951 (*Thelyphonus caudatus*); Moritz, 1957 (Phalangiidae); Legendre, 1959 (Aranea); Weygoldt, 1964, 1965. (Pseudoscorpiones); Pross, 1966 (*Pardosa hortensis*); Weygoldt, 1975 (*Tarantula margine*)]. Only Legendre refers to the evolution of embryonic NSCs. He distinguishes two types of NSCs: A and B. The categorization depends on the location of the cells, either at the surface of a ganglion (type A) or deeper in the ganglion (type B). [Note that the more recent subdivision in A- and B-type NSCs is based on the morphology of the neurosecretory granules (cf. Juberthie, 1983).] Type A cells could be recognized by their size at the earliest stages of differentiation in the 17-day embryo of *Dolomedes*. They are segregated together with metameric neuroblasts from the ectoderm. Type B cells stem from epithelium cells of the coelom (!). They proliferate and are incorporated into neuromere. The origin of type B NSCs from the splanchnopleura, as described by Legendre (1959), is unique. Further embryological studies seem necessary to accept the data. According to Legendre, neither A nor B cells become active during embryogenesis.

The retrocerebral complex of the Aranea includes Schneider's organ I, with the adhering *Tropfenkomplex* (= prosomatic neurohemal organ), where cerebral neurosecretions are released, and Schneider's organ II, which has intrinsic cells, like Schneider's organ I. Whether or not Schneider's organ II also contains neurosecretory fibers from the brain and may act as another neurohemal organ is disputed (see Juberthie, 1983).

The differentiation of these organs has been described in several aranean species: *Dolomedes fimbricatus* (Legendre, 1959); *Pardosa hortensis*

(Pross, 1966); *Pardosa amentata* (Streble, 1966); and *Tarantula margine* (Weygoldt, 1975). The genesis of Schneider's organ II is especially well documented. In *Tarantula*, it can be recognized in the 20- to 25-day-old embryo (Weygoldt, 1975). (Embryonic development can be subdivided into a *protembryonic phase*, which ends with metamerization and an embryonic molt after about 20 days. Then the chorion ruptures and the *deutembryonic phase* starts, in which organogenesis and histogenesis take place; this phase lasts some 70–80 days. Then the prenymph emerges from the brood pouch.) According to Weygoldt (1975), Schneider's organ II is formed by two groups of ganglion cells, which separate from the posterior edge of the subesophageal ganglion. However, this mode of differentiation appears exceptional. Legendre (1959) and Streble (1966) described the separation of the anlagen of Schneider's organ II from the stomodaeum. In *Dolomedes* (Legendre, 1959) two epithelial proliferations from the lateral stomodaeum wall segregate on day 17 of embryonic development (Fig. 1.31). (Neuroblasts differentiate at the same time.) Slightly in front of the anlagen, an unpaired group of cells leaves the stomodaeum roof and give rise to the supraesophageal ganglion. The differentiation of Schneider's organ II takes the same course in *Pardosa amentata* (Fig. 1.32) (Streble, 1966). Legendre and Streble emphasize the striking conformity between the evolution of Schneider's organ II and the CC of insects (see Section 1.3.1.2, above). Both authors consider the glands to be homologous. Streble observed the cytological differentiation of Schneider's organ II by light microscopy and came to the conclusion that secretory activity does not take place in the embryo. But reliable signs of a discharge of neurosecretory material (if one accepts the neurohemal function of Schneider's organ II) can only be obtained by electron microscopic studies, which are as yet lacking.

Schneider's organ I exhibits many similarities to the insect CA and will therefore be treated below. In this context, it should be remembered that the CA of some insect species (e.g., *M. sexta*) also serve as a release site for neurohormones (e.g., PTTH). The fact that the prosomatic neurohemal organ (*Tropfenkomplex*) is part of Schneider's organ I would therefore not speak against a homologization of this organ with insect CA.

The prosomatic neurohemal organ has not been studied during embryogenesis. Thus, it remains open whether neurosecretory activities take place at this stage in this organ or not.

1.3.4.3. JUVENILE HORMONE: CONSIDERATIONS OF ITS SOURCE

The presence of JH or JH-like compounds has never been demonstrated in Chelicerata by physicochemical methods or bioassays to my knowledge. Indirect evidence that JH may occur and exert a function in reproduction of ticks was delivered by Pound and Oliver (1979) using the anti-JH precocene 2. Topical application of several JH analogues to tick eggs blocked embryogenesis, but the mode of action of these substances remained unresolved. An activity of JH on embryogenesis of *Hyalomma dromedarii* was reported by Bassal (1974).

The effect of JH and a number of juvenoids was tested in *Limulus* eggs, embryos, and trilobite larvae (Jegla, 1982). The substances had either a toxic or no effect on larvae. When embryos were subjected to 100 µg/liter JH after the second embryonic molt, the dorsal organ developed abnormally in 65% of the animals. This finding is of interest because an effect of anti-JH on dorsal organ formation was observed in an insect (Dorn, 1982). Nonetheless, Jegla (1982) concluded that JH does not have a role in the morphogenesis of *Limulus*. This statement appears premature. It is certainly not true for ticks and probably not true for other Chelicerata.

Based on comparative embryological studies, Streble (1966) and above all Legendre (1959) pointed out a homology between the CA of Insecta and Schneider's organ I. In *Dolomedes,* the latter differentiates from the ectoderm in front of the appendages of the cheliceran segment (Fig. 1.33). The anlagen appear on day 15 and migrate toward the aorta. Installment of nervous connections, separation of the prosomatic neurohemal organ, and fine structural differentiation have not yet been elucidated. Consequently, questions concerning the function of Schneider's organ I, including that of the prosomatic neurohemal organ, remain unanswered.

1.4.   Summary

Postembryonic growth of Insecta is regulated by three groups of hormones: the neurohormones, which control the peripheral endocrine glands involved in molting; the ecdysteroids, which are produced (in particular the prohormone ecdysone) by the PTGs; and the JHs, synthesized by the CA. Whereas ecdysteroids are present and induce molting in all arthropods, JHs were thought to be confined to Insecta. But recently they were also found in Crustacea (Decapoda). There are indications that they have the same function as in Insecta. Morphogenetic neurohormones probably occur throughout Arthropoda, although they may differ in their chemistry as well as in their action on peripheral hormone glands. Thus, ecdysial glands of Insecta are stimulated by a brain hormone, whereas the corresponding glands in Crustacea and probably Myriapoda are inhibited by brain hormones.

The occurrence of morphogenetic hormones in imagoes and eggs of Insecta and other Arthropoda is currently a center of attention. In Insecta, it is well established that ecdysteroids and most probably morphogenetic neurohormones (i.e., PTTH in *Manduca sexta* and *Bombyx mori,* and diapause hormone in *B. mori*) are present in newly deposited eggs. Whereas for some species JH was reported to occur in newly deposited eggs, this is apparently not the case in others, where JH appears in the course of embryogenesis. It is evident that hormones demonstrated in newly laid eggs must be derived from the mother. Since ovaries (of presumably all insect species) are a site of ecdysteroid synthesis, they can be considered as the source of ecdysteroids found in newly laid eggs. The origin of the neurohormones assayed in newly deposited eggs of *B. mori* and *M. sexta* is presumably the maternal brain (for PTTH) and maternal subesophageal ganglion (for diapause hormone). However, more investigations are needed to allow generalizations. JH, if incorporated into oocytes, stems presumably from the maternal CA. Surprisingly, the serosa of young *Locusta migratoria* embryos was recently described as a JH-synthesizing tissue.

Maternally derived ecdysteroids were monitored in eggs of some Crustacea (Decapoda). In other species, newly deposited eggs included no ecdysteroids; the hormone appeared in these cases during embryonic development. It is not yet clear whether crustacean ovaries (as a rule) produce ecdysteroids or not. Also eggs of Chelicerata (i.e., ticks) contain considerable amounts of ecdysteroids. In newly deposited eggs, they occur either as free hormones or as apolar conjugates. Their source in the female is not quite clear; some observations speak against the ovaries as the production site. The presence of morphogenetic hormones in eggs (newly deposited or embryonated) of other Arthropoda has not yet been investigated.

It has been shown in a number of insect species that the levels of ecdysteroids, JHs, and neurohormones increase during embryonic development (usually concomitantly with or shortly after dorsal closure) in the case of ecdysteroids often correlated with cuticle deposition ("embryonic molts"). The conclusion that embryonic hormone glands may become active seems at hand. But there is only one case where the hormone production of an embryonic gland has been challenged: the CA of *Nauphoeta cinerea* were incubated *in vitro* with radiolabeled precursors of JH and shown to convert them into methyl farnesoate, another precursor of JH III. Whether or not this compound has a biological function remains to be seen. It is interesting to note that methyl farnesoate is apparently the "JH" of Crustacea.

The question as to whether or not embryonic PTGs of insects are a source of ecdysteroids has proved to be difficult to answer. Titer fluctuations of free ecdysteroids can, at least in some cases, be explained with a sequential activity of deconjugating and conjugating enzymes. Nonetheless, several authors suggest a PTG activity in the embryo. An *in vitro* analysis must be awaited to prove this assumption.

There exists a large body of literature on the differentiation of PTGs, which usually takes place somewhat prior to dorsal closure. Few electron microscopical studies have been carried out to find cytological indications for glandular activity. The results have been controversially interpreted. The embryonic CA, on the other hand, which also differentiate shortly before dorsal closure, present fine structural evidence of synthetic activity during late embryogenesis in *Oncopeltus fasciatus* and *Carausius morosus.*

An increase in activity prior to hatching has been shown for several morphogenetic neurohormones (PTTH, eclosion hormone, and diapause hormone) in Insecta. This increase could be due to synthesis of neurosecretions in the brain. Electron microscopical studies revealed neurosecretory processes in two insect species, *O. fasciatus* and *C. morosus.*

Knowledge of embryonic sources of morphogenetic hormones in noninsect arthropods is lacking.

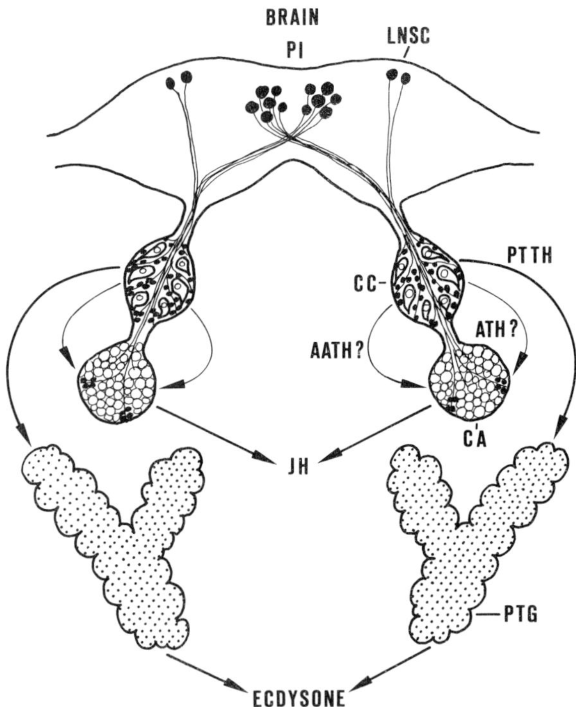

FIGURE 1.1. Diagram showing neuroendocrine apparatus of insects. KEY: AATH = antiallatotropic hormone (allatostatin); ATH = allatotropic hormone (allatotropin); CA = corpus allatum; CC = corpus cardiacum; JH = juvenile hormone(s); LNSC = lateral neurosecretory cells; PI = pars intercerebralis; PTG = prothoracic gland; PTTH = prothoracicotropic hormone.

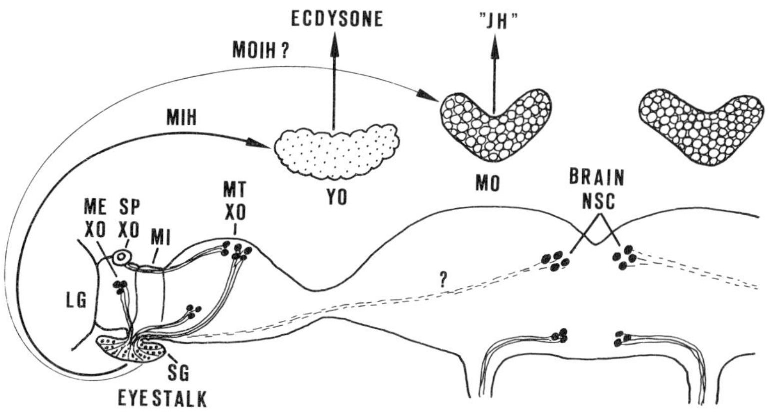

FIGURE 1.2. Diagram showing neuroendocrine apparatus of crustaceans (Decapoda). KEY: "JH" = methyl farnesoate (crustacean "juvenile hormone"); LG = lamina ganglionaris; ME XO = X-organ of medulla externa; MI = medulla interna; MIH = molt-inhibiting hormone; MO = mandibular organ (mandibular gland); MOIH = mandibular-organ-inhibiting hormone; MT XO = X-organ of medulla terminalis; NSC = neurosecretory cells; SG = sinus gland; SP XO = sense pore X-organ (represents a neurohemal organ); YO = Y-organ. (After Gorbman and Bern, 1962; Laufer et al., 1987b.)

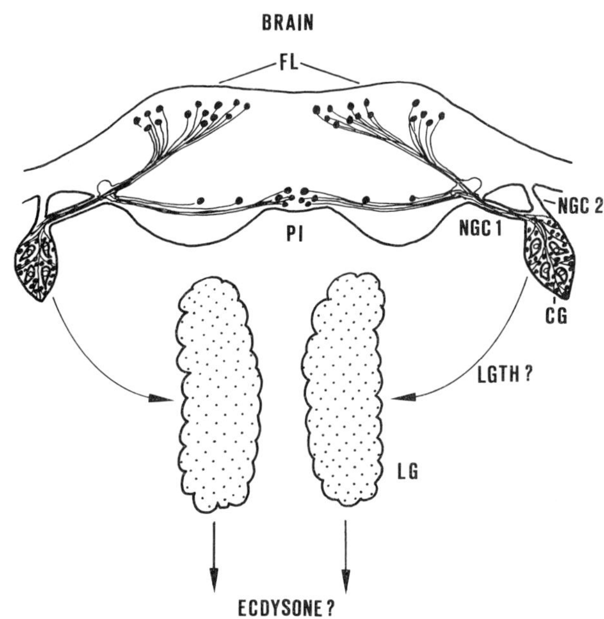

FIGURE 1.3. Diagram showing neuroendocrine apparatus of myriapods. (Chilopoda: *Lithobius forficatus*). CG = cerebral gland; FL = frontal lobe; LG = lymph gland; LGTH = lymph gland stimulating (tropic) hormone; NGC 1, 2 = nervus glandulae cerebralis 1 and 2; PI = pars intercerebralis. (After Scheffel, 1969; Jamault-Navarro and Joly, 1977.)

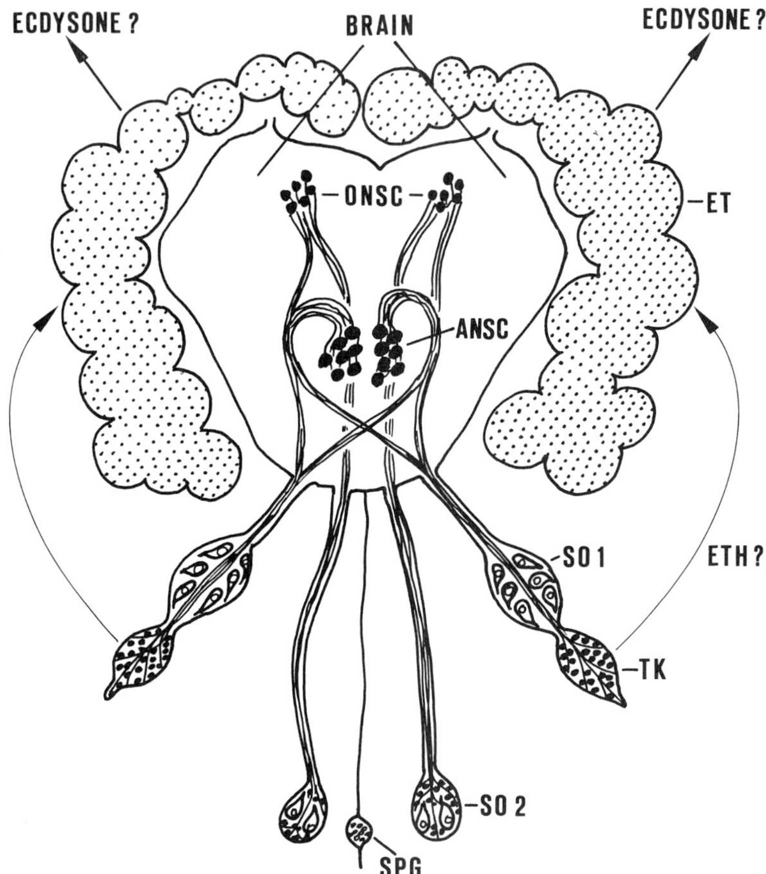

FIGURE 1.4. Diagram showing neuroendocrine apparatus of chelicerates (Araneae). KEY: ANSC = aboral neurosecretory cells; ET = "endocrine tissue"; ETH = "endocrine tissue" stimulating (tropic) hormone; ONSC = oral neurosecretory cells; SO I, II = Schneider's organ I and II; SPG = suprapharyngeal ganglion; TK = Tropfenkomplex (prosomatic neurohemal organ). (After Legendre, 1953; Streble, 1966; Babu, 1973.)

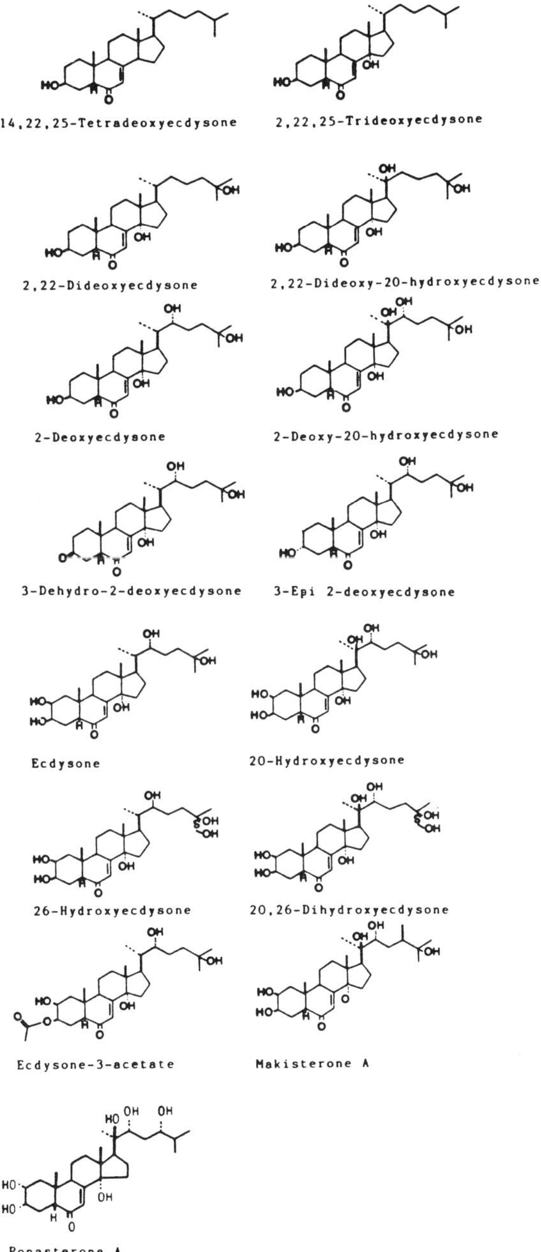

FIGURE 1.5. Free ecdysteroids found in eggs of arthropods (primarily insects). (After several authors; see in particular Hoffmann and Lagueux, 1985.)

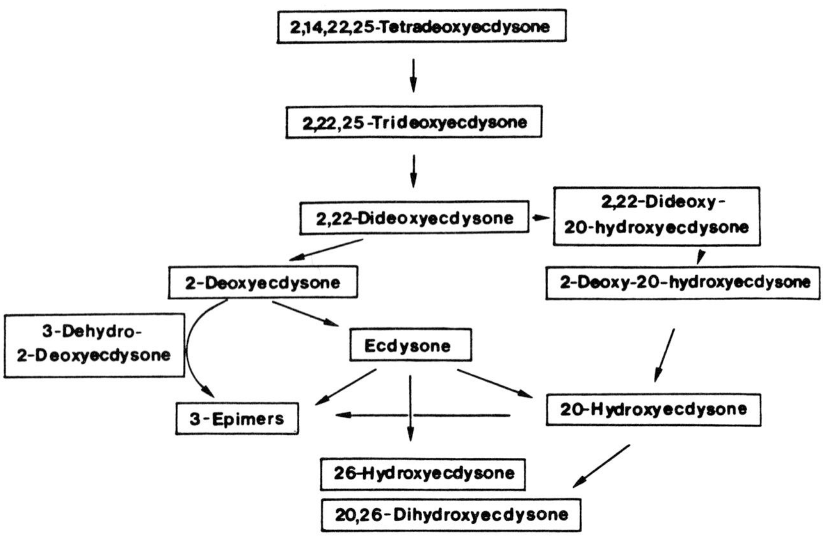

FIGURE 1.6. Biosynthetic and catabolic pathways of ecdysteroids in insect embryos as suggested by Hoffmann and Lagueux (1985).

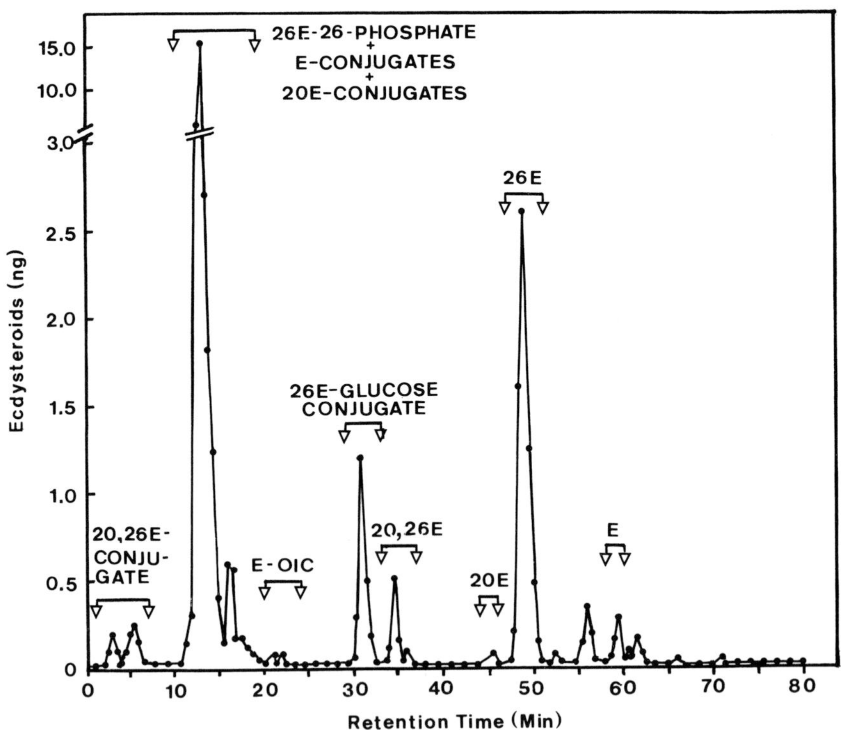

FIGURE 1.7. Ecdysteroids found in the 48-h embryo (shortly after katatrepsis) of *Manduca sexta*. KEY: IS-RPHPLC/RIA analysis; E = ecdysone; 20E = 20-hydroxyecdysone; 26E = 26-hydroxyecdysone; 20,26E = 20, 26-dihydroxyecdysone; E-OIC = ecdysonic acid. (After Warren et al., 1986.)

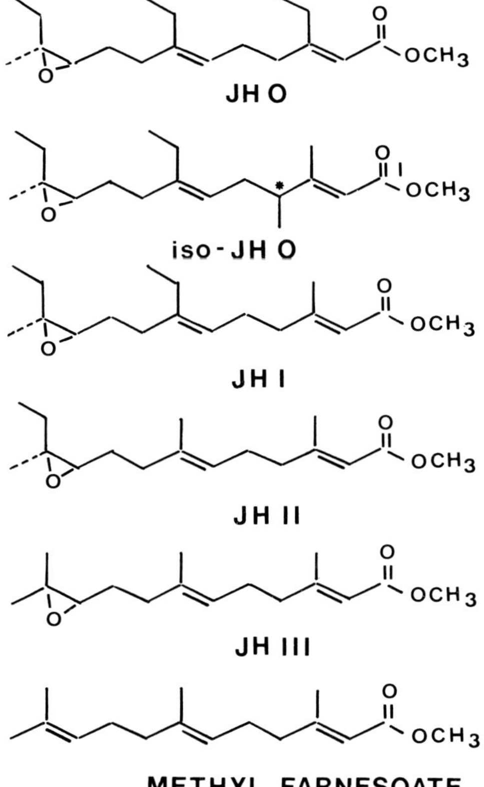

FIGURE 1.8. Juvenile hormones found in arthropod embryos.

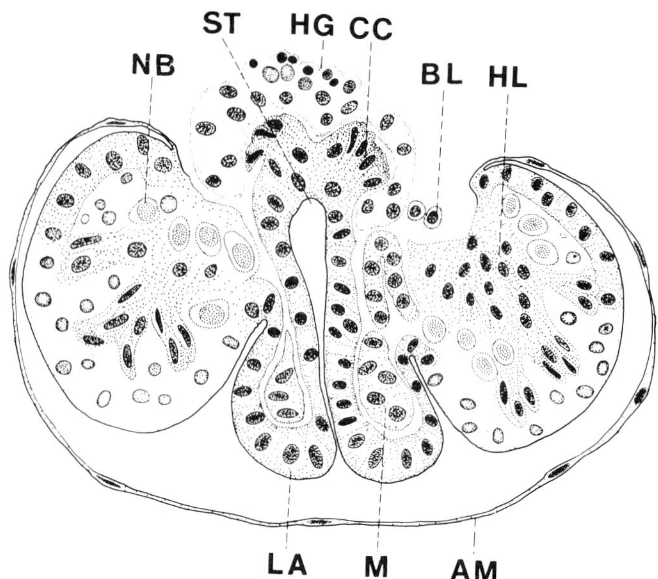

FIGURE 1.9. Differentiation of corpora cardiaca and hypocerebral ganglion from the stomodaeum roof in an insect, *Oncopeltus fasciatus*, 56-h embryo (46% of embryogenesis). Cross section. KEY: AM = amnion; BL = blood cell; CC = differentiating corpus cardiacum anlage; HG = hypocerebral ganglion; HL = head lobe; LA = labium anlage; M = mesoderm; NB = neuroblast; ST = stomodaeum. (From Dorn, 1972.)

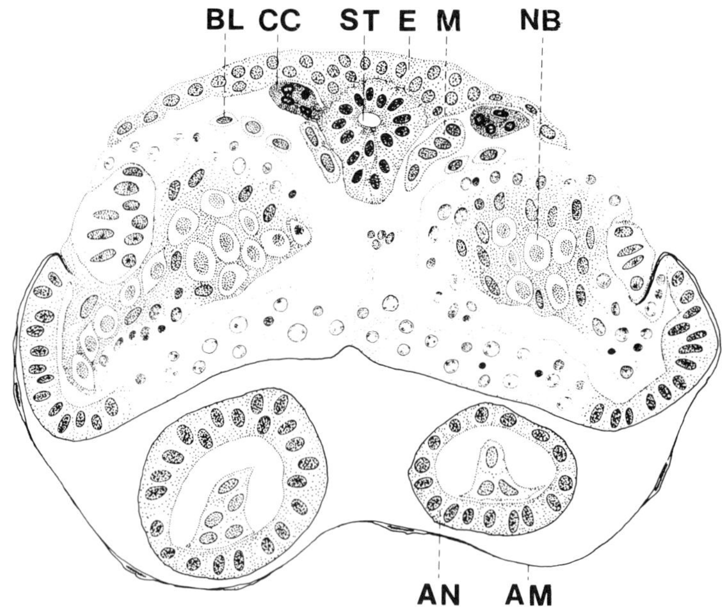

FIGURE 1.10. Differentiation of corpora cardiaca in an insect, *Oncopeltus fasciatus*, 60-h embryo (49% of embryogenesis). Cross section. KEY: AM = amnion; AN = antenna anlage; BL = blood cell; CC = corpus cardiacum anlage; E = entoderm; M = mesoderm; NB = neuroblast; ST = stomodaeum. (From Dorn, 1972.)

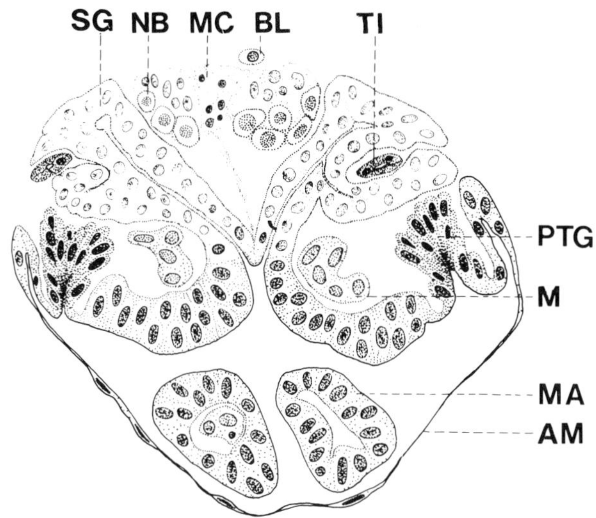

FIGURE 1.11. Differentiation of prothoracic glands in an insect, *Oncopeltus fasciatus*, 62-h embryo (50% of embryogenesis). Cross section. KEY: AM = amnion; BL = blood cell; M = mesoderm; MA = maxilla anlage; MC = median cord; NB = neuroblast; PTG = differentiating prothoracic gland anlage; SG = salivary gland anlage; TI = tubular invagination. (From Dorn, 1972.)

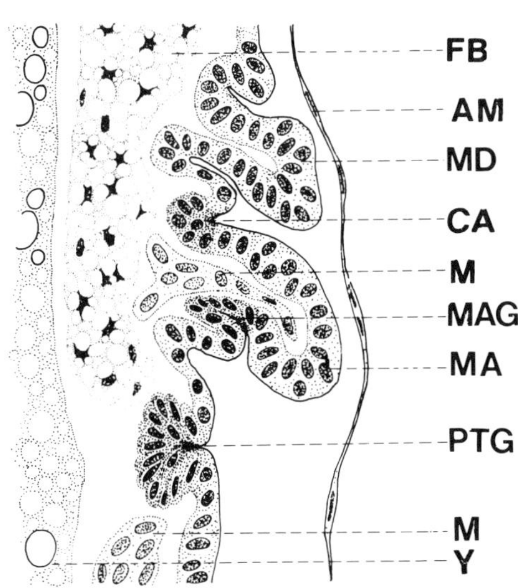

FIGURE 1.12. Differentiation of corpus allatum and prothoracic gland in an insect, *Oncopeltus fasciatus*, 64-h embryo (52% of embryogenesis). Parasagittal section. KEY: AM = amnion; CA = differentiating corpus allatum anlage; FB = fat body; M = mesoderm; MA = maxilla anlage; MAG = differentiating maxillary gland anlage; MD = mandibel anlage; PTG = differentiating prothoracic gland anlage; Y = yolk. (From Dorn, 1972.)

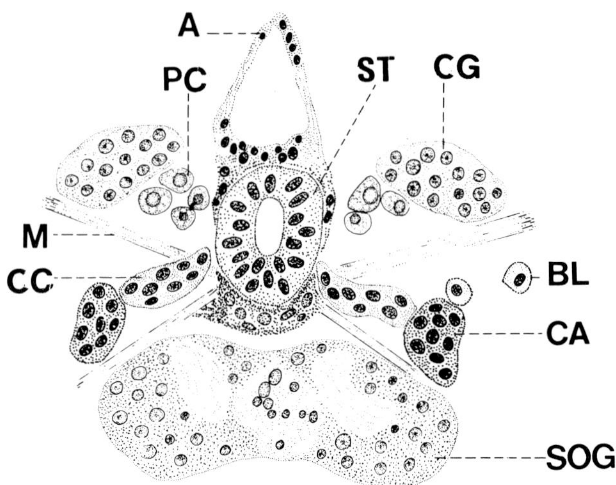

FIGURE 1.13. Corpora allata and corpora cardiaca in the embryo of an insect, *Oncopeltus fasciatus*, 82-h embryo (67% of embryogenesis). Cross section. KEY: A = aorta; BL = blood cell; CA = corpus allatum anlage; CC = corpus cardiacum anlage; CG = cerebral ganglion; M = muscle; PC = pericardial cells; SOG = suboesophageal ganglion; ST = stomodaeum. (From Dorn, 1972.)

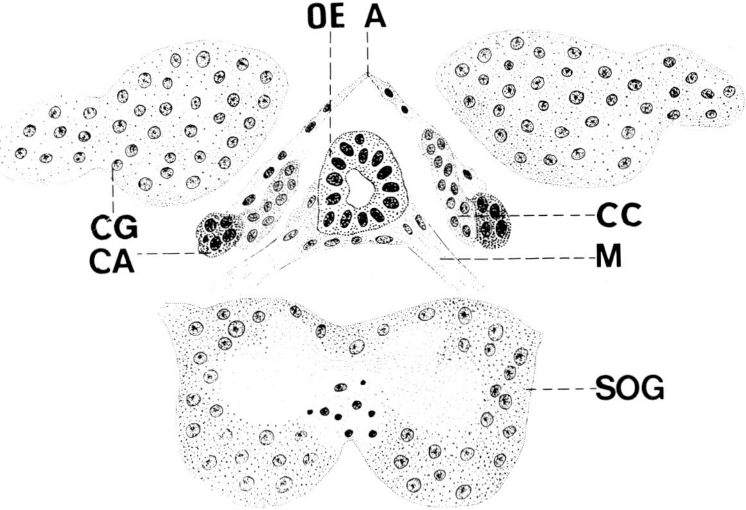

FIGURE 1.14. Corpora allata and corpora cardiaca in the embryo of an insect, *Oncopeltus fasciatus*, 92-h embryo (75% of embryogenesis). Cross section. KEY: A = aorta; CA = corpus allatum; CC = corpus cardiacum; CG = cerebral ganglion; M = muscle; OE = oesophagus; SOG = suboesophageal ganglion. (From Dorn, 1972.)

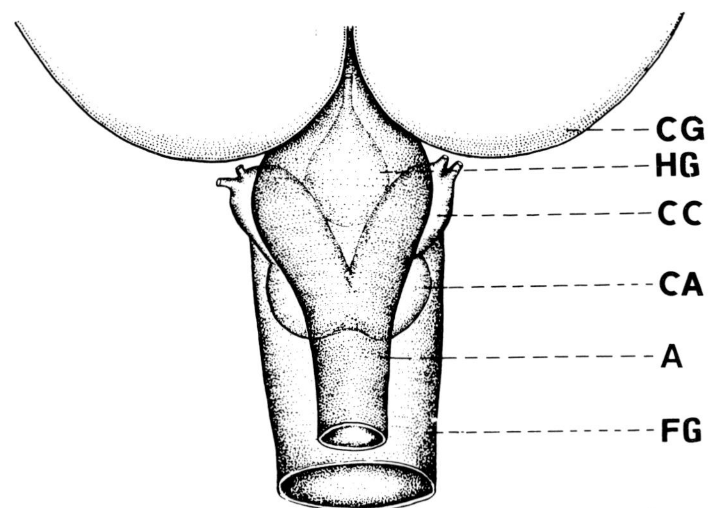

FIGURE 1.15. Organization of the retrocerebral complex in the embryo of an insect, *Oncopeltus fasciatus*, close to hatching. Dorsal view. KEY: A = aorta; CA = corpus allatum; CC = corpus cardiacum; CG = cerebral ganglion; FG = forgut; HG = hypocerebral ganglion. (From Dorn, 1972.)

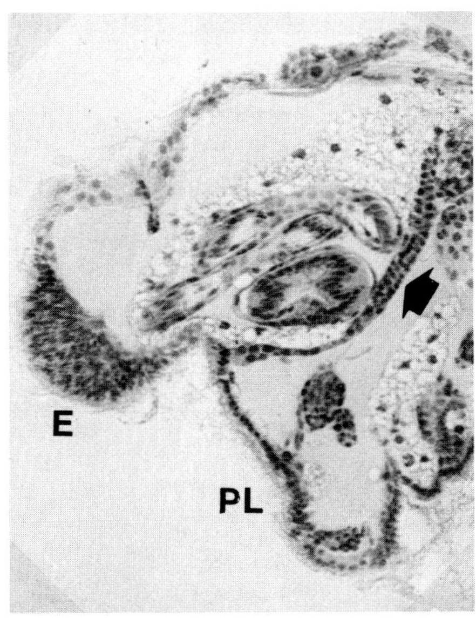

FIGURE 1.16. The prothoracic gland in the embryo of an insect, *Oncopeltus fasciatus*, close to hatching. Para-sagittal section. Arrow points to the prothoracic gland. KEY: E = eye; PL = base of prothoracic leg.

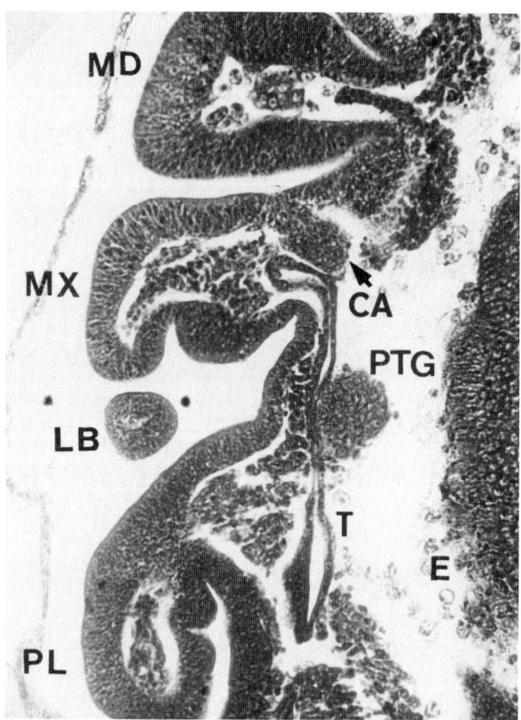

FIGURE 1.17. Differentiation of corpus allatum and prothoracic gland in the embryo of an insect, *Manduca sexta*, 42-h embryo (35% of embryogenesis). Para-sagittal section. KEY: CA = differentiating corpus allatum anlage; E = entoderm; LB = part of the labium anlage; MD = mandibel anlage; MX = maxilla anlage; PL = prothoracic leg anlage; PTG = differentiating prothoracic gland anlage; T = trachea.

FIGURE 1.18. Fine structural organization of embryonic prothoracic gland and ventral gland respectively in insect embryos. KEY: BM = basement membrane; DS = dilated space between adjacent cells; G = glycogen; N = nucleus; PG = pigment granule. (A) Prothoracic gland in the embryo of *Oncopeltus fasciatus*, close to hatching; the gland is considered to be inactive. (From Dorn, 1983.) (B) Ventral gland in the embryo of *Carausius morosus*, stage VI, first phase of presumed endocrine activity. (Drawn from Fig. 1 of Haget et al., 1982).

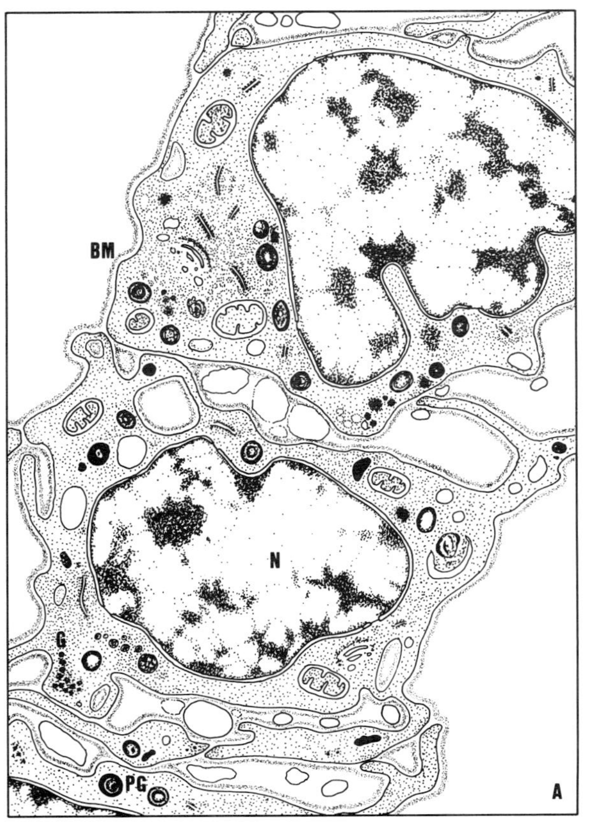

BM
N
G
PG
A

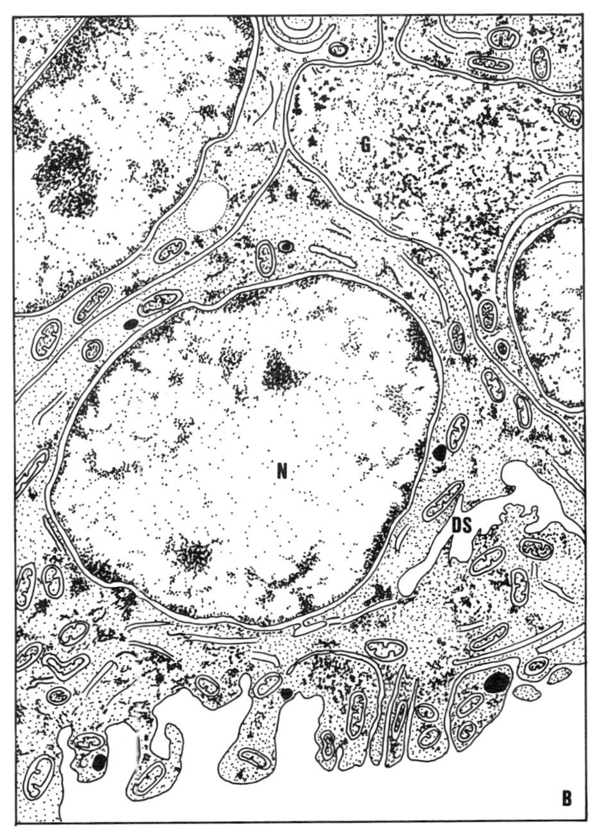

G
N
DS
B

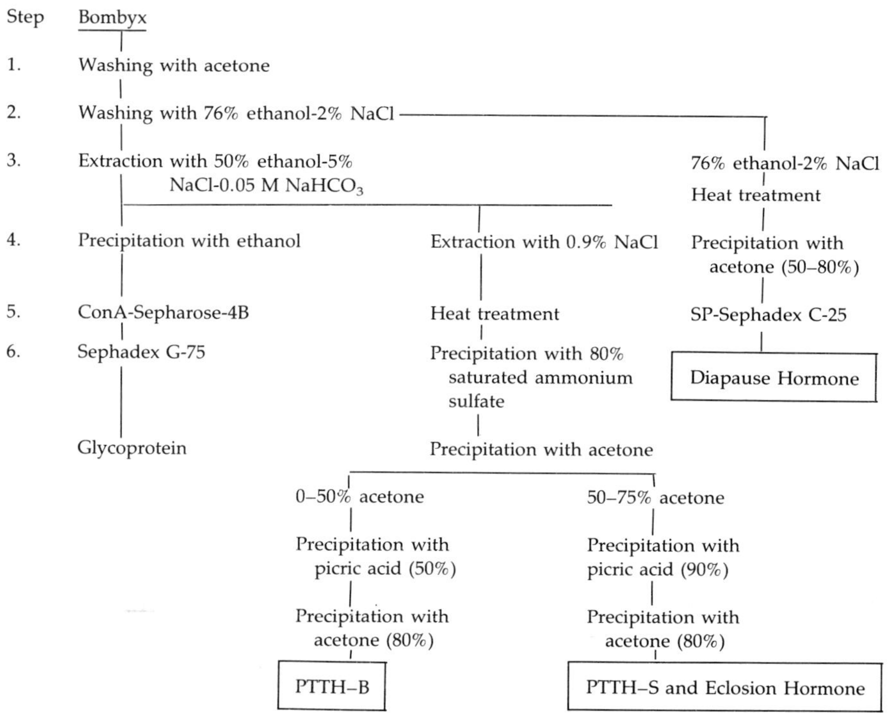

FIGURE 1.19. Purification procedure of neuropeptides from embryos of *Bombyx mori*. (From Fugo, 1987.)

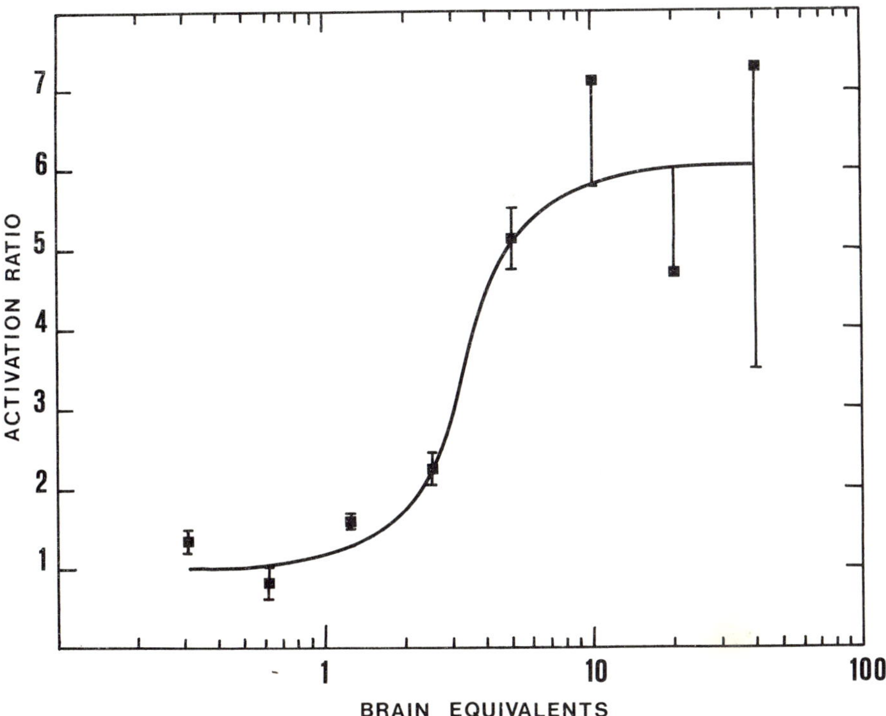

FIGURE 1.20. Prothoracicotropic activity in extracts of embryonic brains of an insect, *Manduca sexta,* dose–response of prothoracic gland activation by an extract of 117-h embryonic brains; activation ratio is the multiple of the basal level of ecdysteroid synthesis by larval prothoracic glands.

FIGURE 1.21. Neurosecretion in the insect embryo. (A) Part of a neurosecretory cell from the pars intercerebralis region in the brain of *Carausius morosus;* embryo close to hatching. ×8000. (B) Part of neurosecretory cell and neuropil in the brain of *Carausius morosus;* embryo close to hatching; note the neurosecretory nerve fibers in the neuropil (arrows). ×14,000. (C) Neurosecretory axons (arrows) in the aorta wall (the neurohemal organ) of *Oncopeltus facsiatus;* embryo close to hatching. ×15,000.

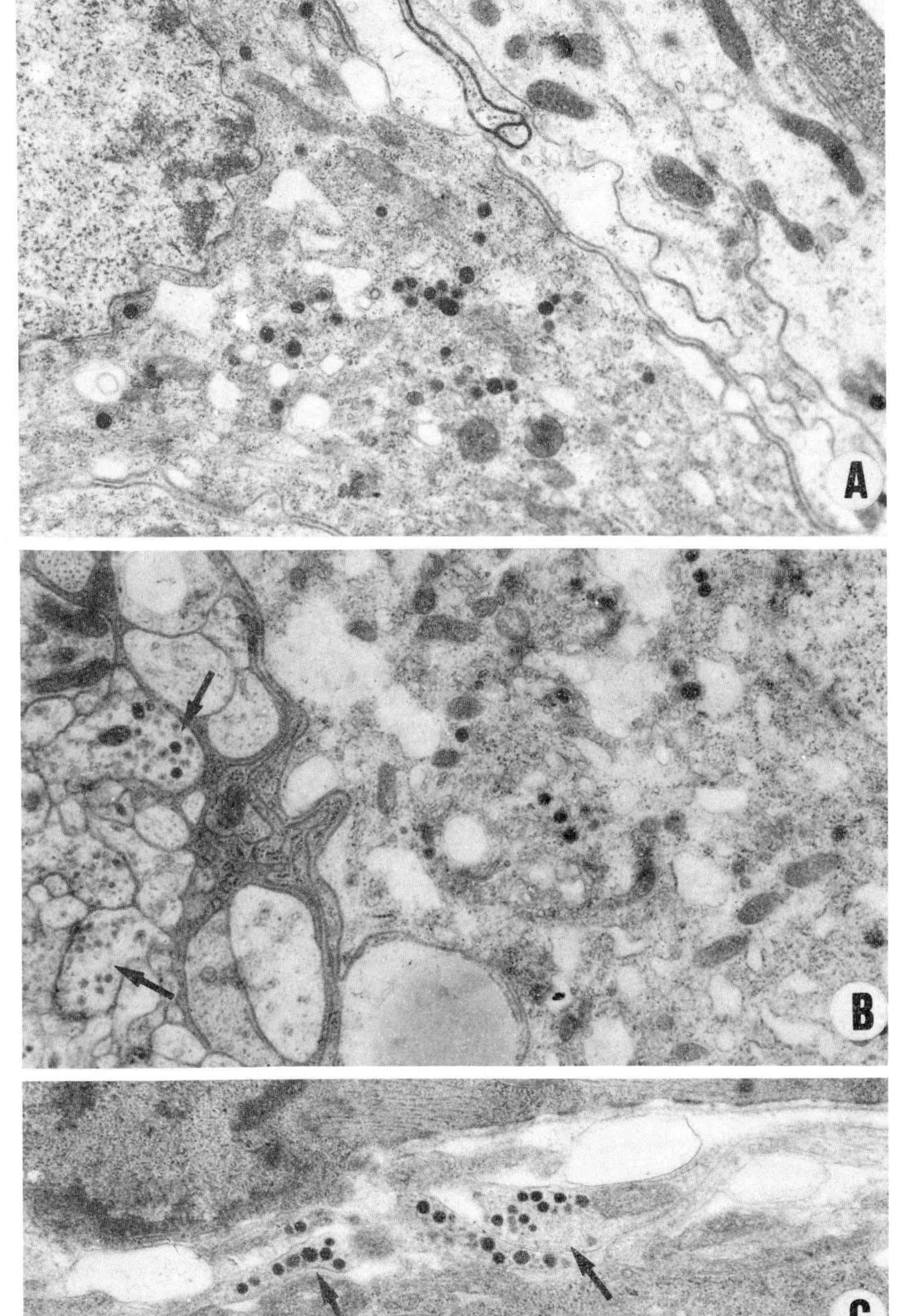

A. Dorn

FIGURE 1.22. Differentiation of the corpus cardiacum in an insect embryo, *Oncopeltus fasciatus*. (From Dorn, 1975b.) (A) The newly formed anlagen of the corpora cardiaca (arrows) at both sides of the stomodaeum (ST), 74-h embryo (60% of embryogenesis). ×1250. (B) Organization of the corpus cardiacum during hatching; note the accumulation of neurosecretory granules in the perikarya. KEY: N = nucleus; OE = oesophagus wall with cuticle against the lumen. ×8,000.

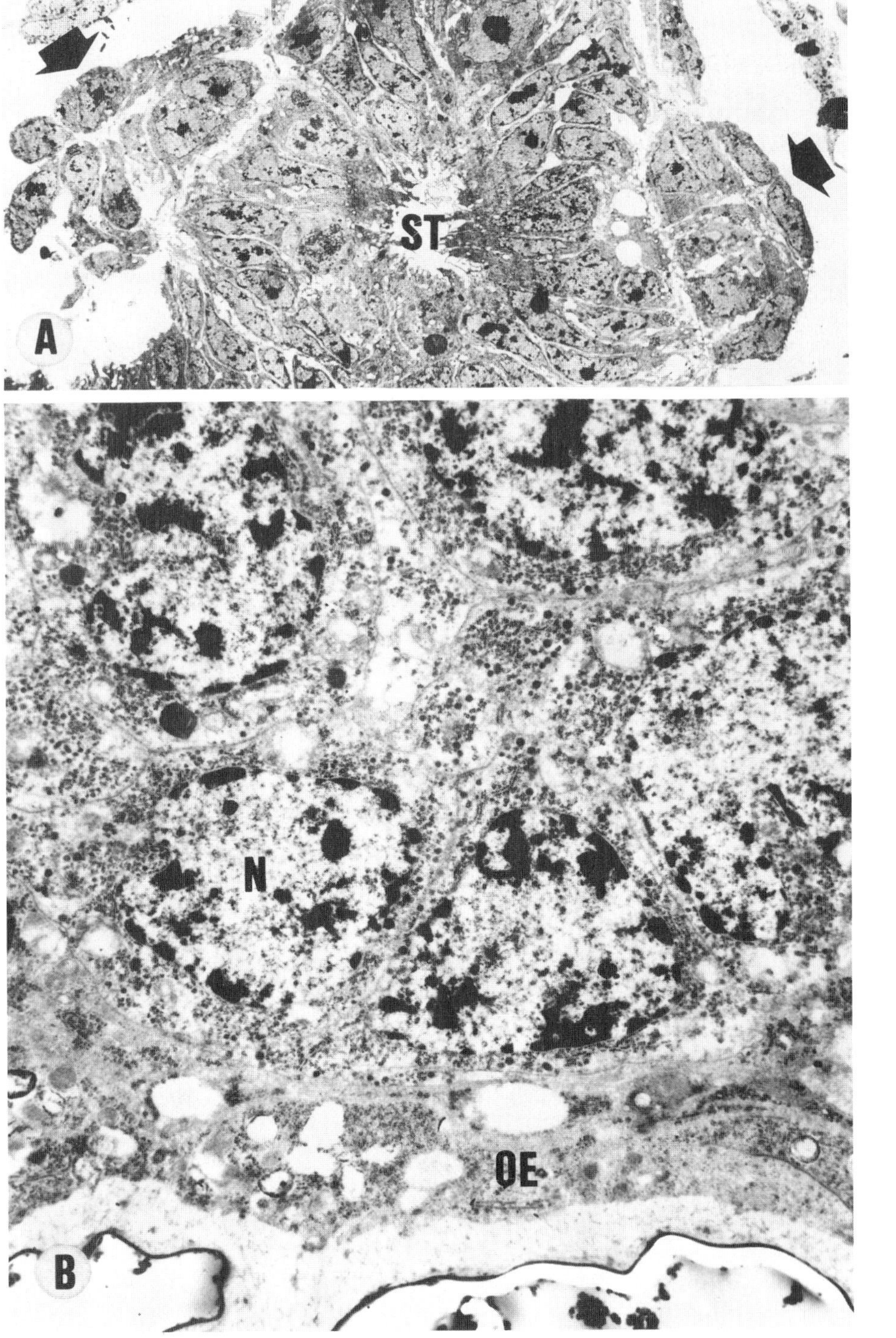

FIGURE 1.23. Neurosecretions in the corpus cardiacum and frontal ganglion of an insect embryo, *Carausius morosus*, close to hatching. (From Dorn, 1985.) (A) Intrinsic and extrinsic elements in the corpus cardiacum. KEY: H = hemocoel; N1 = nucleus of intrinsic neurosecretory cell; N2 = nucleus of glial elements; NA = neurosecretory axon of the extrinsic part of the corpus cardiacum. ×15,000. (B) Neuropil of the frontal ganglion; it is rather distinctly divided into areas with conventional nerve fibers (right half of the micrograph) and such with neurosecretory nerve fibers (left half of the micrograph). ×10,000.

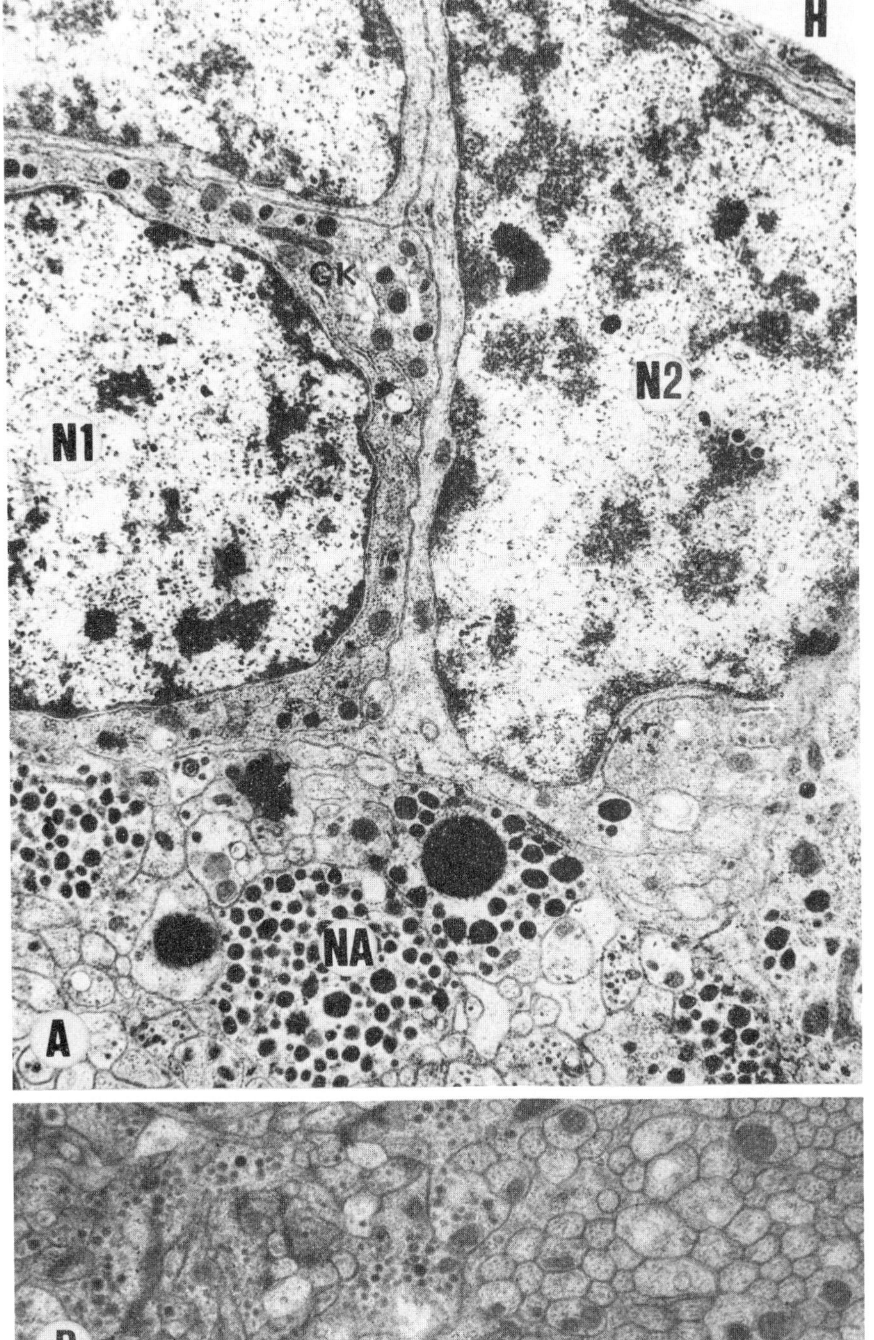

H
CK
N1
N2
NA
A
B

FIGURE 1.24. Differentiation of the corpus allatum in an insect embryo, *Oncopeltus fasciatus*. (From Dorn, 1975a.) (A) Newly formed corpus allatum anlage, 74-h embryo (60% of embryogenesis). ×8500. (B) Part of corpus allatum cell in the 98-h embryo (80% of embryogenesis). KEY: G = glycogen; N = nucleus; PG = pigment granule; RER = concentric whorl of rough endoplasmic reticulum. ×9400.

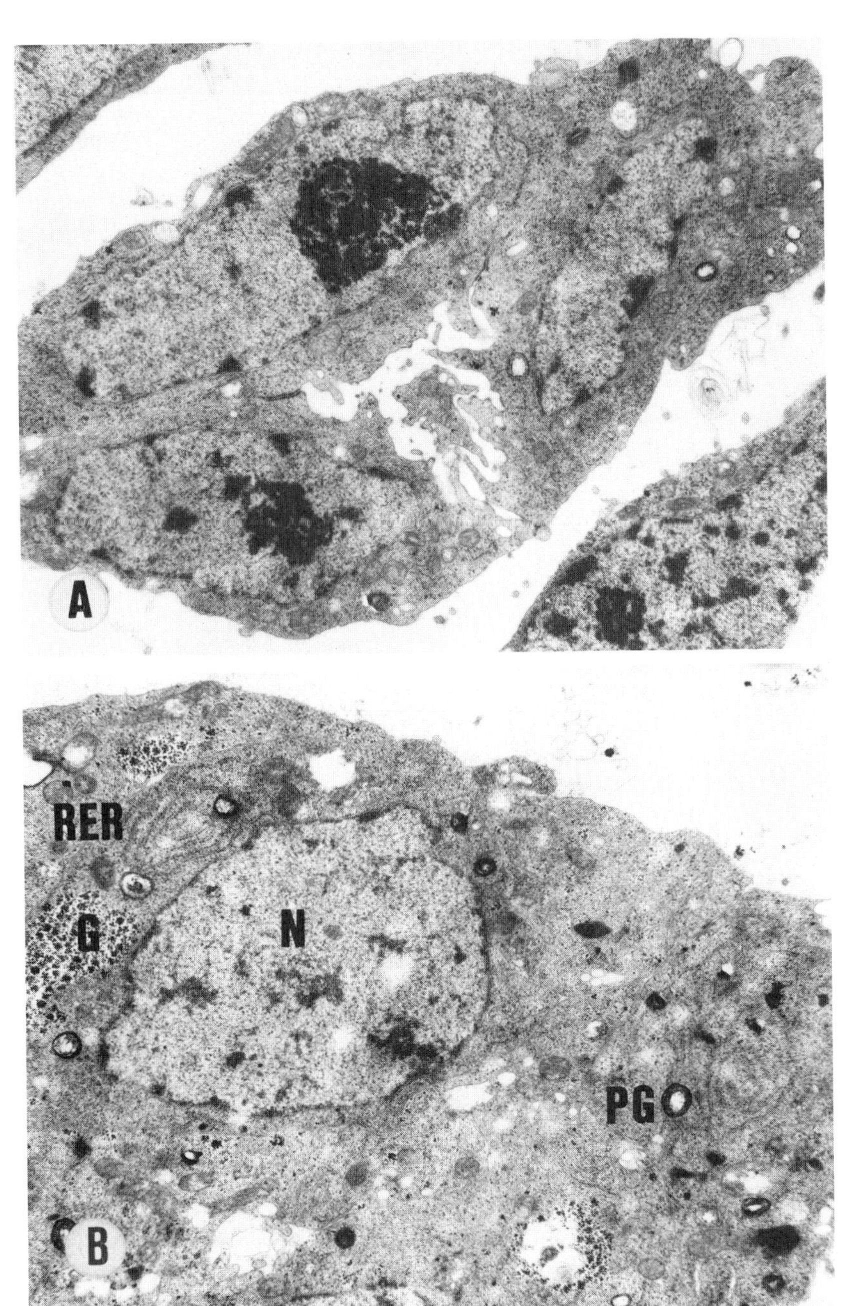

A
B
RER
G
N
PGO

A. Dorn

FIGURE 1.25. The corpus allatum in an insect embryo, *Carausius morosus,* during deposition of larval cuticle. (From Dorn, 1985.) (A) Part of corpus allatum cells facing the hemocoel; note profiles of neurosecretory axons (arrows). KEY: BM = basement membrane; GA = Golgi apparatus; H = hemocoel; L = lysosome; N = nucleus; RER = rough endoplasmic reticulum. ×28,000. (B) Part of corpus allatum cells facing the lumen of the vesicular gland; note profiles of neurosecretory axons (arrows) and the richly interdigitating plasma membranes. KEY: L = lumen of the gland. ×20,000.

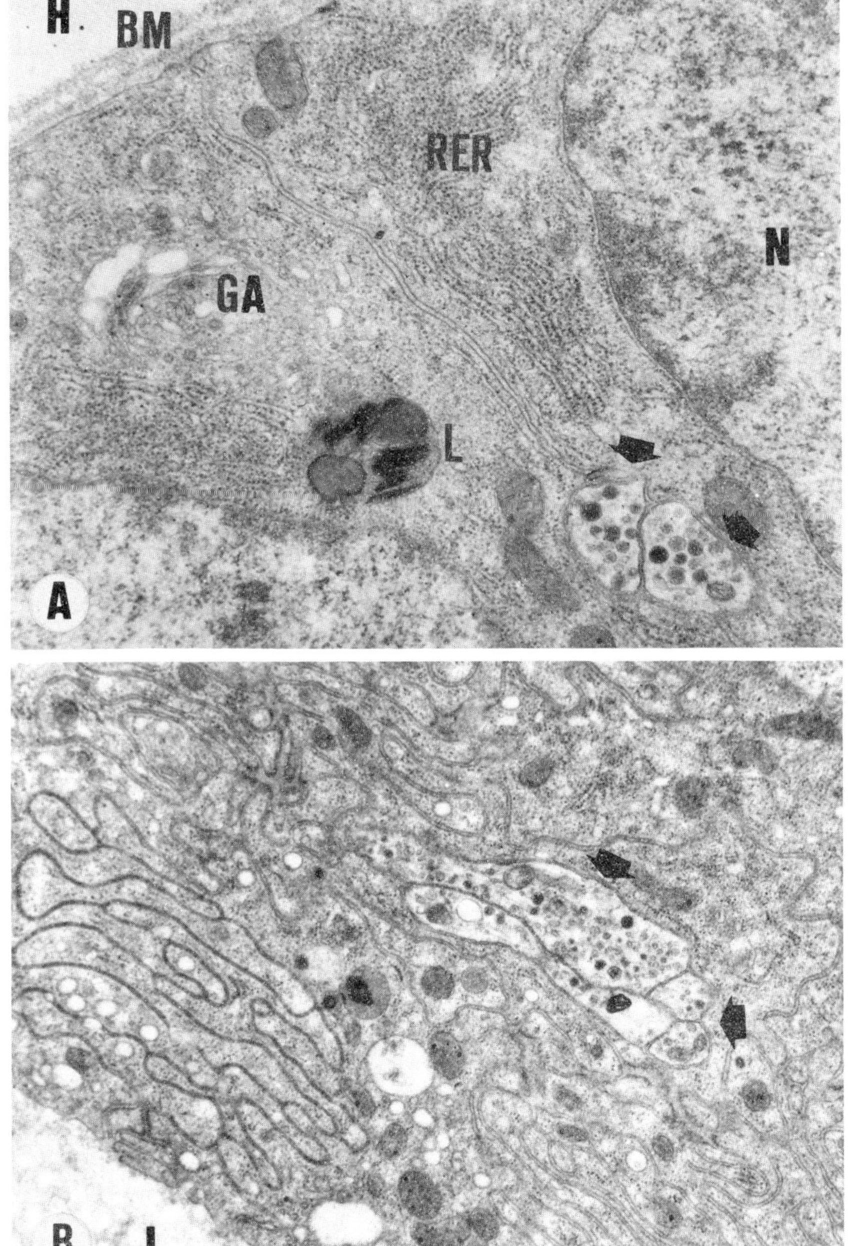
H.
BM
RER
N
GA
L
A
B L

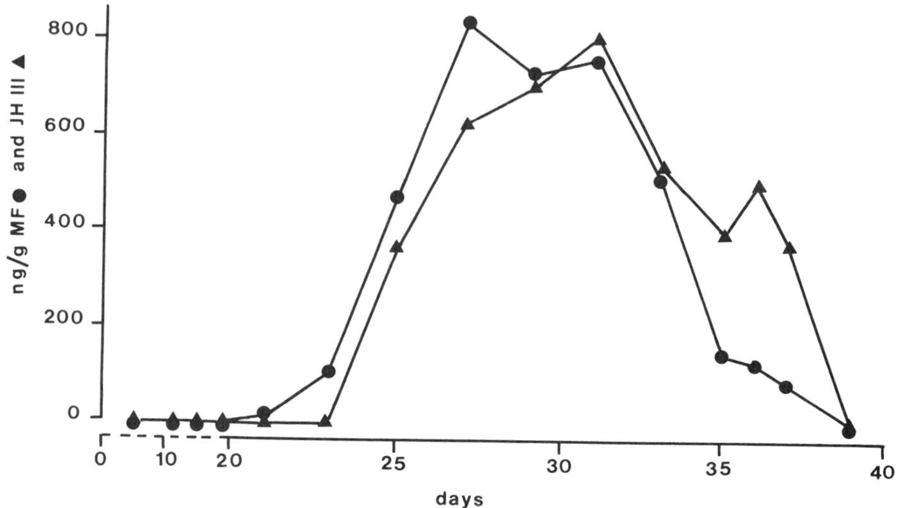

FIGURE 1.26. Titers of JH III (triangles) and of methyl farnesoate (MF) (dots) throughout embryonic development of an insect, *Nauphoeta cinerea*. GC analyses. Abscissa: age of embryos in days. (From Brüning et al., 1985).

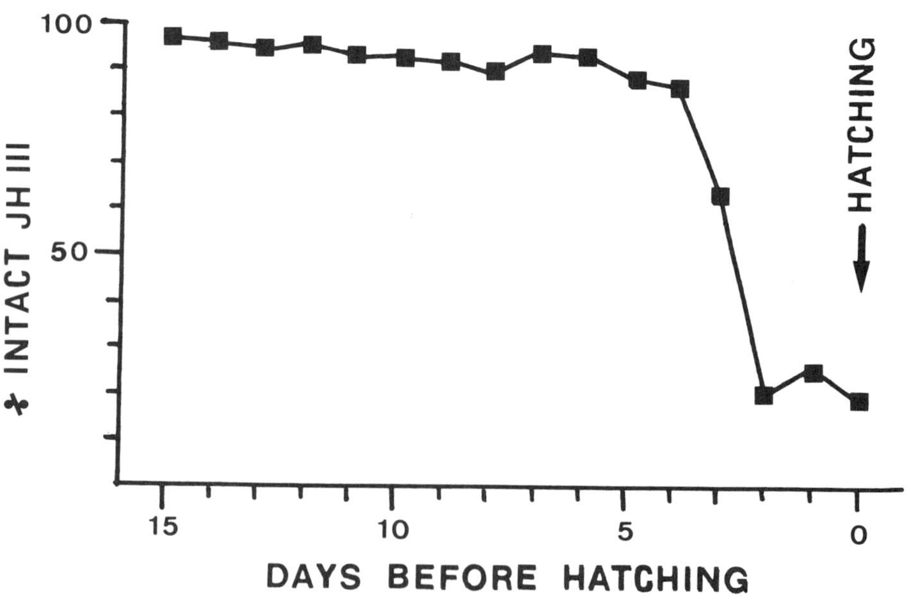

FIGURE 1.27. Degradation of JH III by homogenates of an insect embryo, *Nauphoeta cinerea*; note that only homogenates of embryos close to hatching are active. (From Bürgin and Lanzrein, 1988.)

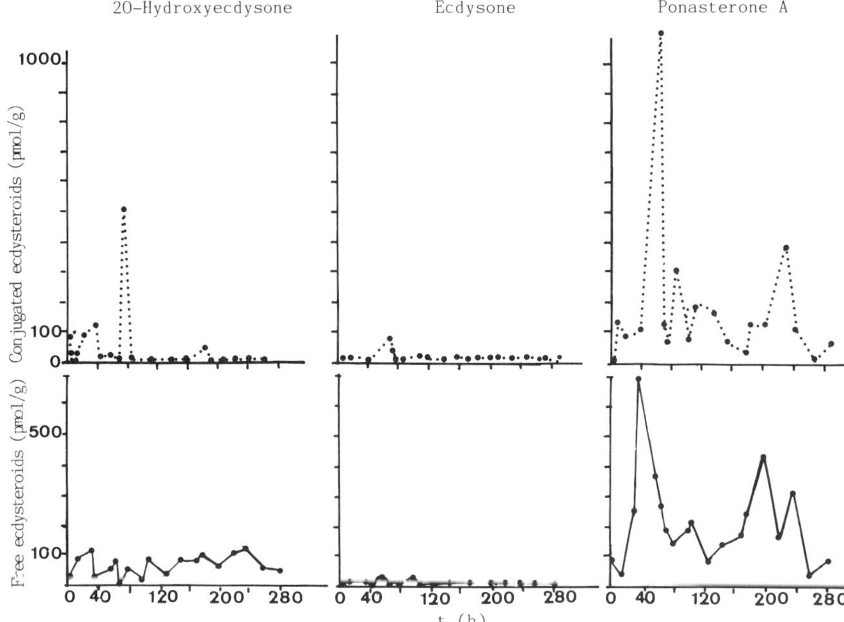

FIGURE 1.28. Ecdysteroid titers during embryonic development of a crustacean, *Carcinus maenas*. Upper graphs: conjugated ecdysteroids. Bottom graphs: free ecdysteroids (20-hydroxyecdysone, ecdysone and ponasterone A). Abscissa: embryonic development in hours. (From Lachaise and Hoffmann, 1982.)

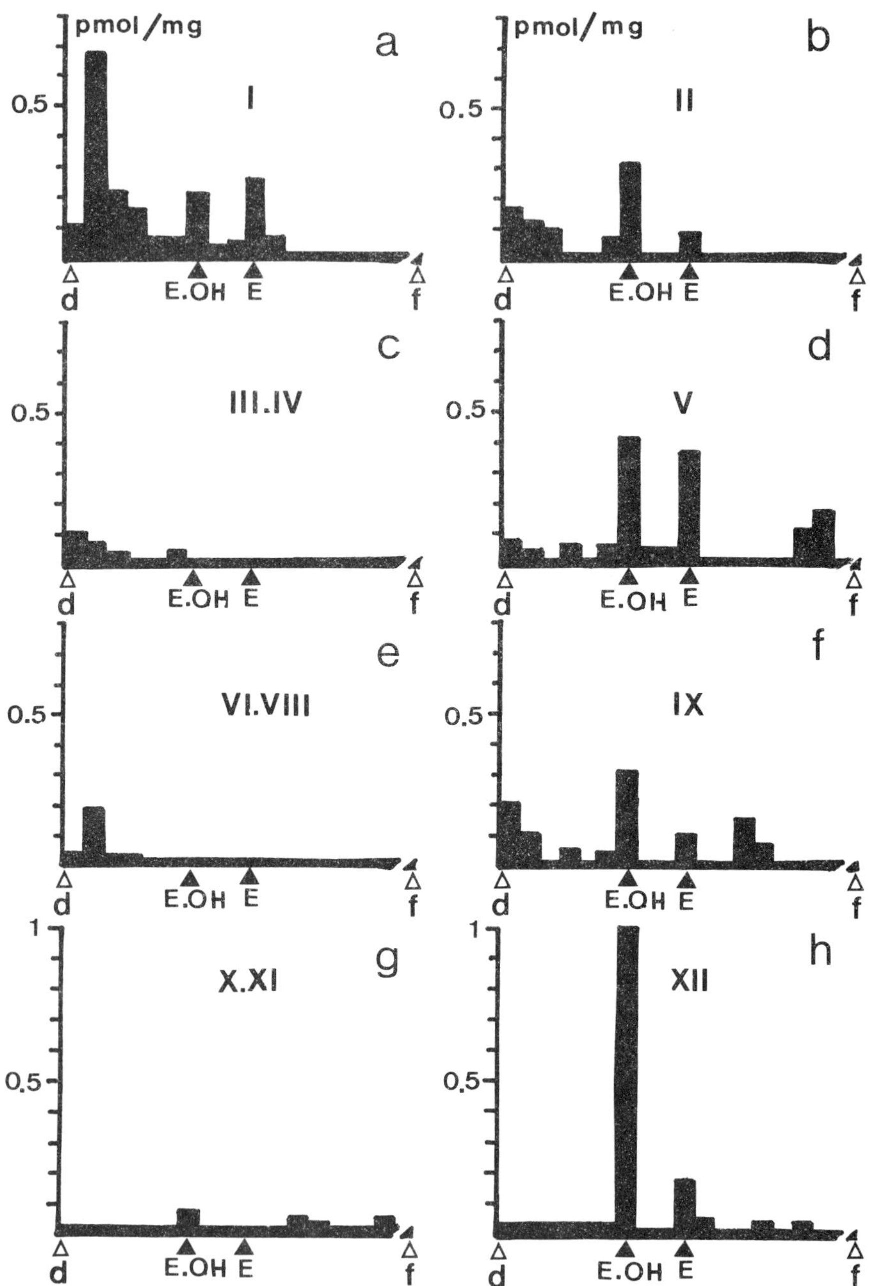

FIGURE 1.29. Ecdysteroid titers during embryonic development of a crustacean, *Acanthonyx lunulatus*. TLC separation and RIA quantification. KEY: I–XII = developmental stages; E = ecdysone; E-OH = 20-hydroxyecdysone; d = start; f = front of chromatogramme. (From Chaix and De Reggi, 1982.)

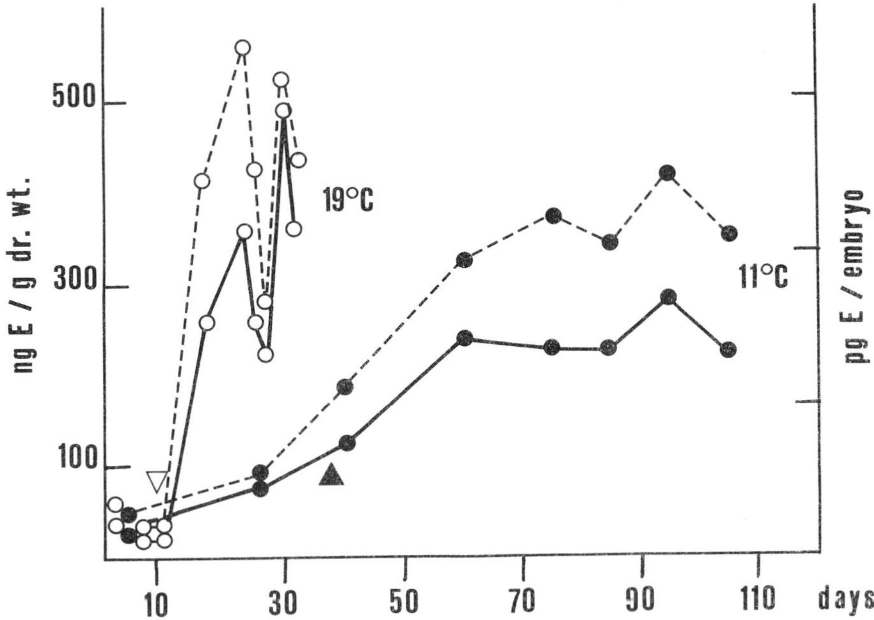

FIGURE 1.30. Ecdysteroid titer during embryonic development of a crustacean, *Palaemon serratus*. Total RIA activity per gram dry weight (———) or per embryo (---) at 11–12°C (●) or 19°C (○). Abscissa: embryonic development in days. The triangles indicate the appearance of the Y-organ (▽ at 19°C; ▲ at 11°C); E = ecdysone equivalents. (From Spindler et al., 1987.)

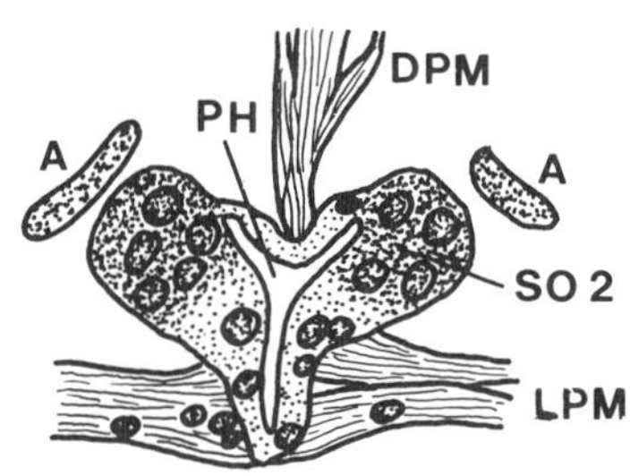

FIGURE 1.31. Differentiation of Schneider's organ II in a chelicerate embryo, *Dolomedes fimbriatus,* from the pharynx; 16-day embryo; note the similarity to the mode of insect corpus cardiacum formation (Fig. 1.9). KEY: A = aorta anlage; DPM = dorsal pharynx muscle; LPM = lateral pharynx muscle; PH = pharynx; SO2 = differentiating anlage of Schneider's organ II. (Redrawn from Legendre, 1959.)

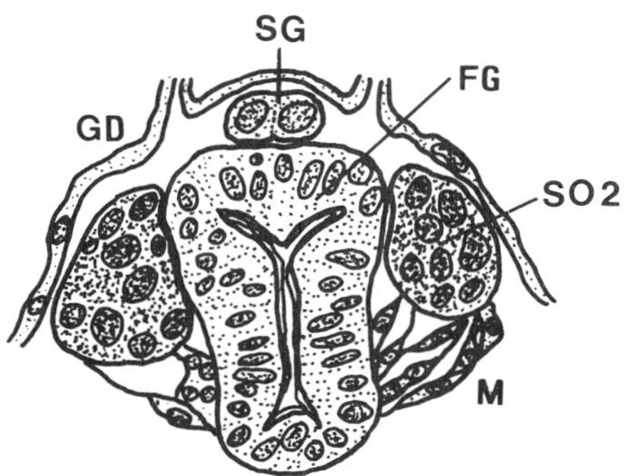

FIGURE 1.32. Anlage of Schneider's organ II in a chelicerate embryo, *Pardosa amentata,* cocoon stage close to hatching; note the similarity of the topographic relations of the organs to insect corpus cardiacum anlage (Fig. 1.10). KEY: FG = foregut; GD = gut diverticle; M = mesoderm; SG = supraoesophageal ganglion; SO2 = anlage of Schneider's organ II. (Redrawn from Streble, 1966.)

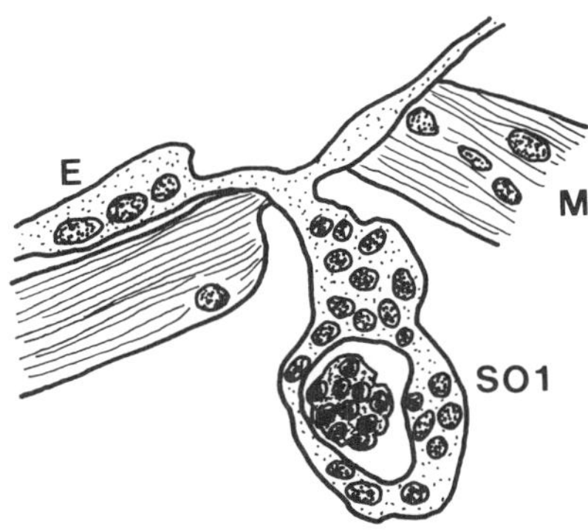

FIGURE 1.33. Differentiation of Schneider's organ I in a chelicerate embryo, *Dolomedes fimbriatus,* from the chelicero-rostral region; 16-day embryo; note the similarity to the mode of insect corpus allatum formation (Figs. 1.12 and 1.17). KEY: E = ectoderm; M = muscle; SO1 = differentiating anlage of Schneider's organ I. (Redrawn from Legendre, 1959).

Acknowledgments

The research work of the author was supported by grants of the Deutsche Forschungsgemeinschaft (Do 163). Ms. E. Sehn and Ms. C. Friederichs gave excellent technical support.

References

Agui, N., W. E. Bollenbacher, N. A. Granger, and L. I. Gilbert. 1980. Corpus allatum is release site for insect prothoracicotropic hormone. Nature (Lond.) 285: 669–670.

Babu, S. 1973. Histology of the neurosecretory system and neurohemal organs of the spider, *Agriope aurantia* (Lucas). J. Morphol. 141: 77–98.

Baker, F. C., L. W. Tsai, C. C. Reuter, and D. A. Schooley. 1988. The absence of significant levels of the known juvenile hormones and related compounds in the milkweed bug, *Oncopeltus fasciatus*. Insect Biochem. 18: 453–462.

Bassal, T. T. M. 1974. Biochemical and physiological studies of certain ticks (Ixodoidea). Activity of juvenile hormone analogues during embryogenesis in *Hyalomma (H.) dromedarii* Koch (Ixodidae). Z. Parasitenkd. 45: 85.

Bellon-Humbert, C. 1983. Développement embryonnaire de *Palaemon serratus*: résultats préliminaires; bases biologiques de l'aquaculture. INFREMER (Inst. Fr. Rech. Exploit. Mer) Actes Colloq. 1: 181–194.

Bergot, B. J., F. C. Baker, D. C. Cerf, G. Jamieson, and D. Schooley. 1981a. Qualitative and quantitative aspects of juvenile hormone titres in developing embryos of several insect species: discovery of a new JH-like substance extracted from eggs of *Manduca sexta*. Pp. 33–45 *in* G. E. Pratt and G. T. Brooks (eds.), *Juvenile Hormone Biochemistry: Developments in Endocrinology*, Vol. 15. Elsevier/North-Holland Publ., Amsterdam and New York.

Bergot, B. J., G. C. Jamieson, M. A. Ratcliff, and D. A. Schooley. 1980. JH zero: new naturally occurring insect juvenile hormone from developing embryos of the tobacco hornworm. Science (Wash., DC) 210: 336–338.

Bergot, B. J., M. Ratcliff, and D. A. Schooley. 1981b. Method for quantitative determination of the four known juvenile hormones in insect tissue using gas chromatography–mass spectroscopy. J. Chromatogr. 204: 231–244.

Bhaskaran, G., G. Jones, and D. Jones. 1980. Neuroendocrine regulation of corpus allatum activity in *Manduca sexta:* sequential neurohormonal and nervous inhibition in the last larva. Proc. Nat. Acad. Sci. USA 77: 4407–4411.

Blanchet, M. F., P. Porcheron, and F. Dray. 1979. Variations du taux des ecdystéroïds au cours du cycle de mue et de vitellogenèse chez le crustace Amphipode *Orchestia gammarellus.* Int. J. Invertebr. Reprod. 1: 133–139.

Bollenbacher, W. E. and N. A. Granger. 1985. Endocrinology of the prothoracicotropic hormone. Pp. 109–151 *in* G. A. Kerkut and L. I. Gilbert (eds.), *Comprehensive Insect Physiology, Biochemistry and Pharmacology,* Vol. 7: *Endocrinology,* Pt. I. Pergamon Press, Oxford and Elmsford, NY.

Bollenbacher, W. E., N. Agui, N. A. Granger, and L. I. Gilbert. 1979. *In vitro* activation of insect prothoracic glands by the prothoracicotropic hormone. Proc. Nat. Acad. Sci. USA 76: 5148–5152.

Bollenbacher, W. E., H. Zvenko, A. K. Kumaran, and L. I. Gilbert. 1978. Changes in ecdysone content during postembryonic development of the wax moth, *Galleria mellonella:* the role of the ovary. Gen. Comp. Endocrinol. 34: 169–179.

Bollenbacher, W. E., E. J. Katahira, M. O'Brien, L. I. Gilbert, M. K. Thomas, N. Agui, and A. H. Baumhover. 1984. Insect prothoracicotropic hormone: evidence for two molecular forms. Science (Wash., DC) 224: 1243–1245.

Boohar, R. and D. H. Bucklin. 1963. Possible ecdysone activity in the embryo of the grasshopper, *Melanoplus differentialis.* Am. Zool. 3: 496.

Borg, T. K., R. A. Bell, and D. J. Picard. 1973. Ultrastructure of neurosecretory cells in the frontal ganglion of the tobacco hornworm, *Manduca sexta* (L.). Tissue Cell 5: 259–267.

Borst, D. W., H. Laufer, M. Landau, E. S. Chang, W. A. Hertz, F. C. Baker, and D. A. Schooley. 1987. Methyl farnesoate and its role in crustacean reproduction and development. Insect Biochem. 17: 1123–1127.

Brüning, E., A. Saxer, and B. Lanzrein. 1985. Methyl farnesoate and juvenile hormone III in normal and precocene treated embryos of the ovoviviparous cockroach, *Nauphoeta cinerea.* Int. J. Invertebr. Reprod. Dev. 8: 269–278.

Bückmann, D., G. Starnecker, K.-H. Tomaschko, E. Wilhelm, R. Lafont, and J.-P. 1986. Isolation and identification of major ecdysteroids from the pycnogonid *Pycnogonum litorale* (Ström) (Arthropoda, Pantopoda). J. Comp. Physiol. B156: 759–765.

Bürgin, C. and Lanzrein. 1988. Stage-dependent biosynthesis of methyl farnesoate and juvenile hormone III and metabolism of juvenile hormone III in embryos of the cockroach, *Nauphoeta cinerea.* Insect Biochem. 18: 3–9.

Cavallin, M. and B. Fournier. 1981. Characteristics of development and variations in ecdysteroid level in *Clitumnus* embryos deprived of their cephalic endocrine glands. J. Insect Physiol. 27: 527–534.

Chaix, J.-C. and M. De Reggi. 1982. Ecdysteroid levels during ovarian development and embryogenesis in the spider crab *Acanthonyx lunulatus*. Gen. Comp. Endocrinol. 47: 7–14.

Charlet, M., F. Goltzene, and J. A. Hoffmann. 1979. Experimental evidence for a neuroendocrine control of ecdysone biosynthesis in adult females of *Locusta migratoria*. J. Insect Physiol. 25: 463–466.

Chaudonneret, J. 1956. Le système nerveux de la région gnathale de l'écrevisse. Ann. Sci. Nat. Zool. Biol. Anim. 18: 33–61.

Chen, J. H., H. Fugo, M. Nakajima, H. Nagasawa, and A. Suzuki. 1986. Neurohormones in developing embryos of the silkworm. *Bombyx mori*. I. The presence of neurohormonal activities in embryos of the silkworm, *Bombyx mori*. J. Seric. Sci. Jpn. 55: 54–59.

Chen, J. H., H. Fugo, M. Nakajima, H. Nagasawa, and A. Suzuki. 1987. Neurohormones in developing embryos of the silkworm, *Bombyx mori*: the presence and characteristics of prothoracicotropic hormone B. J. Insect Physiol. 33: 407–411.

Connat, J.-L. and P. A. Diehl. 1986. Probable occurrence of ecdysteroid fatty acid esters in different classes of arthropods. Insect Biochem. 16: 91–97.

Connat, J.-L., P. A. Diehl, and M. Morici. 1984. Metabolism of ecdysteroids during the vitellogenesis of the tick *Ornithodorus moubata* (Ixodidae, Argasidae): accumulation of apolar metabolites in the eggs. Gen. Comp. Endocrinol. 56: 100–110.

Connat, J.-L., E. M. Dotson, and P. A. Diehl. 1987a. Metabolism of ecdysteroids in the female tick *Amblyomma hebraeum* (Ixodoidea, Ixodidae): accumulation of free ecdysone and 20-hydroxyecdysone in the eggs. J. Comp. Physiol. 157: 689–699.

Connat, J.-L., E. M. Dotson, and P. A. Diehl. 1987b. Comparative study of egg ecdysteroids of several tick species. Abstr. 8th Ecdysone Workshop, Marburg, W. Germany p.40.

Connat, J.-L., P. A. Diehl, H. Gfeller, and M. Morici. 1985. Ecdysteroids in females and eggs of the ixodid tick *Amblyomma hebraeum*. Int. J. Invertebr. Reprod. Dev. 8: 103–116.

Crosby, T., D. Lewis, and H. H. Rees. 1987. Ecdysteroids in adult and embryonic development of the cattle tick, *Boophilus microplus*. Abstr. 8th Ecdysone Workshop, Marburg, W. Germany p. 41.

De Kort, C. A. D. and N. A. Granger. 1981. Regulation of the juvenile titer. Annu. Rev. Entomol. 26: 1–28.

Dennis, R. D. W. 1977. On ecdysone-binding proteins and ecdysone-like material in nematodes. J. Parasitol. 7: 181–188.

Doane, W. W. 1973. Role of hormones. Pp. 291–497 *in* S. J. Counce and C. H. Waddington (eds.), *Developmental Systems: Insects* Vol. 2. Academic Press, Orlando, Florida.

Dorn, A. 1972. Die endokrinen Drüsen in Embryo von *Oncopeltus fasciatus* Dallas (Insecta, Heteroptera): Morphogenese, Funktionsaufnahme, Beeinflussung des Gewebewachstums und Beziehungen zu den embryonalen Häutungen. Z. Morphol. Tiere 74: 52–104.

Dorn, A. 1973. Electron microscopic study on the larval and adult corpus allatum of *Oncopeltus fasciatus* Dallas (Insecta, Heteroptera). Z. Zellforsch. Mikrosk. Anat. 145: 447–458.

Dorn, A. 1975a. Elektronenmikroskopische Studien über das neurosekretorische System von *Oncopeltus fasciatus* Dallas (Insecta, Heteroptera) während der Embryonalentwicklung. Cytobiologie 10: 227–234.

Dorn, A. 1975b. Elektronenmikroskopische Studien über Differenzierung und Funktionsaufnahme der Corpora cardiaca in Embryo von *Oncopeltus fasciatus* Dallas (Insecta, Heteroptera). Cytobiologie 10: 235–248.

Dorn, A. 1975c. Struktur und Funktion des embryonalen Corpus allatum von *Oncopeltus fasciatus* Dallas (Insecta, Heteroptera). Verh. Dtsch. Zool. Ges. 67: 85–89.

Dorn, A. 1977. Hormonal control of egg maturation and embryonic development in insects. Pp. 451–481 *in* K. G. Adiyodi and R. G. Adiyodi (eds.), *Advances in Invertebrate Reproduction*, Vol. 1. Peralam-Kenoth, Karivellur, India.

Dorn, A. 1978a. Ontogenese des endokrinen Systems der Insekten: Untersuchungen an *Oncopeltus fasciatus* (Heteroptera) und *Carausius morosus* (Phasmatodea). Habilitationsschrift. University of Mainz, W. Germany.

Dorn, A. 1987b. Neurosecretion in the frontal ganglion of the stick insect, *Carausius morosus*. Pp. 361–364 *in* W. Bargmann, Oksche, A. Polenov, and B. Scharrer (eds.), *Neurosecretion and Neuroendocrine Activity: Evolution, Structure and Function*. Springer-Verlag, Berlin and New York.

Dorn, A. 1982. Precocene-induced effects and possible role of juvenile hormone during embryogenesis of the milkweed bug, *Oncopeltus fasciatus*. Gen. Comp. Endocrinol. 46: 42–52.

Dorn, A. 1983. Hormones during embryogenesis of the milkweed bug, *Oncopeltus fasciatus* (Heteroptera: Lygaeidae). Entom. Gen. 8: 193–214.

Dorn, A. 1985. Neurosecretion in the insect embryo. Proc. Arthropod Embryol. Soc. Jpn., 1984 pp. 1–16.

Dorn, A. and P. Hoffmann. 1983. Segmentation and differentiation of appendages during embryogenesis of the milkweed bug, *Oncopeltus fasciatus:* a scanning electron microscopical study. Zool. Jahrb. Anat. 109: 277–298.

Dorn, A. and F. Romer. 1976. Structure and function of prothoracic glands and oenocytes in embryos and last larval instar of *Oncopeltus fasciatus* Dallas. Cell Tissue Res. 171: 331–350.

Dorn, A., L. I. Gilbert, and W. E. Bollenbacher. 1987a. Prothroracicotropic hormone activity in the embryonic brain of the tobacco hornworm, *Manduca sexta*. J. Comp. Physiol. B157: 279–283.

Dorn, A., S. T. Bishoff, and L. I. Gilbert. 1987b. An incremental analysis of the embryonic development of the tobacco hornworm, *Manduca sexta*. Int. J. Invertebr. Reprod. Dev. 11: 137–158.

Eckert, M. 1967. Experimentelle Untersuchungen zur Häutungsphysiologie bei Spinnen. Zool. Jahrb. Physiol. 73: 40–101.

El-Hifnawi, E. and G. Seifert. 1972. Elektronenmikroskopische und experimentelle Untersuchungen über die Kragendrüse von *Polyxenus lagurus* (L.) (Diplopoda, Penicillata). Z. Zellforsch. Mikrosk. Anat. 131: 255–268.

Enslee, E. C. and L. M. Riddiford. 1977. Morphological effects of juvenile hormone mimics on embryonic development in the bug *Pyrrhocoris apterus*. Wilhelm Roux' Arch. Entwicklungsmech. Org. 181: 163–181.

Fugo, H. 1987. Partial purification of glycoproteins from heads of adults of the silkworm, *Bombyx mori* (Lepidoptera: Bombycidae) and their physiological activity. Jpn. J. Appl. Entomol. Zool. 31: 194–200.

Fugo, H., J. H. Chen, M. Nakajima, H. Nagasawa, and A. Suzuki. 1987. Neurohormones in developing embryos of the silkworm, *Bombyx mori:* the presence and characteristics of prothoracicotropic hormone-S. J. Insect Physiol. 33: 243–248.

Fugo, H., H. Saito, H. Nagasawa, and A. Suzuki. 1985. Eclosion hormone activity in developing embryos of the silkworm, *Bombyx mori*. J. Insect Physiol. 31: 293–298.

Gabe, M. 1956. Histologie comparée de la glande de mue (organe Y) des crustacées malacostraces. Ann. Sci. Nat. Ser. Zool. Biol. Anim. 18: 145–152.

Gilbert, L. I. and H. A. Schneiderman. 1961. The content of juvenile hormone and lipid in Lepidoptera: sexual differences and developmental stages. Gen. Comp. Endocrinol. 1: 453–472.

Glass, H., H. Emmerich, and K.-D. Spindler. 1978. Immunohistological localization of ecdysteroids in the follicular epithelium of locust oocytes. Cell Tissue Res. 194: 237–244.

Goltzené, F., M. Lagueux, M. Charlet, and J. A. Hoffmann. 1978. The follicle cell epithelium of maturing ovaries of *Locusta migratoria:* a new biosynthetic tissue for ecdysone. Hoppe-Seyler's Z. Physiol. Chem. 359: 1427–1434.

Gorbman, A. and H. A. Bern. 1962. *A Textbook of Comparative Endocrinology.* Wiley, New York.

Goudeau, M. and F. Lachaise. 1983. Structure of the egg funiculus and deposition of embryonic envelopes in a crab. Tissue Cell 15: 47–62.

Guerriero, A. and F. Pietra. 1985. Isolation, in large amounts, of the rare plant ecdysteroid ajugasterone-C from the Mediterranean zoanthid *Gerardia savaglia.* Comp. Biochem. Physiol. 80B: 277–278.

Hagedorn, H. H., J. D. O'Connor, M. S. Fuchs, B. Sage, D. A. Schlaeger, and M. K. Bohm. 1975. The ovary as a source of α-ecdysone in an adult mosquito. Proc. Nat. Acad. Sci. USA 72: 3253–3259.

Hagedorn, H. H., J. P. Shapiro, and K. Hanaoka. 1979. Ovarian ecdysone secretion is controlled by a brain hormone in an adult mosquito. Nature (Lond.) 282: 92–94.

Haget, A., A. Ressouches, and J. Rogueda. 1981. Chronological observations on the activities of the embryonic corpus allatum in *Carausius morosus* Br. (Phasmida, Lonchodidae). Int. J. Insect Morphol. Embryol. 10: 65–81.

Haget, A., A. Ressouches, and J. Rogueda. 1982. Ultrastructural detection of endocrine activities in the ventral glands of the head of *Carausius* (Phasmida: Lonchodidae) during embryonic development. Int. J. Insect Morphol. Embryol. 11: 99–108.

Hammock, B. D. 1975. NADPH-dependent epoxidation of methyl farnesoate to juvenile hormone in cockroach *Blaberus giganteus* L. Life Sci. 17: 323–328.

Hartmann, R., C. Jendrsczok, and M. G. Peter. 1987. The occurrence of a juvenile hormone binding protein and *in vitro* synthesis of juvenile hormone by the serosa of *Locusta migratoria* embryos. Wilhelm Roux's Arch. Dev. Biol. 196: 347–355.

Hoffmann, J. A. and M. Lagueux. 1985. Endocrine aspects of embryonic development in insects. Pp. 435–460 *in* G. A. Kerkut and L. I. Gilbert (eds.), *Comprehensive Insect Physiology, Biochemistry and Pharmacology,* Vol. 1: *Embryogenesis and Reproduction.* Pergamon Press, Oxford and Elmsford, New York.

Hoffmann, J. A., M. Lagueux, C. Hetru, M. Charlet, and F. Goltzene. 1980. Ecdysone in reproductively competent adults and in embryos of insects. Pp. 431–465 *in* J. A. Hoffmann (ed.), *Progress of Ecdysone Research.* Elsevier/North-Holland Publ., Amsterdam and New York.

Horn, D. H. S., J. S. Wilkie, and J. A. Thomson. 1974. Isolation of β-ecdysone (20-hydroxyecdysone) from the parasitic nematode *Ascaris lumbricoides.* Experientia (Basel) 30: 1109–1110.

Hunt, L. M. and D. G. Shapiro. 1973. Larval development following juvenile hormone analogue treatment of eggs of the lesser milkweed bug *Lygaeus kalmii*. J. Insect Physiol. 19: 2129–2134.

Imboden, H. and B. Lanzrein. 1982. Investigations on ecdysteroids and juvenile hormones and on morphological aspects during early embryogenesis in the ovoviviparous cockroach, *Nauphoeta cinerea*. J. Insect Physiol. 28: 37–46.

Imboden, H., B. Lanzrein, J. P. Delbecque, and M. Lüscher. 1978. Ecdysteroids and juvenile hormone during embryogenesis in the ovoviviparous cockroach *Nauphoeta cinerea*. Gen. Comp. Endocrinol. 36: 628–635.

Injeyan, H. S., S. S. Tobe, and E. Rapport. 1979. The effects of exogenous juvenile hormone treatment on embryogenesis in *Schistocerca gregaria*. Can. J. Zool. 57: 838–845.

Ishizaki, H., A. Mizoguchi, M. Fujishita, A. Suzuki, I. Moriya, H. O'oka, H. Kataoka, A. Isogai, H. Nagasawa, S. Tamura, and A. Suzuki. 1983. Species specificity of the insect prothoracicotropic hormone (PTTH): the presence of *Bombyx*- and *Samia*-specific PTTHs in the brain of *Bombyx mori*. Dev. Growth & Differ. 25: 593–600.

Jamault-Navarro, C. and R. Joly. 1977. Localisation et structure des cellules neurosécrétrices protocérébrales chez *Lithobius forficatus* L. (Myriapode Chilopode). Gen. Comp. Endocrinol. 31: 106–120.

Jegla, T. C. 1982. A review of the molting physiology of the trilobite larva of *Limulus*. Pp. 83–101 *in* J. Bonaventura, C. Bonaventura, and S. Tesh (eds.), *Physiology and Biology of Horseshoe Crabs*, Liss, New York.

Joly, R. 1961. Déclenchement expérimental de la mue chez *Lithobius forficatus* L. (Myriapode Chilopode). C. R. Acad. Sci. Paris 252: 1673–1675.

Joly, R. 1962. Les glandes cérébrales, organes inhibiteur de la mue chez les Myriapodes Chilopodes. C. R. Acad. Sci. Paris 254: 1679–1681.

Joly, R. 1966a. Etude expérimental du cycle de mue et de sa régulation endocrine chez les Myriapodes Chilopodes. Gen. Comp. Endocrinol. 6: 519–533.

Joly, R. 1966b. Sur l'ultrastructure de la glande cérébrale de *Lithobius forficatus* L. (Myriapode Chilopode). C. R. Acad. Sci. Paris 263D: 374–377.

Joly, R. 1971. Effet de la destruction de la pars intercerebralis sur l'évolution pondérale chez *Lithobius forficatus* L. (Myriapode Chilopode). C. R. Acad. Sci. Paris 273D: 1208–1209.

Jones, B. M. 1953. Activity in the incretory centres of *Locustana pardalina* during embryogenesis: function of the prothoracic glands. Nature (Lond.) 172: 551.

Jones, B. M. 1956a. Endocrine activity during insect embryogenesis: function of the ventral head glands in locust embryos (*Locustana pardalina* and *Locusta migratoria*, Orthoptera). J. Exp. Biol. 33: 174–185.

Jones, B. M. 1956b. Endocrine activity during insect embryogenesis: control of events in development following the embryonic moult (*Locusta migratoria* and *Locustana pardalina*, Orthoptera). J. Exp. Biol. 33: 685–696.

Juberthie, C. 1983. Neurosecretory systems and neurohemal organs of terrestrial Chelicerata (Arachnida). Pp. 149–203 *in* A. P. Gupta (ed.), *Neurohemal Organs of Arthropods*. Thomas, Springfield, Illinois.

Kästner, A. 1951. Zur Entwicklungsgeschichte von *Thelyphonus caudatus* L. (Pedipalpi). 3. Teil. Die Entwicklung des Zentralnervensystems. Zool. Jahrb. Anat. 71: 1–55.

Kai, H. 1977. The existence of diapause hormone in adult ovaries and mature eggs of the silkworm, *Bombyx mori* L. J. Sericult. Sci. Jpn. 46: 239–245.

Kai, H. and T. Kawai. 1981. Diapause hormone in *Bombyx* eggs and adult ovaries. J. Insect Physiol. 27: 623–627.

Kaplanis, J. N., S. R. Dutky, W. E. Robbins, M. J. Thompson, E. L. Lindquist, D. H. S. Horn, and M. N. Galbraith. 1975. Makisterone A: a 28-carbon hexa-hydroxy molting hormone from the embryo of the milkweed bug. Science (Wash., DC) 190: 681–682.

Kaplanis, J. N., W. E. Robbins, M. J. Thompson and S. R. Dutky. 1973. 26-Hydroxyecdysone: new insect molting hormone from the egg of the tobacco hornworm. Science (Wash., DC) 180: 307–308.

Kopec, S. 1922. Studies on the necessity of the brain for the inception of insect metamorphosis. Biol. Bull. (Woods Hole) 42: 323–342.

Lachaise, F. and J. A. Hoffmann. 1977. Ecdysone et développement ovarien chez un Decapode *Carcinus maenas*. C. R. Acad. Sci. Paris 285D: 701–704.

Lachaise, F. and J. A. Hoffmann. 1982. Ecdysteroids and embryonic development in the shore crab, *Carcinus maenas*. Hoppe-Seyler's Z. Physiol. Chem. 363: 1059–1067.

Lachaise, F., M. Meister, C. Hetru, and R. Lafont. 1986. Studies on the biosynthesis of ecdysone by the Y-organs of *Carcinus maenas*. Mol. Cell. Endocrinol. 45: 253–261.

Lagueux, M., M. Hirn, and J. A. Hoffmann. 1977. Ecdysone during ovarian development in *Locusta migratoria*. J. Insect Physiol. 23: 109–119.

Lagueux, M., P. Harry, and J. A. Hoffmann. 1981. Ecdysteroids are bound to vitellin in newly laid eggs of *Locusta*. Mol. Cell. Endocrinol. 24: 325–338.

Lanzrein, B., H. Imboden, C. Bürgin, E. Brüning, and H. Gfeller. 1984. On titres, origin, and functions of juvenile hormone III, methylfarnesoate, and ecdysteroids in embryonic development of the ovoviviparous cockroach *Nauphoeta cinerea*. Pp. 454–465 *in* J. A.

Hoffmann and M. Porchet (eds.), *Biosynthesis, Metabolism and Mode of Action of Invertebrate Hormones*. Springer-Verlag, Berlin and New York.

Laufer, H., D. Borst, F. C. Baker, C. Carrasco, M. Sinkus, C. C. Reuter, L. W. Tsai, and D. A. Schooley. 1987a. Identification of a juvenile hormone-like compound in a crustacean. Science (Wash., DC) 235: 202–205.

Laufer, H., M. Landau, E. Homola, and D. W. Borst. 1987b. Methyl farnesoate: its site of synthesis and regulation of secretion in a juvenile crustacean. Insect Biochem. 17: 1129–1131.

Legendre, R. 1953. Le système sympathique stomatogastrique ("organe de Schneider") des Araignées du genre Tegenaria. C. R. Acad. Sci. Paris 237: 1283–1285.

Legendre, R. 1959. Contribution a l'étude du système nerveux des Araneïdes. Ann. Sci. Nat. Zool. Biol. Anim. [Ser. 12] 1: 339–473.

Le Roux, A. 1968. Description d'organes mandibulaires nouveaux chez les Crustaces Decapodes. C. R. Acad. Sci. Paris 266D: 1414–1417.

Le Roux, A. 1983. Histogénèse de l'organe Y (glande de mue) chez l'embryon de la crevette *Palaemon serratus* (Pennant) (Crustace Decapode Natantia): bases biologiques de l'aqua culture. IFREMER (Inst. Fr. Rech. Exploit. Mer) 255–262.

Leubert, F., H. Eibisch, H. Kroschwitz, and H. Scheffel. 1982. Ecdysteroid-Biosynthese durch das Ovar von *Lithobius forficatus* (L.) (Chilopoda). Zool. Jahrb. Physiol. 86: 465–476.

Loeb, M. J. and D. K. Hayes. 1980. Neurosecretion during diapause and diapause development in brains of mature embryos of the gypsy moth, *Lymantria dispar*. Ann. Entomol. Soc. Am. 73: 432–436.

Masler, E. P., M. S. Fuchs, B. Sage, and J. D. O'Connor. 1980. Endocrine regulation of ovarian development in the autogenous mosquito, *Aedes atropalpus*. Gen. Comp. Endocrinol. 41: 250–259.

Matz, G. 1980. Variations du taux des ecdystéroïdes au cours de l'embryogenèse de *Leucophaea maderae* Fabr. (Insecte, Dictyoptere). C. R. Acad. Sci. Paris 291D: 501–504.

McCarthy, J. F. 1979. Ponasterone A: a new ecdysteroid from the embryos and serum of brachyuran crustaceans. Steroids 34: 799–806.

McCarthy, J. F. and D. M. Skinner. 1979. Changes in ecdysteroids during embryogenesis of the blue crab, *Callinectes sapidus* Rathbun. Dev. Biol. 69: 627–633.

Mendis, A. W. H., H. H. Rees, and T. W. Goodwin. 1984. The occurrence of ecdysteroids in the cestode, *Moniezia expansa*. Mol. Biochem. Parasitol. 10: 123–138.

Mendis, A. W. H., M. E. Rose, H. H. Rees, and T. W. Goodwin. 1983. Ecdysteroids in adults of the nematode, *Dirofilarai immitis*. Mol. Biochem. Parasitol. 9: 209–226.

Millot, J. 1930. Le tissu reticule du cephalothorax des Araneides et ses dérives: néphrocytes et cellules endocrines. Arch. Anat. Microsc. 26: 43–81. .

Mori. H. and H. Ando. 1983. Embryonic neurohemal organs in arthropods. Pp. 5–16 *in* A. P. Gupta (ed.), *Neurohemal Organs of Arthropods.* Thomas, Springfield, Illinois.

Moritz, M. 1957. Zur Embryonalentwicklung der Phalangiiden (Opiliones, Palpatores) unter besonderer Berücksichtigung der äusseren Morphologie, der Bildung des Mitteldarmes und der Genitalanlage. Zool. Jahrb. Anat. 76: 331–370.

Mueller, N. S. 1962. Control of molting in insect embryos. Ph.D. thesis, University of Wisconsin.

Mueller, N. S. 1963. An experimental analysis of molting in embryos of *Melanoplus differentialis.* Dev. Biol. 8: 222–240.

Nagasawa, H., H. Kataoka, A. Isogai, S. Tamura, A. Suzuki, H. Ishizaki, A. Mizoguchi, Y. Fujiwara, and A. Suzuki. 1984. Amino-terminal amino acids sequence of the silkworm prothoracicotropic hormone: homology with insulin. Science (Wash., DC) 226: 1344–1345.

Neumann-Visscher, S. 1976. The embryonic diapause of *Aulocara elliotti* (Orthoptera, Acrididiae): histological and morphometric changes during diapause development and following experimental termination with juvenile hormone analogue. Cell Tissue Res. 174: 433–452.

Nirde, P., G. Torpier, M. L. De Reggi, and A. Capron. 1983. Ecdysone and 20-hydroxyecdysone: new hormones for the human parasite *Schistosoma mansoni.* FEBS Lett. 151: 223–227.

Novak, V. J. A. 1951. The metamorphosis hormones and morphogenesis in *Oncopeltus fasciatus* Dallas. Vestn. Cesk. Spol. Zool. 15: 1–49.

Novak, V. J. A. 1969. Morphogenetic analysis of the effects of juvenile hormone analogues and other morphogenetically active substances on embryos of *Schistocerca gregaria* (Forskal). J. Embryol. Exp. Morphol. 21: 1–21.

Novak, V. J. A. and S. K. Zambre. 1974. To the problem of structure and function of pleuropodia in *Schistocerca gregaria* Forskal embryos. Zool Jahrb. Physiol. 78: 344–355.

O'Brien, M. A., N. A. Granger, N. Agui, L. I. Gilbert, and W. E. Bollenbacher. 1986. Prothoracicotropic hormone in the developing brain of the tobacco hornworm. *Manduca sexta:* relative amounts of two molecular forms. J. Insect Physiol. 32: 719–725.

Payen, G. and J. D. Costlow. 1977. Effects of a juvenile hormone mimic on male and female gametogenesis of the mud-crab, *Rhithropanopeus harrisii* (Gould) (Brachyura: Xanthidae). Biol. Bull. (Woods Hole) 152: 199–208.

Pflugfelder, O. 1958. *Entwicklungsphysiologie der Insekten,* 2nd ed. Akademische Verlagsgesellschaft Geest & Portig KG, Leipzig, E. Germany.

Pound, J. M. and J. H. Oliver, Jr. 1979. Juvenile hormone: evidence of its role in the reproduction of ticks. Science (Wash. DC) 206: 355–357.

Pratt, G. E., R. C. Jennings, A. F. Hammett, and G. T. Brooks. 1980. Lethal metabolism of precocene I to a reactive epoxide by locust corpora allata. Nature (Lond.) 284: 320–323.

Pross, A. 1966. Untersuchungen zur Entwicklungsgeschichte der Araneae [Pardosa hortensis (Thorell)] unter besonderer Berücksichtigung des vorderen Prosomaabschnittes. Z. Morphol. Ökol. Tiere 58: 38–108.

Raabe, M. 1982. *Insect Neurohormones.* Plenum Press, New York.

Rankin, M. A. and L. M. Riddiford. 1978. Significance of haemolymph juvenile hormone titer changes in timing of migration and reproduction in adult *Oncopeltus fasciatus.* J. Insect. Physiol. 24: 31–38.

Rempel, J. G., B. S. Heming, and N. S. Church. 1977. The embryology of *Lytta viridana* Le Conte (Coleoptera: Meloidae). IX. The central nervous system, stomatogastric nervous system, and endocrine system. Quaest. Entomol. 13: 5–23.

Riddiford, L. M. 1969. Juvenile hormone application to hemipteran eggs: delayed effects on postembryonic development. Am. Zool. 9: 1120.

Riddiford, L. M. 1970. Effects of juvenile hormone on the programming of postembryonic development in eggs of the silkworm. *Hyalophora cecropia.* Dev. Biol. 22: 249–263.

Riddiford, L. M. and C. M. Williams. 1967. The effects of juvenile hormone analogues on the embryonic development of silkworms. Proc. Nat. Acad. Sci. USA 57: 595–601.

Roe, R. M., C. L. Crawford, C. W. Clifford, J. P. Woodring, T. C. Sparks, and B. D. Hammock. 1987a. Characterization of the juvenile hormone esterases during embryogenesis of the house cricket, *Acheta domesticus.* Int. J. Invertebr. Reprod. Dev. 12: 57–72.

Roe, R. M., C. L. Crawford, C. W. Clifford, J. P. Woodring, T. C. Sparks, and B. D. Hammock. 1987b. Role of juvenile hormone metabolism during embryogenesis of the house cricket, *Acheta domesticus.* Insect Biochem. 17: 1023–1026.

Rohdendorf, E. B. and F. Sehnal. 1973. Inhibition of reproduction and embryogenesis in the fire-bat, *Thermobia domestica,* by juvenile hormone analogues. J. Insect Physiol. 19: 37–56.

Romer, F. 1971. Häutungshormone in den Oenocyten des Mehlkäfers. Naturwissenschaften 58: 324–325.

Romer, F. 1979. Ecdysteroids in snails. Naturwissenschaften 66: 471–472.

Romer, F., H. Emmerich, and J. Nowock. 1974. Biosynthesis of ecdysones in isolated prothoracic glands and oenocytes of *Tenebrio molitor in vitro.* J. Insect Physiol. 20: 1975–1987.

Rosenberg, J. 1973. Eine bisher unbekannte endokrine Drüse im Kopf von

Scutigera coleoptrata L. (Chilopoda, Notostigmophora). Experientia (Basel) 29: 690–691.

Roussel, J. P. and R. Aubry. 1981. Présence de substances juvenilisantes dans les oeufs de *Locusta migratoria*, phase solitaire. Arch. Zool. Exp. Gen. 122: 91–97.

Roussel, J. P. and M. Schneider. 1979. Mise en évidence a l'aide d'un dosage biologique d'hormones juvéniles au cours du développement embryonnaire de *Locusta migratoria*. C. R. Acad. Sci. Paris 289D: 29–31.

Rubenstein, E. C., T. J. Kelly, M. G. Schwartz, and C. W. Wood. 1982. *In vitro* synthesis and secretion of ecdysteroids by *Drosophila melanogaster* ovaries. J. Exp. Zool. 223: 305–308.

Sauber, F., M. Reuland, J. P. Berchtold, C. Hetru, G. Tsoupras, B. Luu, M. E. Moritz, and J. A. Hoffmann. 1983. Cycle de mue et ecdystéroïdes chez une sangsue, *Hirudo medicinalis*. C. R. Acad. Sci. Paris 296D: 413–418.

Sbrenna, G. 1974. The fine structure and formation of cuticles during the embryonic development of *Schistocerca gregaria* Forskal (Orthoptera, Acrididae). J. Submicrosc. Cytol. 6: 287–295.

Sbrenna, G., A. Rimondi, and A. Sbrenna Micciarelli. 1983. Fine structure of prothoracic glands in embryo of *Schistocerca gregaria* (Orthoptera). Ultramicroscopy 12: 142.

Sbrenna-Micciarelli. A. and G. Sbrenna. 1972. The embryonic apolyses of *Schistocerca gregaria* (Orthoptera). J. Insect Physiol. 18: 1027–1037.

Scalia, S., A. Sbrenna-Micciarelli, G. Sbrenna, and E. D. Morgan. 1987. Ecdysteroid titres and location in developing eggs of *Schistocerca gregaria*. Insect Biochem. 17: 227–236.

Scalia, S. and E. D. Morgan. 1982. A re-investigation on the ecdysteroids during embryogenesis in the desert locust, *Schistocerca gregaria*. J. Insect Physiol. 28: 647–654.

Scharrer, B. 1983. Current views on the mechanism of release of neurosecretory products. Pp. 602–607 *in* A. P. Gupta (ed.), *Neurohemal Organs of Arthropods*. Thomas, Springfield, Illinois.

Scheer, B. T. 1960. The neuroendocrine system of arthropods. Vitam. Horm. 18: 141–204.

Scheffel, H. 1965. Der Einfluss von Dekapitation und Schnürung auf die Anamorphose der Larven von *Lithobius forficatus* L. (Chilopoda). Zool. Jahrb. Physiol. 71: 359–370.

Scheffel, H. 1969. Untersuchungen über die hormonale Regulation von Häutung und Anamorphose von *Lithobius forficatus* (L.) (Myriapoda, Chilopoda). Zool. Jahrb. Physiol. 74: 436–505.

Schmidt, E. L. and C. M. Williams. 1953. Physiology of insect diapause. V. Assay of the growth and differentiation hormone of Lepidoptera by the method of tissue culture. Biol. Bull. (Woods Hole) 105: 174–187.

Schneiderman, H. A. and L. I. Gilbert. 1958. Substances with juvenile hormone activity in Crustacea and other invertebrates. Biol. Bull. (Woods Hole) 115: 530–535.

Scholl, G. 1963. Embryologische Untersuchungen an Tanaidaceen. Zool. Jahrb. Anat. 80: 500–554.

Sedlak, B. J. 1985. Structure of the endocrine glands. Pp. 25–60 *in* G. A. Kerkut and L. I. Gilbert (eds.), *Comprehensive Insect Physiology, Biochemistry and Pharmacology*, Vol. 7: *Endocrinology*, Pt. I. Pergamon Press, Oxford and Elmsford, NY.

Seifert, G. 1979. Considerations about the evolution of excretory organs in terrestrial arthropods. Pp. 354–372 *in* M. Camatini (ed.), *Myriapod Biology*. Academic Press, Orlando, Florida.

Seifert, G. and J. Rosenberg. 1974. Elektronenmikroskopische Untersuchungen der Häutungsdrüsen ("Lymphstränge") von *Lithobius forficatus* L. (Chilopoda). Z. Morphol. Tiere 78: 263–279.

Slama, K. and C. M. Williams. 1966. "Paper factor" as an inhibitor of the embryonic development of the European bug, *Pyrrhocoris apterus*. Nature (Lond.) 210: 329–330.

Slifer, E. H. 1937. The origin and fate of the membranes surrounding the grasshopper egg; together with some experiments on the source of the hatching enzyme. Q. J. Microsc. Sci. 79: 493–506.

Slinger, A. J., L. N. Dinan, and R. E. Isaac. 1986. Isolation of apolar ecdysteroid conjugates from newly laid oothecae of *Periplaneta americana*. Insect Biochem. 16: 115–119.

Spindler, K.-D., L. Dinan, and M. Londershausen. 1984. On the mode of action of ecdysteroids in crustaceans. Pp. 255–264 *in* J. A. Hoffmann and M. Porchet (eds.), *Biosynthesis, Metabolism and Mode of Action of Invertebrate Hormones*. Springer-Verlag, Berlin and New York.

Spindler, K.-D., A. Van Wormhoudt, D. Sellos, and M. Spindler-Barth. 1987. Ecdysteroid levels during embryogenesis in the shrimp. *Palaemon serratus* (Crustacea Decapoda): quantitative and qualitative changes. Gen. Comp. Endocrinol. 66: 116–122.

Streble, H. 1966. Untersuchungen über das hormonale System der Spinnentiere (Chelicerata) unter besonderer Berücksichtigung des "endokrinen Gewebes" der Spinnen (Araneae). Zool. Jahrb. Physiol. 72: 157–234.

Sturaro, A., A. Guerriero, R. De Clauser, and F. Pietra. 1982. A new, unexpected marine source of a molting hormone: isolation of ecdysterone in large amounts from the zoanthid *Gerardia savaglia*. Experientia (Basel) 38: 1184–1185.

Taghert, P. H., J. W. Truman, and S. E. Reynolds. 1980. Physiology of pupal ecdysis in the tobacco hornworm, *Manduca sexta*. II. Chemistry, distribution and release of eclosion hormone at pupal ecdysis. J. Exp. Biol. 88: 339–349.

Takemi, T. 1963. *In vitro* culture of embryos in silkworms *Bombyx mori*. III. Molt of brainless embryos in culture J. Exp. Biol. 40: 735–739.

Takemoto, T., S. Ogawa, N. Nishimoto, and H. Hoffmeister. 1967. Steroide mit Häutungshormon-Aktivität aus Tieren und Pflanzen. Z. Naturforsch. 22B: 681–682.

Temin, G., M. Zander, and J. P. Roussel. 1986. Physico-chemical (GC-MS) measurements of juvenile hormone III titres during embryogenesis of *Locusta mirgratoria*. Int. J. Invertebr. Reprod. Dev. 9: 105–112.

Tobe, S. S. and B. Stay. 1980. Control of juvenile hormone biosynthesis during the reproductive cycle of a viviparous cockroach. Gen. Comp. Endocrinol. 40: 89–98.

Tombes, A. S. and T. A. Malone. 1977. Ultrastructure of the frontal ganglion of *Chaoborus* and observations on neurosecretory cells during the annual cycle. J. Insect Physiol. 23: 639–648.

Truman, J. W. and L. M. Riddiford. 1970. Neuroendocrine control of ecdysis in silkmoths. Science (Wash., DC) 167: 1624–1626.

Truman, J. W., P. H. Taghert, P. F. Copenhaver, N. J. Tublitz, and L. M. Schwartz. 1981. Eclosion hormone may control all ecdyses in insects. Nature (Lond.) 291: 70–71.

Ude, J., M. Eckert, and H. Penzlin. 1978. The frontal ganglion of *Periplaneta americana* L. (Insecta): an electron microscopic and immunohistochemical study. Cell Tissue Res. 191: 171–182.

Unnithan, G. C., H. A. Bern, and K. K. Nayar. 1971. Ultrastructural analysis of the neuroendocrine apparatus of *Oncopeltus fasciatus* (Heteroptera). Acta Zool. (Stockh.) 52: 117–143.

Unnithan, G. C. and K. K. Nair. 1977. Fine structure of the A cells of the pars intercerebralis of normal and gamma-irradiated female milkweed bugs, *Oncopeltus fasciatus* (Heteroptera, Lygaeidae). J. Morphol. 154: 59–82.

Warren, J. T., W. A. Smith, and L. I. Gilbert. 1984. Simplification of the ecdysteroid radioimmunoassay by the use of protein A from *Staphylococcus aureus*. Experientia (Basel) 40: 393–394.

Warren, J. T., B. Steiner, A. Dorn, M. Pak, and L. I. Gilbert. 1986. Metabolism of ecdysteroids during the embryogenesis of *Manduca sexta*. J. Liquid Chromatogr. 9: 1759–1782.

Welter, V., M. Charlet, M. Reuland, F. Sauber, and J. A. Hoffmann. 1986. Recherches sur les ecdystéroïdes présents dans les cocons de la sangsue *Hirudo medicinalis* au cours de l'embryogenèse. Int. J. Invertebr. Reprod. Dev. 9: 321–331.

Weygoldt, P. 1960. Embryologische Untersuchungen an Ostrakoden. Die

Entwicklung von *Cyprides litoralis* (G. S. Grady) (Ostracoda, Podocoda, Cytheridae). Zool. Jahrb. Anat. 78: 369–426.

Weygoldt, P. 1964. Vergleichend-embryologische Untersuchungen an Pseudoscorpionen (Chelonethi). Z. Morphol. Ökol. Tiere 54: 1–106.

Weygoldt, P. 1965. Vergleichend-embryologische Untersuchungen an Pseudoscorpionen. III. Die Entwicklung von *Neobisium muscorum* Leach (Neobisiinea, Neobisiidae). Mit dem Versuch einer Deutung der Evolution des embryonalen Pumporgans. Z. Morphol. Ökol. Tiere 55: 321–382.

Weygoldt, P. 1975. Untersuchungen zur Embryologie und Morphologie der Geisselspinne *Tarantula marginemaculata* C. L. Koch (Arachnida, Amblypygi, Tarantulidae). Zoomorphologie 82: 137–199.

Whitehead, D. L., E. W. Osir, F. D. Obenchain, and L. S. Thomas. 1986. Evidence for the presence of ecdysteroids and preliminary characterization of their carrier proteins in the eggs of the brown ear tick *Rhipicephalus appendiculatus* (Neumann). Insect Biochem. 16: 121–133.

Whitehead, D. L. and K. Sellheyer. 1982. The identification of ecdysterone (20-hydroxyecdysone) in 3 species of molluscs (Gastropoda: Pulmonata). Experientia (Basel) 38: 1249–1251.

Whiting, P. and L. Dinan. 1987. The presence and formation of apolar ecdysteroid conjugates in the eggs and ovaries of the house cricket, *Acheta domesticus*. Abstr. 8th Ecdysone Workshop, Marburg, W. Germany p. 124.

Wigglesworth, K. P., D. Lewis, and H. G. Rees. 1985. Ecdysteroid titre and metabolism to novel apolar derivatives in adult female *Boophilus microplus* (Ixodidae). Arch. Insect Biochem. Physiol. 2: 39–54.

Yamashita, O. and K. Hasegawa. 1985. Embryonic diapause. Pp. 407–434 *in* G. A. Kerkut and L. I. Gilbert (eds.), *Comprehensive Insect Physiology, Biochemistry and Pharmacology*, Vol. 1: *Embryogenesis and Reproduction*, Pergamon Press, Oxford and Elmsford, New York.

Yin, C.-M. and G. M. Chippendale. 1975. Insect frontal ganglion: fine structure of its neurosecretory cells in diapause and nondiapause larvae of *Diatraea grandiosella*. Can. J. Zool. 53: 1093–1100.

Zhu, X. X., H. Gfeller, and B. Lanzrein. 1983. Ecdysteroids during oogenesis in the ovoviviparous cockroach, *Nauphoeta cinerea*. J. Insect Phyiol. 29: 225–235.

Zilch, R. 1974. Die Embryonalentwicklung von *Thermosbaena mirabilis* Monod (Crustacea, Malacostraca, Pancarida). Zool. Jahrb. Anat. 93: 462–576.

POSTEMBRYONIC SOURCES

Morphology, Histology, and Ultrastructure of JH-Producing Glands in Insects

2

PIERRE CASSIER

2.1. Introduction

Insect *corpora allata* (CA) (Heymons, 1897a,b, 1899), which were previously called paired posterior visceral ganglia (Hofer, 1887), ganglia
allata (Heymons, 1895), or corpora incertae (Meinert, 1861), are located in
the posterior regions of the head or, in rare instances, in the thorax.

They were first identified in the ant (Meinert, 1861; Forel, 1874); however, their endocrine function was first suggested by Nabert (1913) and
later by Ito (1918) and Müller (1829), and was experimentally demonstrated
by Wigglesworth (1935), Bounhiol (1936a,b, 1937a,b,c, 1938a,b,c,d,e,f),
Pflugfelder (1937a,b), Piepho (1938a,b,c,d), Bodenstein (1938a,b,c), and
Weed-Pfeiffer (1936a,b). Since these early experiments, CA have been
found in many insect species. In his comprehensive topographic histological study, which is still often referred to, Cazal (1948) presented both
the available data and his own observations on more than 130 species (see
also Cazal and Guerrier, 1946). He pointed out that he had no information
on these structures in Grylloblattidae, Zoraptera, Strepsiptera,
Raphidioptera, and Plecoptera. His preliminary observations on Apterygota were later completed by other workers (Chaudonneret, 1949;
Bitsch, 1962; Watson, 1964a,b; Rohdendorf, 1965; Cassagnau and
Juberthie, 1967; Rohdendorf and Watson, 1969; Palevody and Grimal, 1975;
Palevody, 1976).

Although the secretion of CA is designated by the generic name
juvenile hormone (JH), five different JHs have been identified (Fig. 2.1).

The identification of JH I (C_{18}JH) [methyl (2E,6E)-(10R-11S)-10,11-ep-
oxy-7-ethyl-3,11-dimethyl-2,6-tridecadienate] as the JH of *Hyalophora
cecropia* by Röller et al. (1967) and Dahm et al. (1968, 1976) allowed progress
in the JH metabolism studies (biosynthesis, degradation, mode of action of
the hormone, regulation of physiological JH-dependent processes). In
fact, the CA of adult male *H. cecropia* produce only JH I acid, which is subsequently methylated in the accessory gland (Shirk et al., 1976; Peter et al.,
1981; Dahm et al., 1981) with the intermediate formation of S-adenosylmethionine (SAM) (Weirich and Culver, 1979; Metzler et al., 1971, 1972).

JH II (C_{17}JH) [methyl (2E,6E)-(10R,11S)-10,11-epoxy-3,7,11-tri-
methyl-2,6-tridecadienoate] was subsequently identified in *H. cecropia*
(Meyer et al., 1968, 1970).

JH III (C_{16}JH) [methyl (2E,6E)-(10R)-10,11-epoxy-3,7,11-trimethyl-2,6-
dodecadienoate] was found in the hemolymph of many species (Judy et
al., 1973a,b, 1975; Trautmann et al., 1974b, 1976) and also in cultures of
CA (Judy et al., 1973b, 1975; Peter and Dahm, 1975; Müller et al, 1974;
Dahm et al., 1976). In *Galleria mellonella*, the imaginal wing disks of mobile
prepupa exhibit the ability to methylate JH-acid. Therefore, some eleva-
tions of JH III titer in the hemolymph after the second day of seventh

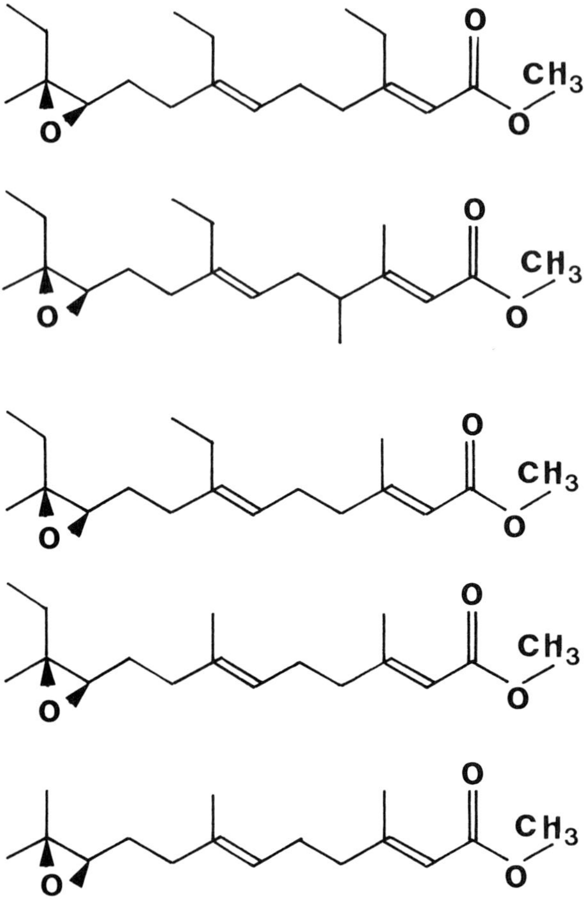

$JHO(C_{19}JH)$

4-Methyl JH I (iso-JHO)

$JHI(C_{18}JH)$

$JH\ II(C_{17}JH)$

$JH\ III(C_{16}JH)$

FIGURE 2.1. Chemical structures of identified juvenile hormones.

instar are due in part to JH-acid methyltransferase activity in the imaginal disks.

JH 0 and 4-methyl–JH I (Fig. 2.1) were recently extracted from embryos of *H. cecropia* (Bergot et al., 1981) and *Manduca sexta* (Bergot et al., 1981). Sites of secretion of these embryonic juvenile hormones are yet to be identified.

With one exception (see Lanzrein et al., 1975; Lüscher and Lanzrein, 1976) it is not yet clear whether any of the three JHs produced by the CA has a particular morphogenetic or gonadotropic effect. For each JH-dependent physiological process, there may be different activity thresholds or different induction capacities for these three hormones (Röller and

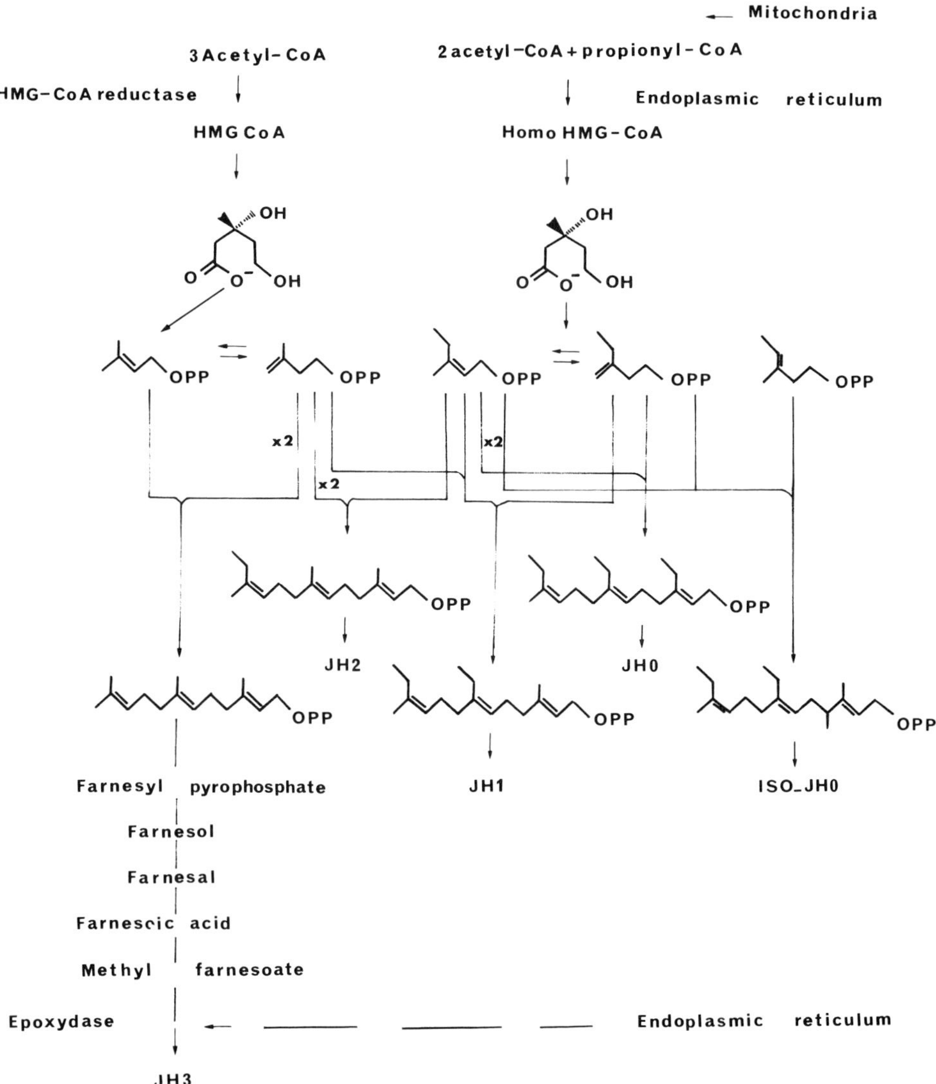

FIGURE 2.2. Major pathways of JH(s) biosynthesis.

Dahm, 1968). JH I and II have been identified only in Lepidoptera (Röller et al., 1967; Meyer et al., 1968; Dahm et al., 1976; Dahm and Röller, 1970; Röller and Dahm, 1970; Schooley et al., 1973; Judy et al., 1973a; Jennings et al., 1975a,b; Tobe and Stay, 1985). JH III alone was found in orthopteran, coleopteran, and hymenopteran species (Judy et al., 1973b; Trautmann et al., 1974b, 1976) and also in cultures of orthopteran and coleopteran CA (Judy et al., 1973b, 1975; Peter and Dahm, 1975; Müller et al., 1974; Dahm et al., 1976). These findings suggest that the original JH of insects is JH III and that JH I and JH II are special evolutionary achievements in Lepidoptera (Fig. 2.2). These JHs are transported in the hemolymph, free or linked to specific JH-binding proteins (Tobe and Stay, 1985).

The endocrine activity of the CA (Fig. 2.3) is related to morphogenesis (inhibition of metamorphosis), activation of follicular cells and sexual accessory glands, control of polymorphism, general physiology (synthesis of proteins, respiration, effect on fat body, oenocytes, water balance, diapause, coloration, pheromone secretion), behavior and so on. The different aspects of classic endocrinology have been extremely well documented in several reviews, which the reader should consult (Gilbert, 1963, 1964, 1974, 1976; de Wilde, 1964; Wigglesworth, 1964, 1965; Novak, 1966; Cassier, 1967, 1979; Engelmann, 1968, 1970; Menn and Beroza, 1972; Wy-

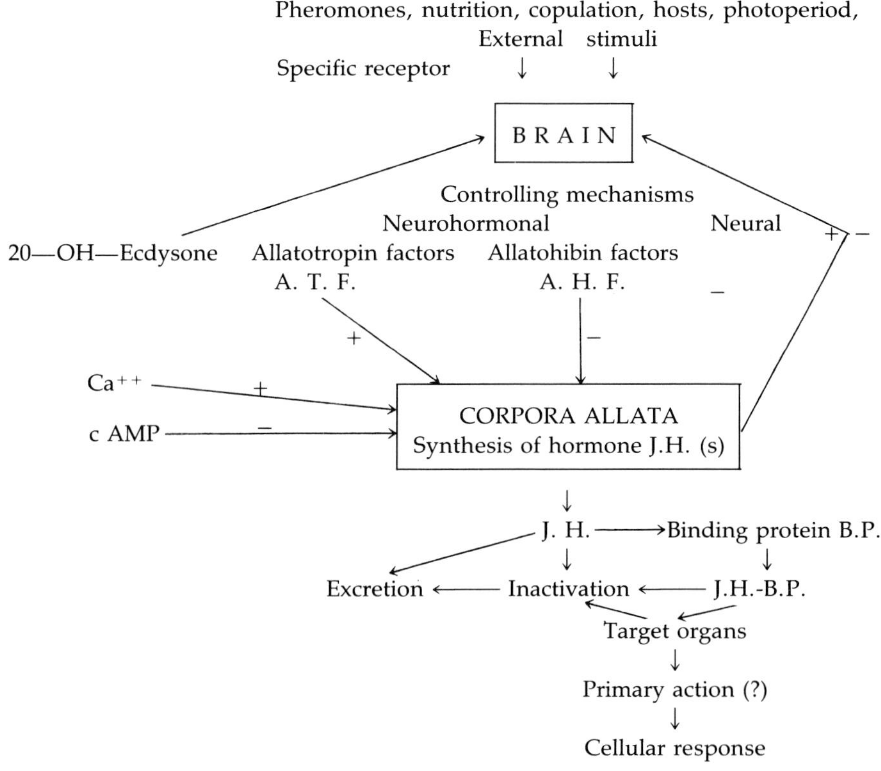

FIGURE 2.3. Mechanisms of control of CA activity.

att, 1972; Gilbert and King, 1973; Doane, 1973; de Wilde and de Loof, 1973; Slama et al., 1974; Willis, 1974; Steel, 1975; Kerkut and Gilbert, 1984). The interest in this field of research is further demonstrated by the amount of research carried out on juvenoids, growth regulators, and JH mimetics with specific short-range action on economically important insects (Henrick et al., 1976; Zurflueh, 1976).

Histological and cytological observations are presented here in the hope that they will provide a better understanding of CA activities. The latter, as well as the JH titer, are influenced by a variety of external and neural or humoral factors (Fig. 2.3).

2.2.   The Embryonic Origin of Corpora Allata

The CA are ectodermal in origin and arise early in embryonic development. This was established by Heymons in 1895 and confirmed by Haget (1977). The site of appearance of the buds or original infoldings (Fig. 2.4), at first paired and symmetrical, differs slightly from species to species. The CA may be formed by invaginations at the anterior part of the mandibular somite (*Corynodes:* Paterson, 1936), at its posterior part (*Pieris:* Eastham, 1930), between the mandibular and the maxillary segments (*Silpha:* Smereczynski, 1932; *Locusta:* Roonwal, 1936, 1937; Maltête, 1962; *Carausius:* Weismann, 1926; Pflugfelder, 1937a) or even in the maxillary somite (*Forficula* and *Gryllus,* Heymons, 1895). In *Apis,* they appear to be derived from lateral outgrowths of the epithelium of the transverse bar of the tentorium (Pflugfelder, 1937a; Nelson, 1915).

The conspicuous differences are due (1) to the degree of morphogenetic movements, which affect the ventral part of the gnathal region, and (2) to the timing of bud formation or appearance. The available data enabled Haget (1977) to confirm that the CA are intersegmental organs originating in a region anterior to the maxillary segment.

Subsequently, the CA buds form two coherent cellular masses, which then migrate in dorsal and mesal directions toward the ventrolateral angles of the coelomic sacs of the antennal segment.

Throughout the various orders of insects, the more highly evolved species is, the more marked is the migration of the CA. Thus, in Thysanura, the CA or corps jugaux (Chaudonneret, 1946, 1949; Bitsch, 1962) still occupy the ventral position. In Odonata, they are anterior and ventral (Hanström, 1940a,b; Cazal, 1948; Schaller, 1968), in contact with the circumesophageal connectives. In the majority of Heterometabola, they flank the esophagus laterally and remain distinct. However, in numerous species of Dermaptera, Hemiptera and Holometabola, the CA fuse on the ventral surface of the aorta. Finally, in higher Diptera, they form a single

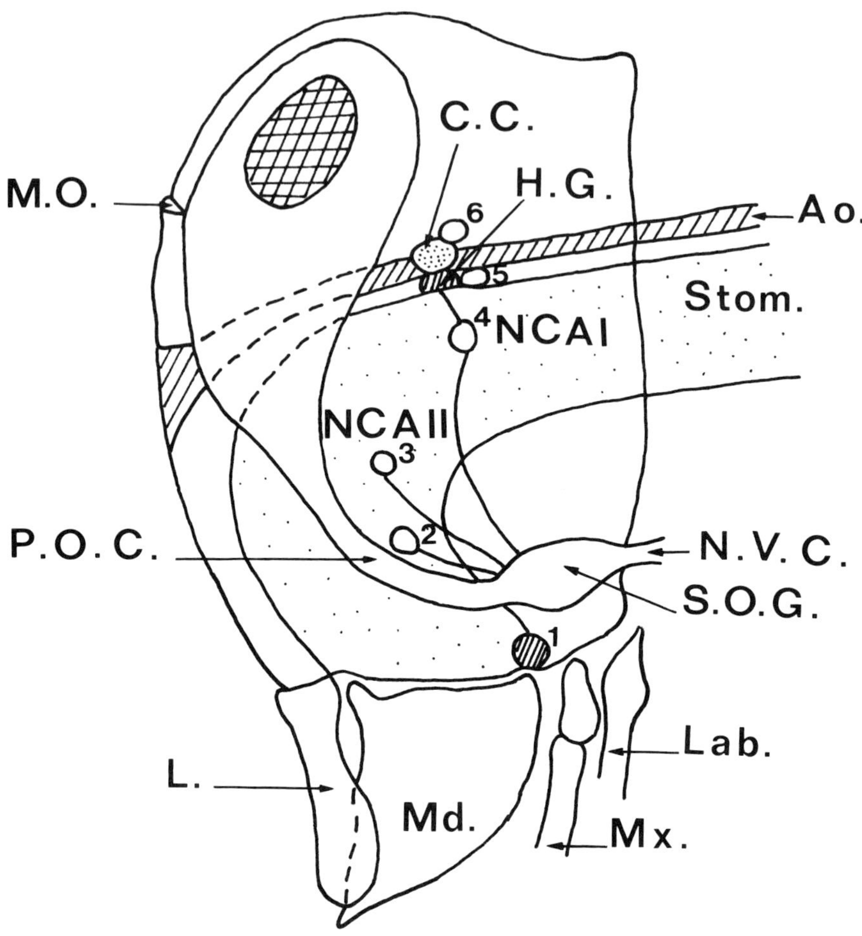

FIGURE 2.4. Schematic drawing, showing the origin (1) of the CA, their innervation (NCA I; NCA II) and their final position in Thysanura (2), Odonata (3), various Holometabola (4), Hemiptera (5) and higher Diptera (6). KEY: Ao = aorte; CA = corpora allata; CC = corpora cardiaca; HG = hypocerebral ganglion; L = labrum; Lab = labium; MD = mandibles; MO = median ocellus; Mx = maxillae; POC = perioesophageal connectives; SOG =suboesophageal ganglia; Stom = stomodaeum; NVC = nervous central cord.

mass above the aorta (Fig. 2.4). In some rare instances, the CA are situated extremely posteriorly. Thus, in *Sialis, Japyx,* and the larvae of *Lampyris,* they are located in the prothorax. In the embryo of *Carausius* (Wiesmann, 1926), the CA buds very rapidly become vesiculated. In *Oncopeltus fasciatus,* the original central vesicle resulting from the invagination disappears gradually during the course of the development (Dorn, 1975). In their final position, the CA, after dorsal closure of the embryo, frequently become associated with the corpora cardiaca (CC); this was confirmed in embryos of *Locusta migratoria* (Maltête, 1962, cited in Haget, 1977).

2.3.   Morphological Types: The Evolutionary Process

At the moment of hatching, the CA occupy different positions in various groups or species; their positions with regard to CC also vary considerably. Based on his observations, Cazal (1948) described the following five morphological types (Fig. 2.4):

The lateralized type.   This is the most common type and is found in the primitive forms of most groups. It is characterized by the presence of two CA, symmetrically arranged on each side of the digestive tube, connected by completely individualized allatocardiac nerves (NCA I). Beginning with this type, their evolution always follows the same pattern. The CA reach the dorsal surface of the esophagus, join with each other, and fuse either beneath or above the aorta. Thus, two CA may still be observed beneath the aorta in Isoptera, Heteroptera, Cryptocera, and Cicadidae.

The distal lateralized type.   This type is seen in several Diptera and Coleoptera, and differs from the lateralized type by the attachment of homolateral CC and CA.

The semicentralized type.   This type, found in Paleoptera and primitive Neoptera (Dictyoptera, Orthoptera), is characterized by a sagittal fusion of the CC (or corpus paracardiacum), which are attached directly to the aorta and the hypocerebral ganglia; the CA remain distinct, while the NCA I may (Orthoptera) or may not (Dictyoptera) remain separate, depending on the closeness of the relationship between the CC and CA.

The centralized type.   In this type (Embioptera, Dermaptera, Psocoptera, various gymnocerate Heteroptera, Cicadidae, Aphididae), there is a hint of the semicentralized type, since the CA fuse and form a separate mass located beneath the aorta, in direct contact with the CC so that it seems as if they were actually inside the CC. In *Oncopeltus*, the incomplete fusion of the CA may give rise to a bilobed structure (Novak, 1951; Unnitham et al., 1971).

The annular type.   The annular type (ring-shaped type), characteristic of higher Diptera (Brachycera, Cyclorrhapha), results from a fusion above the aorta of the CA, which, together with the ventral CC and the lateral peritracheal glands, form a ring around the aorta known as Weismann's ring. Formation of this ring takes place in the larval stage, and the NCA I are not visible; they become apparent during metamorphosis, from the onset of the imaginal molt as the ecdysial glands (peritracheal glands) degenerate. Furthermore, the CA, in contact with the brain of the larvae, migrate posteriorly to the prothorax of the imago.

2.4.   Innervation and Tracheal Supply of the CA

The innervation of the CA is double and most complex (Fig. 2.5). These endocrine glands are connected both to the brain by the superior allatocardiac nerve NCA I via the CC and NCC and to the subesophageal ganglion (NCA II), either directly or by mean of the tritocerebral paracardiac nerve (NCC IV). Since the origin of the CA is ventral, the relationship with the superior centers seems to have been acquired secondarily.

2.4.1.   *Subesophageal Innervation (NCA II)*

Subesophageal innervation is primitive and exists only in Ephemeroptera (Hanström, 1940a; Cazal, 1948; Bounhiol et al., 1953), Thysanura (Cazal, 1948), Collembola (Cassagnau and Juberthie, 1967), and in young stages of Odonata, where the cerebral innervation appears only in older larvae (Cazal, 1948). In *Thermobia domestica*, a fine nerve from the CC is also found (Bitsch and Lapalisse, 1984). In *Locusta*, the NCA II consists of neurosecretory fibers that emerge from the subesophageal cells (Chalaye, 1965, 1966), and the pattern is the same in Blattidae and Culicidae (Füller, 1960; Harker, 1960). In Thysanura (Chaudonneret, 1949), the *corps jugaux* are entirely covered with maxillary nerves but are directly innervated by the subesophageal ganglion. In Machilidae (Bitsch, 1962), the CA are innervated both by maxillary and mandibulary neuromers. In *Periplaneta americana* (Pipa and Novak, 1979), each NCA II contains axons from two groups of somata—group A (2 cells) and group B (5 cells)—located on the ipsilateral side of the subesophageal ganglion.

2.4.2.   *Cerebral Innervation*

At the other extreme are insects in which the innervation arises exclusively from the brain, e.g., in *Leptinotarsa decemlineata* (Schooneveld, 1970) and *Hydrous piceus* (Coleoptera) (de Lerma, 1956), as well as in some Diptera (Juberthie and Cassagnau, 1971). The *cerebral innervation* of the CA is provided by the superior allatocardiaca nerve (NCA I), composed of ordinary and neurosecretory fibers that arise in protocerebrum and traverse the CC.

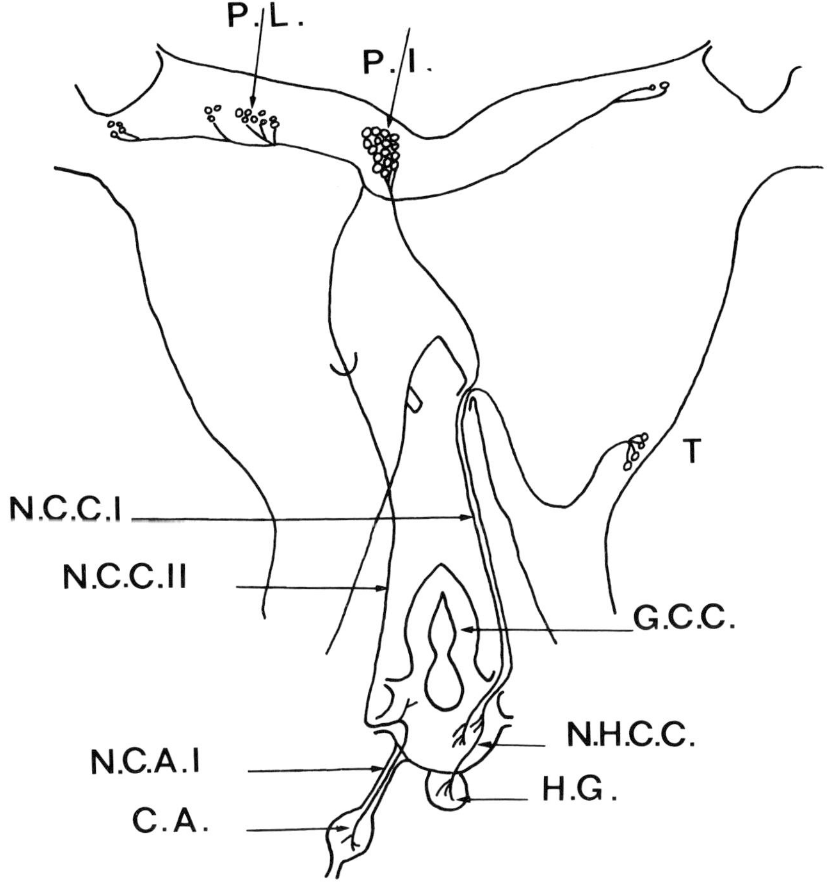

FIGURE 2.5. Schematic reconstruction showing the innervation of the CA of *Locusta* by cells located in the pars lateralis (PL) via the external nerve of the corpora cardiaca (NCC II), the neurohemal part of the corpora cardiaca (NCC II), the neurohemal part of the corpora cardiaca (NHCC), and the superior nerve of the CA (NCA I). Axons from cells of the pars intercerebralis (PI) and the tritocerebrum (T) terminate in the NHCC and the hypocerebral ganglia (HG) via NCC I; GCC = glandular part of the corpora cardiaca (CC).

The relationship between the pars intercerebralis (a neurosecretory center), the CC (a neurohemal organ), and the CA (an endocrine gland) is frequently compared to those in the hypothalamic–hypophyseal complex, in vertebrates, particularly at the level of adenohypophysis.

In insects with the lateralized type and some semicentralized type (Paleoptera, Orthoptera), the NCA I are distinct. They enter the CA at the

level of the hilum, then lose their sheath, and their fibers ramify among
the glandular cells. However, in Dictyoptera and in species with the centralized and annular types, the NCA I are morphologically nondiscerni-
ble because of the coalescence of the CA and CC. It is only by histological
studies that the fibers can be recognized. They frequently form a periglandular plexus, and then enter the gland by several hila and ramify
among the glandular cells (Schultz, 1960; Fukuda et al., 1966; Tombes and
Smith, 1970; Dorn, 1973; Baehr et al., 1973; Melnikova and Panov, 1975;
Palevody and Grimal, 1975; Morohoshi et al., 1976a). In *Hyalophora* (Waku
and Gilbert, 1964), only one part of the CA is covered by the neurosecretory fibers.

The afferent and efferent nerves are surrounded by a thick (3–4 μm)
stratified neural sheet and by a perilemma consisting of a layer of thin (1–
3 μm) glial cells (the lemmoblasts of Edwards et al., 1958; the Schwann
cells of Hess, 1958; the neuroglial cells of Trujillo-Cenoz, 1962). Extension
of these cells enclose the isolated fibers or bundles of fibers. They are
united by septate desmosomes. Their ovoid nucleus (4–6 μm in diameter)
generally contains only one nucleolus; the fragmented chromatin clumps
are attached to the nuclear membrane.

Since the brain figures largely in the regulation of the CA, it will clearly
be necessary to identify the cells that are the direct regulators of the CA
and to trace the integrator pathway relationship. However, simply know-
ing the origin, content, and termination sites of nerves that arrive at the
CA is insufficient to demonstrate an interaction with the CA cells. It is
possible that the CA are regulated by neurosecretions, released at a more
distant site, e.g. the CC, and that the neurosecretory release sites "synaptoids" (Scharrer, 1968), seen in CA of many insects (Cassier, 1979), regulate functions other than those of the CA. For example, current evidence
points to the release of prothoracotropic hormone (PTTH) in neurons
ramifying in the CA of larval Lepidoptera (Gibbs and Riddiford, 1977;
Agui et al., 1980; Carrow et al., 1981, 1984). Nevertheless, details of the
morphology and topography of the nerve cells and neurosecretory cells
(NSC) that project to the retrocerebral complex (CC and CA) are essential
in designing and interpreting experiments relevant to understanding the
interactions between the nervous system and the CA (Tobe and Stay,
1985).

The cells contributing to the nerves entering the CC and/or the CA
have been established by several complementary methods of vizualiza-
tion, including silver stains, stains of neurosecretory material, immunochemical staining, retrodiffusion of cobalt chloride, or treatment with
peroxidase (Raabe, 1982; Pipa, 1983). The principal nerves and their cells
of origin are shown in Fig. 2.5. There are many variations on this plan;
e.g., tracts of nervi corporis cardiaci (NCC) I and II join within the brain of

Manduca sexta (Nijhout, 1975) and *L. decemlineata* (Khan et al., 1984), and the NCA I are distinct from the CC in Orthoptera (Mason, 1973) and contiguous with the CC in Dictyoptera (Fraser and Pipa, 1977; Tobe and Stay, 1985).

Retrograde filling of a transected CA shows all cells projecting to the cut, and consequently does not distinguish between those axons that simply traverse the CA and those that may terminate there. Also, it is not possible to distinguish between those axons that terminate on the surface and those that ramify within the CA. Combined retrograde and orthograde filling and sectioning at different locations have solved some of these problems. For instance, Pipa's study (1978) in *P. americana* showed that the medial and lateral cells of the protocerebrum were filled from the NCA I. However, orthograde filling of the NCC I (derived from contralateral medial cells) showed fibers traversing the CC and ramifying in the CA, whereas orthograde filling of the NCC II (derived from ipsilateral cells) showed fibers only in the CC (Gunde and Penzlin, 1980). Thus, in *P. americana*, medial cells are likely those innervating the CA, whereas axons of lateral cells terminate on the anterior surface of the CA in the extension of CC described by Pipa and Novak (1979) as a cap of neurosecretory endings.

In the Colorado potato beetle, *Leptinotarsa decemlineata* (Khan et al., 1984), the CA after application of horseradish peroxidase (HRP) appear to be innervated by a single group of about eight cells. These cells belong to the L-type lateral neurosecretory cells. The axons from each of these cells split into two pathways: one pathway arborizes extensively in the dorsal part of the brain adjacent to the pars intercerebralis (PI), whereas the other passes unbranched through the neuropile of the protocerebral lobe and CC into the CA. Additional cell groups in the PI were stained when the CC were contaminated with HRP. Lateral neurosecretory cells are involved in the regulation of JH. In *M. sexta*, both medial and lateral cells were shown to project to the CA by retrograde filling with cobalt chloride (Nijhout 1975; Buys and Gibbs, 1981).

The injection of individual cells resolved the projections more specifically. By injecting HRP—or alternatively Lucifer yellow—into brain cells that showed the opalescent blue interference colors of neurosecretory granules, Carrow et al. (1984) distinguished two groups of lateral cells with axons that ramify within the CA: one group (IIa) was contralateral, and the other (IIb) was ipsilateral to the CA. One of the two cells of group IIa is presumably the primary source of PTTH (Agui et al., 1979; Carrow et al., 1981).

In *Schistocerca cancellata* (= *paranensis*) (Strong, 1965a,b), the neurosecretory fibers reaching the CA arise exclusively from lateral protocerebral neurosectretory cells via the NCC II. In *Locusta migratoria migratorioides* (Cassier and Fain-Maurel, 1970), they arise also from median protocerebral neurosecretory cells (pars intercerebralis) along the slope of the internal paracardial nerves (NCC I).

In *L. migratoria* (Savio strain), retrograde diffusion of cobalt chloride (Poras, 1983) shows that each CA is innervated by 15 ipsilateral cells and 2 contralateral cells located in the lateral part of the protocerebrum, but the bundles of fibers form a loop in the area of the pars intercerebralis and then are included in the NCC II; they are involved in the inhibition of CA activity during imaginal diapause (Poras et al., 1983). Conversely, axons coming from the pars intercerebralis never project into the CA.

In *Diploptera punctata*, some lateral and medial cells terminate either in or on the CA (Lococo and Tobe, 1984). Further, intracellular recording and dye injection were used to study the structure and electrophysiological properties of individual neurons that project to the CA of this cockroach. Neurons in the pars intercerebralis generate long-duration, tetradotoxin-sensitive action potentials. Dye injection revealed two cell types: one extends axons to the contralateral nervi corporis cardiaci I, some of which innervate the CA; the other extends a major axon down each of the circumoesophageal connectives. Neurons in the pars intercerebralis also generate long-duration potentials. These neurons extend axons to the ipsilateral NCC II, which continue on to terminate in the CC and the CA. Small groups of all the above neuronal types are dyed and electrically coupled. Penetration and dye injection into nerve terminals in the CA and CC confirmed the innervation of the CA by neurons located in the PI and pars lateralis and revealed a third class of neurons that have terminals in the CA: intrinsic neurons of the CC (Thompson et al., 1987).

Filling of single cells with Co^{2+} shows that axonal collateral may project into both hemispheres of the brain and into proto-, deuto-, and tritocerebra (Zaretsky and Loher, 1983; Carrow et al., 1984).

Immunoreactivity to some vertebrate hormones has been demonstrated in the CA of *Leucophaea maderae*, including β-endorphin, luteinizing hormone-releasing factor, and substance P (Hansen et al., 1982), and in *M. sexta*, enkephalin, insulin, somatostatin, substance P, vaso-intestinal polypeptide, and pancreatic polypeptide (El-Salhy et al, 1983). This immunoreactivity to vertebrate hormone may or may not have functional significance in insects, but antisera to these hormones may be useful for mapping of neurosecretory cells (NSC). Stefano and Scharrer (1981) have found high-affinity binding of an analogue of an enkephalin in the brain of *L. maderae*. Since 30% more binding occurred in pregnant females than in males or in larvae of both sexes, it seems plausible that insects may utilize peptides similar to those of vertebrates for regulatory purposes.

2.4.3. *The Tracheal Supply of the CA*

The CA are abundantly supplied with tracheae; their distribution seems to follow the same pattern as that of the nerve fibers. Thus, where NCA I are distinct (e.g., in *Locusta*), they run side by side with the nerve and enter the gland at the level of the hilum and ramify among the gland cells. However, when these nerves lose their individuality, there is a network of tracheae surrounding the gland, which then ramify and run into the glandular parenchyma (Cazal, 1948; Busselet, 1968; Baehr et al., 1973; Bitsch and Lapalisse, 1984). In the latter example, the tracheae are often enclosed between the basal laminae and the allata cells.

2.5. Histological Types of CA

The characteristic shape of the CA is ovoid to round, but they may be elongate as in large larvae and adults of *Libellula depressa* (Odonata) or polylobed as in *Hydrous piceus* (Coleoptera) (Cazal, 1948). The size of the glands is frequently about the diameter of the aorta or smaller. However, there is much variation among species and within a species, and the size differs with age, sex, polymorphism, and activity cycle of the glands.

2.5.1. *Histological Types*

The structural study of CA shows that there are at least six distinguishable types of cells, which correspond mainly to glandular cells [undifferentiated, normal, or polyploid (Özbas, 1957; Joly, 1976; Mendes, 1948)] and to associated elements (axons, neurosecretory fibers, glial cells, tracheoblasts, or peripheral conjunctive cells). Furthermore, these glands are often surrounded by fat body tissue and pericardial cells. The distribution of these different cellular types, their abundance, and their morphometric and histological characteristics provide so many criteria that Cazal (1948) was able to identify four principal types of CA on their bases, as described in the following subsections.

2.5.1.1. THE PSEUDOLYMPHOID (OR LYMPHOID) TYPE

This type is found in Paleoptera (Ephemeroptera, Odonata) and is characterized by the presence of many closely packed cells. They have a reduced amount of cytoplasm, which is slightly basophilic, and do not have any apparent secretory activity. The small nuclei (7–9 μm in diameter) are round or oval. There are a few mitochondria; the Golgi apparatus is not well developed, and it is difficult to distinguish any secretory products or secretion-related phenomena.

2.5.1.2. THE SMALL-CELL TYPE

The small-cell type is fairly common, appearing in Blattidae, Mantidae, Orthoptera, Dermaptera, Paraneoptera, and in numerous Oligoneoptera. It is characterized by the presence of many gland cells, which are small, irregularly shaped, and particularly rich in mitochondria, Golgi apparatus, and secretory grains.

In Acrididae and Coccinellidae, the peripheral cells are more or less palisade, whereas the central cells are irregularly shaped; many intercellular cavities may also be seen whose size varies according to the physiological state of the insect and the size of the gland.

Fusion of these intercellular cavities sometimes gives rise to a large central cavity. This is the case in senescent males of *Locusta*, where this phenomenon is even more marked because of degeneration of many of the central cells (Cassier, 1965b; Fain-Maurel and Cassier, 1969, 1970).

The nuclei are generally average sized (10–12 μm in diameter), oval or elongated, and contain many chromatin blocks. The cytoplasm is clear and basophilic. The well-developed chondrion consists of granular or filiform mitochondria. Each cell contains a 5- to 12 ring thick or scale-like Golgi area.

Recent fine structural analysis of CA from selected Ephemeroptera at different stages of the life cycle (Kaiser, 1980) would suggest that there is no radical difference between "active cells" of CA of Paleoptera and those of other small-celled CA. Some CA have only small cells, which is undoubtedly correlated with the tiny size of the insect, e.g., collembolans (Cassagnau and Juberthie, 1967; Palevody and Grimal, 1975; Bitsch and Lapalisse, 1984) and aphids (Elliot, 1976).

2.5.1.3. THE MACROCELL TYPE

The macrocell type observed in *Panorpa* is characteristic of Trichoptera, Lepidoptera, Hymenoptera, and some cyclorraphous Diptera. It differs greatly from the aforementioned types and is apparently the most highly evolved. The allata cells are few in number and are large: *Hyphantria*, 20–25 cells (Melnikova and Panov, 1975); *Aphis craccivora*, 10–12 cells (Elliot, 1976); *Speophyes* and *Diaprysius*, 16–18 cells (Deleurance and Charpin, 1971); *Troglodromus bucheti gaveti*, 50 cells (Deleurance and Charpin, 1972); *Hypera postica*, 85 cells (Tombes and Smith, 1970); *Drosophila melanogaster*, 20–24 cells (Poulson, 1945; King et al., 1966a). In *Bombyx mori* (Fukuda et al., 1966), the peripheral cells are the largest (50 × 90 μm) and are certainly more active than those in the center (30 × 50 μm) of the organ. Occasionally, the limited number of cells may be correlated with the small size of the insect: *Folsomia candida* (Collembola), 3 cells (Palevody and Grimal, 1975; Palevody, 1976). The nuclei are hypertrophied and branched in appearance. The cytoplasm is abundant and contains many inclusions.

2.5.1.4.   THE VESICULAR TYPE

The vesicular type is noteworthy for an epithelium-like arrangement of the glandular cells in a single layer or several layers surrounding a central cavity. Many investigators consider this arrangement to be a reappearance of primitive characteristics. This type is found in Phasmidae, where an epithelium lines a subspherical cavity containing a PAS (periodic acid–Schiff)-positive mass, the central body apparently being formed by five or six chitinous lamellae that are concentrically disposed. Each lamella may correspond to an exuvium removed during each of the molts, which occur during postembryonic development in Phasmidae. These exuvial structures, which are inevitably confined, persist during the imaginal stage (Joly, 1976; Haget et al., 1981). Such an interpretation deserves careful study.

The CA of Psyllidae, Mallophagidae, Anoplura (Cazal, 1948), Thysanura (Chaudonneret, 1949; Bitsch and Lapalisse, 1984), Machilidae (Bitsch, 1962), and Blattidae (*Diploptera punctata*) (Engelmann, 1970; Johnson et al., 1985) are also of the vesicular type. In *Psocoptera* (Badonnel, 1934), each CA contains three cavities and three central bodies; however, these are not present in Mallophaga or in Anoplura.

In *Japyx*, the nuclei of allata cells are grouped on the periphery of the organ in a type of envelope. In Embioptera, the cells have a radial arrangement. In Isoptera, the glandular cells are spindle shaped. The outer extremity is connected to the basal laminae; the inner one, which is more elongated and chromophilic, is wound in a whorl at the center of the CA (Cazal, 1948).

2.5.2.   *Activity and Volumetric Variations of the CA*

2.5.2.1.   EMBRYONIC PERIOD

The possibility that embryonic CA become differentiated and synthetize JH has been recently proposed from histological and fine structure analyses of the CA and from titers of JH in the developing embryos.

In *Eurygaster integriceps*, CA appear to be active in the last third of embryogenesis in that they have ring-shaped nuclei (Polivanova and Bocharova, 1974, 1982), which are found in the glands of mature females but not in hibernating individuals (Panov and Bassurmanova, 1970). Such morphological criteria are only suggestive, and measurements of JH titers would be necessary to confirm their activity.

It was recognized early in studies on titers of JH that eggs contained substances with JH bioactivity and that these were in part the products of the developing embryos. Eggs produced by allatectomized females of *Hyalophora cecropia* lacked JH in early development but showed normal amounts of hormone later, presumably the product of developed CA (Gilbert and Schneiderman, 1961). Recent chemical identification of JHs in embryos of *M. sexta* have demonstrated the presence of JH I, JH II, and the previously unknown JH 0. JH I, the most prominent, increased in titer before the embryonic molt (day 5.5) and subsequently decreased (Bergot et al., 1980, 1981).

Changing JH titers, which suggest activity of embryonic CA, have also bee observed by bioassay in other species. In *Nauphoeta cinerea*, two peaks of JH, determined by *Galleria* assays, were found between dorsal closure and hatching (Imboden et al., 1978). In *Oncopeltus fasciatus*, a large increase in JH titer was shown to occur by *Tenebrio* bioassay after the CA appeared differentiated at the end of blastokinesis (Dorn, 1975). However, the extent of the titer increase is uncertain because, although Bergot et al. (1981) verified the presence of JH III in *O. fasciatus* embryos, no more than 0.09 ng/g was found by physicochemical methods, whereas the bioassay suggested about 1 μg JH I equivalents/g at this stage of development. Nevertheless, Dorn (1975) presents other evidence of CA activity. The fine structure is consistent with that of active glands (Dorn, 1975), and chemical destruction or inhibition of CA with precocene resulted in an embryonic defect (failure of dorsal closure) that was prevented in 30% of the cases by simultaneous treatment with JH (Dorn, 1982).

In *Carausius morosus*, ultrastructural data suggest two periods of endocrine activity by embryonic CA (Haget et al., 1981). These active periods coincided with dorsal closure in the prothorax and the darkening of the mandibles. They were characterized in part by decreased density of the mitochondrial matrix, increased extracellular spaces, and a sequential change in the endoplasmic reticulum.

The first chemical identification of JH in insect embryos was performed in *M. sexta*. In addition to JH I, two JHs were found: JH 0 (Bergot et al., 1980) and 4-methyl–JH I (Bergot et al., 1981b). The titers of JH 0 and 4-methyl-JH I were higher than those of JH I and JH II; all the JHs were undetectable on day 0, appeared during embryogenesis in varying titers, and—with the exception of JH I—disappeared by hatching time. To demonstrate that JH biosynthesis was occurring in these embryos, these authors inhibited JH biosynthesis with fluoromevalonate (Quistad et al., 1981). Not only did such treatment produce larvae with symptoms of JH deficiency, it also resulted in reduced titers of JH 0, JH I, and 4-methyl–JH I in the older embryos (Bergot et al., 1981b). Thus, it seems that not only do CA of embryo produce JH but they synthesize a form specific to embryos.

In 4- to 5-day-old embryos of *L. migratoria*, at the end of blastokinesis, serosal epitheliae contain an immunohistologically detectable cytosolic protein (molecular weight $\sim$ 240 kD) which is related to the JH carrier protein in the hemolymph of the same species, and which binds triated JH III ($1 kD \sim 10^{-8} M$). At this early stage of development, the CA of the embryos are not yet fully differentiated and do not synthesize JH III in organ cultures. The earliest detectable JH III production by CA in isolated heads is on day 6. On the other hand, serosal epithelium of 4- and 5-day-old embryos provide JH III in organ cultures, as has been shown by methylation of [10-^{3}H]JH III acid to [10-^{3}H]JH III and by incorporation of triated CH_3 from L-[*methyl*-^{3}H]methionine into JH III. Isolated heads and abdomens of the embryos used as donors for the serosal preparations did not show methyltransferase activity responsible for JH III biosynthesis. The serosal cells represent a hitherto unrecognized source of methyltransferase activity and of JH III production. Degradation of JH III to JH III–acid was also observed (Hartmann et al., 1987).

Embryos of *Nauphoeta cinerea* contain very high quantities of methyl farnesoate and JH III at stages between dorsal closure and hatching. The increase in the titer of JH III could be prevented by applying ethoxy-precocene shortly after dorsal closure (Brüning et al., 1985). [10-^{3}H] methylfarnesoate injected into embryos is epoxydized only to an extremely small extent into JH III and hardly metabolized at all. Furthermore, in embryos treated after dorsal closure with a dose of ethoxy-precocene, Brüning and Lanzrein (1987) observed a certain delay in development, malformation of the eye, and lack of certain larval characteristics of the cuticle such as pigmentation and formation of hairs and bristles; the fat body disintegrated and the midgut did not differentiate normally, thus impeding yolk resorption. When a physiological dose of JH III was applied on ethoxy-precocene-treated embryos, the formation of the midgut was normalized and the disintegration of the fat body was less pronounced. Investigators thus have direct evidence that, in embryos, JH III plays a role in the differentiation of the midgut epithelium, and the development of the fat body, and that the hormone controls the appearance of larval characteristics of the cuticle.

2.5.2.2 POSTEMBRYONIC DEVELOPMENT AND IMAGINAL LIFE

During the postembryonic development and imaginal life of insects, the CA are subject to two types of obvious changes in volume frequently associated with differences in gland activity: allometric growth and periodic growth. Bioassays have been relied upon for estimation of activity of the glands. One can use gland size to estimate glandular activity by comparing *in vitro* rates of JH biosynthesis with gland size.

Allometric Growth

This regular increase, independent of any physiological event, may or may not be accompanied by modifications in form and is related to insect growth. An allometric type of growth follows the mitotic events (hyperplasia) and that of cell enlargement (hypertrophy) whose relative importance varies according to species.

In the majority of species (e.g., Blattidae) these two phenomena coexist during their larval and imaginal stages; in Acrididae, particularly in *Locusta*, this coexistence occurs only during the larval stage, whereas during the imaginal stage it results only from an increase in the cytoplasmic mass, from dilatation of intercellular spaces, and from endopolyploidy in a relatively limited number of cells (Özbas, 1957; Cassier, 1965b; Joly, 1968).

In small insects, the CA contain a limited number of cells (e.g., *Folsomia candida*, three cells), and growth depends exclusively on cytoplasmic and nuclear hypertrophy, which may or may not be associated with endopolyploidy (Palevody and Grimal, 1975; Palevody, 1976). However, in *Thermobia domestica*, growth is associated with an increase in the number of cells (Rohdendorf and Watson, 1969; Bitsch and Lapalisse, 1984). In *O. fasciatus*, an increase occurs 3 to 4 days after the imaginal molt and continues during the egg-laying period (from day 8 to 38), thus the rate of increase is 30. The increase in size is only partially due to divisions (Novak, 1951; Johansson, 1958b). Inactive CA may contain pycnotic cells (Cassier, 1979; Bitsch and Lapalisse, 1984).

In *Calliphora erythrocephala* (Lea and Thomsen, 1969; Thomsen and Thomsen, 1970), the volume of the CA increased three- or fourfold between the first ($200 \times 10^3 \mu m^3$) and the fourth day of imaginal life, resulting exclusively from growth of various cells. In flies fed on a protein-deficient diet, the CA volume was reduced ($100 \times 10^3 \mu m^3$).

In general, resumption of CA activity after the imaginal molt, during sexual maturation, or after diapause is associated with an increase in their volume. This has been observed in *Anacridium* (Girardie and Granier, 1973), *Pterostichus nigrita* (Hoffmann, 1970), *Nebria brevicollis* (Ganagarajah, 1965), *Macrodytes marginalis* (Joly, 1945), *Carabus nemoralis* (Klug, 1959), *Galeruca tanaceti* (Siew, 1965), *Leptinotarsa* (de Wilde and de Boer, 1969), and *Aulacophora foveicollis* (Saini, 1966).

Periodic Growth

In most insects, the CA undergo cyclical changes of variable importance during larval development and in adults. These changes, which result in fluctuations of CA activity, affect in particular the nucleus–cytoplasm ratio.

In larvae, these changes are synchronous with the molt cycles; this has been established in Holometabola and Heterometabola (Novak, 1954). Thus, in *Oncopeltus* (Novak, 1951, 1954; Johansson, 1958b), at the onset of each larval stage, the CA volume is reduced, but it progressively increases and reaches maximal values in the middle of the intermolt cycle, which is the phase of intense endocrine activity; later, the volume decreases in parallel to the slowing down of this endocrine activity. Although these glands are considered nonactive during the last stage, their volume still increases, but to a smaller degree than during the earlier stages; this certainly corresponds only to allometric growth.

In *B. mori* (Legay, 1950), the increase in volume occurs at each larval molt and to a lesser extent during the intermolt.

The synchronous changes in the molt cycles depend partly on hyperplasia (mitotic activity), which has been established in *Carausius* (Pflugfelder, 1937b) and in *Rhodnius* (Wigglesworth, 1936, 1948), and more particularly on cytoplasmic hypertrophy, which is shown by a high reduction in the nucleus–cytoplasm ratio.

Most authors agree that increase in volume and cell number of CA during larval development is more often associated with somatic growth than with increased activity of the CA. This has been confirmed by direct measurements of rates of JH biosynthesis and CA volume for selected stages of *M. sexta* (Granger et al., 1979) and in more extensive measurements in the last two larval stadia of *S. gregaria* (Injenyan and Tobe, 1981), and *D. punctata* (Szibbo et al., 1982). In *D. punctata* nymphs, both volume and cell number increased in the second half of the penultimate and last stadia, but there were no significant changes in volume per nucleus, which reflected the observed changes in rates of JH biosynthesis. In both stadia, the rate of JH biosynthesis per cell was higher early in the stadium than later in the stadium (Szibbo et al., 1982).

In the *imaginal stage,* the cyclic fluctuations are synchronous with ovarian cycles and are manifested by variations of the gonadotropic activity of the CA (Thomsen, 1947; Wigglesworth, 1948, 1964; Engelmann, 1957, 1959, 1970; Johansson, 1958; Nayar, 1958; Scharrer and von Harnack, 1958; Strangeways-Dixon, 1961, 1962; Highnam, 1962a,b, 1964; Ganagarajah, 1965; Siew, 1965; Strong, 1965a; Cassier, 1967; Wilkins, 1969; Pratt and Davey, 1972a,b,c; Weaver et al., 1974; Dahm et al., 1981).

Cazal (1948) observed frequent mitoses in the CA of adults, as well as of larvae, and hypothesized that mitotic cycles in adult CA might accompany cycles in activity of the glands. In *Musca domestica* (Adams et al., 1968; Adams, 1970), a small CA releases the JH whereas a large one stores it; this may depend on the action of an oostatic hormone, which is contrary to Lea's ideas (1975).

These phenomena have been particularly well studied in *Leucophaea*

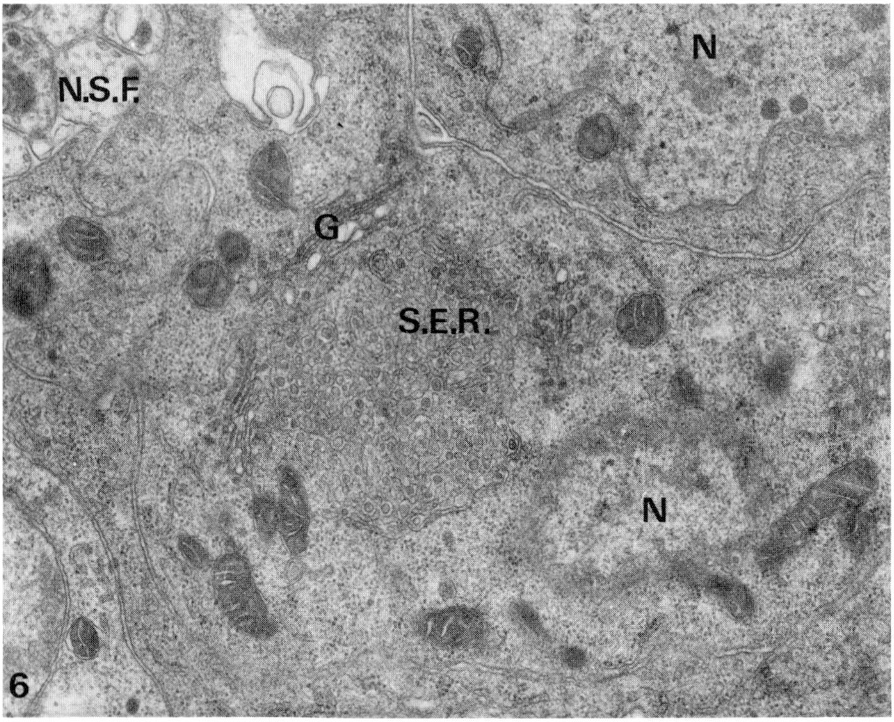

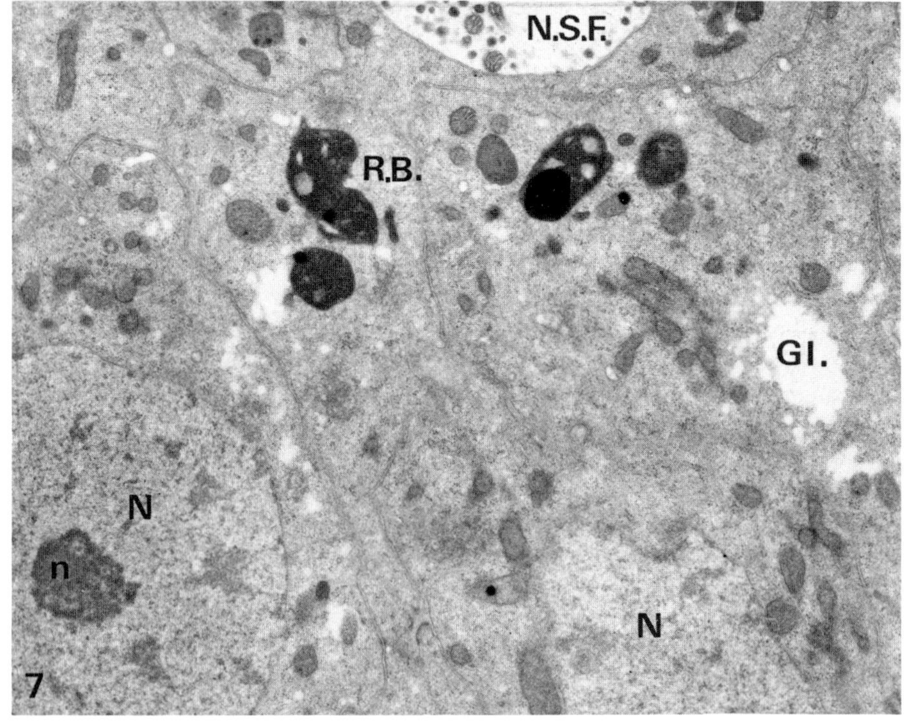
N.S.F.
N
G
S.E.R.
N
6
N.S.F.
R.B.
GI.
N
n
N
N.S.F.
7

maderae, a viviparous cockroach (Scharrer and von Harnack, 1958; von Harnack, 1958; Engelmann, 1957, 1958). Glandular activity (vitellogenesis, a period of rapid growth) is associated with an increase in the number of cells (mitosis); during nonactive periods of gestation, the number of cells diminishes following pycnosis. These two processes (mitosis, pycnosis) do in fact coexist, but their relative importance varies, mitosis being predominant during vitellogenesis, and pycnosis being characteristic of gestation periods.

However, the volume of the gland increased to a greater extent than the number of cells, so that the number of nuclei per area of gland decreased at the highest gland volume, reflecting the increase in cytoplasmic volume of the cells. The change in the nucleus–cytoplasm ratio was found to be the most conspicuous reflection of the JH-dependent egg laying in this cockroach (Engelman, 1957; Scharrer and von Harnack, 1958). Unfortunately, rates of JH biosynthesis are not available to compare this well-documented morphometric study. However, morphometric measurements and *in vitro* synthetic activity for CA of *D. punctata* do exist. Cyclic changes in cell number occur during the vitellogenesis cycle (Engelmann, 1959; Szibbo and Tobe, 1981a), but the increase in cell number does not account for the increase in synthetic capacity of the glands, i.e., activity of a pair of CA expressed as rate per cell increase, reflecting the change in rate for the entire gland pair. A comparison of morphometry (Engelman, 1959; Szibbo and Tobe, 1981a) with rates of JH biosynthesis (Tobe, 1977; Szibbo and Tobe, 1981a) showed that the nucleus–cytoplasm ratio began to increase before, and decreased after the rates of biosynthesis. A similar discrepancy between gland volume and gland activity was demonstrated in *N. cinerea* (Lanzrein et al., 1978) and between gland volume and JH titer (by RIA) in *Labidura riparia* (Figs. 2.6 and 2.7) (Baehr et al., 1982). Thus, there are instances of reasonable, although not exact, correlation between gland activity and volume. Yet, even in the same species, good correlation between volume and activity during one physiological state may not hold for another physiological state. This was clearly shown for the wasp *Polistes gallicus,* in which the rate of JH biosynthesis reflects the volume of the CA in egg-maturing females but not

FIGURES 2.6–2.7. CA of an adult female of *Labidura riparia* (Dermaptera). 2.6 (*above*). First cycle of vitellogenesis. Active CA cells contain—in the vicinity of the nucleus (N) and the Golgi apparatus (G)—characteristic structures of associated tubules and vesicles of smooth endoplasmic reticulum (SER); NSF = neurosecretory fibers ×32,000.
2.7 (*below*). First egg-care period. In the inactive CA cells, SER structures (cf. Fig. 2.6) generate residual bodies (RB). KEY: Gl = glycogen; NSF = neurosecretory fibers; n = nucleus; N = nucleus. ×13,600.

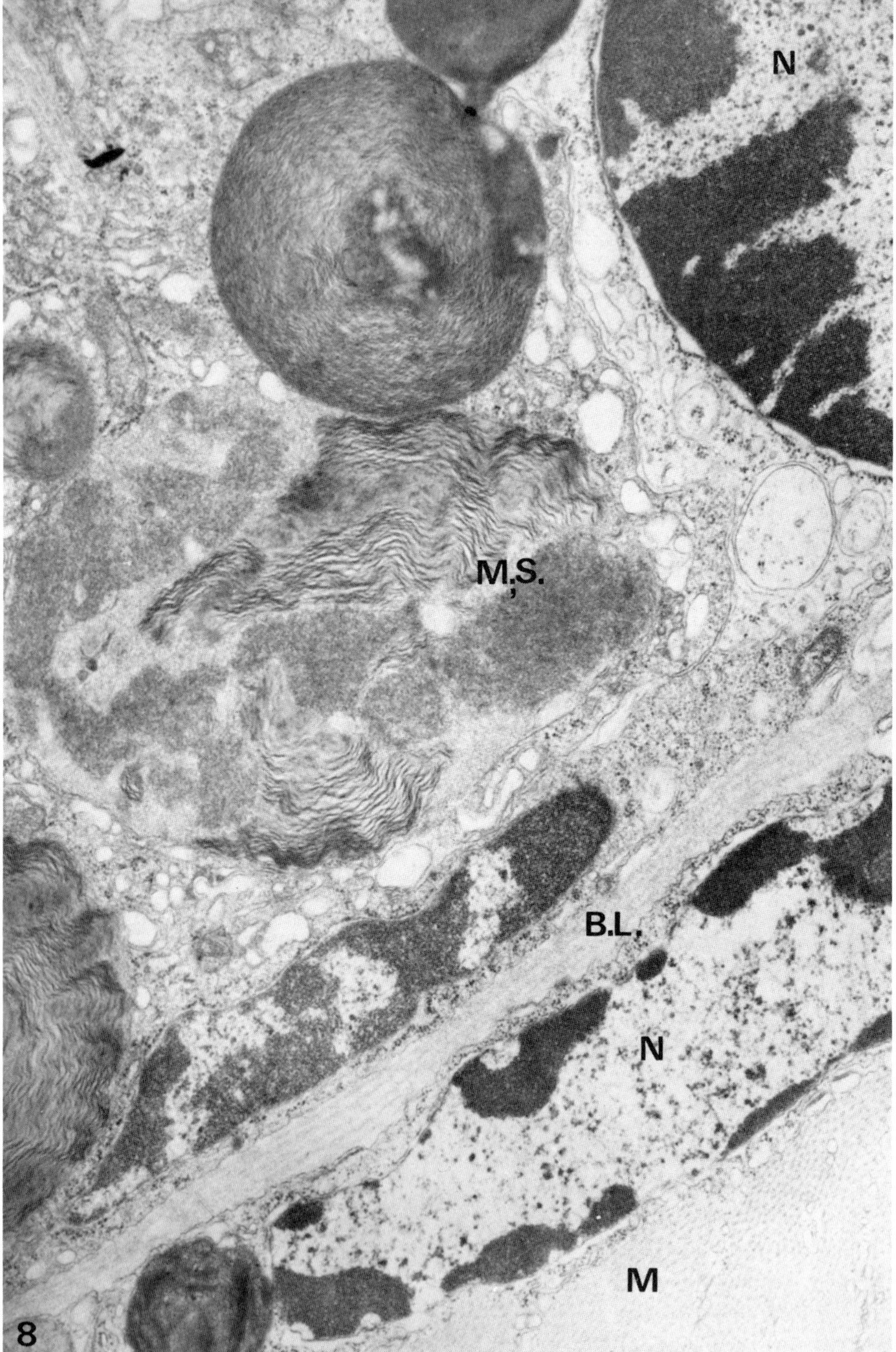

N
M,S.
B.L,
N
M
8

in overwintering females with quiescent ovaries, nor in ovariectomized females (Röseler et al., 1980).

In *Teleogryllus commodus* (Ruegg et al., 1986), the mean rate of *in vitro* JH release was undetectable on the day of emergence, rising to approximately 0.80 pmol h^{-1} per pair of CA on day 6. During activation of the CA (day 2.5), a second highly active population of CA appeared, showing a mean rate of biosynthesis approximately 1 order of magnitude greater than the basal rate. The increase in the rate of release correlated with the growth of oocytes.

The CA of a mated female of *D. punctata* showed an increase in cell number from 6000 cells/CA on the day of emergence to a maximum of about 9000 cells/CA by day 5. The number of cells per CA then declined, and by day 8 the CA was composed of about 6000 cells. In the normal mated animal, the previously established increase in biosynthetic activity of the CA correlated with the increase in number of cells in the CA. In virgin females, no significant change in cell number was observed during the period, and this correlates with the demonstrated low and constant rates of JH biosynthesis. Following ovariectomy, however, an increase in cell number occurred (to 10,000 cells/CA) by day 5 and cell number remained elevated until at least day 8. Low rates of JH biosynthesis have been observed previously in ovariectomized females, and thus the increase in cell number observed in these females is not accompanied by a corresponding increase in biosynthetic activity. The cell number of the CA in ovariectomized female was reduced by injection of 20-OH-E. Tobe et al. (1984) believe that either mating or surgical removal of inhibitory signals from cerebral neurosecretory cells permit the proliferation of the cells of the CA. These authors also proposed that factors from the ovary are responsible for both the initiation of the cycle of JH biosynthesis and the reduction of CA cell number at the end of the gonadotrophic cycle.

Similar phenomena were observed in the *corps jugaux* (= CA) of adult *Petrobius maritimus* (Fig. 2.8) (Cassier, 1979), in *Acridium* (Girardie and Granier, 1973; Warkievi-Granier and Leonide, 1971), and *T. domestica* (Bitsch and Lapalisse, 1984).

The factors contributing to sexual maturation, egg laying, and increase in fecundity (mating, enforced activity, feeding, excitatory pheromones, wounds, amputations) also stimulate the postimaginal growth of the CA. Conversely, factors that delay sexual maturation and cause a decline in fecundity (gestation, parasitism, inhibitory pheromones, imaginal diapause, overcrowding, presence of mature eggs in the ovarioles) are marked by reduced activity and size of the CA (Cassier, 1967).

FIGURE 2.8. Inactive CA of *Petrobius maritimus* (Apterygota). Glandular cells contain several fingerprint-like and myeloid structures (MS). KEY: BL = basal laminae; M = muscle; N = nucleus. ×10,300.

The CA of ovariectomized female *Acheta domesticus* (Huignard, 1964), *Melanoplus differentialis* (Pfeiffer, 1940, 1945), *L. maderae* (Scharrer and von Harnack, 1961), *R. prolixus* (Wigglesworth, 1936; Baehr et al., 1982), *Calliphora vicina* (Thomsen, 1940), and *D. melanogaster* (Vogt, 1942; Doane, 1961) and of sterile females (the *Drosophila fes* mutant) (Vogt, 1942; King et al., 1961, 1966b) are abnormally large, with few exceptions (Highnam, 1962a,b; Joly, 1964; Loher, 1966; Wilhelm and Lüscher, 1947).

2.5.2.3. VOLUMETRIC CHANGES AND POLYMORPHISM

In Acrididae (phase polymorphism) the CA of solitary locusts are larger than those of gregarious individuals (Nickerson, 1954; P. Joly, 1949, 1967; L. Joly, 1960; Carlisle and Ellis, 1959; Staal, 1961). Their hypertrophy is associated with increased activity and contributes to the development of phase polymorphism (chromotropic, gonadotropic, and morphogenetic effects: Cassier, 1967, 1974; Injeyan and Tobe, 1981; Dale and Tobe, 1986). In *Schistocerca* the level of circulating JH causes the differentiation and functioning of integumentary glands (Cassier and Delorme, 1973, 1974; Cassier and Delorme-Joulie, 1975, 1976a,b,c; Cassier, 1977a,b), which are responsible for the secretion of a stimulatory specific sexual pheromone (Norris, 1952, 1954; Loher, 1958, 1960, 1961).

In the honey bee *Apis mellifera* (social polymorphism), the queens possess large CA whereas those of the workers are average sized and those of the drones are small. Even in the larval stages, the differences between queens and workers are discernible; they become even more noticeable during the pupal and adult stages (Lukoschus, 1955a). Presumptive queen larvae of honey bees have larger CA than presumptive worker larvae (Wirtz, 1973); although rates of JH biosynthesis are not known for these glands, the higher JH titer in the presomptive queen larvae (Lensky et al., 1978) suggests that the larger glands may be more active.

In adult workers (sterile females), there are periods of considerable growth of the CA that are, in particular, related to the activity of the hypopharyngeal glands (6–12 days), achievement of the first orientation flight (9–10 days), secretion of wax glands (14–15 days), and acquisition of behavior in field bees. In this last case, the sudden increase in the level of JH causes degeneration of the hypopharyngeal glands and a fall in the level of vitellogenic and hemolymphatic proteins (Rutz et al., 1976). Subsequently, the CA activity decreases (Pflugfelder, 1937a,b, 1948, 1958; Hanström, 1942).

The behavioral polymorphism in egg-maturing females of the wasp *P. gallicus* is accompanied by correlated differences in size and rates of JH biosynthesis: dominant females have larger glands and higher rates of JH biosynthesis than do subordinate females (Röseler et al., 1980).

In the bumble bee, *Bombus terrestris*, CA from workers, maintained in the presence of queens, synthesized JH titers at low rates; removal of the queens resulted in an increased rate of JH biosynthesis and egg maturation by the queenless workers (Röseler and Röseler, 1978; see also Chapter 10 by Röseler in Part 3). Presumably, the volume of CA also increased, as Röseler (1977) has shown a positive correlation between CA volume and JH dependent egg growth in queenless workers of this species. A similar phenomenon has been observed in the neotenic reproductives of the termite, *Zootermopsis angusticollis*, which upon removal of the king and queen undergo an increase in volume of the CA, as well as increased rates of JH biosynthesis, as measured by *in vitro* RCA (Greenberg and Tobe, 1985).

In *Postelectrotermes nayari* (Isoptera, Kalotermitidae) (Varma, 1979), the nymphal CA are smaller in size, whereas in pseudergates and reproductives, and mainly in soldiers, they are larger.

The polymorphism of Aphids is also correlated with CA activity (Hardie, 1987).

Several workers, having considered the aforementioned observations, have concluded that the degree of activity of the CA may be assessed by measuring their size and nucleus–cytoplasm ratio (see Highnam, 1964; Wigglesworth, 1964; P. Joly and L. Joly, 1970a). Such a correlation has been confirmed in many cases but is not a general rule, and in the literature there are various examples of cases in which no relationship exists between the size of the CA and the development of oocytes (Johansson, 1958a; Mordue, 1967; Engelmann, 1968; Lea and Thomsen, 1969; Tobe and Pratt, 1975a,b; Gillott and Dogra, 1972; Brousse-Gaury, 1971c, 1972; Belles and Piulachs, 1983). Thus, bilateral sectioning of antennae between the scape and the pedicel in *Blabera fusca* (Brousse-Gaury et al., 1973; Brousse-Gaury and Cassier, 1975) causes a retardation of ovulation and reduced ovarian activity, but results in an increase in the size of the CA.

In *L. decemlineata*, during egg-laying, little correlation could be found between CA volume and rates of JH biosynthesis (Khan et al., 1982). This was also true for gonadotrophic female locusts, e.g., *S. gregaria* (Tobe and Pratt, 1975b; Injeyan and Tobe, 1981) and *L. migratoria* (Ferenz and Kaufner, 1981). Nonsynchronized fluctuating biosynthesis of JH is one possible explanation of this lack of correlation. (Khan et al., 1982). In all of these species, a high degree of variability in rates of JH biosynthesis between individuals has been observed.

2.5.3. *Sexual Dimorphism*

Male and female CA are generally comparable during the juvenile or larval stage. However, following the imaginal molt, the process of allometric growth results in a difference in their size, independent of any morphometric differences. This sexual dimorphism results in either hypertrophic male or hypertrophic female CA.

2.5.3.1. HYPERTROPHIC CA OF FEMALES

This is the most common type of sexual dimorphism. It has been described in Orthoptera (Fain-Maurel and Cassier, 1969b), in Coccidae (Homoptera) (Cazal, 1948; Pflugfelder, 1936), in Isoptera (Holmgren, 1909; Pflugfelder, 1937a,b; Strindberg, 1913), in *Aphenogaster* (Coleoptera) (Cazal, 1948), and has been particularly well studied in *B. mori* (Lepidoptera) (Fukuda et al., 1963, 1966). In the last-mentioned case, the CA cells of the female may be distinguished from those of the male by the presence of many osmiophilic vacuoles of varying size (0.3–14.0 μm in diameter), which are particularly large (12 μm) and abundant in the circumferential cells. These inclusions are not formed when the CA are separated from the brain (Fukuda and Takeuchi, 1965). They are associated with lamellar structures or rod-shaped mitochondria with well-developed cristae of the tubular type.

In males, the osmiophilic bodies are infrequent and small, and the chondriosomes (cup or ring shaped) have many lamellar cristae; particularly well-developed endoplasmic reticulum may be seen. Degenerating cells may be observed in the center of the CA.

The sexual differences are basically evident immediately after the imaginal molt, but are well established some 50 h later. In *Anacridium* (Orthoptera) (Girardie and Granier, 1973), sexual dimorphism during the active period concerns the size and the mitotic index (female CA are larger and show more mitoses than those of males). An example of small male CA, synthesizing hormone at lower rates than the larger female CA, has been demonstrated in *D. punctata* (Szibbo and Tobe, 1981b). The volume and number of nuclei of male glands are less than half those of female glands during periods of low JH biosynthesis. The male glands remain relatively constant in activity, cell number, and volume, whereas the female CA undergo large cyclic changes in both during a vitellogenic cycle. Also, denervated male glands, whether transplanted into females or left in males, maintained a low rate of biosynthesis only twice the normal rate (Szibbo and Tobe, 1981a). This was a small increase relative to the female CA, which increase some 8- to 10-fold in a vitellogenic cycle (Stay and Tobe, 1981).

2.5.3.2. HYPERTROPHIC CA OF MALES

This type of sexual dimorphism is less common than that in females; it has, however, been reported in various Lepidoptera: *Ephestia kuhniella* (Schrader, 1938), *Hyalophora cecropia* (Gilbert and Schneiderman, 1961b), and *Galleria mellonella* (Rehm, 1951). In *H. cecropia*, the glands of the adult male, which have fivefold greater wet weight than female glands (Gilbert and Schneidermann, 1961), appear to differ from female glands not only in quantity but also in quality of product: the male gland does not have the methyltransferase necessary to complete biosynthesis of JHs from expoxy acids, whereas the female gland does (Dahm et al., 1981; Sparagana et al., 1984). Although it is not certain whether the female gland secretes JH acid and/or JH *in vivo*, male glands produced 100-fold greater quantity of JH acid *in vitro* than those of females (Dahm et al., 1981). An *in vitro* radiochemical assay for JH accumulation using male accessory reproductive organs showed that the CA of female are 10 times more active in JH acid production than are male CA (Shirk et al., 1983).

2.6. Cytological and Infrastructural Characteristics of the CA

Only a few species have been studied by electron microscopy; in addition, these observations do not form a cohesive picture, since in some species only juvenile forms and larvae have been studied whereas in others only adults have been examined. This variation is largely dependent on whether the researcher's principal concerns were problems related to postembryonic development or those of reproductive activity. There are very few complete systematic studies, and even in these the results are sometimes difficult to compare because different techniques were used (osmium fixation, or double fixation with glutaraldehyde–osmium or paraformaldehyde–osmium), the difference in inclusion media markedly affects the preservation of organelles, especially the endoplasmic reticulum. Preservation of tissue remains a source of difficulty in the fine structural analysis of the CA, mainly when cytotoxic activity of various allatostatic compounds are to be tested (precocene, dihydroprecocene, or methyl farnesoate).

Combination of glutaraldehyde and osmium tetroxide gave the best results mainly when osmotic pressure was respected (Brousse-Gaury et al., 1973; Odhiambo, 1966a; Joly et al., 1968; Papillon et al., 1976; Fain-Maurel and Cassier, 1969a,b; Guelin and Darjo, 1974; Melnikova and Panov, 1975; Tobe et al., 1976; Tobe and Saleuddin, 1977). According to these authors, this procedure retains in the tissues, to some extent, radiolabeled farnesoic acid, methyl farnesoate, and JH III. Glutaraldehyde plus paraformaldehyde (Karnovsky, 1965), modified by Bradley and Edwards (1979), has been used. A modification of Karnovsky's fixative containing picric acid (Ito and Karnovsky, 1968), which is reported to fix SER in steroid-secreting cells, was used by Kaiser (1980), Feyereisen et al. (1981c), and Johnson et al., (1985) with reasonable results.

Nevertheless, the available results have a common denominator: the fact that research was carried out to find fine cytological criteria that act as markers for the synthetic activity of CA and the organelles involved in the production of JH. The studies are carried out indirectly. CA removed in the middle of the intermolt phase, in mature males, or in females during vitellogenesis are considered active according to the classic criteria (Scharrer, 1961, 1964a; Fukuda et al., 1966; Odhiambo, 1966b; Fain-Maurel and Cassier, 1969; Melnikova and Panov, 1975). However, CA removed at the beginning or at the end of an intermolt during the last larval stage, immediately after the imaginal molt, during egg laying, in viviparous females during gestation, in sterile castes (workers or soldiers), and in insects undergoing an imaginal diapause are considered inactive (Waku and Gilbert, 1964; Odhiambo, 1966c; Panov and Bassurmanova, 1970; Tombes and Smith, 1970; Guelin and Darjo, 1974). These categories may be completed by including insects in which the CA are inactive because of a mutation (e.g., the *fes* mutant in *D. melanogaster:* King et al., 1966a) or because of experimental intervention such as parsectomy (Joly et al., 1967, 1969; Baehr et al., 1973), fasting (Thomsen and Thomsen, 1969, 1970; Brousse-Gaury et al., 1973), modifications of the sensory afferents (Brousse-Gaury et al., 1973; Brousse-Gaury and Cassier, 1975), or castration (Cassier, 1979).

Only those elements of the CA (see Section 2.5) that have functional significance are discussed (glandular cells, axons, and neurosecretory fibers); the associated elements (glial cells, tracheoblasts, and conjunctive cells) are not considered here.

2.6.1. *Basal Lamina*

The noncellular layer enclosing the endocrine glands of insects has been given various names: the mesenchymatic envelope (Cazal, 1948), the conjunctive tissue (Willey and Chapman, 1960; Scharrer, 1963), the stroma (Scharrer, 1964a; Normann, 1965; Thomsen and Thomsen, 1970), the basal membrane (Odhiambo, 1966b), the tunica (King et al., 1966a), the cytoplasmic membrane (Fukuda et al., 1966), and the tunica propria (Beaulaton, 1964; Busselet, 1969). The term *basal lamina* seems to be the most appropriate and does not lead to presumptions regarding its origin or function.

The presence of the basal lamina is a constant feature of the CA; it is involved in glandular cohesion. It is continuous with that of the allatocardiac nerves (NCA I) when they are distinct, and with that of the CC when the NCA I are joined. In the latter case, the distribution of nerves, neurosecretory fibers, and tracheae (see Section 2.4) is such that the basal lamina envelops these structures also (Baehr et al., 1973). According to Cazal (1948) the basal lamina is rarely continuous, and thus some of the glandular cells are in direct contact with the hemolymph. No discontinuity has been established at the infrastructural level, although hemocytes are capable of traversing the membrane (Busselet, 1969; Fukuda et al., 1966).

The basal lamina frequently has, on its exterior, a discontinuous layer of flattened mesenchymal cells. These cells have a dense nucleus and a few organelles. The cells or lemnoblasts contain short rough endoplasmic reticular cisternae, free ribosomes, mitochondria and dictyosomes, and, in general, have no smooth endoplasmic reticulum. The thickness of the basal lamina varies greatly from one species to another: from 0.6 to 1 or 2 μm in *Antheraea* (Busselet, 1969); from 0.02 to 2 μm in larvae and from 0.1 to 0.2 μm in adults of *Rhodnius* (Busselet, 1969; Baehr et al., 1973); 0.1 μm in *Calliphora* (Thomsen and Thomsen, 1970); 0.25 μm in *Aphis* (Elliot, 1976); and from 0.1 to 1 μm in *Hyalophora* (Waku and Gilbert, 1964). The thickness of the basal lamina also varies with the physiological state of the insect; as a general rule, it is relatively thin, slightly tortuous, and directly apposed to the cytoplasmic membrane in active CA. However, it is thicker, has scallop shapes, which may be more or less marked, and may be separated from the glandular cells by a 3- or 4-μm-wide space in nonactive CA (Busselet, 1969; Baehr et al., 1973; Thomsen and Thomsen, 1970).

The basal lamina of the CA of *Leucophaea, Blaberus craniifer* (Harper et al., 1967), *B. fusca* (Brousse-Gaury et al., 1973), *Schistocerca* (Odhiambo, 1966b,c) and *Thermobia domestica* (Bitsch and Lapalisse, 1984) contains collagen fibers with characteristic periodic transverse striations. These fibers are rich in hydroxyproline and hydroxylysine.

The texture of the basal lamina is also variable. It may be homogeneous and granulofilamentous, as in *Aphis* (Elliot, 1976), *Folsomia* (Palevody and Grimal, 1975; Palevody, 1976), and *Hyalophora* (Waku and Gilbert, 1964), or composed of layers of lamellae each 0.02–0.2-μm thick, as in *Antheraea* and *Rhodnius* (Busselet, 1969; Baehr et al., 1973).

Finger-like projections of the basal lamina penetrate to a greater or lesser depth into the cavities between the cells of the glandular parenchyma (Scharrer, 1961, 1963, 1964b; Thomsen and Thomsen, 1970; Willey, 1961; Waku and Gilbert, 1964; Odhiambo, 1966c; Melnikova and Panov, 1975; Brousse-Gaury et al., 1973; Brousse-Gaury and Cassier, 1975). Scharrer (1961, 1962a,b, 1963, 1964b) stressed, in particular, the importance of this structure, which—in the absence of a circulatory apparatus—seems to provide an effective means of supplying nutrients and removing the secretory product, JH, from the CA.

The basal lamina, which may be stained with aniline blue (Cazal, 1948) or PAS, is composed of neutral and acid mucopolysaccharides associated with a protein component. No lipid has been detected. It may contain inclusions of varying sizes that are homogeneous in *Calliphora* (Thomsen and Thomsen, 1970) and in nymphs of *Rhodnius* (Busselet, 1969), in which they measure from some 120–150 to 600–650 nm in diameter, and more frequently from 200 to 250 nm. In neither case has the origin of these inclusions or their significance been elucidated. In *Leucophaea* (Scharrer, 1971), dense, "cribriform" pleomorphic inclusions measuring up to 1.6 μm have been seen. They are formed in the Golgi apparatus (C-body material) and released by exocytosis into intercellular spaces, although part of the C-body material may be captured by the multivesicular bodies. Cribriform bodies are particularly numerous in the extracellular stroma of inactive CA, which may simply suggest a relationship with the endocrine functions of these glands. The junctional modification of the plasma membranes shutting on the external and trabecular basal lamina are hemidesmosomes.

2.6.2.   *Gland Cells*

2.6.2.1.   LIGHT AND DARK CELLS

Examination of CA by electron microscopy shows two types of glandular cells, distinguished by their different electron densities, namely, light and dark cells. According to Busselet (1969), Thomsen and Thomsen (1969), Johnson (1966), and Deleurance and Charpin (1971, 1972), these distinctions have no functional or structural significance; they are an artifact due to variable penetration of fixatives. However, Dorn (1973) considered that the differences of electron density observed in *Oncopeltus* had a functional significance. He proposed that the dark cells, which predominate during the last nymphal stage and in young adults, are inactive. The light cells, which are characteristic of males and mature females, are active; they are larger than the dark cells, and their nuclei have more indentations.

During the moulting and reproductive cycles, which coexist in adult females of the primitive insect, *T. domestica,* the CA undergo periodic changes; their volume and number of cells are lower at the beginning and at the end of each molting cycle, and higher about the middle. However, the nucleus–cytoplasm ratio does not undergo significant variations. The glandular cells, which are all similar, undergo two distinct stages: (1) the so-called clear cells present at the beginning and the end of every molting cycle, and (2) the so-called dense cells in the middle of the cycles. If there is no insemination at the beginning of each intermolting period, the terminal oocytes are totally resorbed and most of the CA cells are in an intermediate stage between "clear" and "dense" cells. These data are correlated with probable variations in the physiological activity of the CA during normal adult cycle (Bitsch and Lapalisse, 1984).

2.6.2.2. PLASMA MEMBRANES AND
MEMBRANE DIFFERENTIATION

Plasma Membranes

One of the essential contributions of electron microscopy studies has been the demonstration of plasma membranes. Previous observations using the light microscope have led to a hypothesis suggesting the syncytial nature of CA, or the existence of multinuclear cells (see Cazal, 1948, for references). Plasma membranes are universally distributed, and this rule has no known exception in the case of CA. The difficulties encountered in their study are essentially due to their meandering path since the allata cells have extremely complex forms (branched or star shaped), particularly those located at the center of large, compact glands. The plasma membrane in contact with the peripheral basal lamina and its finger-like processes (see Section 2.6.1, above) itself projects numerous fine or thick digitations, which are supported by microtubules oriented parallel to their longidudinal axis or reinforced at their extremities by hemidesmosomes. In *Schistocerca*, independently of these digitations (Odhiambo, 1966a,b), there also exist crypts provided with microvillii that are formed by invaginations of the plasma membrane of certain cells, or developed in the zone of junction of several glandular cells (Papillon et al., 1976).

Generally, the course of the plasma membrane varies according to the physiological condition of the CA. Thus, in *Aphis craccivora* (Elliot, 1976), *Hypera postica* (Tombes and Smith, 1970, 1972), *H. cecropia* (Waku and Gilbert, 1964), *Schistocerca gregaria* (Odhiambo, 1966a), *R. prolixus* (Busselet, 1969; Baehr et al., 1973), *Antheraea pernyi* (Busselet, 1969), *L. maderae* (Scharrer, 1964a), and *Celerio lineata* (Schultz, 1960), the plasma membranes of active CA are much less sinuous than those of inactive CA. The increase in size of the cytoplasmic mass associated with cellular activity causes distension of the membranes. Thus, in *Schistocerca* (Papillon et al., 1976), the aforementioned crypts are clearly visible in the inactive gland immediately after the imaginal molt but are subsequently found less frequently and are less well developed.

Intercellular Junctions (Figs. 2.9 and 2.10)

Glandular cohesion of the CA is essentially ensured by the peripheral basal lamina (see Section 2.6) and the interdigitations of the projection of adjacent cells. The differentiations of junctional membranes are not well developed. In the region of contact of the basal lamina and the finger-like processes of the glandular cells, the extremities of these cellular digitations are generally reinforced by hemidesmosomes associated with microtubules (Girardie and Granier, 1974a; Tombes and Smith, 1970;

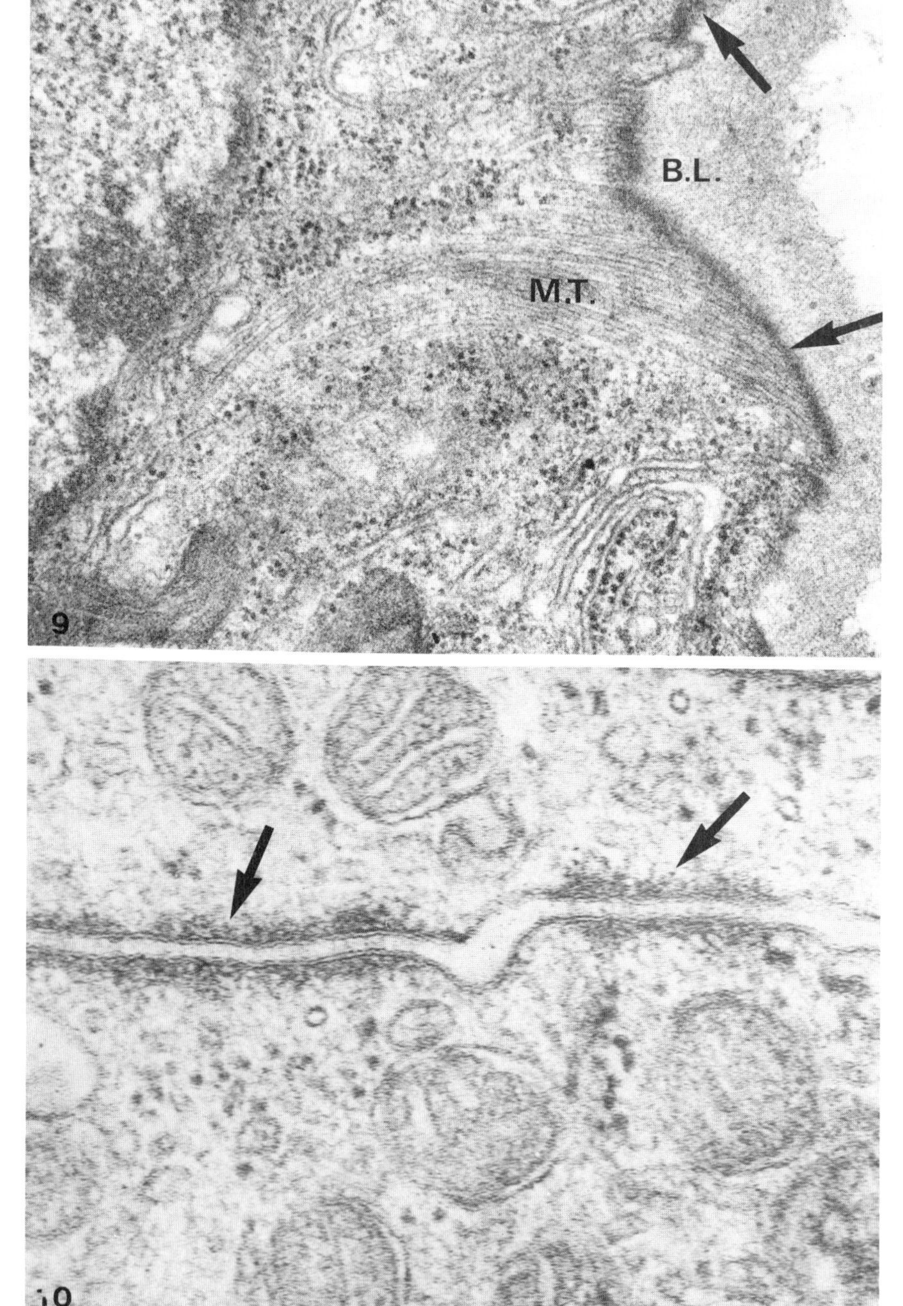

B.L.
M.T.
9
10

Melnikova and Panov, 1975; Bitsch and Lapalisse, 1984). In *B. fusca*, the hemidesmosomes may be up to 2.5 μm long. In *H. cecropia*, the accumulation of dense material corresponding to the hemidesmosomes is visible only in active glands; this may be related to the secretion of JH (Waku and Gilbert, 1964). In *Hyphantria* (Melnikova and Panov, 1975), no specialized system of junctions between the glandular cells has been observed. In *B. mori* (Fukuda et al., 1966), larval stages of *Rhodnius* (Busselet, 1969), *Locusta migratoria* (Fain-Maurel and Cassier, 1969b), and *Oncopeltus* (Dorn, 1973), the junctional system consists of occasional desmosomes of the zonula or macula adherens type; it consists of numerous macula adherens in *A. pernyi* (Busselet, 1969). In *Anacridium* (Girardie and Granier, 1974a) and *Drosophila* (King et al., 1966a,b), gap junctions (tight junctions) and desmosomes are associated with microtubules. In *R. prolixus*, during imaginal life (Baehr et al., 1973) and in *B. fusca* (Brousse-Gaury et al., 1973), this junctional system is particularly complex, because it consists of gap junctions, maculae adherens, and septate desmosomes.

In the vesicular CA of *T. domestica* (Bitsch and Lapalisse, 1984), distal parts of the plasma membrane of the CA cells show long septate desmosomes continued by short *zonulae adherens*.

The presence of gap junctions suggests that the CA cells are coupled, and indeed this has been demonstrated by dye injections and intercellular recording in CA of *D. punctata*. The amplitude of electrotonic potentials elicited by negative current injection varied intensely with the rate of JH biosynthesis by the CA. Forskolin and IBMX induced elevation of cyclic AMP (that is, adenosine 3′,5′-cyclic monophosphate), increased the electrotonic potential, but reduced JH release by day 8 from the CA (Lococo et al., 1986).

Pinocytotic Vacuoles

Pinocytotic vacuoles notable for the radial ornamentation of their cytoplasmic surface have frequently been observed during their formation both when they are associated with or are close to the plasma membrane and when they are scattered throughout the cytoplasm (coated vesicles). They are particularly large (100–200 mn in diameter) in *Rhodnius* (Baehr et al., 1973) and in *Blabera* Brousse-Gaury et al., 1973).

FIGURE 2.9. CA of *Pyrrhocoris apterus* (Heteroptera). In the vicinity of the basal laminae (BL), plasma membranes are provided with hemidesmosomes (arrow) associated with numerous microtubules (MT). ×27,000.

FIGURE 2.10. CA of *Blabera fusca* (Dictyoptera). Glandular cells are associated by numerous *maculae adhaerens* (arrow). Note mitochondria. ×56,172. (new magnification due to reduction)

2.6.2.3. INTERCELLULAR SPACES AND MICROTUBULES

Intercellular Spaces

The intercellular spaces deserve particular attention, since in the CA they constitute an important pathway not only for the supply of structural or energetic material and hormonal precursors but also for the transport of JH into the hemocoel. In resting glands or glands where the level of activity is low, the cells are contiguous, the plasma membranes are apposed, and the intercellular spaces are narrow and moniliform. However, during the peak of hormonal activity, the intercellular spaces become much wider (Schultz, 1960; Waku and Gilbert, 1964; Scharrer, 1964a; Pasteels and Deligne, 1965; King et al., 1966a; Odhiambo, 1966b; Joly, 1968; Brousse-Gaury et al., 1973; Brousse-Gaury and Cassier, 1975; Baehr et al., 1973; Haget et al., 1981).

The Intercellular Microtubules

In the fourth instar larvae of *B. mori* (Morohoshi et al., 1976a), the intercellular spaces contain numerous microtubules that form connections between the glandular cells and axons and even penetrate the basal lamina. These unusual structures may participate in the transport of intercellular material—or even of the brain hormone, which in this species enters the hemolymph in the region of the CA (Morohoshi et al., 1975; Morohoshi and Shimada, 1975).

In *L. migratoria migratorioides* (Cassier, 1979), the intercellular spaces frequently contain an electron-dense substance that is usually amorphous and does not have the appearance of the entangled fibrils (4 nm in diameter) characteristic of the basal lamina; it is able to reorganize into macrofilaments, as is indicated by the coexistence of these two forms. The origin and significance of the intercellular material and tubules is still conjectural: they may play a role in (1) glandular cohesion, (2) secretion, or (3) intercellular transport. The first hypothesis is the most likely one, since the material is similar to the intercellular oriented fibrillar material that appears during the reaggregation of dissociated embryonic cells (Overton, 1969).

2.6.2.4. THE NUCLEUS

Although allata cells are uninuclear, their nucleus has not been the subject of much research. In CA of the lymphoid (see Section 2.5.1.1) or microcellular type (Section 2.5.1.2), the very numerous nuclei are regular in shape, being spherical or ovoid, and of small dimensions: 7–9 and 10–12 μm in diameter, respectively. However, in the macrocellular type, the nuclei are less frequent but are large, lobed, and branched or star shaped.

The size and appearance of the nuclei, regardless of the type, vary according to the level of glandular activity. Cellular activity of the CA is accompanied by nuclear swelling, dilatation of the perinuclear space, and dispersion of chromatic clumps (Panov and Bassurmanova, 1970; Elliot, 1976; Thomsen and Thomsen, 1970; Waku and Gilbert, 1964; Baehr et al., 1973). For example, in *Leucophaea* (Scharrer, 1964a), the nuclear dimensions increase from 11.7 × 8.8 μm in inactive glands to 15 × 11 μm in the active state.

The dilatations of the perinuclear space, which measures 15–20 nm at rest, are frequently accompanied by a proliferation of the external rough endoplasmic reticular layer of the nuclear sheath, thus increasing the amount of endoplasmic reticulum. This phenomenon, comparable to that resulting in the production of the annulated lamella, has been reported in *Hyphantria* (Melnikova and Panov, 1975), *Drosophila* (Aggarwal and King, 1969), *Calliphora erythrocephala* (Thomsen and Thomsen, 1970), and Bathyscinae (Coleoptera) (Deleurance and Charpin, 1971, 1972); it may be the source of reticular pseudocrystalline structures of "twisted areas" (Deleurance and Charpin, 1971, 1972).

The number of nucleoli is a specific characteristic, varying according to species from 1 (*Rhodnius, Triatoma, Eurygaster*), to 2 (*Leucophaea*), or 3–5 (*Antheraea, Bombyx, Hyalophora*). In inactive glands, the nucleoli are large and compact; in active glands, they have a tendency to fragment (e.g., *Rhodnius:* Baehr et al., 1973) or to take on a very characteristic annular configuration (e.g., *Eurygaster:* Panov and Bassurmanova, 1970; *B. mori:* Fukuda et al., 1966).

The presence of nuclear inclusions has been reported. In *A. pernyi* the single inclusion is large (1.75 × 1.75 μm), is Feulgen positive, and contains a large quantity of RNA and DNA (Busselet, 1969); in *Leucophaea*, it is a spherical body (700 nm in diameter) of an unknown nature.

2.6.2.5. MITOCHONDRIA

Mitochondria are an important element in a cell, and their study can provide information about the structure and the activity of the CA.

In inactive glands, the mitochondria are generally scarce and are represented principally by globular (0.15–0.3 μm in diameter) or rod-shaped (0.5–1.0 μm × 0.3–0.5 μm) elements that are uniformly distributed in the cytoplasm (on average, 10.4 per 25 μm in *Hyalophora*). These organelles contain lamellar cristae with no preferential orientation in a dense matrix and zero, one, or two opaque bodies 100–150 nm in diameter (Waku and Gilbert, 1964; Scharrer, 1964a; Odhiambo, 1966b; Panov and Bassurmanova, 1970; Thomsen and Thomsen, 1970; Baehr et al., 1973; Papillon et al., 1976). Several cup-shaped elements have also been observed in *Hyphantria* (Melnikova and Panov, 1975). In *B. fusca*, the mitochondria are frequently found close to the nucleus (Brousse-Gaury et al., 1973).

Important modifications, correlated with the cycle of activity have

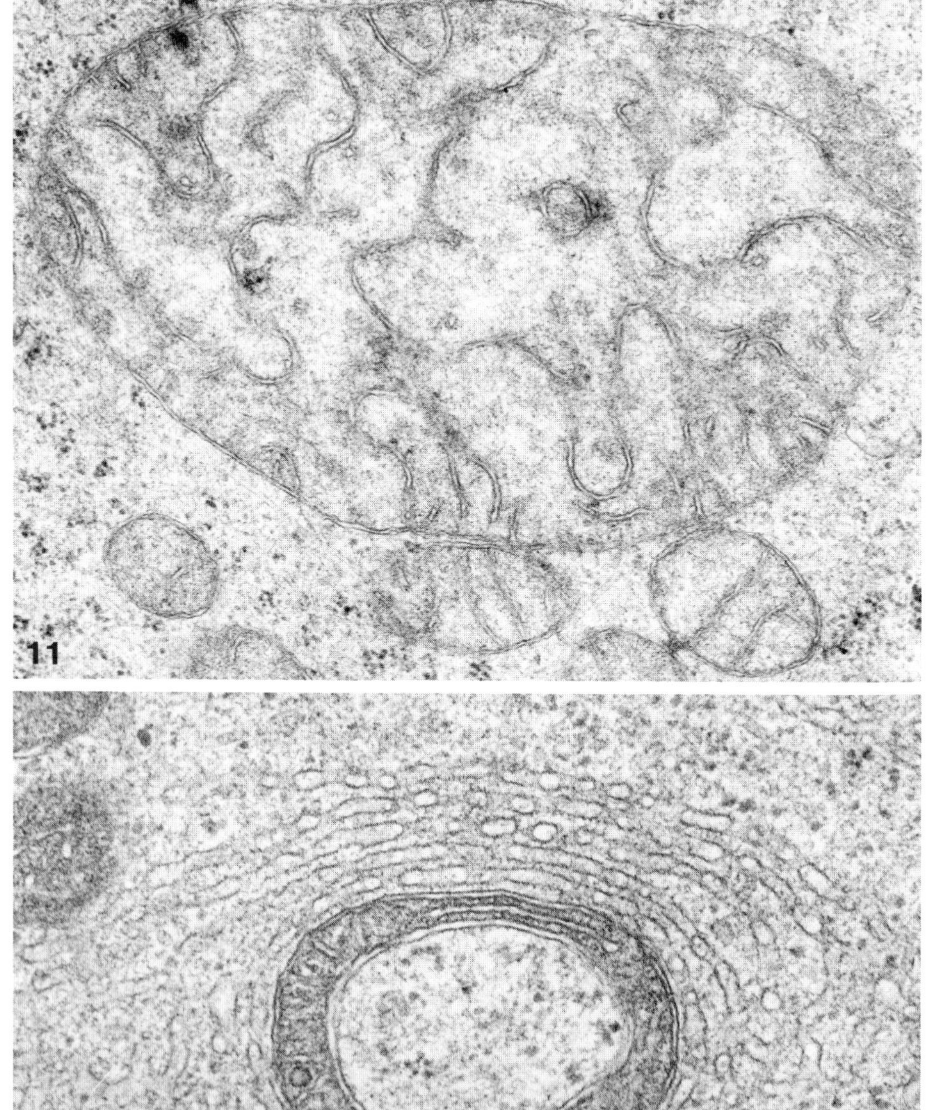

frequently been observed; these generally consist of a considerable increase in the mitochondrial population (39.7 per 25 μm^2 in *Hyalophora*) (Waku and Gilbert, 1964) and of an increase in the size of the mitochondria, which also frequently undergo a change in shape, becoming dumbbell, cup, or ring shaped, or give rise to macromitochondria as a result of the matrix becoming opaque and the number of cristae increasing (Scharrer, 1964a; Fukuda et al., 1966; Odhiambo, 1966b; Joly et al., 1968; Baehr et al., 1973; Brousse-Gaury et al., 1973; Brousse-Gaury and Cassier, 1975; Dorn, 1973; Melnikova and Panov, 1975; Joly, 1976). Mitochondria frequently appear as a globular element (0.1–0.2 μm in diameter) arranged in small chains, which is a sign of mitochondrial division (Busselet, 1969; Baehr et al., 1973; Brousse-Gaury et al., 1973; Brousse-Gaury and Cassier, 1975).

Cup- or ring-shaped mitochondria frequently form concentric structures in which they regularly alternate with cisternae of smooth endoplasmic reticulum (e.g., *Bombyx:* Fukuda et al., 1966; *Locusta:* Fain-Maurel and Cassier 1969b; *Schistocerca:* Odhiambo, 1966b). During the transition from an active phase to a resting phase, many mitochondria disintegrate or are incorporated into lytic bodies (Scharrer, 1964a).

Locusta migratoria migratorioides deserves particular mention. In this species, sexual dimorphism is associated with mitochondrial pleomorphism (Fain-Maurel and Cassier, 1969b). After the imaginal molt, each sexual type (male or female) and even each physiological state (females during maturation and mature females) may be defined and individualized by the type of mitochondria. In males, the activity of the CA is continuous and the mitochondrial population is homogeneous. The mitochondria are characterized by their swollen appearance and their large size, by the increasing transparency of the matrix material, and by the large number and regular arrangement of the cristae.

In females, the activity of the CA is cyclic and the mitochondrial population is heterogeneous (Figs. 2.11–2.14). It consists of (1) a strain of small mitochondria with an electron-dense matrix substance that are permanent source elements from which various strains develop in the glandular cells and (2) a strain characteristic of each physiological stage. During sexual maturation, a line of chondriosomes differentiates from the permanent line. These mitochondria contain a clear matrix substance with a

FIGURE 2.11–2.12. CA of a mature female of the African migratory Locust, *Locusta migratoria migratorioides* (Orthoptera).
2.11 (*above*). Very active cells of the CA of a green solitarious female, containing giant mitochondria with clear matrix and wavy cristae. ×20,500.
2.12 (*below*). Gregarious female. At the end of each vitellogenic cycle, elongated or annular mitochondria are frequently associated with concentrically disposed smooth endoplasmic reticulum. ×31,900.

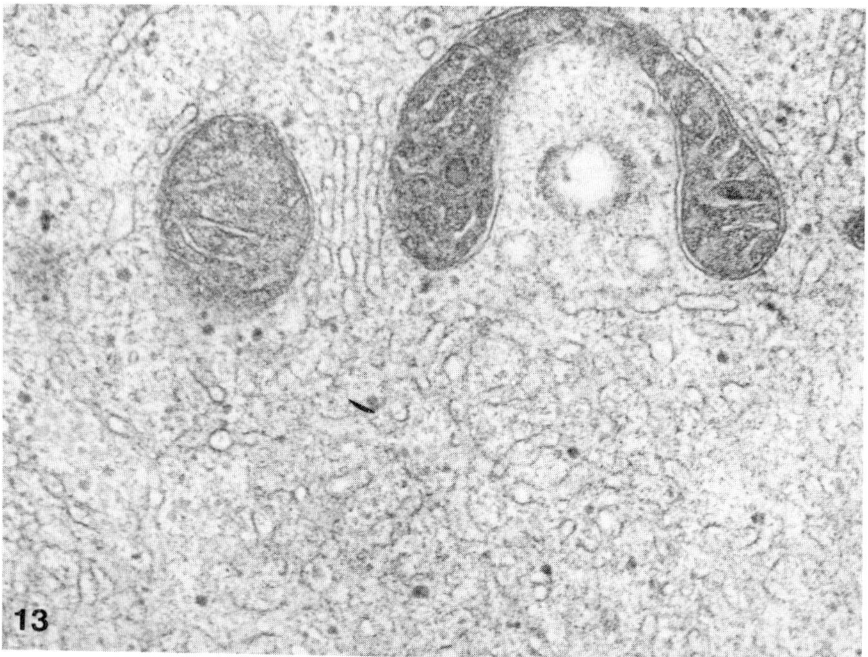

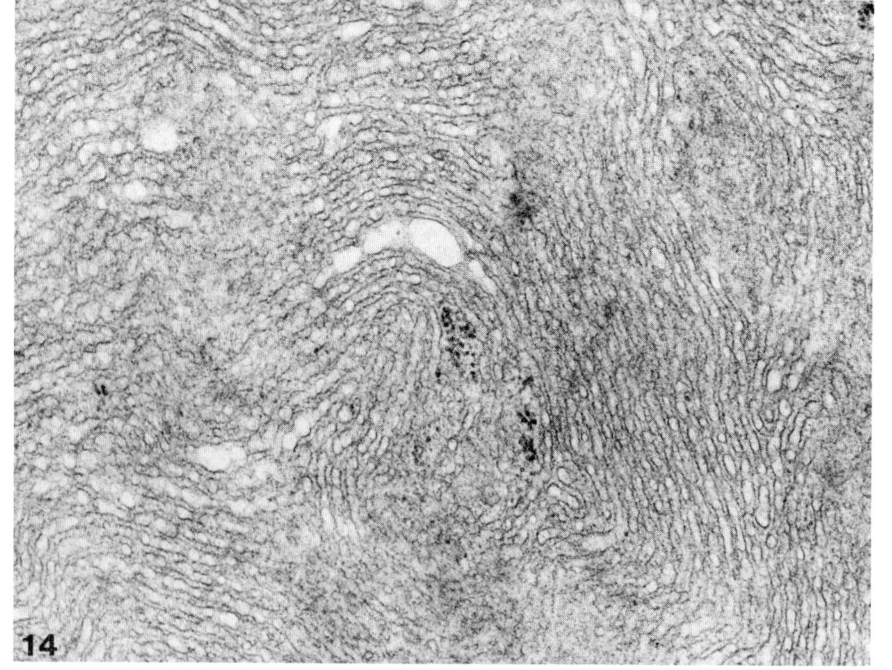

small number of fine lamellar rectilinear cristae; some of these mitochondria develop into giant mitochondria (1.5–2.0 μm × 2.0–3.0 μm), characterized by marked proliferation of their lamellar cristae, which form loops and rings (see also *Anacridium:* Girardie and Granier, 1974a). However, during ovarian cycles, individualization of the mitochondria occurs at the expense of the permanent line: very elongate mitochondria containing an electron-opaque matrix substance will become dumbbell shaped or later cup shaped, and take part, by their association, in the construction of a large mitochondrial complex—the chondriosphere of De Robertis and Sabatini (1958). They are frequently found in organs involved in the synthesis of steroids (Christensen and Chapman, 1959; Idelman, 1964, 1970), as well as in the prothoracic glands (Beaulaton, 1964, 1968a,b; Cassier and Fain-Maurel, 1970, 1971a,b).

2.6.2.6. ENDOPLASMIC RETICULUM

Of all the organelles in allata cells, the endoplasmic reticulum has attracted most attention of cytologists who engaged in the early ultrastructural investigations. The endoplasmic reticulum could, because of its abundance, provide an exact index of the level of glandular activity, which in turn could provide investigators with a means of evaluating the consequences of their experiments and thus the mechanisms that control the activity of CA. But although JHs are lipid-soluble sesquiterpenoids (Röller et al., 1967; Meyer et al., 1968; Judy et al., 1973a,b), their synthesis is not necessarily linked to the formation of an abundant smooth endoplasmic reticulum (Table 2.1). Thus, in insects belonging to the first group, during peaks of hormonal activity, glandular cells have shown evident enrichment in rough endoplasmic structures (Figs. 2.15 and 2.16). However, in insects belonging to groups II to V (Table 2.1), under the same conditions, the study of CA reveals an increase in smooth endoplasmic reticulum, which in the extreme case (Figs. 2.17 and 2.18) (group V) may invade the cell and form structures that may be pseudocrystalline, lenticular, whorled, or concentric. Moreover, on examining Table 2.1, it can be seen that there is no direct relationship between the type of endo-

FIGURES 2.13–2.14. CA of a mature female of the African migratory Locust, *Locusta migratoria migratorioides* (Orthoptera). Evolution of the endoplasmic reticulum during an ovarian cycle of a crowded female.
2.13 (*above*). During vitellogenesis, smooth endoplasmic reticulum progressively invades cytoplasmic area. ×14,880. (new magnification)
2.14 (*below*). At the moment of egg laying, concentrically disposed smooth endoplasmic reticulum generates fingerprint-like structures and then myelinic structures associated with lipid droplets or mitochondria. ×20,500.

TABLE 2.1. Nature of the Endoplasmic Reticulum and of the JH Secreted in Active Corpora Allata Cells

Order	Species	Stage[a]	Nature of JH	References
Group I: Smooth endoplasmic reticulum absent; hypertrophy of the rough endoplasmic reticulum				
Phasmoptera	*Carausius morosus*	L,A,E	JH III	Smith, 1968; Cassier, 1979; Meyer and Pflugfelder, 1958; Haget et al., 1981
Heteroptera	*Eurygaster integriceps*	A	ND	Panov and Bassurmanova, 1970
	Rhodnius prolixus	L,A	JH III	Busselet, 1969 : Baehr et al., 1973
Coleoptera	*Diaprysius serullazy*	L	ND	Deleurance and Charpin, 1971
	D. fagei	L	ND	Deleurance and Charpin, 1971
Lepidoptera	*Antheraea pernyi*	L	ND	Busselet, 1969
Group II: Smooth endoplasmic reticulum slightly developed; predominant rough endoplasmic reticulum				
Heteroptera	*Oncopeltus fasciatus*	L,A	ND	Dorn, 1973
	Pyrrhocoris apterus	L,A	ND	P. Cassier, unpublished data
Homoptera	*Aphis craccivora*	L,A	ND	Elliot, 1976
	Myzus persicae	A	ND	Bowers and Johnson, 1966
Coleoptera	*Hypera postica*	A	ND	Tombes and Smith, 1970, 1972; Tombes, 1972

Group III: Smooth endoplasmic reticulum and rough endoplasmic reticulum present

Thysanura	*Petrobius maritimus*	A	ND	Cassier, 1979
Collembola	*Folsomia candida*	A	ND	Palevody and Grimal, 1975; Palevody, 1976; Cassagnau and Juberthie, 1967
	Hypogastrura tullbergi	L,A	ND	Lauga-Reyrel, 1984a,b
Blattodea	*Leucophaea maderae*	L,A	JH III	Scharrer, 1964a, 1971
	Diploptera punctata	A	JH III	Johnston et al., 1985; Szibbo and Tobe, 1981a,b; Tobe et al., 1985
	Blabera fusca	A	JH III	Brousse-Gaury et al., 1973; Brousse-Gaury and Cassier, 1975
	Periplaneta americana	A	JH III	Gundel and Penzlin, 1980; Pipa and Novak, 1979
Ephemeroptera	*Ephemera danica*	L,A	ND	Kaiser, 1980
Hymenoptera	*Apis mellifera*	L	JH III	Wirtz, 1973
	Polistes gallicus	L,A	JH III	Delfino et al., 1981
Diptera	*Chironomus riparius*	A	ND	Credland and Scales, 1981
	C. melanotus	L	ND	Kummel, 1969

(continued)

TABLE 2.1. (*Continued*)

Order	Species	Stage[a]	Nature of JH	References
Group IV: Smooth endoplasmic reticulum preponderant but unorganized				
Thysanura	*Thermobia domestica*	A	ND	Bitsch and Lapalisse, 1984
Lepidoptera	*Monema flavescens*	L	ND	Takeda, 1977
Group V: Rough endoplasmic reticulum negligible; smooth endoplasmic reticulum preponderant and organized				
Orthoptera	*Anacridium aegyptium*	L,A	ND	Girardie and Granier, 1974a
	Locusta migratoria	L,A	JH III	Joly et al., 1967, 1968; Fain-Maurel and Cassier, 1969b; Cassier and Fain-Maurel, 1970; Guelin and Darjo, 1974; Poras et al., 1985
	Schistocerca gregaria	A	JH III	Odhiambo, 1966b; Papillon et al., 1976
	Acheta domesticus	A	ND	Bradley and Edwards, 1979
Heteroptera	*Oncopeltus fasciatus*	L,A	ND	Unnitham et al., 1971
Dermaptera	*Labidura riparia*	A	ND	Cassier, 1979; Caussanel and Cassier, 1978; Baehr et al., 1982
Coleoptera	*Speophyes lucidulus*	L,A	ND	Deleurance and Charpin, 1971, 1973
	Troglodomus brucheti gaveti	A	ND	Deleurance and Charpin, 1972

	Isereus colasi	A	ND	Deleurance and Charpin, 1972
	Bathysciola schiodtei	A	ND	Deleurance and Charpin, 1972
	Choleva lucidulus	L,A	ND	Deleurance and Charpin, 1973
	Xyleborus ferrugineus	A	ND	Chu and Norris, 1979; Chu, 1979
	Leptinotarsa decemlineata	L,A	JH III	Deleurance and Charpin, 1971; Khan et al., 1984; Schooneveld, 1970
	Choleva cistiloides	A	ND	Charpin, 1973
Diptera	*Drosophila melanogaster*	L,A	ND	King et al., 1969a,b; Aggarwal and King, 1969
	Calliphora erythrocephala	P	ND	Thomsen and Thomsen, 1970
	Calliphora stygia	A	ND	Johnson, 1966
Lepidoptera	*Hyalophora cecropia*	P	JH I, II, JH I acid	Waku and Gilbert, 1964
	Pectinophora gossypiella	L	ND	Raina and Borg, 1980
	Celerio lineata	A	ND	Schultz, 1960
	Bombyx mori	L,A	ND	Fukuda et al., 1966; Morohoshi et al., 1976a,b
	Hyphantia cunea	L	ND	Melnikova and Panov, 1975
	Miscellaneous (*Galleria, Ostrinia, Diatraea, Plodia*)		ND	Nishiitsutsuji-Uwo, 1960, 1961
	Manduca sexta	L	JH I, II, III	Sedlak, 1982; Sedlak et al., 1983; El-Salhy et al., 1983

[2]Stage: L = larvae; P = pupa; A = adult; E = embryo; ND = not determined.   [3]From Tobe and Stay (1985).

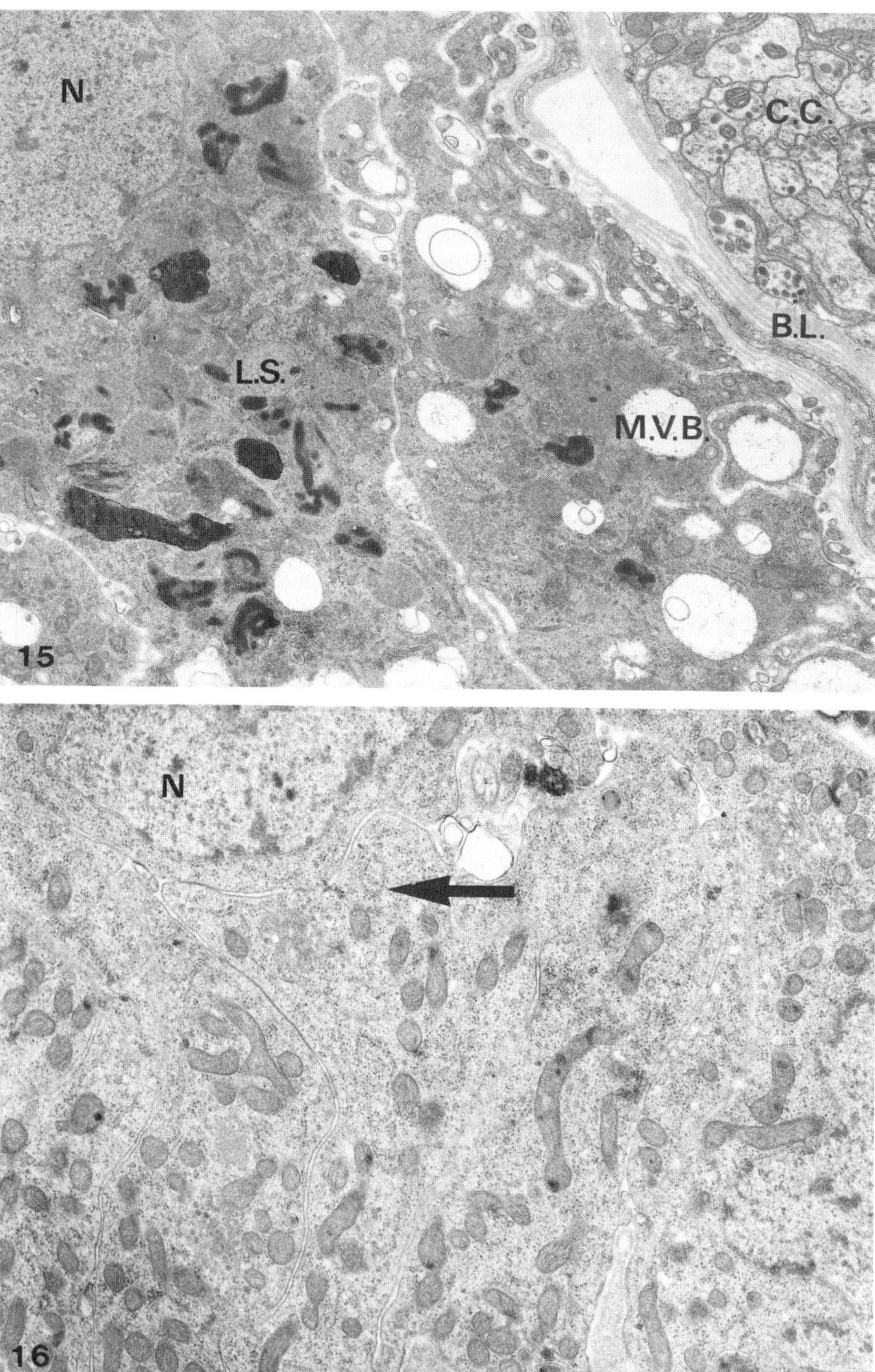
N
C.C.
B.L.
L.S.
M.V.B.
15
N
N
16

plasmic reticulum, the systematic position, or the physiological state of an insect (nymph, pupa, or adult). It is worth noting that insects belonging to groups IV and V and, to a certain extent, group III, are the most common, are large, or have a holometabolic type of postembryonic development. Those belonging to group I are heterometabolic or larvae of holometabolic insects.

According to Deleurance and Charpin (1971, 1972), the structural differences in CA of primitive species having small eggs (extended development) and those of the most evolved species having large eggs (direct development) are essentially dependent on the amount of endoplasmic reticulum present. In the coleopterans *Bathysciola* and *Speophyes*, species that have small eggs, the granular reticulum is abundant, while the smooth endoplasmic reticulum is represented only by tubules and scarce or grouped vesicles. And in *Troglodromus* and *Isereus*, which have large eggs, the rough endoplasmic reticulum is not well developed, but the smooth reticulum is abundant and formed of lamellae that organize to form well-individualized, lenslike structures (6 × 3 μm in diameter).

The presence of ribosomes, either scattered or associated in polysomes, has been reported in the CA in all species studied. Their number increases in parallel with the level of glandular activity and with the development of the cytoplasmic mass (Scharrer, 1964a; Waku and Gilbert, 1964; Fukuda et al., 1966; Odhiambo, 1966b; King et al., 1966a; Busselet, 1969; Fain-Maurel and Cassier, 1969b; Thomsen and Thomsen, 1970; Melnikova and Panov, 1975). However, in *Eurygaster* (Panov and Bassurmanova, 1970), the number of ribosomes decreases during the active phase but, in parallel, the ergastoplasmic membrane system begins a considerable development.

The chronological succession of different forms of endoplasmic reticulum, which represent stages of the cycle of activity associated with the development of larval stages or with ovarian cycles, has been confirmed (Scharrer, 1964a; Odhiambo, 1966b; Joly et al., 1967; Fain-Maurel and Cassier, 1969a; Brousse-Gaury et al., 1973; Brousse-Gaury and Cassier, 1975; Melnikova and Panov, 1975; Cassier, 1979).

FIGURES 2.15–2.16. CA of a 15-day old adult female of *Rhodnius prolixus* (Hemiptera).

2.15 (*above*). Starved female. Inactive glandular cells contain various dense lytic structures (L.S.) and multivesicular bodies (MVB). KEY: BL = basal lamina; CC = corpora cardiaca; N = nucleus. ×18,000.

2.16 (*below*). Blood-fed female. Active glandular cells deprived of lytic structures contain numerous mitochondria and a well-developed rough endoplasmic reticulum (arrow). ×24,000.

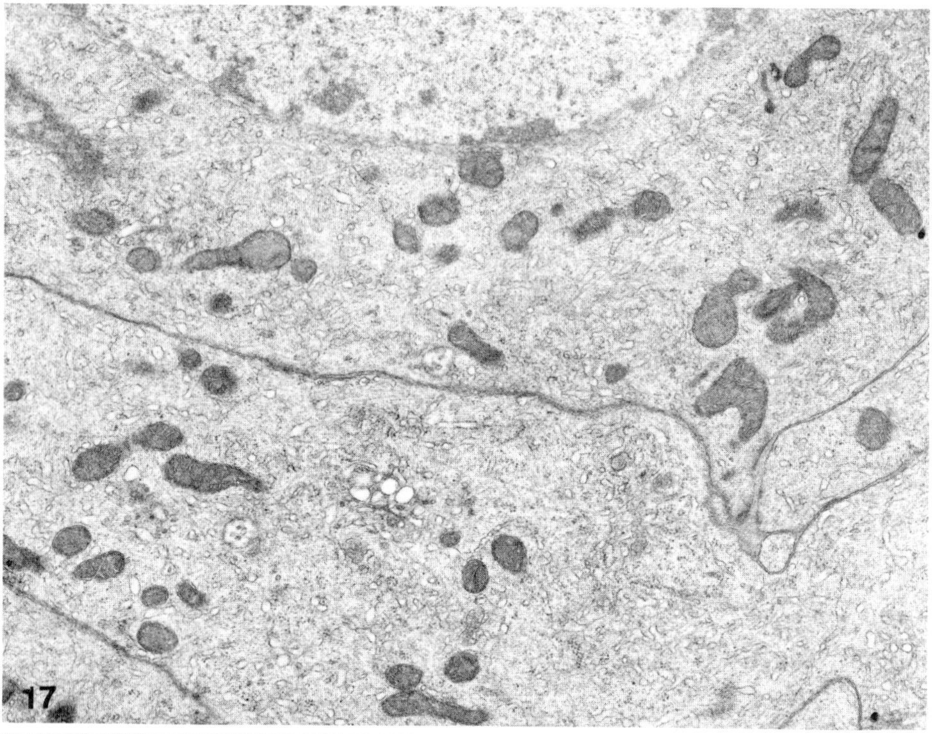

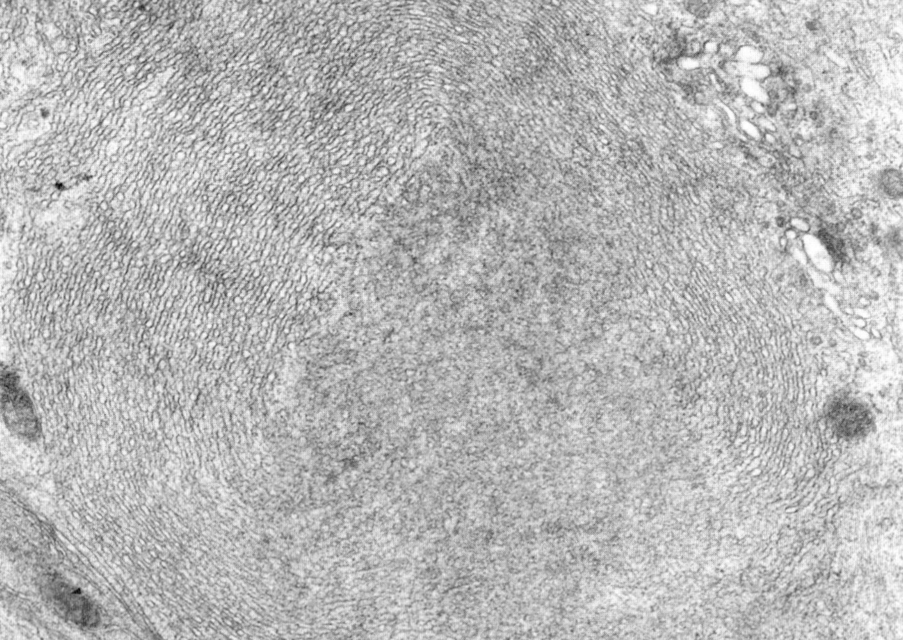

In *Anacridium* (Girardie and Granier, 1974a), the ultrastructural appearance of the CA does not change during oocyte maturation. The sequence of different ultrastructural appearances of the endoplasmic reticulum is not temporal but spatial; it occurs simultaneously in each CA regardless of the stage of vitellogenesis. So oocyte maturation results from a succession of secretory cycles, which continue asynchronously in all the cells of the organ.

In *L. migratoria migratorioides* (Kazalinsk strain), Guelin and Darjo (1974) observed a distinct distribution of various forms of reticulum in each allata cell, which enabled them to define three zones: a perinuclear A zone, rich in mitochondria, tubules, ribosomes, polysomes, and short rough endoplasmic reticulum tubules, containing a crown of dictyosomes; an internal lobed B zone specialized with secretory processes and having abundant smooth reticulum associated with the dictyosomes; and an external C zone composed of a group of digitations that serve to anchor the cell—a zone in which there is no smooth reticulum and only a few dictyosomes.

Such polarity may be observed during a period of activity in the allata cells of *B. fusca* (Brousse-Gaury et al., 1973; Brousse-Gaury and Cassier, 1975) (Figs. 2.19–2.22). The progressive increase in gonadotropic activity of the CA is manifested by an increase in the glandular cells of ergastoplasmic formations, which, in the perinuclear zone, are involved in the formation of large rolls of membranes and also of agranular tubular reticulum scattered inside the peripheral cytoplasmic digitations (Figs. 2.21 and 2.22).

There may be a regional distinction. Thus, the peripheral allata cells of *Calliphora* (Thomsen and Thomsen, 1970), *Bombyx* (Fukuda et al., 1966), *Schistocerca* (Odhiambo, 1966b,c), *Hyphantria* (Melnikova and Panov, 1975) and *Locusta* (Cassier and Fain-Maurel, 1970; Fain-Maurel and Cassier, 1969b, 1970) are richer in reticulum and inclusions and more active than the central cells. The opposite is seen in *Anacridium* (Girardie and Granier, 1974a).

Experimental data suggest a strict relationship between the level of activity of the CA and the development of the endoplasmic reticulum. In

FIGURES 2.17–2.18. CA of a female adult of *Locusta migratoria* (Savio-strain) (Orthoptera).

2.17 (*above*). Inactive CA of a 25-day-old diapausing female (L/D: 16/11). ×6800.

2.18 (*below*). Very active CA of a 25-day-old diapausing female (L/D = 16/11), 10 days after severance of NCC II. Concentrically disposed smooth endoplasmic reticulum. ×13,000.

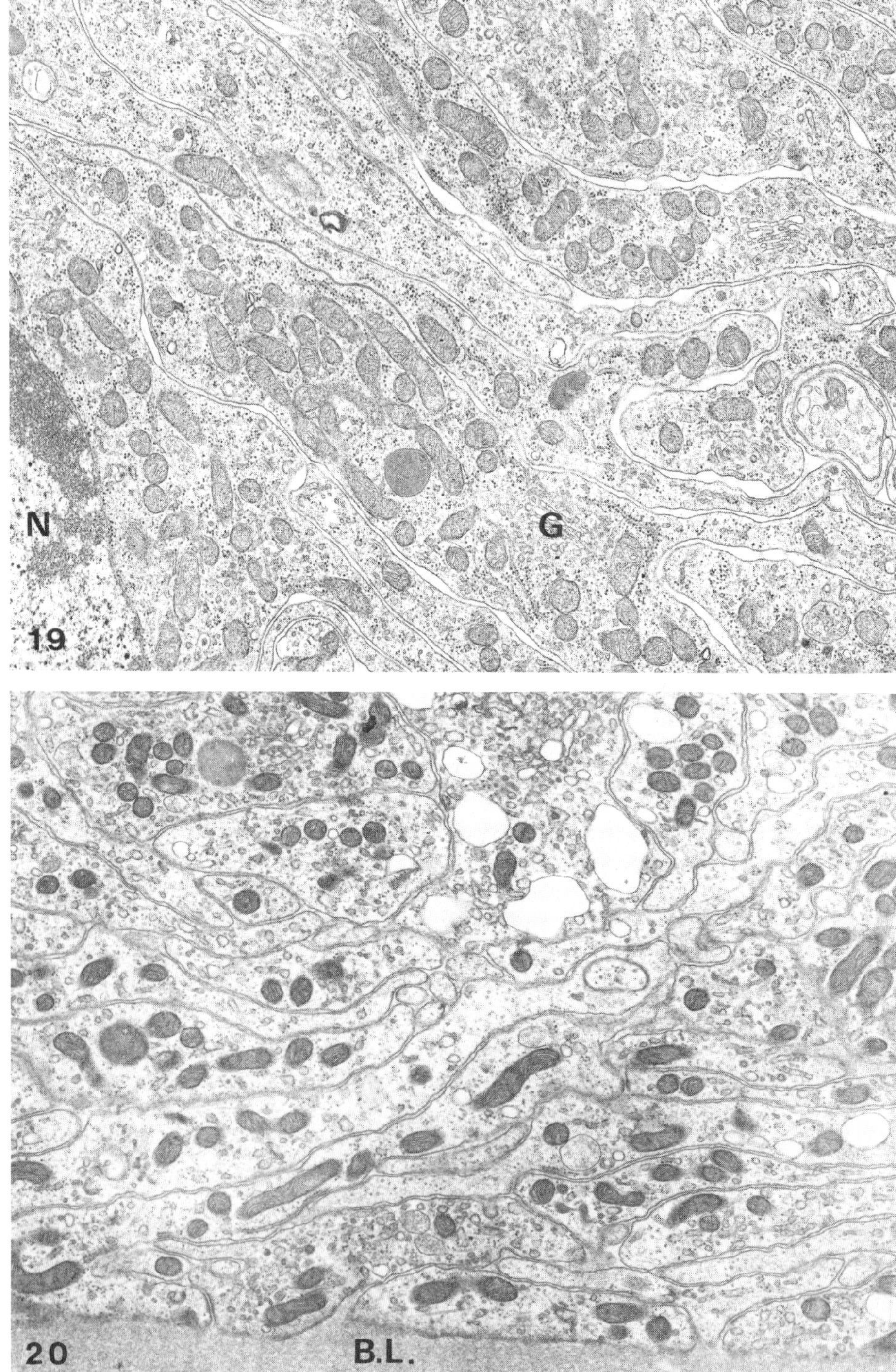

N
G
19
20
B.L.

Blabera fusca (Brousse-Gaury et al., 1973) (Figs. 2.23–2.26) cauterization of
the right ocellus in isolated females results in a large and significant increase in gonadotropic activity associated with hypertrophy of the right
CA; the allata cells contain long, flexible, curved, or monoliform
ergastoplasmic cisternae and a large number of large mitochondria. In
Locusta, the disconnection of male active CA rich in endoplasmic reticulum or their implantation in larvae or females results in less than 8
days in aplasia of the gland, its inactivation, and disappearance of the reticulum. A reduction in the diurnal temperature from 33° to 28°C in the
breeding colonies of *Schistocerca* (Papillon et al., 1972; Cantacuzène et al.,
1972) (Figs. 2.27 and 2.28) results in male sterility and a decline in female
fecundity, a reduction in volume of CA, and a noticeable reduction in the
amount of different forms of endoplasmic reticulum (Papillon et al., 1976).
Finally, in *Locusta*, ovariectomy causes hypertrophy of CA, associated
with proliferation of the endoplasmic reticulum. In *Eurygaster* (Panov et
al., 1970), the CA of females parasitized by *Clytomia* have the appearance
of inactive and degenerate glands. This last finding was also confirmed by
Girardie and Granier (1974b) in females of *Anacridium* parasitized by
Metacemyia calloti during nondiapausing periods; nevertheless, certain
cells maintained the criteria of activity.

The transition from an active phase to a period of inactivity is associated with elimination of the endoplasmic reticulum in the form of lamellar
bodies or pseudomyelinic structures (Fain-Maurel and Cassier, 1969a,b;
Deleurance and Charpin, 1972; Palevody and Grimal, 1975; Girardie and
Granier, 1974a).

In insects belonging to the groups IV and V (Table 2.1), the proliferation of smooth endoplasmic reticulum may give rise to structures of variable size and appearance: lenslike bodies in coleopteran Bathyscinae
(Deleurance and Charpin, 1971, 1972); allatum bodies in *Drosophila* (King
et al. 1966a,b); paracrystalline bodies in *Hyphantria* (Melnikova and Panov, 1975); concentrically disposed areas in *Locusta* and *Schistocerca*; twisted
areas in *Leucophaea* (Scharrer, 1964a) and *B. fusca* (Brousse-Gaury et al.,
1973); and aggregates or networks of tubules in *Calliphora* (Thomsen and
Thomsen, 1970), *Hyalophora* (Waku and Gilbert, 1964), and *Labidura* (Figs.

FIGURES 2.19–2.20. Inactive CA of a virgin female of *Blabera fusca*
(Dictyoptera) Compare with Figs. 2.21–2.22.
2.19 (*above*). Perinuclear part of a glandular cell contains only short cisternae of rough endoplasmic reticulum. Note numerous mitochondria. KEY:
N = nucleus; G = Golgi apparatus. ×8200.
2.20 (*below*). Peripheral cytoplasmic infoldings. KEY: BL = basal laminae.
×8200.

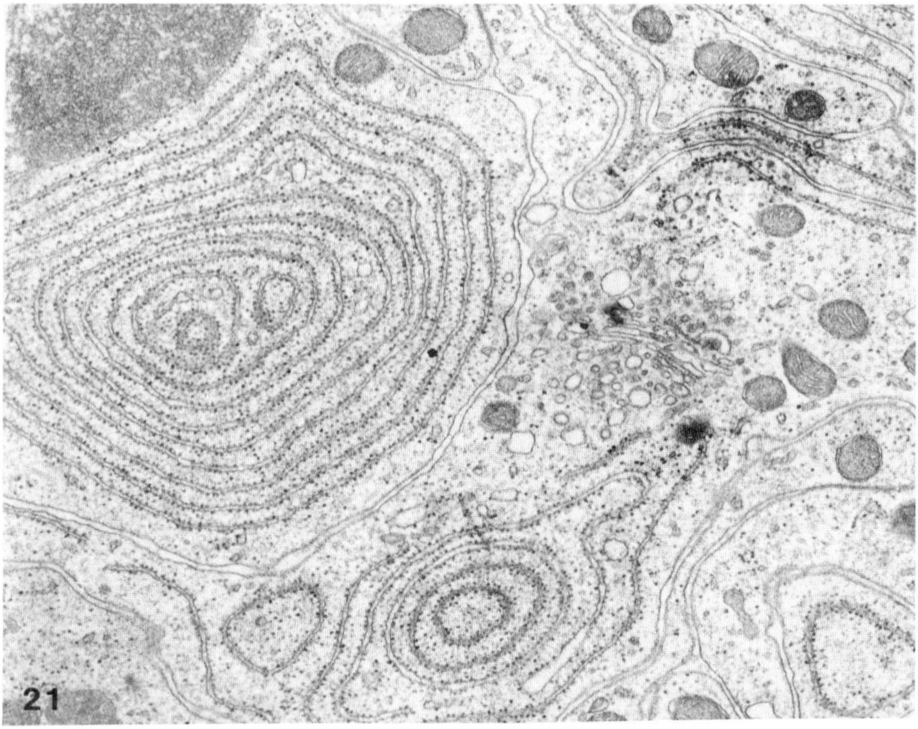

21

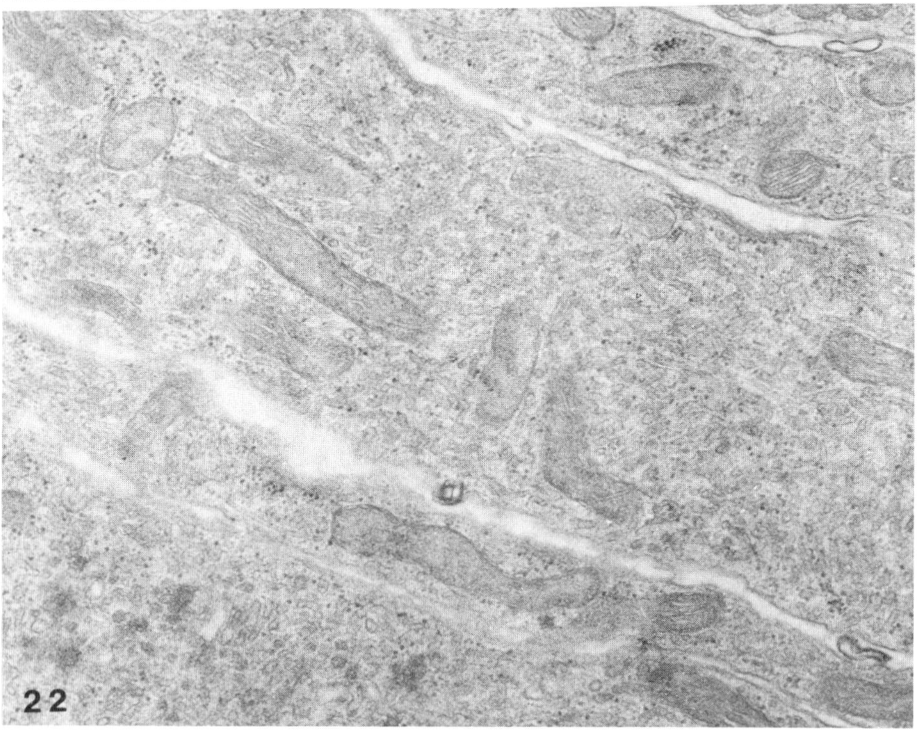

22

2.6 and 2.7). These structures characterize active CA. In addition: (1) they are frequently associated with organelles (mitochondria, rough endoplasmic reticulum, Golgi apparatus), inclusions (lipid droplets, osmiophilic bodies) and even the nucleus, whose granular outer membrane seems to be involved in their formation; (2) an area of sinuous tubules, vacuoles, or vesicles with a clear content frequently surrounds the structures and seems to migrate toward the plasma membrane (Thomsen and Thomsen, 1970; Melnikova and Panov, 1975); (3) these structures are not specific to CA but seem to be correlated to the synthesis of lipid or small molecules, and similar structures have been observed in the steroidogenic organs of vertebrates (Berchtold, 1969, 1970a,b; Picheral, 1968; Christensen, 1965), the seminal vesicles of *Schistocerca* (Cantacuzène et al., 1972), the oenocytes of *Locusta* (Fain-Maurel and Cassier, 1969a, 1972), and the digestive tubes of *Petrobius* (Fain-Maurel and Cassier, 1972).

In conclusion, morphological analysis of CA demonstrates that, in spite of the chemical identity and similarity of the synthetized JH, there is great disparity in the development of the organelles involved, particularly of the endoplasmic reticulum. The qualitative (rough or smooth) and quantitative differences in the endoplasmic reticulum seem to be related to the level of activity of the CA, which depends upon the relative size of insects, and the intensity of physiological processes (morphogenesis, vitellogenesis). The use of fine-analytical methods for the assay of cellular fractions, or immunocytological or radiographic techniques, should enable workers to localize JH synthesis and to appreciate the function of the endoplasmic reticulum.

The incorporation of tritiated farnesic acid in the form methyl farnesoate and of C_{16}JH (JH III) into the CA of *Schistocerca* adult females and examination by autoradiographic methods (Tobe et al., 1976) show that the granules of silver were almost entirely localized upon the surface of the smooth endoplasmic reticulum and in the intercellular spaces. These data demonstrate the rapid conversion of the precursor into JH and the excretion of the hormone. Mitochondria are involved in the production of acetyl CoA ATP, NAD^{+}, and SAM (*S*-adenosylmethionine) (Tobe and Stay, 1985). Furthermore, the use of [*methyl*-^{14}C]methionine *in vitro* demonstrated that there was no satisfactory correlation between the size of the CA and their level of activity; the rate of secretion varied proportionally

FIGURES 2.21–2.22. Active CA of a vitellogenic female of *Blabera fusca* (Dictyoptera) Compare with Figs. 2.19 and 2.20.
2.21 (*above*). Perinuclear part of glandular cells contain concentrically disposed cisternae of rough endoplasmic reticulum. ×13,000.
2.22 (*below*). Peripheral cytoplasmic infoldings contain numerous vesicles or tubules of smooth endoplasmic reticulum. ×13,000.

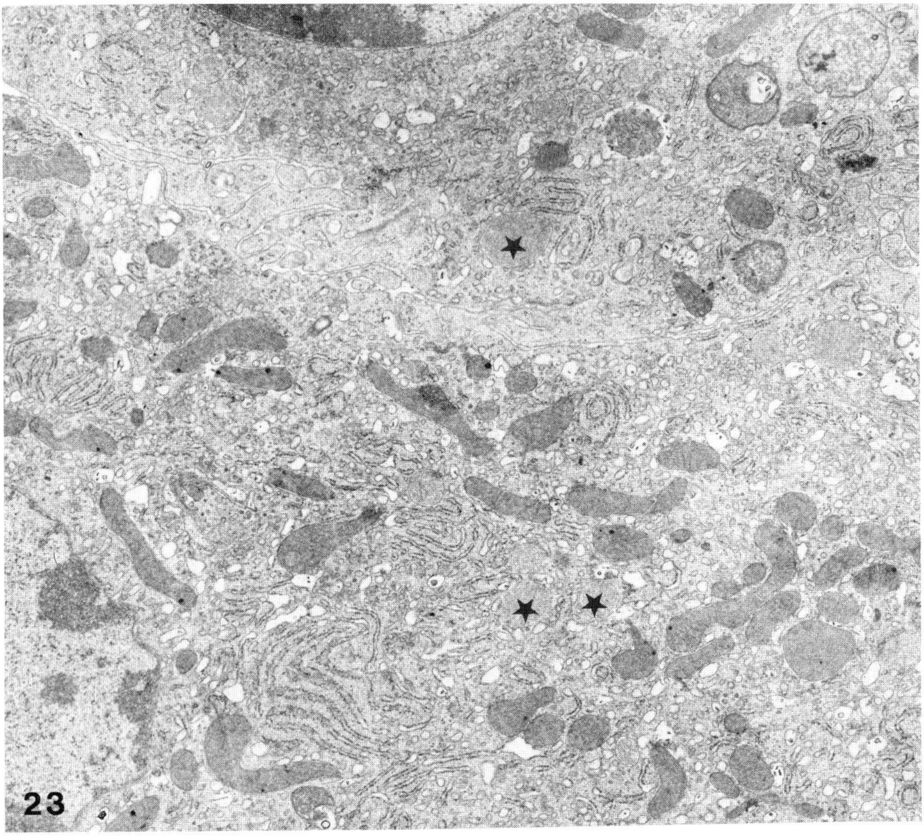

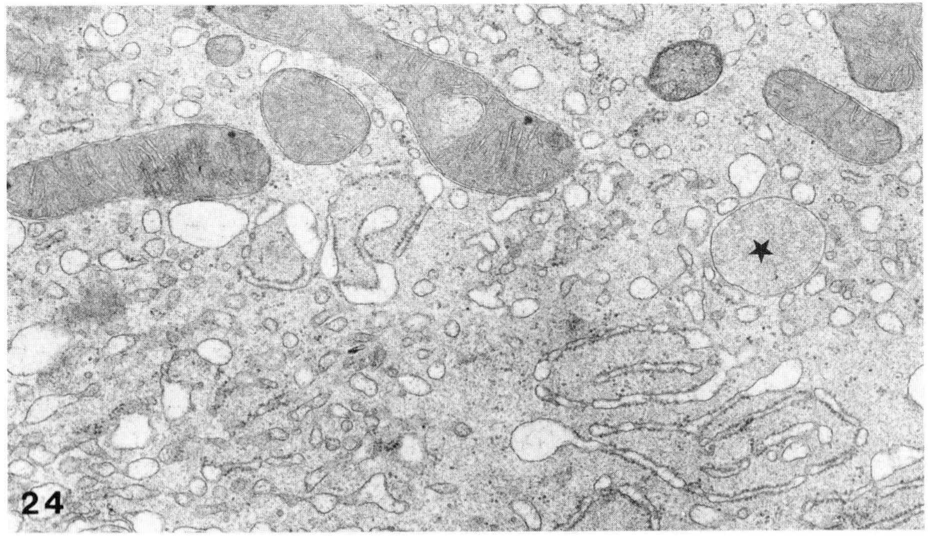

with the synthetic capacity, which excludes the possibility of an accumulation of JH (Tobe and Pratt, 1974, 1975a,b; Pratt and Tobe, 1974a,b). It seems, therefore, that the activity of CA results from a modulation of synthesis rather than of secretion.

2.6.2.7. GOLGI APPARATUS

The dictyosomes are frequently unobstrusive structures, simple masses of vesicles or aggregates of rare, short, flattened saccules and vesicles (60–120 nm in diameter) with a smooth or coated wall (transition vesicles, multivesicular bodies); their appearance does not vary greatly (Scharrer, 1964a,b; Fain-Maurel and Cassier, 1969a,b; Thomsen and Thomsen, 1970; Panov and Bassurmanova, 1970; Dorn, 1973; Brousse-Gaury et al., 1973; Brousse-Gaury and Cassier, 1975; Elliott, 1976; Bitsch and Lapalisse, 1984).

In male adult *Schistocerca*, sexual maturation is accompanied by growth of the dictyosomes from day 5 of imaginal life (Odhiambo, 1966b). In *Celerio*, the Golgi apparatus is very active (Schultz, 1960). In *Leucophaea* (Scharrer, 1971), it participates in the elaboration of "cribriform structures." In *Rhodnius* (Baehr et al., 1973), opaque granules 0.1 μm in diameter) were seen associated with the Golgi apparatus, scattered in the cytoplasm, or even attached to the plasma membrane; although their number increased with the activity of the CA (triggered by nutrition, severing nerves, extirpation of the pars intercerebralis, or ovariectomy), their functional significance remains unknown. They may be hypothetically grouped with the primary lysosomes, which participate in the degradation of structures implicated in the synthesis of JH.

In *Carausius morosus* (Cassier, 1979) (Figs. 2.29 and 2.30), the secretory activity of the allata cells is very noticeable but is irregularly distributed inside the cells lining these vesicular organs. Numerous large granules composed of glycoproteins are released into the central cavity.

2.6.2.8. THE SECRETION

Both the chemical nature of JHs and their solubility in organic solvents make it impossible for them to be preserved and visualized at the ultrastructural level. No morphological parameter can be used to localize the hormone in the allata cells or to determine exactly the mechanism of its

FIGURES 2.23–2.24. Hyperactive CA of *Blabera fusca* (Dictyoptera) after right ocellus cauterization.
2.23 (*above*). Hypertrophied cytoplasmic areas contain numerous ergastoplasmic structures, giant mitochondria and enigmatic granules (*). ×5600.
2.24 (*below*). Detail of a cytoplasmic area in Fig. 2.23. ×12,500.

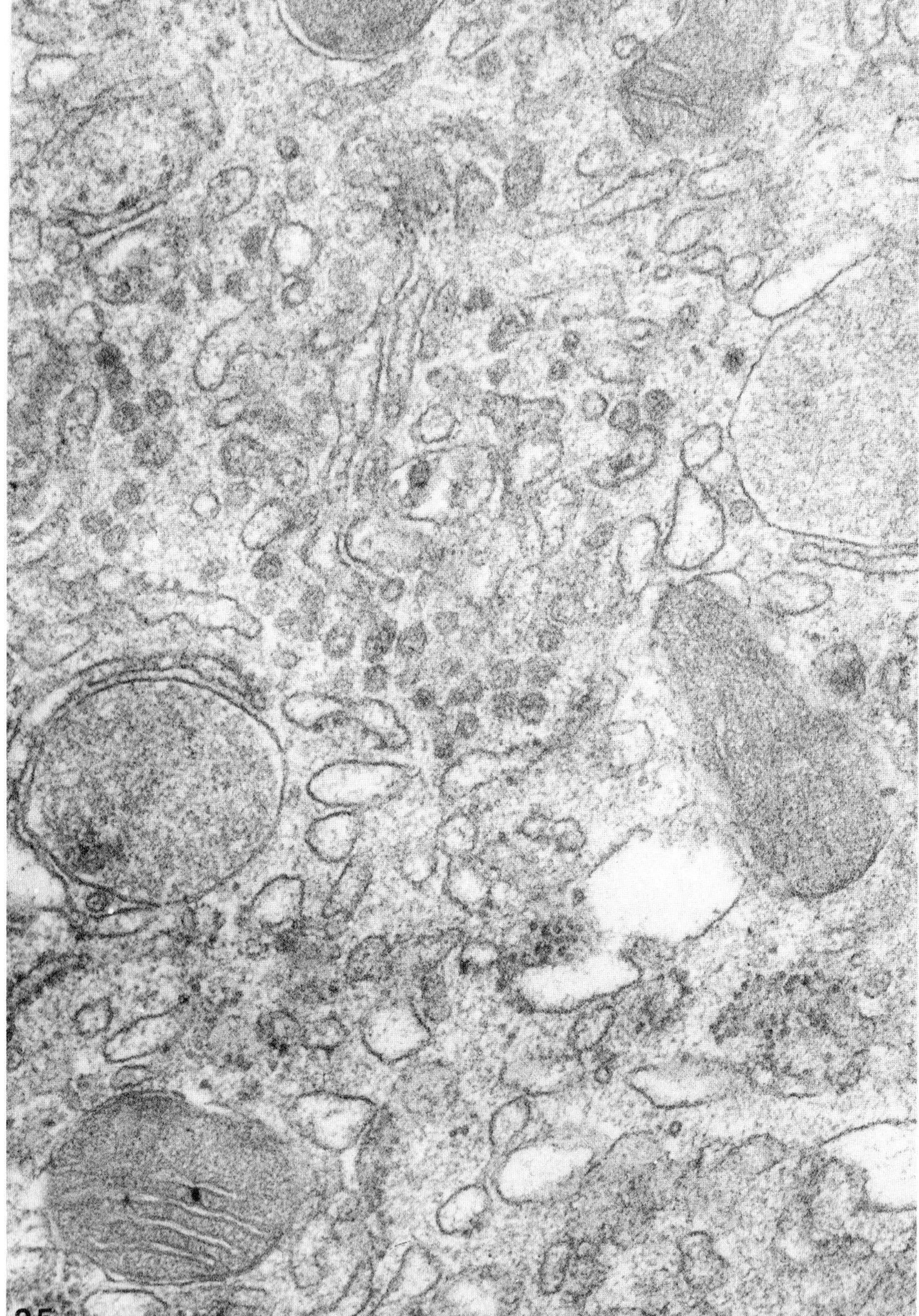

25

release into the hemolymph (vesicles or clear vacuoles) (Ito, 1918; Mendes, 1948; Nayar, 1956a,b; Highnam, 1958; Elliot, 1976). Moreover, during the phase of normal activity, JH is released in proportion to its production; this has been experimentally demonstrated *in vitro* in *Schistocerca* (Tobe and Pratt, 1974). Since carrier proteins for JH are known in hemolymph (Goodman, 1983), it seems likely that these facilitate movement of hormone away from the extracellular proteins that act as carriers for JH precursors. The rough endoplasmic reticulum (RER), in continuity with the SER (smooth endoplasmic reticulum) (Scharrer, 1964a; Fain-Maurel and Cassier, 1969a), would provide for carriers within the endoplasmic reticulum. Electron-dense material can be seen within the SER (Deleurance and Charpin, 1978; Baehr et al., 1982; Bitsch and Lapalisse, 1984). Segments of ER frequently appear in proximity to the plasma membrane of CA cells (Scharrer, 1964a; Kaiser, 1980; Johnson et al., 1985), suggesting that release of JH may depend upon such juxtaposition.

In light microscopical studies, a secretion product may be observed. According to Cazal (1948), the use of hematoxylin after osmium fixation clearly shows that the secretory product begins at the center of the gland and moves progressively toward the periphery. Hematoxylinophilic globules appear near the granular mitochondria and at the extremities or in the concavities of filamentous mitochondria. The secretion progressively fills the allata cells, becomes osmiophilic, and passes into the intercellular spaces or diffuses into the hemocoel.

No ultrastructural observation has demonstrated the presence of secretion either free or in granules. The secretory processes observed in *Leucophaea* (Scharrer, 1971) and *Rhodnius* (Busselet, 1969; Baehr et al., 1973) are negligible. However, one may assume that the description, summarized by Cazal (1948), indicates that the allata cells are rich in lipids, lipoproteins, lipofuchsins, and chromoproteins, which correspond to the lipids, reticular structures, lysosomes, and residual bodies observed with electron microscopy (Pasteels and Deligne, 1965).

2.6.2.9. RESERVE SUBSTANCES: GLYCOGEN AND LIPIDS

Glycogen

The high glycogen content of the CA depends essentially upon the nutritional condition of the insect; it does not correlate well with the cycle of activity (Thomsen and Thomsen, 1970; Dorn, 1973; Scharrer, 1971). Melnikova and Panov (1975) found high glycogen levels in the allata cells

FIGURES 2.25. Golgi area in a glandular cell of an hyperactive CA of *Blabera fusca* (Dictyoptera) after right ocellus cauterization. ×31,900.

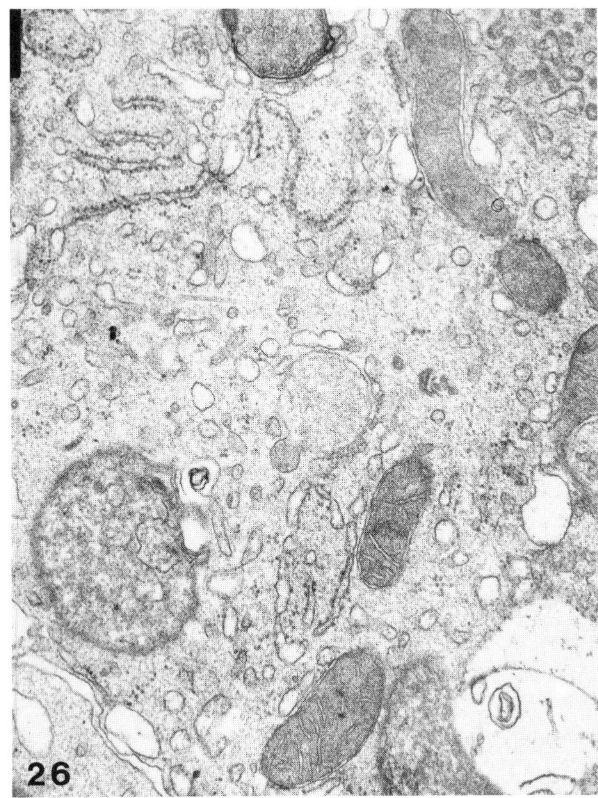

FIGURE 2.26 (*above*). Cytoplasmic and Golgi areas of an hyperactive CA of *Blabera fusca* (Dictyoptera) after right ocellus cauterization. Lytic activities. ×12,500.

FIGURES 2.27–2.28 (*opposite*). CA of a 23-day-old adult female of the desert locust, *Schistocerca gregaria* (Orthoptera). Nocturnal temperature: 20°C. Effect of the diurnal rearing temperature.
2.27 (*above*). Active CA with well-developed smooth endoplasmic reticulum. Diurnal temperature: 33°C. ×13,000
2.28 (*below*). Inactive CA with reduced endoplasmic reticulum. Diurnal temperature: 28°C ×13,000.

of *Hyphantria* during nonactive periods, whereas during active periods these cells contained no glycogen. In *Rhodnius* (Busselet, 1969), no glycogen was found in fasting larvae, but it appeared after a meal.

Lipids (Fig. 2.32)

In *Calliphora* (Thomsen and Thomsen, 1970), glandular cells during the active periods show numerous spherical lipid droplets (0.5–1.5 μm in

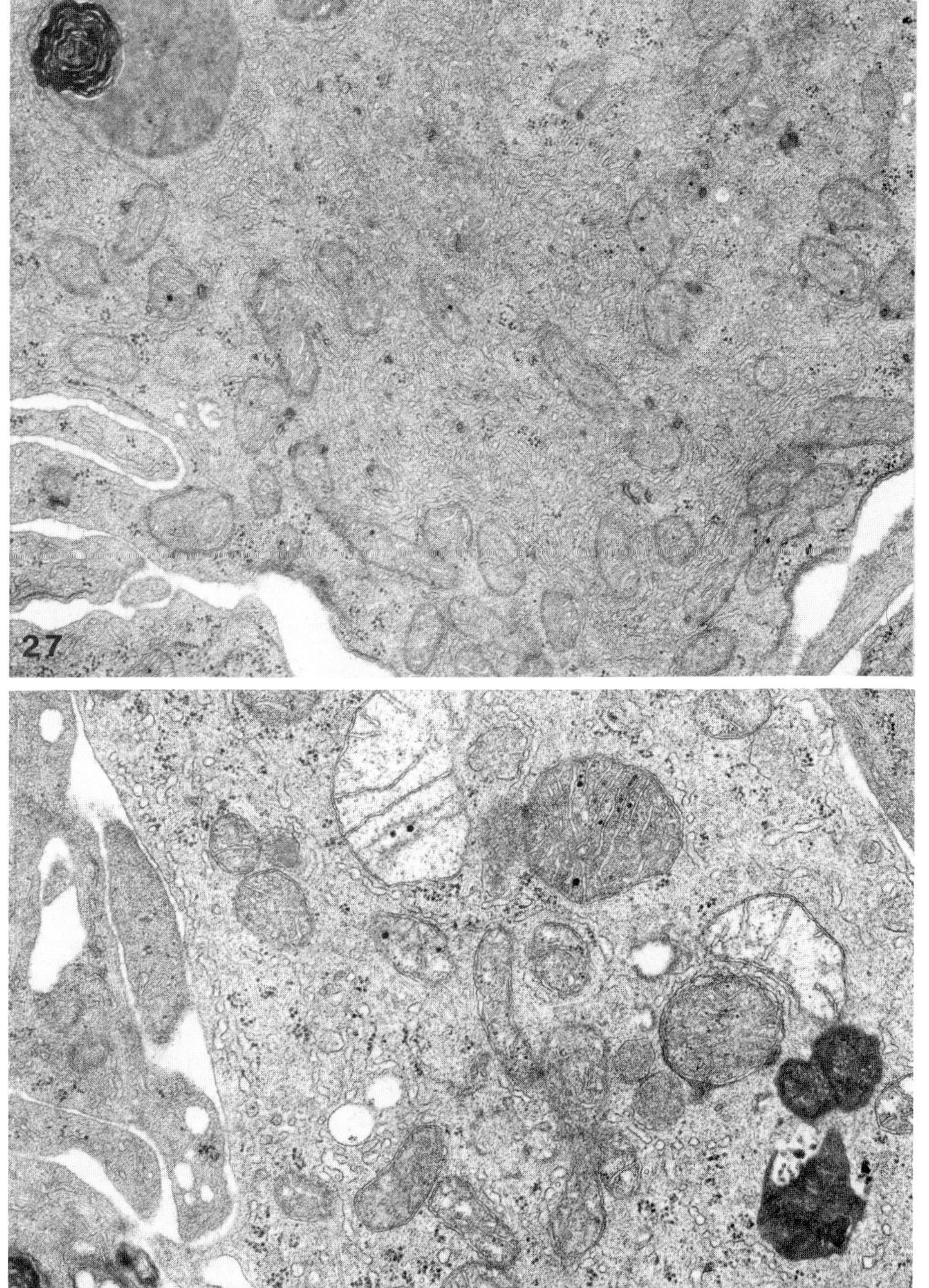

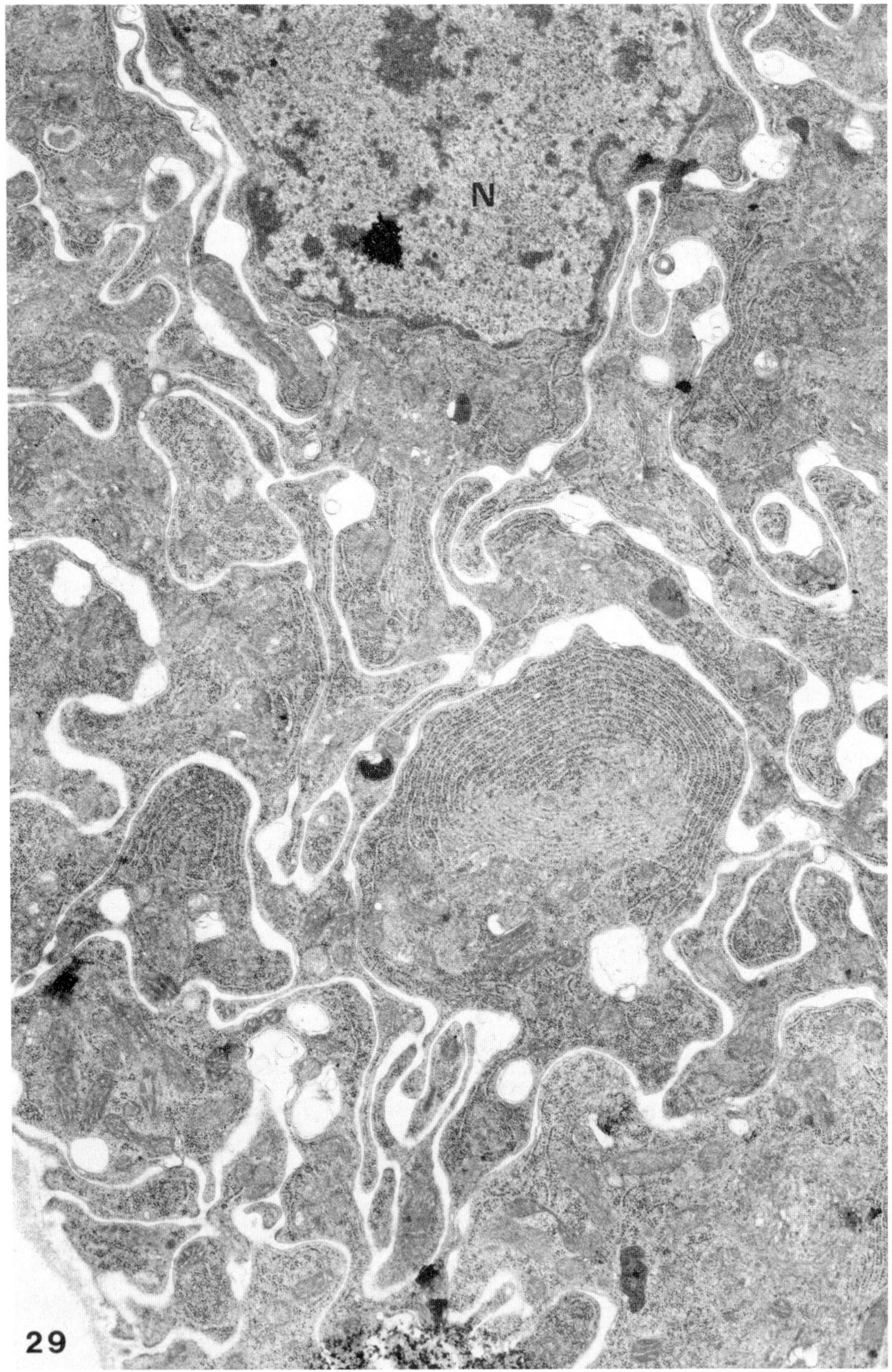

N
29

diameter), which are particularly abundant at the periphery of the gland. During nonactive periods, they are small, rare, or absent. Similar observations have been made in *Hyalophora* (Waku and Gilbert, 1964), where lipid droplets disappear in diapausing animals, in active females of *Bombyx* (Fukuda et al., 1966), in *Antheraea* (Busselet, 1969), and in *Diatraea grandiosella* (Yin and Chippendale, 1977). In Acrididae, the central part of concentric fields of endoplasmic reticulum are frequently occupied by a lipid droplet.

2.6.2.10. MICROTUBULES

Microtubules (25–30 nm in diameter) are abundant in allata cells; in general they are randomly distributed, without any preferential orientation (Busselet, 1969; Thomsen and Thomsen, 1970; Tombes and Smith, 1970; Scharrer, 1971; Deleurance and Charpin, 1972; Dorn, 1973; Baehr et al., 1973; Brousse-Gaury et al., 1973; Girardie and Granier, 1974a; Morohoshi et al. 1976a,b), except at the level of the cytoplasmic peripheral infolding (Odhiambo, 1966b,c; Guelin and Darjo, 1974). In the fusiform allata cells of *Folsomia* (Palevody and Grimal, 1975; Palevody, 1976), the microtubules are arranged parallel to the major axis of the cells.

2.6.2.11. LYTIC STRUCTURES

Independently of the lamellar formations, which participate, at the end of each active cycle, in the elimination of reticulum involved in JH synthesis, the allata cells, like any active cells, contain various lytic structures (primary lysosomes, cytosegresomes or autophagic vacuoles, multivesicular bodies, residual bodies). Lysosomes (dense bodies) are frequently found in *Schistocerca* (Odhiambo, 1966), *Calliphora* (Thomsen and Thomsen, 1970), *Leucophaea* (Scharrer, 1964a), *Drosophila* (King et al., 1966a,b), *Aphis* (Elliot, 1976), and *Hypera* (Tombes and Smith, 1970). In *Oncopeltus* (Dorn, 1973), they were particularly frequent after the imaginal molt and in senescent adults. Similar results were found during the active period in *Hyalophora* (Waku and Gilbert, 1964). The multivesicular bodies, which have classic structures, have been studied in particular in *Rhodnius* (Busselet, 1969; Baehr et al., 1973) and in *Spheophyes lucidulus* (Deleurance and Charpin, 1971).

2.6.2.12. ULTRASTRUCTURE AND CA ACTIVITY

Although the use of structural parameters has given useful information about the state of glandular activity, it has been impossible by this means alone to obtain quantitative data. Structural analysis must be coupled

FIGURE 2.29. Active CA of the stick insect *Carausius morosus* (Phasmida) showing a well-developed and concentrically disposed rough endoplasmic reticulum. KEY: N = nucleus. ×6800.

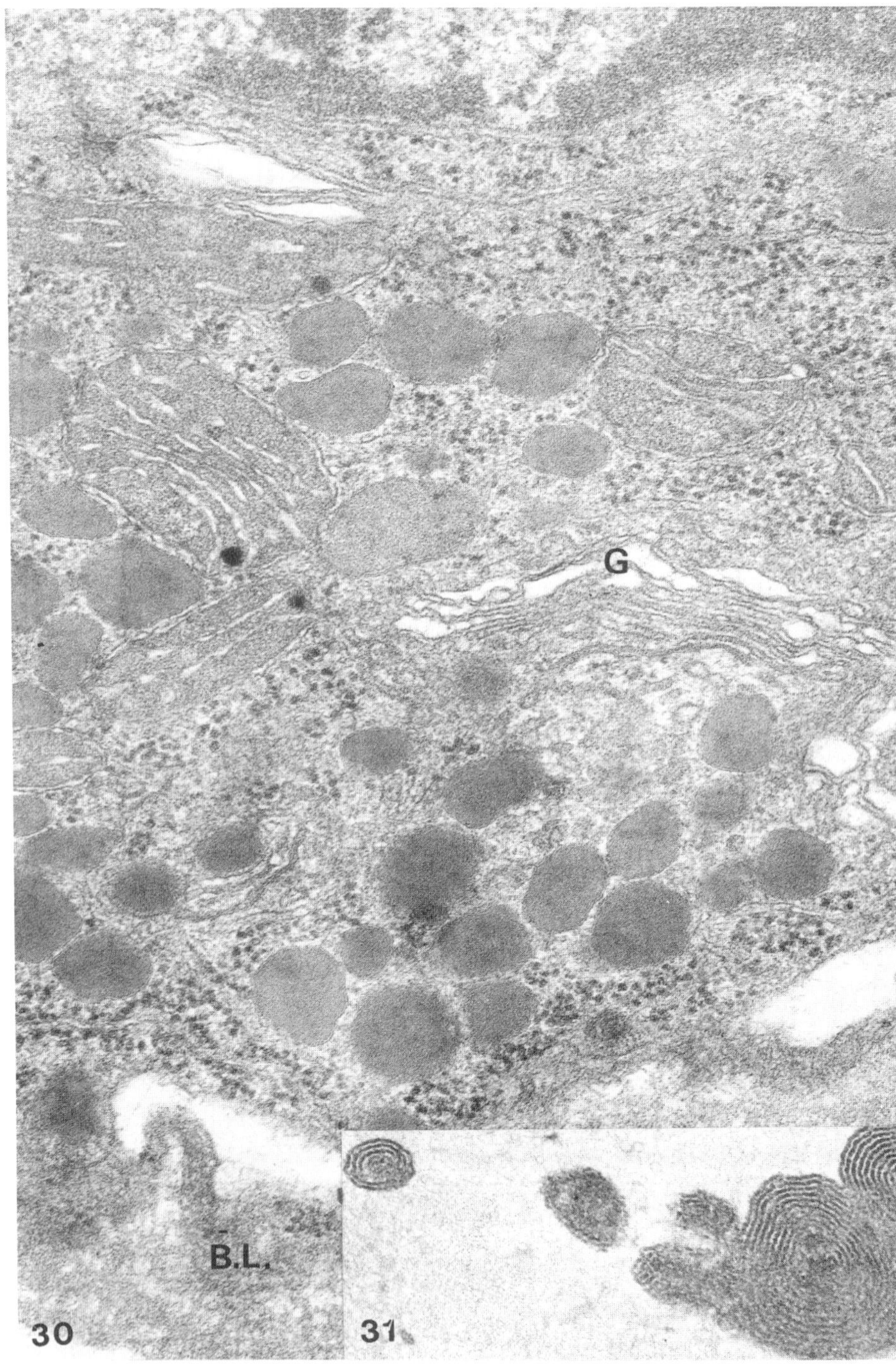

G
B.L.
30
31

with radioimmunoassay (RIA) or radiochemical assay (RCA) measurements. In *M. sexta*, rates of JH biosynthesis have been determined during the last two stadia by RIA (Granger et al., 1979, 1982a,b) and by RCA (Kramer and Law, 1980); they reflect JH titer measurements (Fain and Riddiford, 1975; Riddiford, 1980). JH biosynthesis is high during the first half of the penultimate larval stadium, and it declines before ecdysis to the last stadium. During the early days of the last larval stadium, JH biosynthesis is high again; it falls in midstadium and may rise again before the pupal molt. Rates are low in pupa. But Sedlak et al. (1983) were not able to discern great structural changes.

In *D. grandiosella*, structure of CA (Yin and Chippendale, 1979a) and titers of JH (Yin and Chippendale, 1979a; Bergot et al., 1976) have been investigated. SER was most abundant as vesicles in prediapausing larvae (highly active glands), as stacked forms in middiapause (declining activity), and as whorled forms in late diapause (inactive). SER of all forms was least abundant in nondiapausing larvae (inactive glands).

In the male of *S. gregaria*, the rates of JH biosynthesis by CA during sexual maturation (Avruch and Tobe, 1978) allowed comparisons with ultrastructural data (Odhiambo, 1966b). As in females of *L. migratoria* (Girardie et al., 1981) and *S. gregaria* (Tobe and Pratt, 1975a; Tobe, 1977), variations between animals is great; thus, it is difficult to relate fine structural changes to rate changes with precision. Odhiambo (1966) reported more SER and mitochondria of complex shapes in CA of animals developing toward sexual maturity than in those just after adult emergence. Papillon et al. (1976) also reported the earlier appearance and greater abundance of fields of vesicles of ER in sexually mature animals (reared at 33°C), compared with animals that did not mature sexually (reared at 28°C) (Figs. 2.27 and 2.28).

In *L. migratoria*, correlation of oocytes size with CA structure (Joly et al., 1968; Fain-Maurel and Cassier, 1969a,b; Guelin and Darjo, 1974) has shown changes—especially in the ER and mitochondria—that appear to be correlated with changing activity of the glands. The fine ultrastructure parameters (Fain-Maurel and Cassier, 1969a) indicate that the CA are most active when oocytes are relatively early in vitellogenesis; this concurs approximately with the findings of Joly et al. (1968) and Guelin and

FIGURES 2.30–2.31. Part of the vesicular CA of the stick insect *Carausius morosus* (Phasmida).

2.30. Fourth larval instar. Rough endoplasmic reticulum and Golgi apparatus (G) elaborate numerous dense granules extruded in the central cavity. KEY: BL = basal laminae. ×33,000.

2.31 (*inset*). Characteristic "cribriform structures" enclosed in the central cavity. ×13,000.

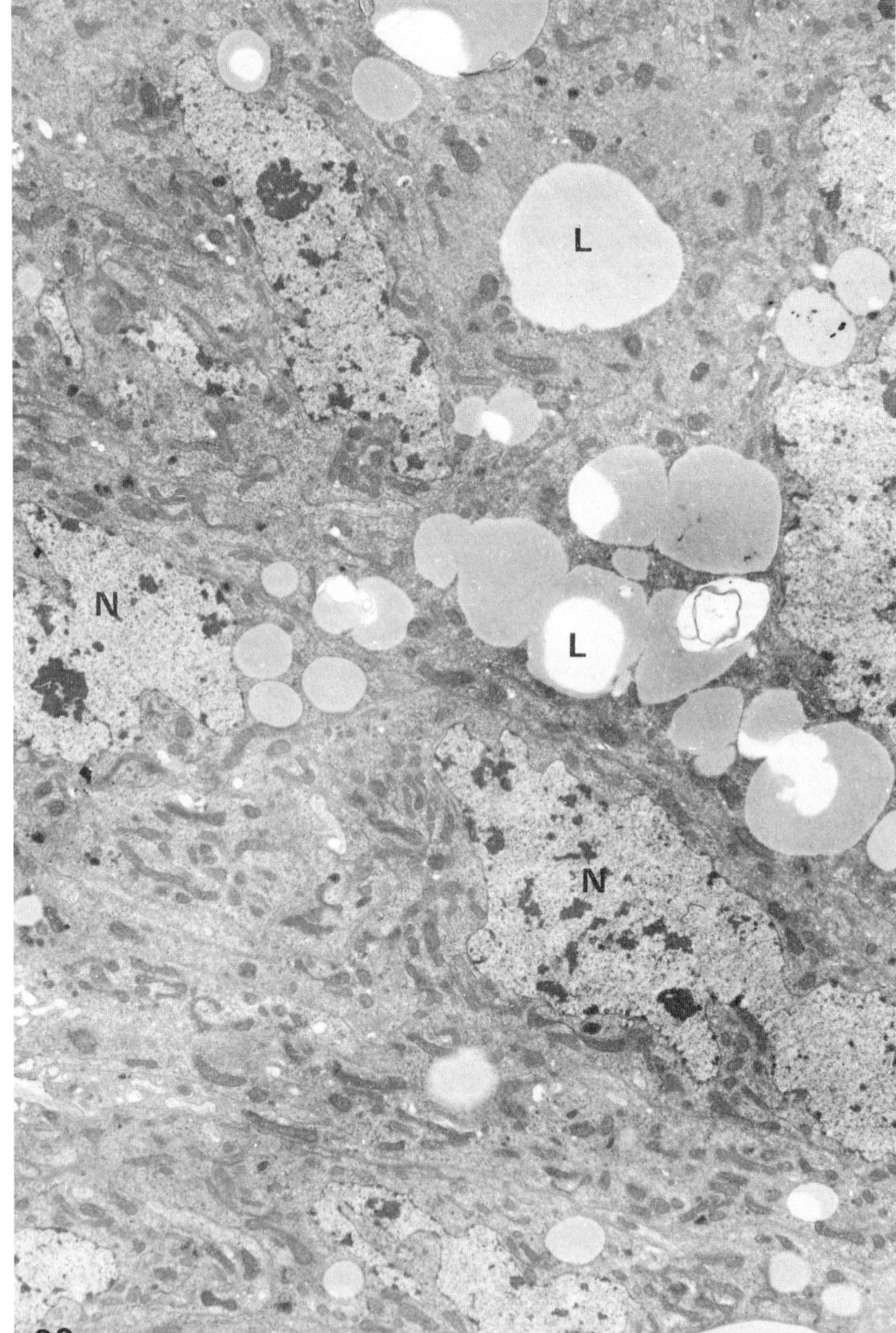

L
L
N
N
32

Darjo (1974). Direct measurement of the rate of JH biosynthesis by CA (Girardie et al., 1981; Ferenz and Kaufner, 1981) have revealed two peaks of JH biosynthesis for each gonotrophic cycle; so the fine structural diagnosis of activity agrees only roughly with the direct measurements (Tobe and Stay, 1985) (Figs. 2.33 and 2.34).

In adult females of *D. punctata*, rates of JH biosynthesis and ultrastructure of CA can be compared (Feyereisen et al., 1981c; Johnson et al., 1985). The rates of biosynthesis of individual glands in a pair are similar (Szibbo and Tobe, 1981a). In glands of highest activity, SER, RER, and mitochondria became obscure; newly formed autophagic vacuoles were seen. Glands of decreasing rates of biosynthesis showed decreased width and irregular shape of mitochondria with dense matrix; the SER was again easily seen as tubules; the Golgi complexes were larger and more conspicuous; and the RER remained in curved forms.

Activity of CA following ovariectomy has been demonstrated to be low in *D. punctata* and *N. cinerea* (Stay and Tobe, 1978; Lanzrein et al., 1981), but cytological parameters might easily be construed as indicating the contrary (Tobe et al., 1985) (Figs. 2.35 and 2.36). In *P. gallicus*, the synthetic rates of glands in ovariectomized animals were low, yet the volume of the CA was characteristic of that of glands with increasing activity (egg-maturing females) (Röseler et al., 1980). Similarly, in *Acheta domesticus*, CA of ovariectomized females synthetized JH at lower rates than those of normal females (Strambi, 1981).

In *Labidura riparia*, the alternation of JH-dependent vitellogenic cycles in the ovaries with JH-independent periods of egg guarding has been utilized by Baehr et al. (1982) for the fine structural analysis of active and inactive CA. The hemolymph titer of JH was determined by RIA (Baehr et al., 1976, 1979, 1981). JH titers increased during vitellogenesis, and active CA exhibited well-developed SER, cisternae of RER, and elongated mitochondria. During the inactive periods, JH titers were low and CA were provided with autophagic vacuoles containing "structured bodies" of anastomosed SER; mitochondria became globular and dense; glycogen appeared in the cytoplasm (Figs. 2.6 and 2.7). Ovariectomy results in an abnormally high titer of hemolymph JH; the CA displayed a steady increase of about twice the normal maximal volume observed during vitellogenesis; nuclei of cells were swollen, the RER and "structured body" were abundant, and glycogen had accumulated (Baehr et al., 1982).

In *L. decemlineata* (Leclercq-Smekens and Naisse, 1986) during imaginal diapause, rates of JH biosynthesis are negligible and the glandular cells of

FIGURE 2.32. Active CA of adult female of *Pyrrhocoris apterus* (Heteroptera) contain numerous lipid droplets (L). KEY: N = nucleus. ×3300.

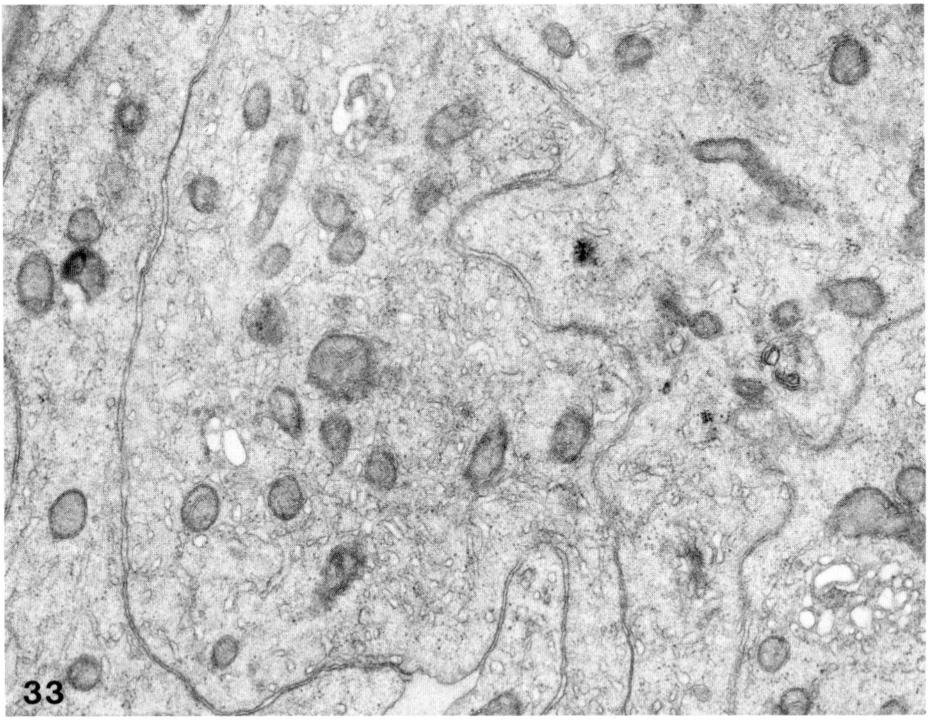

33

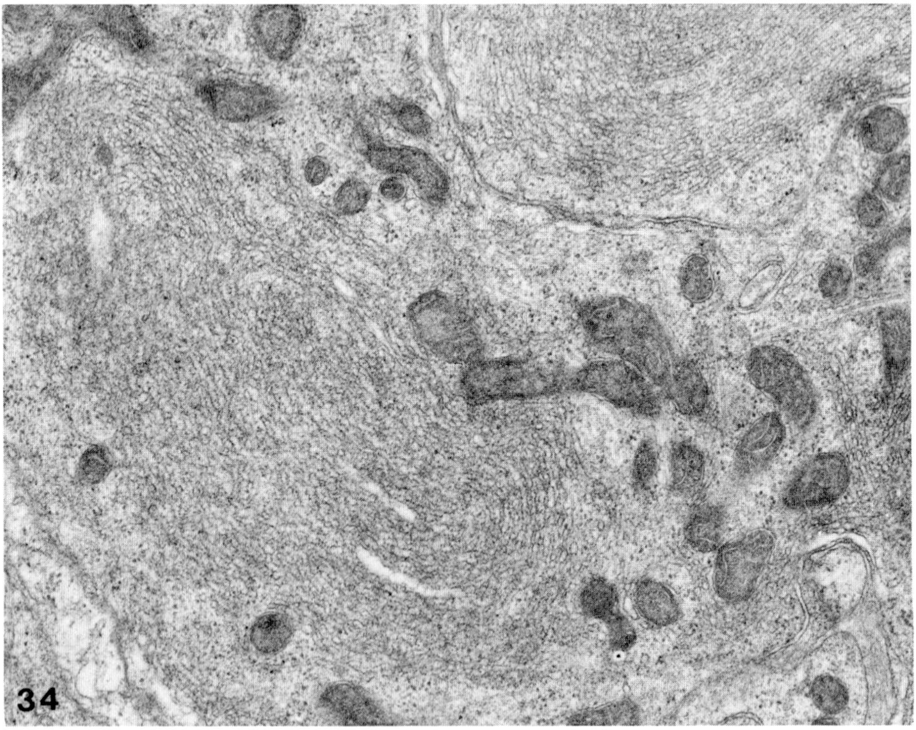

34

CA show all characteristics of inactive cells—reduced cytoplasm, infoldings of the nuclear membrane, rare and disperse mitochondria, and conspicuous Golgi apparatus; the RER and lipid droplets are not well developed. After the breakdown of diapause state, the rate of JH increases proportionally to the length of the photoperiod, as does the number of glandular cells showing symptoms of activity: the cytoplasm mass, the number of mitochondria and liposomes, the size of the Golgi apparatus, and the development of RER and SER all increase.

2.6.3. *Neurosecretory Fibers*

The presence of neurosecretion and neurosecretory fibers in the CA has been established many times by light microscopy (see Cassier, 1967, for references). Electron microscope studies have confirmed these results, which have been found in all cases examined. (Figs. 2.37 and 2.38). The neurosecretory fibers penetrate the CA through one or several hila, and then ramify at the center of the glandular parenchyma and rapidly lose their glial sheaths. The naked, dilated endings are characterized by the presence of neurosecretory granules that are free or attached to the plasma membrane with synaptic (35–60 nm) and pinocytic vesicles. The process of pinocytosis has been observed by several investigators (Cassier and Fain-Maurel, 1970; Busselet, 1969; Brousse-Gaury et al., 1973; Girardie and Granier, 1974a,b; Thomsen and Thomsen, 1970; Fukuda et al., 1966).

Apparently only one type of granule is present in *Anacridium* (100–200 nm in diameter) (Guelin and Darjo, 1974), *Oncopeltus fasciatus* (Unnithan et al., 1966), *Manduca sexta* (Sedlak, 1981), *B. mori* (75–200 nm in diameter) (Fukuda et al., 1966; Morohoshi et al., 1976a,b), *Troglodromus* and *Cytodromus dapsoides* (90 nm) (Deleurance and Charpin, 1972; Deleurance, 1967), *Hyalophora* (50–150 nm) (Waku and Gilbert, 1964), and *Calliphora* (110–200 nm in diameter) (Thomsen and Thomsen, 1970); however, two, three, or more types of granules of variable diameter and density exist in *Hyphantria* (Melnikova and Panov, 1975), *Euborellia annulipes* (Awasthi, 1979), *A. craccivora* (Elliot, 1976), *Celerio* (100–300 nm in diameter) (Schultz, 1960), *Schistocerca* (80–185 nm in diameter) (Odhiambo, 1966b), *Rhodnius*

FIGURES 2.33–2.34. CA of a 61-day-old adult female of *Locusta migratoria* (Savio strain).
2.33 (*above*). Inactive CA of a diapausing female (L/D = 16/11) with reduced endoplasmic reticulum. ×10,000.
2.34 (*below*). Active CA of a normal female (L/D = 11/16) with well-developed and concentrically disposed smooth endoplasmic reticulum. ×13,000.

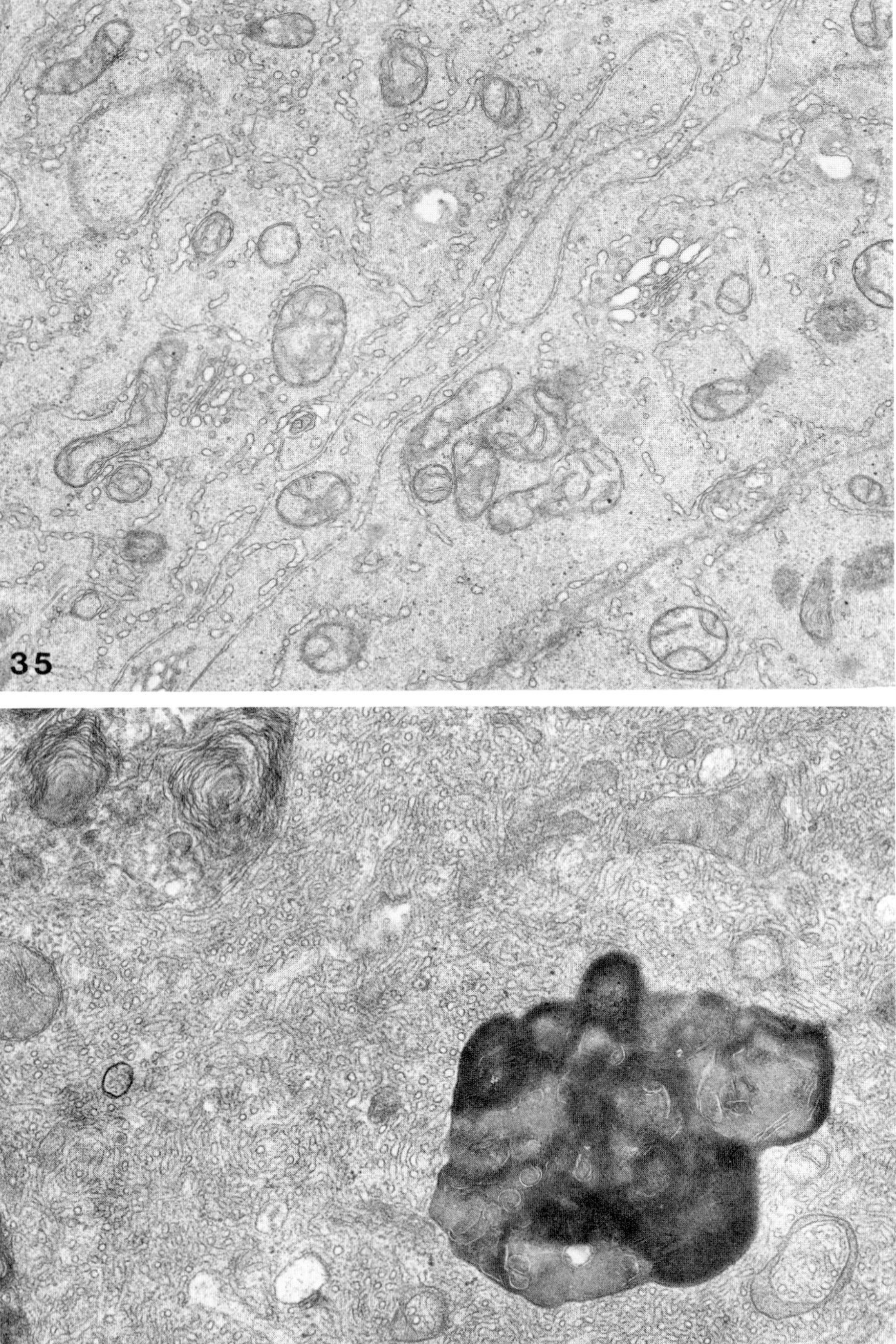

(Busselet, 1969), *Oncopeltus* (Dorn, 1973), *Blabera* (Brousse-Gaury, 1973; Brousse-Gaury and Cassier, 1975), and *Xyleborus ferrugineus* (Chu and Norris, 1979). In *Locusta* (Fain-Maurel and Cassier, 1970; Cassier and Fain-Maurel, 1970; Cassier, 1979), four fiber types have been identified, three of which are also present in the internal (NCC I) and external (NCC II paracardiac nerves, the CC, and the allatocardiac nerves (NCA I).

From the examination of these data, it seems obvious that neurosecretory granules (100–2500 nm in diameter) are well represented in insect CA. It is possible that they are aminergic in nature and are involved in the inhibitory or stimulatory mechanisms regulating the activity of the CA. Indeed, in *Locusta*, in the CC, which contain the same types of fibers as the CA, Lafon-Cazal (1976) and Lafon-Cazal and Arluison (1976), using a technique of induced fluorescence (according to Falck and Owman, 1965) and high-resolution autoradiography, demonstrated the presence of indolamines (serotonin), catecholamines, and dopamine.

In *L. decemlineata* (Khan and Buma, 1985), the exocytosis of neurosecretory granules in the CA was visualized and quantified in beetles kept under two different photoregimes. The number of exocytosis figures in the neurosecretory synapses of beetles kept under short days was significantly higher than that of beetles kept under long-day conditions; furthermore, under short-day conditions, CA are more richly innervated and the CA activity is at least partly restrained by neurally mediated factors.

In *M. sexta* (Sedlak, 1981), two areas of the CA of last-instar larvae contain neurosecretory axons: (1) the gland proper, which consists of typical stellate glandular cells, and (2) the sheath, in which groups of neurosecretory axons are separated by the stromal sheath. In midinstar larvae, the neurosecretory axons within the gland are quite large in comparison with those of the pupa. This size fluctuation during development and the proximity of these axons to gland cells suggest that these fibers may influence JH synthesis and/or secretion. The population of neurosecretory axons in the sheath, on the other hand, does not change throughout postembryonic development. The products of these fibers have easy access to the general circulation; thus, the CA sheath may serve as a neurohemal area for prothoracicotropic hormone (PTTH), since this tropic hormone is released from the CA during development (Agui et al.,

FIGURES 2.35–2.36. CA of a 17-day-old adult female of the desert locust, *Schistocerca gregaria* (Orthoptera).
2.35 (*above*). CA of normal female. Smooth endoplasmic reticulum is reduced. ×32,000.
2.36 (*below*). CA of castrated female. Hypertrophy of the smooth endoplasmic reticulum. ×32,000.

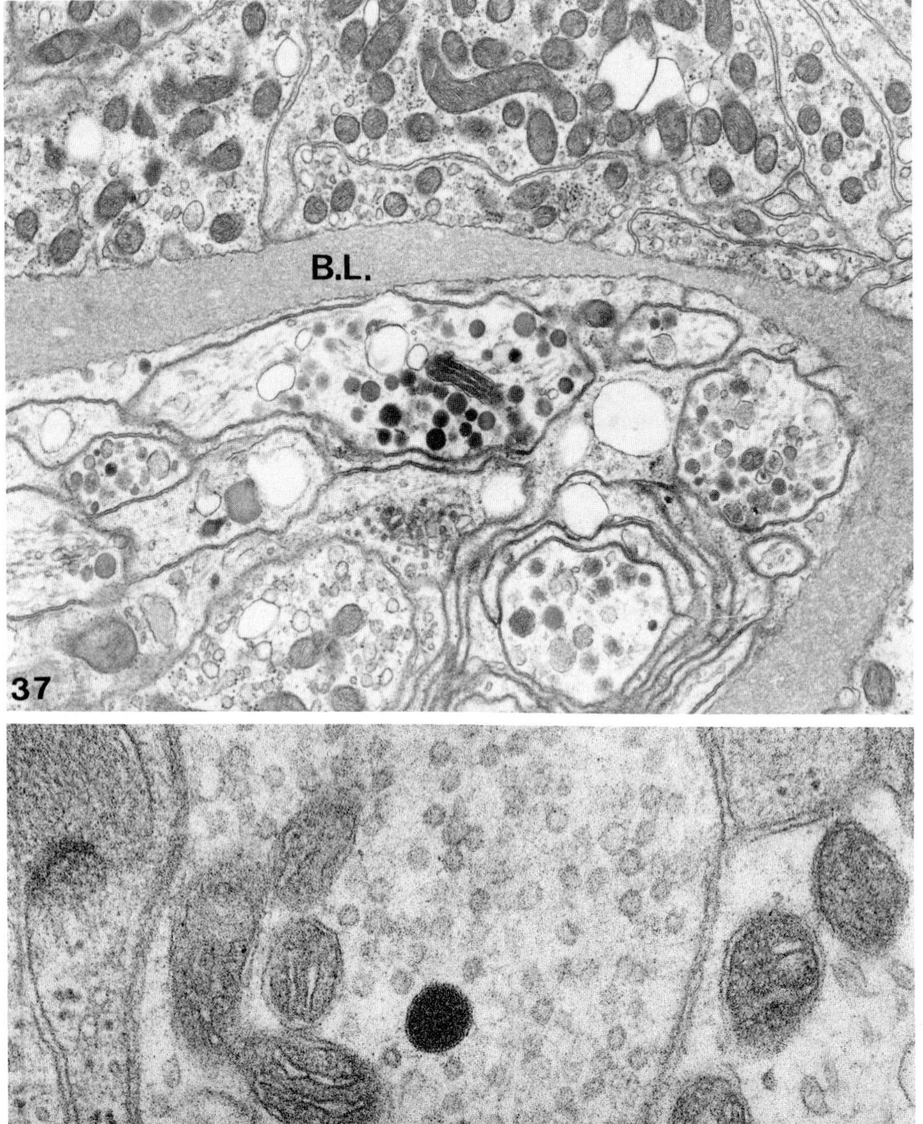

B.L.
37
38

1979, 1980). In a variety of lepidopterans, synaptoids release sites are found in both areas. Common nerve fibers that have no neurosecretory-type dense granules are associated with neurosecretory fibers.

With certain exceptions, the integrity of the innervation of the CA is indispensable for the maintenance and regulation of their activity, as well as for their postimaginal growth. Thus, in *Schistocerca*, Highnam et al. (1963), Pener (1965, 1967), and Strong and Amerasinghe (1977) found that implanted CA were incapable of inducing sexual receptivity and ovarian development in allatectomized females. In *L. migratoria migratorioides*, restoration of these faculties required either a massive implantation (Albrecht and Cassier, 1964) or repeated implantations (Cassier, 1964a,b, 1965a, b, 1967; Cassier and Papillon, 1968). A similar situation exists in *L. migratoria cinerascens* (Girardie 1967b). In the African migratory locust, after severance of the NCA I, the rate of JH biosynthesis *in vitro* by disconnected CA is low (Couillaud et al., 1984, 1985, 1987), but JH-mediated physiological events appear to occur normally.

Conversely, the ablation of a CA induced compensating hypertrophy of the symmetrical gland (Cassier, 1965a,b, 1966a,b,c). Similar results have been reported by Quo Fu (1965), Müller (1965a), and Johansson (1958).

A number of substances that react with antisera to bioactive peptides have been demonstrated in the CA of various species. So a gastrin/CCK(cholecystokinin)-like substance has been revealed in CA of *L. maderae* (Hansen et al., 1987), *C. erythrocephala* (Duve and Thorpe, 1984), and *Manduca sexta* (El Salhy et al., 1983). The CA of *L. maderae* contain ACTH-MSH (adrenocorticotropic hormone–melanocyte-stimulating hormone) (Hansen et al., 1982) and gastrin (Kramer et al., 1977). Similarly, CA of *P. americana* (Verhaert et al., 1984) contain bovine neurophysin I, arginine vasopressine, and oxytocin. The functions of those immunoreactive products are as yet unclear. In the Colorado beetle, *L. decemlineata*, the lateral neurosecretory cells, which are immunoreactive with antisera to β-endorphin and ovin prolactin, may regulate CA activity (Veenstra et al., 1985).

FIGURES 2.37–2.38. Neurosecretory fibers in the CA of *Leucophaea maderaea* (Dictyoptera).
2.37 (*above*). Bundle of fibers surrounded by glial infoldings at the level of a hilum. KEY: BL = basal laminae. ×10,000.
2.38 (*below*). Detail of a synaptoid area (arrow). ×66,000.

2.6.4. Ultrastructural Effects of
Inhibitors of JH Biosynthesis

Many compounds interfere with the activity of CA and with the normal developmental process governed by JH in insects. These insect growth regulators (IGRs; see Staal, 1975) are JH mimics (juvenoids, JH agonists, JH analogues) or JH antagonists, i.e., anti-JH agents (Camps and Belles, 1986). The former group of IGRs (reviewed by Heinrich, 1982) includes the "paper factor" juvabione (Bowers et al., 1976a,b), farnesol (Schmialek, 1963), hydroprene, methoprene (Heinrich et al., 1973), and fenoxycarb (Ro 13-5223; see Masner et al., 1981). Anti-JH agents include the precocenes (Bowers et al., 1976), 3'-fluomevalonolactone (Quistad et al., 1981), terpenoid imidazoles (Kuwano et al., 1983, 1984), and dichloroalkyl hexanoate, as well as their analogues (Quistad et al., 1985) (cf. Staal, 1986). Such compounds produce prothetily (e.g., premature pupation) in treated larvae and disrupt other physiological processes regulated by JH; typically the effects of anti-JH compounds can be restored by co-application of compounds with JH activity.

A few compounds appear to possess both juvenoid and antagonist activity, at least on certain insect species (e.g., ethyl-4-[2-(t-butylcarbonyloxy)butoxy]benzoate (ETB); Staal, 1977a,b; Horn et al., 1983).

2.6.4.1. PRECOCENES

Precocenes (ageratochromenes) were originally isolated from the house plant *Ageratum houstonianum* (Alertsen, 1955; Kasturi and Manithomas, 1967) and were subsequently demonstrated to cause precocious metamorphosis in *Oncopeltus fasciatus* (Bowers et al., 1976a,b) and selected members of the Hemiptera and Orthoptera. *Locusta migratoria* is a species very sensitive to precocene treatment (Pener et al., 1978; Schooneveld, 1979). Precocenes are rapidly converted to the 3,4-dihydrodiols; they serve as substrates for epoxidase within the CA (Pratt and Bowers, 1977; Pratt et al., 1980). The formation of highly reactive precocene epoxides results in extensive alkylation of macromolecules, followed by cell death and ultimately the destruction of much of the CA tissue (Feyereisen et al.,

FIGURES 2.39–2.40. CA of the adult female of *Blatella germanica* (Dictyoptera). *In vitro* culture: 6 h.
2.39 (*above*). Control. Culture in a normal Grace's medium. The general organization of the CA is well preserved. KEY: N = pycnotic nucleus. ×9000.
2.40 (*below*). Culture in a Grace's medium supplemented with $10^{-3}M$ precocene II. Degenerative glandular cells with hypertrophied mitochondria and pycnotic nuclei (N). ×9000.

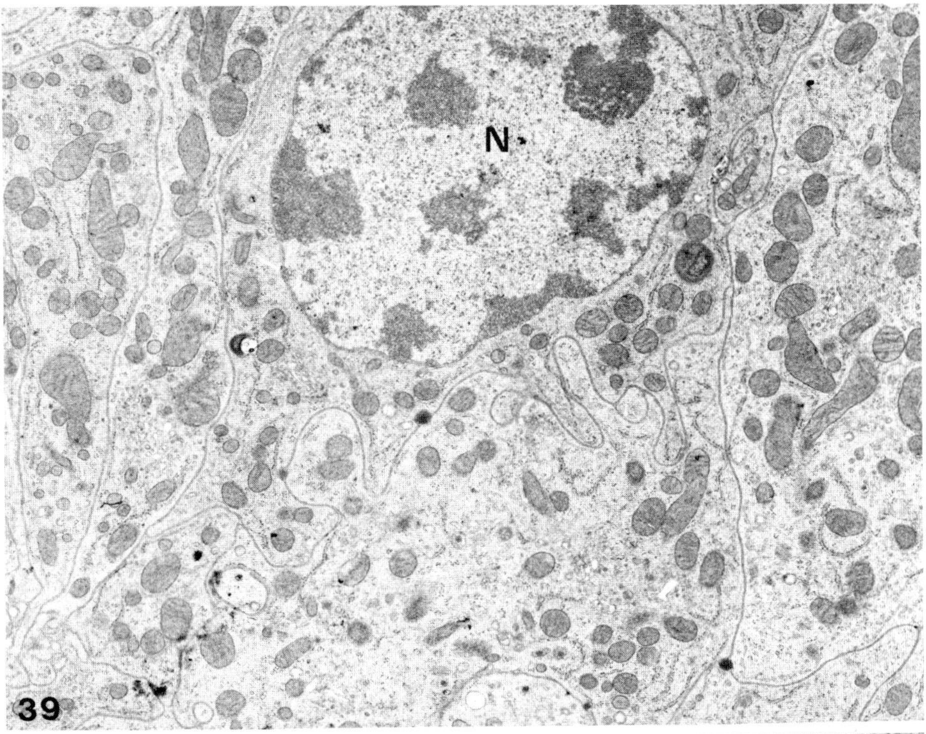

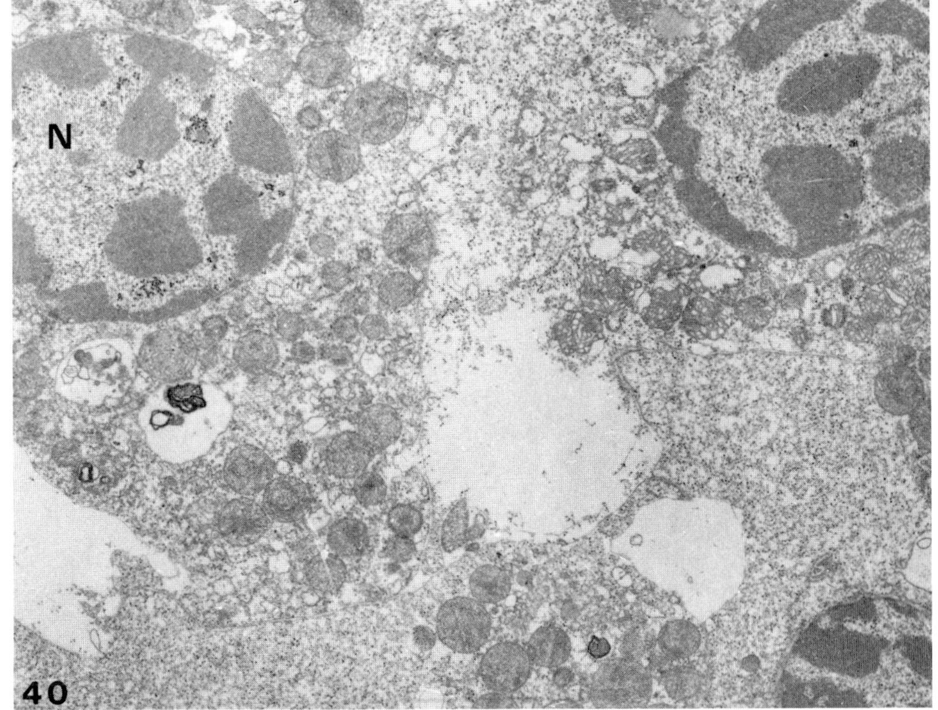
N
N
39
40

1981d). Thus, the precocenes appear to be "suicide substrates" for the epoxidase within the CA.

The ultrastructure of precocene-treated glands did show that with long exposure to precocene the CA underwent irreversible degeneration. The cells were filled with lysosome-like bodies, autophagic vacuoles, and multivesicular bodies (Unnitham et al., 1971). In *D. punctata*, a short-term exposure to precocene II results in aggregation of the SER (Feyereisen et al., 1981c). Precocene action can be also appreciated *in vitro* (Figs. 2.39 and 2.40). Since precocenes appear to exert their effects directly on the CA by disrupting JH biosynthesis, they should be regarded as proallatocidins.

2.6.4.2. FLUOMEVALONATE

Fluomevalonate, a vertebrate hypocholesterolaenic agent, had considerable anti-JH activity in *M. sexta* and several other Lepidoptera (Quistad et al., 1981), and resulted in a marked reduction in levels of endogenous JH I, JH II, and JH III (Edwards et al., 1983). Fluomevalonate probably acts by inhibiting JH biosynthesis, presumably by interfering with the enzymatic conversion of mevalonate or homomevalonate to their corresponding isopentyl derivatives. Co-applications of JH or high levels of mevalonic acid counteracted fluomevalonate action (Quistad et al., 1982). JH levels in *M. sexta* were also suppressed in a dose-dependent fashion following topical treatment with the potent anti-JH agent fluomevalonolactone (Baker et al., 1986).

In adult females of *P. americana*, topical treatment with fluomevalonate delays the formation of the first posttreatment ootheca; this effect is dose related and can be counteracted by simultaneous application of a JH analogue (hydroprene) or by a fivefold excess of mevalonic acid. However, the treatment with fluomevalonate did not affect either the carrying period of ootheca, or their weight or viability (Edwards et al., 1985).

2.6.4.3. COMPACTIN

Compactin (mevastin), a fungal metabolite (Endo, 1981), inhibits 3-hydroxy-3-methylglutaryl (HMG) CoA reductase in vertebrates and acts as potent hypocholesterolemic agent. In *M. sexta* (Monger et al., 1982), *P. americana* (Edwards and Price, 1983), *D. punctata, B. germanica* (Belles et al., 1987) compactin is an effective inhibitor of *in vitro* JH biosynthesis but appears to be relatively ineffective *in vivo*.

2.6.4.4. ALLYLIC ALCOHOLS

Allylic alcohols acted on a preliminary step of JH biosynthesis, the condensation of dimethylallypyrophosphate with two units of isopentenylpyrophosphate to give farnesylpyrophosphate (cf. Staal, 1986).

2.6.4.5.   IMIDAZOLES

Imidazoles are potent anti-JH agents, but their mode of action is unknown; they may act as inhibitors of epoxidation. So far, their action has been restricted to the commercial silkworm, *Bombyx mori* (Staal, 1986).

2.7.   Conclusions

Histological and ultrastructural studies of JH-producing structures (CA, embryonic serosal, accessory glands, etc.) have yielded useful information on cellular organellae involved in JH biosynthesis and on their interrelationships. Such techniques provide potent tools enabling researchers to appreciate the activity of inhibitors of JH biosynthesis and ascertain their significance as control agents of insects of economic importance. Furthermore, anti-JH research is a fascinating field in the experimental endocrinology of insects and has raised a great many questions of fundamental biological and biochemical nature that invite further study (Staal, 1986).

In future, such approaches will be extended to other arthropods; a JH-like compound has been identified in a crustacean, *Libinia emarginata* (Laufer et al., 1986). Hemolymph JH III and methyl farnesoate are secreted by mandibular organs in relation to female reproduction, with the highest rates near the end of the ovarian cycle when oocyte growth and vitellogenesis are greatest.

2.8.   Summary

Corpora allata are endocrine glands originating in an ectodermal region anterior to the maxillary segment. The CA buds form two coherent cellular masses and then migrate in dorsal and mesal directions toward the coelomic sacs of the antennal segment. Throughout the various orders of Insecta, the more highly evolved the species is, the more marked is the migration of CA. In their final position, the CA frequently become associated with the corpora cardiaca. Five morphological and histological types are described.

Generally, the CA are connected both with the brain by the NCA I, via the CC and NCC, and with the suboesophageal ganglion, via NCA II. They contain various axons and neurosecretory fiber types. Their relationship with the superior centers seems to have been acquired secondarily.

Although the secretion of CA is designated by the generic name *juvenile hormone* (JH), five different JHs have been identified. They are involved in various processes: morphogenesis, development, reproduction, polymorphism, general physiology, behavior, and so on. Activity of

CA is associated with volumetric variations during the embryonic period, postembryonic development (allometric and periodic growth), and imaginal life (ovarian cycles). Volumetric changes are also associated with polymorphism and sexual dimorphism.

Cytological and ultrastructural analysis of CA demonstrates that, in spite of the chemical identity and similarity of the synthesized JH, there is a great disparity in the development of the organelles, particularly of the endoplasmic reticulum and of the mitochondria that are directly involved in the successive steps of the JH biosynthesis. The amount of mitochondria and the qualitative (rough or smooth) and quantitative differences in the endoplasmic reticulum seem to be directly related to the level of activity of the CA, which depends upon the relative size of insects and the intensity of physiological processes (morphogenesis, vitellogenesis). RIA and RCA measurements allow us to correlate ultrastructure and CA activity.

Acknowledgments

I sincerely appreciate the help of Mesdames J. Bigler, N. Coudouel, N. Diot, and A. Janicot for their technical help in the preparation of the manuscript.

References

Adams, T. S. 1970. Ovarian regulation of the corpus allatum in the housefly *Musca domestica*. J. Insect Physiol. 16: 349–360.

Adams, T. S., A. M. Hintz, and J. G. Pomonis. 1968. Oöstatic hormone production in houseflies *Musca domestica* with developping ovaries. J. Insect Physiol. 14: 983–993.

Aggarval, S. K. and R. C. King. 1969. A comparative study of the ring glands from wild type and 1(2) gl mutant *Drosophila melanogaster*. J. Morphol. 129: 171–200.

Agui, N., W. E. Bollenbacher, N. A. Granger, and L. I. Gilbert. 1980. Corpus allatum is release site for insect prothoracicotropic hormone. Nature (Lond.) 285: 669–670.

Agui, N., N. A. Granger, L. I. Gilbert, and W. E. Bollenbacher. 1979. Cellular localization of the insect prothoracicotropic hormone: *in vitro* assay of a single neurosecretory cell. Proc. Natl. Acad. Sci. USA 76(11): 5694–5698.

Albrecht, F. O. and P. Cassier. 1964. Modifications de la fécondité et de la descendance chez *Locusta migratoria migratorioides* (R. et F.), phase solitaire, après implantation de nombreux corps allates. C. R. Acad. Sci. Paris 259: 3375–3377.

Alertsen, A. R. 1955. Ageratochromene, a heterocyclic compound from the essential oils of *Ageratum* species. Acta Chem. Scand. 9: 1725–1726.

Avruch, L. I. and S. S. Tobe. 1978. Juvenile hormone biosynthesis by the corpora allata of the male desert locust, *Schistocerca gregaria*, during sexual maturation. Can. J. Zool. 56: 2097–2102.

Awasthi, V. B. 1979. The ultrastructure of corpus allatum of the earwig *Euborella annulipes* Lucas (Dermaptera: Labiduridae). Z. Mikrosk. Anat. Forsch. 93: 982–991.

Badonnel, A. 1934. Recherches sur l'anatomie des Psoques. Bull. Biol. Fr. Belg. 18: 240.

Baehr, J. C., P. Cassier, and M. A. Fain-Maurel. 1973. Contribution expéri-mentale et infrastructurale à l'étude de la dynamique du corpus al-latum de *Rhodnius prolixus* Stål: influence de la nutrition, de l'activité ovarienne, de la pars intercerebralis et de ses connexions. Arch. Zool. Exp. Gen. 114(4): 611–626.

Baehr, J. C., P. Pradelles, and F. Dray. 1979. A radioimmunological assay for naturally occurring juvenile hormones using iodinated tracers: its use in the analysis of biological samples. *Ann. Biol. Anim. Bio-chem. Biophys.* 19: 1827–1836.

Baehr, J. C., J. P. Caruelle, P. Porcheron, and P. Cassier. 1981. Quantifica-tion of juvenile hormones using a radioimmunological assay in the analysis of biological samples. Pp. 47–57 *in* G. E. Pratt and G. T. Brooks (eds.), *Juvenile Hormone Biochemistry: Action, Agonism and An-tagonism*. Elsevier, Amsterdam and New York.

Baehr, J. C., P. Cassier, Cl. Caussanel, and P. Porcheron. 1982. Activity of corpora allata, endocrine balance and reproduction in female *Labidura riparia* (Dermaptera). Cell Tissue Res. 225: 267–282.

Baehr, J. C., P. Pradelles, P. Cassier, and F. Dray. 1977a. Aspects méth-odologiques et critiques du dosage radioimmunologique des hor-mones juvéniles. Proc. 4th Sem. Physiol. Insecte, Bordeaux, May 1977 pp. 300–301.

Baehr, J. C., P. Pradelles, P. Cassier, and F. Dray. 1977b. Contribution à l'étude de l'hormone juvénile par des techniques immunologiques. Bull. Soc. Zool. Fr. Suppl. 1: 77.

Baehr, J. C., P. Pradelles, C. Lebreux, P. Cassier, and F. Dray. 1976. A sim-ple and sensitive radioimmunoassay of Insect Juvenile Hormone using an iodinated tracer. FEBS (Fed. Eur. Biochem. Soc.) Lett. 69(1): 123–128.

Baker, F. C., E. Lee, B. J. Bergot, and D. A. Schooley. 1981. Isomerization of isopentenyl pyrophosphate and hemoisopentenyl pyrosphosphate by *Manduca sexta* corpora cardiaca–corpora allata homogenates. Pp. 67–80 *in* G. E. Pratt and G. T. Brooks (eds.), *Juvenile Hormone Biochemistry: Action, Agonism and Antagonism*. Elsevier, Amsterdam and New York.

Baker, F. C., C. A. Miller, L. W. Tsai, G. C. Jamieson, D. C. Cerf, and D. A. Schooley. 1986. The effect of juvenoids, anti-juvenile hormone agents, and several intermediates of juvenile hormone biosynthesis or the *in vivo* juvenile hormone levels in *Manduca sexta* larvae. Insect Biochem. 16: 741–747.

Beaulaton, J. 1964. Etude des ultrastructures de la glande prothoracique du Ver à soie (*Antheraea pernyi* Guer.). Thesis of the third cycle, Cell. Biol., University of Clermont-Ferrand, France.

Beaulaton, J. 1968a. Etude des glandes prothoraciques de Vers à soie: ultrastructure et cytochimie. D.S.N. thesis, University of Clermont-Ferrand, France; Arch. orig. Centre Document, CNRS.

Beaulaton, J. 1968b. Modifications ultrastructurales des cellules sécrétrices de la glande prothoracique de vers à soie au cours des deux derniers âges larvaires. I. Le chondriome, et ses relations avec le reticulum agranulaire. J. Cell Biol. 39(3): 501–525.

Belles, X. and M. D. Piulachs. 1983. Dessarrollo de los corpora allata, oocitos y glándulas colaterales durante el primer ciclo gonótrofico de *Blatella germanica* (L.). Rev. Espanol. Fisiol. 39: 149–154.

Belles, X., J. Casas, A. Messeguer, and M. D. Piulachs. 1987. *In vitro* biosynthesis of JH III by the corpora allata of adult females of *Blattella germanica*. Insect Biochem. 17: 1007–1015.

Benford, H. H. and J. T. Bradley. 1986. Early detection and juvenile hormone-dependence of cricket vitellogenin. J. Insect Physiol. 32(2): 109–116.

Ben Khay, A., L. Gourdoux, R. Moreau, and J. Dutrieu. 1987. Effects of intestinal insulin-like peptide on glucose catabolism in male adult *Locusta migratoria*. Arch. Insect Biochem. Physiol. 4: 223–239.

Berchtold, J. P. 1969. Contribution à l'étude ultrastructurale des cellules interrénales de *Salamandra salamandra* L. (Amphibien, Urodèle). I. Conditions normales. Z. Zellforsch. 102: 357–375.

Berchtold, J. P. 1970a. Sur un aspect particulier du reticulum endoplasmique agranulaire dans les cellules interrénales de *Salamandra salamandra* L. stimulées par l'A.C.T.H. endogène. C. R. Acad. Sci. Paris 270: 626–628.

Berchtold, J. P. 1970b. Contribution à l'étude ultrastructurale des cellules interrénales de *Salamandra salamandra* L. (Amphibien Urodèle). II. Action de l'ACTH endogène. Z. Zellforsch. 110: 517–539.

Bergot, B. J., G. C. Jamieson, M. A. Ratcliff, and D. A. Schooley. 1980. JH

Zero: new naturally occurring insect juvenile hormone from developing embryos of the tobacco hornworm. Science (Wash., DC) 210: 336–338.

Bergot, B. J., D. A. Schooley, E. M. Chippendale, and C. M. Yin. 1976. Juvenile hormone titer determinations in the southwestern corn borer, *Diatraea grandiosella*, by electron capture–gas chromatography. Life Sci. 18: 811–820.

Bergot, B. J., F. C. Baker, D. C. Cerf, G. Jamieson, and D. A. Schooley. 1981. Qualitative and quantitative aspects of juvenile hormone titers in developing embryos of several insect species: discovery of a new JH-like substance extracted from eggs of *Manduca sexta*. Pp. 33–45 *in* G. E. Pratt and G. T. Brooks (eds.), *Juvenile Hormone Biochemistry*. Elsevier, Amsterdam and New York.

Bitsch, J. 1962. Le complexe nerveux hypocérébral et les corpora allata des Machilides (Insectes, Thysanura). C. R. Acad. Sci. Paris 254: 1501–1504.

Bitsch, J. 1983. Recherches récentes sur le développement et la reproduction des Thysanoures (Insecta, Apterygota). Pedobiologia 25: 287–292.

Bitsch, J. and J. Lapalisse. 1984. Modifications volumétriques et ultrastructurales des corpora allata du Lépisme *Thermobia domestica* (Pacard) (Thysanura: Lepismatidae) au cours des cycles biologiques. Int. J. Insect Morphol. Embryol. 13(1): 37–49.

Bodenstein, D. 1938a. Untersuchungen zum Metamorphose-problem. I. Kombinierte Schnürungs und Transplantationsexperimente an *Drosophila*. Wilhelm Roux' Arch. Entwicklungsmech. Org. 137: 1–18.

Bodenstein, D. 1938b. Untersuchungen zum Metamorphose-problem. II. Entwicklungsrelationen in verschmolzenen Puppenteilen. Wilhelm Roux' Arch. Entwicklungsmech. Org. 139: 636–60.

Bodenstein, D. 1938c. Untersuchungen zum Metemorphose-problem. III. Über die Entwicklung der ovarien in Thoraxlosen puppenabdomen. Biol. Zentralbl. 58: 329–332.

Bounhiol, J. J. 1936a. Métamorphose après ablation des corpora allata chez le Ver à soie. C. R. Acad. Sci. Paris 203: 388–391.

Bounhiol, J. J. 1936b. Dans quelles limites l'écérébration des larves de Lépidoptères est-elle compatible avec leur nymphose. C. R. Acad. Sci. Paris 203: 1182–1184.

Bounhiol, J. J. 1937a. La métamorphose des Insects serait inhibée dans leur jeune âge par les corpora allata. C. R. Soc. Biol. Paris 126: 1189–1191.

Bounhiol, J. J. 1937b. Mue surnuméraire chez le Ver à soie élevé dans les conditions normales. C. R. Soc. Biol. Paris 124: 1223–1226.

Bounhiol, J. J. 1937c. Métamorphose prématurée après ablation des corpora allata chez le jeune Ver à soie. C. R. Acad. Sci. Paris 205: 175–178.

Bounhiol, J. J. 1938a. Rôle possible du ganglion frontal dans la métamorphose de *Bombyx mori*. C. R. Acad. Sci. Paris 206: 773–774.

Bounhiol, J. J. 1938b. Recherches expérimentales sur le déterminisme de la métamorphose chez les Lépidoptères. Bull. Biol. Fr. Belg. (Suppl.) 24: 1–200.

Bounhiol, J. J. 1938c. Les fonctions des corps allates. Arch. Zool. Exp. Gén. 81: 54–64.

Bounhiol, J. J. 1938d. Les conceptions modernes sur la métamorphose des Insectes. Rev. Zool. Agr. Appl. Bordeaux 3: 1728–1732.

Bounhiol, J. J. 1938e. Nymphose partielle localisée chez des Vers à soie divisés en trois parties par deux ligatures. C. R. Acad. Sci. Paris 217: 203.

Bounhiol, J. J. 1938f. Les glandes mandibulaires jouent-elles un rôle dans la métamorphose chez *Bombyx mori*. C. R. Acad. Sci. Paris 17: 237–238.

Bounhiol. J. J., M. Gabe, and L. Arvy. 1953. Données histophysiologiques sur la neurosécrétion chez *Bombyx mori* L. et sur ses rapports avec les glandes endocrines. Bull. Biol. Fr. Belg. 87: 1–11.

Bowers, B. and B. Johnson. 1966. An electron microscope study of the corpora cardiaca and secretory neurons in the aphid, *Myzus persicae* (Sulz.) Gen. Comp. Endocrinol. 6(2): 213–230.

Bowers, W. S. 1983. The Precocenes. Pp. 517–523 *in* R. G. H. Downer and H. Laufer (eds.), *Endocrinology of Insects*. Liss, New York.

Bowers, W. S., T. Ohta, J. S. Cleere, and P. A. Marsella. 1976a. Discovery of insect anti-juvenile hormones in plants. Science (Wash., DC) 193: 542–547.

Bowers, W. S., T. Ohta, J. S. Cleere, and P. A. Marsella. 1976b. Discovery of insect anti-hormones: *in vitro* biosynthesis of JH III by *Dysdercus fasciatus*. J. Exp. Zool. 228: 555–559.

Bradley, J. T. and J. S. Edwards. 1979. Ultrastructure of the corpus cardiacum and corpus allatum of the house cricket *Acheta domesticus*. Cell Tissue Res. 198: 201–208.

Brousse-Gaury, P. 1971a. Influence de stimuli externes sur le comportement des Blattes. I. Les organes sensoriels céphaliques, point de départ de réflexes neuro-endocriniens. Ann. Sci. Nat. Zool. [12] 13: 181–332.

Brousse-Gaury, P. 1971b. Influence de stimuli externes sur le comportement neuro-endorcrinien des Blattes. II. Histophysiologie des voies réflexes neuro-endocriniennes. Ann. Sci. Nat. Zool. Biol. [12] 13: 333–350.

Brousse-Gaury, P. 1971c. Influence de stimuli externes sur le comportement neuro-endocrinien des Blattes. Thesis, Faculty of Science, University of Paris, No. 5073.

Brousse-Gaury, P. 1972. Innervation cérébrale et sous-oesophagienne des formations endocrines rétrocérébrales chez *Blabera fusca* Br. et *Periplaneta americana* L. Bull. Biol. Fr. Belg. 4: 315–336.

Brousse-Gaury, P. and P. Cassier. 1975. Contribution à l'étude de la dynamique des corpora allata chez *Blabera fusca:* influence du jeûne, du groupement et du régime photopériodique. Bull. Biol. Fr. Belg. 59(3/4): 253–277.

Brousse-Gaury, P., P. Cassier, and M. A. Fain-Maurel. 1973. Contribution expérimentale et infrastructurale à l'étude de la dynamique des corpora allata chez *Blabera fusca:* influence du groupement visuel, des afférences ocellaires et antennaires. Bull. Biol. Fr. Belg. 107: 143–169.

Brüning, E., A. Saxer, and B. Lanzrein. 1985. Methyl farnesoate and juvenile hormone III in normal and precocene treated embryos of the ovoviviparous cockroach, *Nauphoela cinerea.* Int. J. Invertebr. Reprod. Dev. 8: 269–278.

Brüning, E. and B. Lanzrein. 1987. Function of juvenile hormone III in embryonic development of the Cockroach, *Nauphoeta cinera.* Int. J. Invertebr. Reprod. Dev. 12: 29–44.

Busselet, M. 1968. Données ultrastructurales sur les corps cardiaques d' *Antheraea pernyi* Guer. (Lepidoptera, Attacidae). C. R. Acad. Sci. Paris 267(26): 2337–2340.

Busselet, M. 1969. Données histochimiques et ultrastructurales sur les corps allates de *Rhodnius prolixus* Stål. et *Antheraea pernyi* Guer. Bull. Soc. Zool Fr. 94(3): 372–503.

Buys, C. M. and D. Gibbs. 1981. The anatomy of neurons projecting to the corpus cardiacum from the larval brain of the tobacco hornworm, *Manduca sexta* L. Cell Tissue Res. 215: 505–513.

Camps, F. and X. Belles. 1986. Inhibidores de les hormonas juveniles de Insectos. Bioquim. Biol. Mol. 2: 70–75.

Cantacuzène, A. M., S. Lauverjat, and M. Papillon. 1972. Influence de la température d'élevage sur les caractères histologiques de l'appareil génital de *Schistocerca gregaria.* J. Insect Physiol. 18: 2077–2093.

Carlisle, D. B. and P. E. Ellis. 1959. La persistance des glandes ventrales céphaliques chez les Criquets solitaires. C. R. Acad. Sci. Paris 249: 1059–1060.

Carrow, G. M., R. L. Calabresse, and C. M. Williams. 1981. Spontaneous and evoked release of prothoracicotropin from multiple neurohemal organs on the tobacco hornworm. Proc. Natl. Acad. Sci. USA 78: 5866–5870.

Carrow, G. M., R. L. Calabrese, and C. M. Williams. 1984. Architecture and physiology of insect cerebral neurosecretory cells. J. Neurosci. 4(4): 1034–1044.

Cassagnau, P. and C. Juberthie. 1967. Structures nerveuses, neurosécrétions et organes endocrines chez les Collemboles. II. Le complexe cérébral des Entomobryomorphes. Gen. Comp. Endocrinol. 8: 489–502.

Cassier, P. 1963. Action des implantations de corps allates sur la réactivité phototropique de *Locusta migratoria migratorioïdes* (R. et F.) phase grégaria. C. R. Acad. Sci. Paris 257: 4048–4049.

Cassier, P. 1964a. Etude et interprétation des effets à long terme des implantations de corps allates sur la réactivité phototropique de *Locusta migratoria migratorioïdes* (R. et F.), phase grégaire. C. R. Acad. Sci. Paris 258: 723–725.

Cassier, P. 1964b. La réactivité phototropique de *Locusta migratoria* (R. et F.), phase gregaria, après implantation abdominale de corps allates. Insectes Soc. 11(2): 131–140.

Cassier, P. 1964c. Effets immédiats et transmis des implantations de corps allates sur la fécondité et la descendance des femelles de *Locusta migratoria migratorioïdes* (R. et F.), phase grégaire. C. R. Acad. Sci. Paris 259: 2706–2708.

Cassier, P. 1965a. Déterminisme endocrine de quelques caractéristiques phasaires chez *Locusta migratoria migratorioïdes*. Insectes Soc. 12(1): 71–80.

Cassier, P. 1965b. Le comportement phototropique du Criquet migrateur (*Locusta migratoria migratorioïdes* R. et F.): bases sensorielles et endocrines. Ann. Sci. Nat. Zool. Paris 7: 213–358.

Cassier, P. 1966a. L'activité corps allates et la reproduction du Criquet migrateur africain, *Locusta migratoria migratorioïdes* R. et F. *Bull. Soc. Zool. Fr.* 91(2): 133–148.

Cassier, P. 1966b. Effets de l'ablation d'un corps allate et de la section du nerf allato-cardiaque symétrique sur la fécondité des femelles isolées du Criquet migrateur (*Locusta migratoria migratorioïdes*) et sur les caractéristiques de leur descendance. C. R. Acad. Sci. Paris 262(11): 1276–1279.

Cassier, P. 1966c. Effects de l'ablation d'un corps allate sur la fécondité et la descendance des femelles isolées du Criquet migrateur (*Locusta migratoria migratorioïdes* R. et F.) (Insecte, Orthoptéroïde, Acrididae). Insectes Soc. 13(1): 17–27.

Cassier, P. 1967. La reproduction des Insectes et la régulation de l'activité des corps allates. Ann. Biol. Fr. 6(11/12): 595–670.

Cassier, P. 1974. Der Phasenpolymorphismus der Wanderheuschreken. Pp. 110–151 *in* G. H. Schmidt (ed.), *Sozialpolymorphismus bei Insekten: Problem der Kastenbildung in Tierreich*. Wissenschaftliche Verlagsgesellschaft, Stuttgart.

Cassier, P. 1977a. The imaginal development of the integument in the desert locust, *Schistocerca gregaria* Forsk: influence of the sex and phasic states. Adv. Insect Reprod. 1: 344–355.

Cassier, P. 1977b. La différenciation imaginale du tégument chez le Criquet pélerin, *Schistocerca gregaria* Forsk. IV. Les étapes de la morphogenèse des unités glandulaires. Arch. Anat. Microsc. Morph. Exp. 66: 145–161.

Cassier, P. 1979. The corpora allata of Insects. Int. Rev. Cytol. 57: 1–73.

Cassier, P. and C. Delorme. 1973. Différenciation des structures tégumentaires chez les imagos de *Schistocerca gregaria* Forsk: données infrastructurales et expérimentales. Proc. 2nd Sem. Physiol. Insects, Rennes.

Cassier, P. and C. Delorme, 1974. La sécrétion de la phéromone sexuelle mâle chez le Criquet pélerin, *Schistocerca gregaria* Forsk: étude expérimentale et ultrastructurale de l'évolution des formations tégumentaires au cours de la maturation sexuelle. *Ann. Zool. Ecol. Anim.* 6(2): 153–208.

Cassier, P. and C. Delorme-Joulie. 1975. Modalités de la différenciation imaginale du tégument de *Schistocerca gregaria* Forsk, phase grégaire. Proc. 3rd Sem. Physiol. Insects, Strasbourg.

Cassier, P. and C. Joulie-Delorme. 1976a. La différenciation imaginale du tégument chez le Criquet pélerin, *Schistocerca gregaria*. I. L'évolution post-imaginale et son déterminisme. Arch. Zool. Exp. Gen. 117: 95–116.

Cassier, P. and C. Delorme-Joulie. 1976b. La différenciation imaginale du tégument chez le Criquet pélerin, *Schistocerca gregaria* Forsk. II. L'évolution au cours de la mue imaginale et son déterminisme chez les mâles grégaires. Ann. Sci. Nat. Zool. [12] 18(3): 295–309.

Cassier, P. and C. Delorme-Joulie. 1976c. La différenciation imaginale du tégument chez le Criquet pélerin, *Schistocerca gregaria* Forsk. IV. Les différences phasaires et leur déterminisme. Insectes Soc. 23(2): 179–198.

Cassier, P. and M. A. Fain-Maurel. 1970. Contribution à l'étude infrastructurale du système neurosécréteur rétrocérébral chez *Locusta migratoria migratorioïdes* (R. et F.). II. Le transit des neurosécrétions. Z. Zellforsch. Mikrosk. Anat. 111: 483–492.

Cassier, P. and M. A. Fain-Maurel. 1971a. Formations endocrines et stéroïdogènèse: étude comparée chez les Insectes ptérygotes et aptérygotes. Colloq. stéroïdogènèse, Clermont-Ferrand, Association pour le Développement de l'Enseignement et de la Recherche (ADER).

Cassier, P. and M. A. Fain-Maurel. 1971b. Modalités de l'évolution et du renouvellement du chondriome au cours des cycles d'activité des glandes de mue de *Petrobius maritimus* Leach (Insecte aptérygote). Arch. Zool. Exp. Gen. 112(3): 457–470.

Cassier, P. and M. Papillon. 1968. Effets des implantations de corps allates sur la reproduction des femelles groupées de *Schistocerca gregaria* (Forsk) et sur le polymorphisme de leur descendance. C. R. Acad. Sci. Paris 266(10): 1048–1051.

Cazal, M. 1971a. Les corpora cardiaca chez *Locusta migratoria* L. et leurs fonctions. Thesis, Academy of Montpellier, CNRS.

Cazal, P. 1948. Les glandes endocrines rétrocérébrales des Insectes. Bull. Biol. Fr. Belg. Suppl. 32: 1–227.

Cazal, P. and Y. Guerrier. 1946. Recherches sur les glandes endocrines rétrocérébrales des Insectes. I. Les Orthoptères. Arch. Zool. Exp. Gen. 84: 303–334.

Chalaye, D. 1965. Recherches histochimiques et histophysiologiques sur la neurosécrétion dans la chaîne nerveuse ventrale du Criquet migrateur, *Locusta migratoria*. C. R. Acad. Sci. Paris 260(26): 7010–7013.

Chalaye, D. 1966. Recherches sur la destination des produits de neurosécrétion de la chaîne nerveuse ventrale du Criquet migrateur, *Locusta migratoria*. C. R. Acad. Sci. Paris 262(1): 161–164.

Charpin, P. 1973. Etude ultrastructurale des corps allates chez les femelles de *Choleva cisteloïdes* Fröl (Colépotères Catopidae de la sous famille des Capitonae) au cours de la diapause ovarienne. C. R. Acad. Sci. Paris 277: 2181–2184.

Charpin, P. 1975. Evolution ultrastructurale des corps allates au cours du dernier stade larvaire chez *Choleva augustata* Fab. (Coléoptères, Catopidae de la sous-famille des Catopidae). C. R. Acad. Sci. Paris 280D: 1997–2000.

Chaudonneret, J. 1946. Sur la présence d'une glande neuroendocrine dans la maxille de *Thermobia domestica* Pack. C. R. Acad. Sci., Paris 223: 292–293.

Chaudonneret, J. 1949. A propos du corps jugal des Thysanoures. Bull. Soc. Zool. Fr. 74: 164–167.

Christensen, A. K. 1965. The fine structure of testicular interstitial cells in guinea pigs. J. Cell. Biol. 26: 911–935.

Christensen, A. K. 1975. Leydig cells. Handb. Physiol. (Sect. 7) 5: 57–94.

Christensen, A. K. and G. B. Chapman. 1959. Cup-shaped mitochondria in interstitial cells of the albino rat testis. Exp. Cell. Res. 18: 576–579.

Chu, H. M. 1979. Ultrastructural features of the corpus allatum of the newly transformed versus ovipositing female beetle, *Xyleborus ferrugineus* (Coleoptera: Scolytidae). Diss. Abstr. Int. B Sci. Eng. 40: 1050.

Chu, H. M., and D. M. Norris. 1979. Comparative morphology and ultrastructure of the corpora allata in newly emerged and sexually mature female *Xyleborus ferrugineus* (Fabr.) (Coléoptera: Scolytidae). Int. J. Insect Morphol Embryol. 8: 359–374.

Couillaud, F., B. Mauchamp, and A. Girardie. 1985. Regulation of juvenile hormone titer in African locust. Experientia (Basel) 41: 1165–1167.

Couillaud, F., B. Mauchamp, and A. Girardie. 1987. Biological, radio-chemical and physicochemical evidence for the low activity of disconnected corpora allata in locust. J. Insect Physiol. 33(4): 223–228.

Couillaud, F., J. Girardie, S. S. Tobe, and A. Girardie. 1984. Activity of disconnected corpora allata in *Locusta migratoria*: juvenile hormone biosynthesis *in vitro* and physiological effects *in vivo*. J. Insect. Physiol. 30(7): 551–556.

Credland, P. F. and M. D. C. Scales 1981. Ultrastructure of the retro-cerebral complex in the adult midge, *Chironomus riparius* Mg (Diptera: Chironomidae). Int. J. Insect Morphol. Embryol. 10: 451–461.

Dahm, K. H. and H. Röller. 1970. The juvenile hormone of the giant silkmoth *Hyalophora golveri* (Strecker). Life Sci. 9: 1397–1400.

Dahm, K. H., H. Röller, and B. M. Trost. 1968. The juvenile hormone. IV. Stereochemistry of juvenile hormone and biological activity of some of its isomers and related compounds. Life Sci. 7(4): 129–137.

Dahm, K. H., G. Bhaskaran, M. G. Peter, P. D. Shirk, K. R. Seshan, and H. Röller. 1976. On the identify of the juvenile hormone in insects. Pp. 19–47 *in* L. I. Gilbert (ed.), *The Juvenile Hormones*. Plenum Press, New York.

Dahm, K. H., G. Bhaskaran, M. G. Peter, P. D. Shirk, K. R. Seshan, and H. Röller. 1981. The juvenile hormones of *Cecropia*. Pp. 183–198 *in* C. F. Sehnal, A. Zabza, J. J. Menn, and B. Cymborowski (eds.), *Regulation of Insect Development and Behaviour*, Pt. I. Wrocław Technical University Press., Wrocław, Poland.

Dale, J. F. and S. S. Tobe. 1986. Biosynthesis and titre of juvenile hormone during the first gonotrophic cycle in isolated and crowded *Locusta migratoria* females. J. Insect. Physiol. 32(9): 763–769.

de Lerma, B. 1956. Corpora cardiaca et neurosécrétion protocérébrale chez le Coléoptère *Hydrous piceus* L. Ann. Sci. Nat. Zool. 18: 235–250.

Deleurance, S. 1967. La neurosécrétion chez les Coléoptères cavernicoles: imago. C. R. Acad. Sci. Paris 264 (2): 392–394.

Deleurance, S. and P. Charpin. 1971. Sur les corps allates des *Bathysciinae* (Coléoptères cavernicoles). C. R. Acad. Sci. Paris 273: 177–180.

Deleurance, S. and P. Charpin. 1972. Aspects comparatifs des corps allates chez les Bathysciinae (Coléoptères cavernicoles) imago. C. R. Acad. Sci. Paris 274: 405–408.

Deleurance, S. and P. Charpin. 1973. Recherches expérimentales sur les mues larvaire et nymphale chez le *Choleva* sp. (Coléoptères Catopidae, sous-famille des Catopidae). C. R. Acad. Sci. Paris 277: 65–67.

Deleurance, S. and P. Charpin. 1975. Evolution ultrastructurale des corps allates de la larve du dernier stade de *Choleva angustata* Fab. (Catopidae): action du jeûne, en présence ou non d'ecdystérone. C. R. Acad. Sci. Paris 280D: 2113–2116.

Deleurance, S. and P. Charpin. 1978. Ultrastructural dynamics of the corpus allatum of *Choleva angustata* Fab. (Coleoptera, Catopidae). Cell. Tissue Res. 191: 151–160.

Delfino, G., M. T. Marino Piccioli, and C. Calloni. 1981. Ultrastructure of the corpora allata in *Polistes gallicus* (L) (Hymenoptera: Vespidae). Monit. Zool. Ital. 15: 29–52.

De Robertis, E. and D. Sabatini. 1958. Mitochondrial changes in the adreno cortex of normal hamsters. J. Biophys. Biochem. Cytol. 4: 667–671.

De Wilde, J. 1964. Reproduction. Pp. 10–58 *in* M. Rockstein (ed.), *Physiology of Insecta*, Vol. 1. Academic Press, Orlando, Florida.

De Wilde, J. and J. De Boer. 1969. Humoral and nervous pathways in photoperiodic induction of diapause in *Leptinotarsa decemlineata*. J. Insect. Physiol. 15: 661–675.

De Wilde, J. and A. De Loof. 1973. Reproduction: endocrine control. Pp. 97–157 *in* M. Rockstein (ed.), *Physiology of Insecta* Vol. 1. Academic Press, Orlando, Florida.

Doane, W. W. 1961. Developmental physiology of the mutant sterile (2) adipose of *Drosophila melanogaster*. III. Corpus allatum complex and ovarian transplantations. J. Exp. Zool. 146: 275–298.

Doane, W. W. 1973. Rôle of hormones in insect development. Pp. 291–497 *in* S. J. Counce and C. H. Waddington (eds.), *Developmental Systems: Insects*, Vol. 2. Academic Press, Orlando, Florida.

Dorn, A. 1973. Electron microscopic study on the larval and adult corpus allatum of *Oncopeltus fasciatus* Dallas (Insecta: Heteroptera). Z. Mikrosk. Anat. Forsch. 145: 447–458.

Dorn, A. 1975. Struktur und Funktion des embryonalen Corpus allatum von *Oncopeltus fasciatus* Dallas (Insecta, Heteroptera). Verh. Dtsch. Zool. Ges., 1974 pp. 85–89.

Dorn, A. 1982. Precocene-induced effects and possible role of juvenile hormone during embryogenesis of the milkweed bug, *Oncopeltus fasciatus*. Gen. Comp. Endocrinol. 46: 42–52.

Duve, H. and A. Thorpe. 1984. Immunocytochemical mapping of gastrin/CCK-like peptides in the neuroendocrine system of the blowfly *Calliphora vomitoria* (Diptera). Cell Tissue Res. 237: 305–320.

Eastham, L. E. S. 1930. The formation of germ layers in insects. Biol. Rev. Camb. Phil. Soc. 5: 1–20.

Edwards, J. P. and N. R. Price. 1983. Inhibition of juvenile hormone III biosynthesis in *Periplaneta americana* with the fungal metabolite compactin (ML-236 B). Insect Biochem. 13(2): 185–189.

Edwards, J. P., B. J. Bergot, and G. B. Staal. 1983. Effects of three compounds with anti-juvenile hormone activity and a juvenile hormone analogue on endogenous juvenile hormone levels in the tobacco hornworm, *Manduca sexta*. J. Insect Physiol. 29(1): 83–89.

Edwards, J. P., D. C. Cerf, and G. B. Staal. 1985. Inhibition of oötheca production in *Periplaneta americana* (L.) with the anti–juvenile hormone fluoromevalonate. J. Insect Physiol. 31(9): 723–728.

Edwards, C. A., H. Ruska, and E. De Harven. 1958. Electron microscopy of peripheral nerves and neuromuscular junctions in the wasp leg. J. Biophys. Biochem. Cytol. 4: 107–114.

Elliot, H. J. 1976. Structural analysis of the corpus allatum of an aphid. *Aphis craccivora*. J. Insect Physiol. 22: 1275–1279.

El Salhy, M., S. Falkner, K. J. Kramer, and R. D. Speirs. 1983. Immunohistochemical investigations of neuropeptides in the brain, corpora cardiaca and corpora allata of an adult lepidopteran insect, *Manduca sexta* (L.). Cell Tissue Res. 232: 295–317.

Endo, A. 1981. Biological and pharmacological activity of inhibitors of 3-hydroxy-3-methylglutaryl coenzyme A reductase. Trends Biochem. Sci. 6: 10–12.

Engelmann, D. 1957. Die Steuerung der Ovarfunktion bei der Ovoviviparen Schabe *Leucophaea maderae* (Fabr.). J. Insect. Physiol. 1: 257–278.

Engelmann, F. 1958. The stimulation of the corpora allata in *Diploptera punctata* (Blattaria). Anat. Rec. 132(3): 432–433.

Engelmann, F. 1959. The control of reproduction in *Diploptera punctata* (Blattaria). Biol. Bull. (Woods Hole) 116: 406–419.

Engelmann, F. 1968. Endocrine control of reproduction in insects. Annu. Rev. Entomol. 13: 1–26.

Engelmann, F. 1970. *The Physiology of Insect Reproduction*. Pergamon Press, Oxford and Elmsford, New York.

Fain-Maurel, M. A. and P. Cassier. 1969a. Etude infrastructurale des corpora allata de *Locusta migratoria migratorioides* (R. et F.), phase solitaire, au cours de la maturation sexuelle et des cycles ovariens. C. R. Acad. Sci. Paris 268 D: 2721–2724.

Fain-Maurel, M. A. and P. Cassier. 1969b. Pléomorphisme mitochondrial dans les corpora allata de *Locusta migratoria migratorioïdes* (R. et F.) au cours de la vie imaginale. Z. Zellforsch. Mikrosk. Anat. 102: 543–559.

Fain-Maurel, M. A. and P. Cassier. 1970. Contribution à l'étude infrastructurale du système neurosécréteur rétrocérébral chez *Locusta migratoria migratorioïdes* (R. et F.). I. Les corpora cardiaca. Z. Zellforsch. Mikrosk. Anat. 111: 471–482.

Fain-Maurel, M. A. and P. Cassier. 1972. Sur une nouvelle modalité de l'agencement en "cotte de mailles" du reticulum endoplasmique. J. Microsc. (Paris) 14(1): 121–124.

Fain, M. J. and L. M. Riddiford. 1975. Juvenile hormone titers in the hemolymph during late larval development of the tobacco hornworm *Manduca sexta* (L.). Biol. Bull. (Woods Hole) 149: 506–521.

Falck, K. B. and C. Owman. 1965. A detailed methodological description of the fluorescence method for the cellular demonstration of biogenic monoamines. Acta Univ. Lund 2: 1–23.

Fawcett, D. W. 1961. The membranes of the cytoplasm. Lab. Invert. 10: 1162–1180.

Feldlaufer, M. F., J. A. Svoboda, and E. W. Herbert Jr. 1986. Makisterone A and 24-methylenecholesterol from the ovaries of the honey bee, *Apis mellifera* L. Experientia (Basel) 42: 200–201.

Ferenz, H. J. and Kaufner, I. 1981. Juvenile hormone synthesis in relation to oogenesis in *Locusta migratoria*. Pp. 135–145 *in* G. E. Pratt and G. T. Brooks (eds.), *Juvenile Hormone Biochemistry*. Elsevier, Amsterdam and New York.

Feyereisen, R. and S. S. Tobe 1981. A rapid partition assay for routine analysis of juvenile hormone release in insect corpora allata. *Anal. Biochem.* 111: 372–375.

Feyereisen, R., T. Friedel, and S. S. Tobe. 1981a. Farnesoic acid stimulation of C_{16} juvenile hormone biosynthesis by corpora allata of adult female *Diploptera punctata*. Insect Biochem. 11(4): 401–409.

Feyereisen, R., J. Koener, and S. S. Tobe. 1981b. *In vitro* studies with C_2, C_6 and C_{15} precursors of C_{16}JH biosynthesis in the corpora allata of adult female *Diploptera punctata*. Pp. 81–92 *in* G. E. Pratt and G. T. Brooks (eds.), *Juvenile Hormone Biochemistry: Action, Agonism and Antagonism*. Elsevier, Amsterdam and New York.

Feyereisen, R., G. Johnson, J. Koener, B. Stay, and S. S. Tobe. 1981c. Precocenes as pro-allatocidins in adult female *Diploptera punctata*: a functional and ultrastructural study. J. Insect Physiol. 27(12): 855–868.

Forel, A. 1874. Les fourmis de la Suisse. Neue Denkschr. Allgem. Schweiz. Ges. Gersant. Naturwiss., Zurich.

Fraser, J. and R. Pipa. 1977. Corpus allatum regulation during the metamorphosis of *Periplaneta americana*: axon pathways. J. Insect Physiol. 23: 975–984.

Füller, H. B. 1960. Morphologische und experimentelle Untersuchungen über die neurosekretorischen Verhaltnisse im Zentralnerven system von Blattiden und Culiciden. Zool. Jahrb. Abt. Allg. Zool. Phys. 69: 223–250.

Fugo, H., J. H. Chen, M. Nakajima, H. Nagasawa, and A. Suzuki. 1987. Neurohormones in developing embryos of the silkworm, *Bombyx mori:* the presence and characteristics of prothoracicotropic hormone. J. Insect Physiol. 33: 243–248.

Fukuda, S. and S. Takeuchi. 1965. On the origin of osmiophilic bodies contained in the corpus allatum cells of female moth of the silkworm *Bombyx mori.* (In Japanese.) Zoll. Mag. (Tokyo) 74: 392.

Fukuda, S., G. Eguchi, and S. Takeuchi. 1963. Electron microscopical studies on the sexual difference on the corpora allata in the silkworm. Zool. Mag. (Tokyo) 72: 345.

Fukuda, S., G. Eguchi, and S. Takeuchi. 1966. Histological and electron microscopical studies on sexual differences in structure of the corpora allata of the moth of the silkworm, *Bombyx mori.* Embryology 9(2): 123–158.

Ganagarajah, M. 1965. The neuro-endocrine complex of adult *Nebria brevicollis* and its relation to reproduction. J. Insect. Physiol. 11: 1377–1387.

Gibbs, D. and L. M. Riddiford. 1977. Prothoracicotropic hormone in *Manduca sexta:* localization by a larval assay. *J. Exp. Zool.* 66: 255–256.

Gilbert, L. I. 1963. Hormones controlling reproduction and molting in invertebrates. *Comp. Endocrinol.* 2: 1–46.

Gilbert, L. I. 1964. Physiology of growth and development: endocrine aspects. Pp. 150–214 *in* M. Rockstein (ed.), *Physiology of Insecta*, Vol. 2. Academic Press, Orlando, Florida.

Gilbert, L. I. 1974. Endocrine action during insect growth. Pp. 347–390 *in* Roy O. Greep (ed.), *Recent Progress in Hormone Research.* Academic Press, Orlando, Florida.

Gilbert, L. I. (ed.) 1976. *The Juvenile Hormones.* Plenum Press, New York.

Gilbert, L. I. and D. S. King. 1973. Physiology of growth and development: endocrine aspects. Pp. 249–370 *in* M. Rockstein (ed.), *The Physiology of Insects*, Vol. 1. Academic Press, New York.

Gilbert, L. I. and H. A. Schneiderman. 1961a. Some biochemical aspects of insect metamorphosis. Am. Zool. 1: 11–51.

Gilbert, L. I. and H. A. Schneiderman. 1961b. The content of juvenile hormone and lipid in Lepidoptera: sexual differences and developmental changes, *Gen. Comp. Endocrinol.* 1(5/6): 453–472.

Gilbert, L. I., W. B. Bollenbacher, and N. A. Granger. 1980. Insect endocrinology: regulation of endocrine glands, hormone titer, and hormone metabolism, Annu. Rev. Physiol. 42: 493–510.

Gillott, C. and G. S. Dogra. 1972. Neurosecretory cell and corpus allatum activity during production of successive egg batches in virgin *Melanoplus sanguinipes* (Fab.) Gen. Comp. Endocrinol. 18: 126–132.

Girardie, A. 1967a. La pars intercerebralis chez *Locusta migratoria* L. (Orthoptère) et son rôle dans le développement. D.S.N. thesis, University of Strasbourg; Arch. Orig. Centre Documentation CNRS, No. 1308.

Girardie, A. 1967b. Contrôle neuro-hormonal de la métamorphose et de la pigmentation chez *Locusta migratoria cinerascens* (Orthoptère). Bull. Biol. Fr. Belg. 101,(2): 79–114.

Girardie, A. and S. Granier. 1973. Système endocrine et physiologie de la diapause imaginale chez le criquet égyptien *Anacridium aegyptium*. J. Insect. Physiol. 19: 2341–2358.

Girardie, J. and S. Granier. 1974a. Ultrastructure des corps allates d'*Anacridium aegyptium* (Insecte Orthoptère) à l'avant-dernier stade larvaire et durant la vie imaginale. Arch. Anat. Microsc. Morphol. Exp. 63: 251–267.

Girardie, J. and S. Granier. 1974b. Rôle des corps allates dans la castration parasitaire d'*Anacridium aegyptium* (Insecte Orthoptère) infesté par *Metacemyia calloti* (Insecte: Diptère). Arch. Anat. Microsc. Morphol. Exp. 63(3): 269–280.

Girardie, J., S. S. Tobe, and A. Girardie. 1981. Biosynthèse de l'hormone juvénile C_{16} (JH III) et maturation ovarienne chez le Criquet migrateur. C. R. Acad. Sci. Paris 293: 443–446.

Goodman, W. G. 1983. Hemolymph transport of ecdysteroids and juvenile hormone. Pp. 147–159 in R. G. H. Downer and H. Laufer (eds.), *Endocrinology of Insects*. Liss, New York.

Granger, N. A., S. M. Niemiec, L. I. Gilbert, and W. E. Bollenbacher. 1982a. Juvenile hormone III: Biosynthesis by the larval corpora allata of *Manduca sexta*. J. Insect Physiol. 28(4): 385–391.

Granger, N. A., S. M. Niemiec, L. I. Gilbert, and W. E. Bollenbacher. 1982b. Juvenile hormone synthesis *in vitro* by larval and pupal corpora allata of *Manduca sexta*. Mol. Cell. Endocrinol. 28: 587–604.

Granger, N. A., W. E. Bollenbacher, R. Vince, L. I. Gilbert, J. C. Baehr, and F. Dray. 1979. *In vitro* biosynthesis of juvenile hormone by the larval corpora allata of *Manduca sexta*: quantification by radioimmunoassay. Mol. Cell. Endocrinol. 16: 1–17.

Greenberg, S. and S. S. Tobe 1985. Adaptation of a radiochemical assay for juvenile hormone biosynthesis to study caste differentiation in a primitive termite. J. Insect. Physiol. 31: 347–352.

Guelin, M. and A. Darjo. 1974. Etude ultrastructurale des corpora allata en relation avec le contrôle photopériodique de leur fonction gonadotrope chez *Locusta migratoria migratorioïdes*. C. R. Acad. Sci. Paris 278: 491–494.

Gundel, M. and H. Penzlin. 1980. Identification of neuronal pathways between the stomatogastric system and the retrocerebral complex of the cockroach *Periplaneta americana* (L.). Cell Tissue Res. 208: 283–297.

Haget, A. 1977. Embryologie des Insectes. Pp. 1–387 *in* P. P. Grassé (ed.), *Traité de zoologie*, Vol. 8, pt. V. Masson, Paris.

Haget, A., A. Ressouches, and J. Rogueda. 1981. Chronological and ultra-structural observations on the activities of the embryonic corpus allatum in *Carausius morosus* Br. (Phasmida: Lonchodidae). Int. J. Morphol. Embryol. 10(1): 65–81.

Hamilton, M. D., R. R. Rojas, and J. G. Baust. 1986. Juvenile hormone modulation of cryoprotectant synthesis in *Eurosta solidaginis* by a component of the endocrine system. J. Insect. Physiol. 32(11): 971–979.

Hammock, B., J. Nowock, W. Goodman, V. Stamoudis, and L. I. Gilbert. 1975. The influence of hemolymph binding protein on juvenile hormone stability and distribution in *Manduca sexta* fat body and imaginal discs *in vitro*. Mol. Cell. Endocrinol. 3: 167–181.

Hansen, B. L., G. N. Hansen, and B. Scharrer. 1982. Immunoreactive material resembling vertebrate neuropeptides in the corpus cardiacum and corpus allatum of the insect *Leucophaea maderae*. Cell Tissue Res. 225: 319–329.

Hansen, B. L., G. N. Hansen, and B. Scharrer. 1986. Immuno-cytochemical demonstration of a material resembling vertebrate ACTH and MSH in the corpus cardiacum–corpus allatum complex of the insect *Leucophaea*. Pp. 213–222 *in* G. B. Stefano (ed.), *Handbook of Comparative Aspects of Opioid and Related Neuropeptide Mechanisms*, Vol. 1. CRC Press, Boca Raton, Florida.

Hansen, G. N., B. L. Hansen, and B. Scharrer. 1987. Gastrin/CCK-like immunoreactivity in the corpus cardiacum–corpus allatum complex of the cockroach *Leucophaea maderae*. Cell Tissue Res. 248: 595–598.

Hanström, B. 1940a. Inkretorische Organe, Sinnesorgane und Nerven-system des Kopfes einiger niederer Insektenordungen. K. Sven. Vetenskapsakad. Handl. 3(8): 1–266.

Hanström, B. 1940b. Die Chromatophoraktivierende Substanz des Insektenkopfes K. Fysiogr. Sällsk. Handl. Lund Univ. Arsskr. [1] 51(36): 1–20.

Hanström, B. 1942. Die Corpora cardiaca und Corpora allata der Insekten. Biol. Gen. 15: 485–531.

Hardie, J. 1987. The corpus allatum, neurosecretion and photo-periodically controlled polymorphism in an aphid. J. Insect Physiol. 33(3):201–205.

Harker, J. E. 1960. Endocrine and nervous factors in insects circadian rythms. Cold Spring Harbor Symp. Quant. Biol. 25: 229–287.

Harper, E., S. Seifter, and B. Scharrer. 1967. Electron microscopic and biochemical characterization of collagen in blattarian insects. J. Cell Biol. 33(2): 385–394.

Hartmann, R., C. Jendrsczok, and M. G. Peter. 1987. The occurrence of a juvenile hormone binding protein and *in vitro* synthesis of juvenile hormone by the serosa of *Locusta migratoria* embryos. Wilhelm Roux's Arch. Dev. Biol. 196: 347–355.

Heinrich, C. A. 1982. Juvenile hormone mimics. Pp. 315–340 *in* J. R. Coats (ed.), *Insecticides: Mode of Action*. Academic Press, Orlando, Florida.

Heinrich, C. A., G. B. Staal, and J. B. Siddall. 1973. Alkyl 3,7,11-Trimethyl-2,4-dodecadienoates, a new class of potent insect growth regulators with juvenile hormone activity. J. Agric. Food. Chem. 21: 354–359.

Heinrich, C. A., G. B. Staal, and J. B. Siddall. 1976. Structure–activity relationships in some juvenile hormone analogs. Pp. 48–60 *in* L. I. Gilbert (ed.), *The Juvenile Hormones*. Plenum Press, New York.

Hess, A. 1958. The fine structure of nerve cells and fibers neuroglia, and sheaths of the ganglion chain in the cockroach *Periplaneta americana*. J. Biophys. Biochem. Cytol. 4: 731–742.

Heymons, R. 1895. *Die Embryonalentwicklung von Dermapteren und Orthopteren unter besonderer Berücksichtigung der Keimblätterbildung*. Fischer, Jena.

Heymons, R. 1897a. Entwicklungsgeschichte Untersuchungen an *Lepisma saccharina*. Z. Wiss. Zool. 62: 583–631.

Heymons, R. 1897b. Ueber die Organisation und Entwicklung von *Bacillus rossii*. Sitzungsber. Akad. Wiss. Berlin 2: 368–373.

Heymons, R. 1899. Ueber blaschenformige Organe bei den Gespenstheuschrecken. Sitzungsber. Akad. Wiss. Berlin 4: 363–373.

Highnam, K. C. 1958. Activity of corpora allata during pupal diapause in *Mimias tiliae* (Lepidoptera). Q. J. Microsc. Sci. 99(2): 171–180.

Highnam, K. C. 1962a. Neurosecretory control of ovarian development in *Schistocerca gregaria*. Q. J. Microsc. Sci. 103(61): 57–72.

Highnam, K. C. 1962b. Neurosecretory control of ovarian development in the desert locust. Mem. Soc. Endocrinol. 12: 379–390.

Highnam, K. C. 1964. Endocrine relationships in insect reproduction. Proc. 3rd Symp. R. Entomol. Soc. Lond. pp. 26–42.

Highnam, K. C., O. Lusis and L. Hill. 1963. The role of the corpora allata during oocyte growth in the desert locust, *Schistocerca gregaria* Forsk. J. Insect Physiol. 9: 587–596.

Hofer, B. 1887. Untersuchungen über den Bau der Speicheldrüsen und des dazu gbörenden Nervenapparats von *Blatta*. Nova Acta Leopold. Carol. Dtsch. Akad. 51: 345–395.

Hoffmann, H. J. 1970. Neuro-endocrine control of diapause and oöcyte maturation in the beetle, *Pterostichus nigrita*. *J. Insect Physiol*. 16: 629–642.

Holmgren, N. 1909. Termitenstudien, I. K. Sven. Vetenskapsakad. Handl. 44: 1–215.

Horn, D. H. S., R. H. Nearn, J. B. Siddall, G. B. Staal, and D. C. Cerf.

1983. Synthesis of the two optical isomers of the insect anti–juvenile hormone agent ethyl p-[2-(t-butylcarbonyloxy)butoxy]benzoate (ETB) and their biological activities. Aust. J. Chem. 36: 1409–1417.

Huignard, J. 1964. Recherches histophysiologigues sur le contrôle hormonal de l'ovogenèse chez *Gryllus domesticus*. C. R. Acad. Sci. Paris 259: 1551–1560.

Idelman, S. 1964. Mitochondries et liposomes: description d'une transformation mitochondriale observée dans la cortico-surrénale du Rat. J. Microsc. (Paris) 3: 437–446.

Idelman, S. 1970. Ultrastructure of the mammalian adrenal cortex. Int. Rev. Cytol. 27: 181–282.

Imboden, H., B. Lanzrein, J. P. Delbecque, and M. Lüscher. 1978. Ecdysteroids and juvenile hormone during embryogenesis in the ovoviviparous cockroach *Nauphoeta cinerea*. Gen. Comp. Endocrinol. 36: 628–635.

Injeyan, H. S. and S. S. Tobe. 1981. Phase polymorphism in *Schistocerca gregaria*: assessment of juvenile hormone synthesis in relation to vitellogenesis. J. Insect Physiol. 27: 203–210.

Ito, H. 1918. On the glandular nature of the corpora allata of the Lepidoptera. Bull. Imp. Univ. Tokyo, Seric. 1: 64–103.

Ito, S. and M. Karnovsky. 1968. Formaldehyde–glutaraldehyde fixatives containing trinitro compounds. J. Cell Biol. 39: 168a.

Jennings, R. C., K. J. Judy, and D. A. Schooley. 1975a. Biosynthesis of the homosesquiterpenoid juvenile hormone JH II methyl ($2E,6E,10Z$) -10,11-epoxy-3,7,11-trimethyltridecadienoate from 5-^3H-homomevalonate in *Manduca sexta*. J. Chem. Soc. Chem. Commun. pp. 21–22.

Jennings, R. C., K. J. Judy, A. Schooley, M. S. Hall, and J. B. Siddall. 1975b. The identification and biosynthesis of two juvenile hormones from the tobacco budworm moth (*Heliothis virescens*). Life Sci 16: 1033–1040.

Johansson, A. S. 1958a. Hormonal regulation of reproduction in the milkweed bug, *Oncopeltus fasciatus* (Dallas). Nature (Lond.) 181: 198–199.

Johansson, A. S. 1958b. Relation of nutrition to endocrine reproductive function in the milkweed bug, *Oncopeltus fasciatus*. Nytt. Mag. Zool. (Oslo) 7: 1–132.

Johnson, B. 1966. Ultrastructure of probable sites of release of neurosecretory materials in an insect, *Calliphora stygia* Fabr. (Diptera). Gen. Comp. Endocrinol. 6(1): 99–108.

Johnson, G. D., B. Stay, and S. M. Rankin. 1985. Ultrastructure of corpora allata of known activity during the vitellogenic cycle in the Cockroach *Diploptera punctata*. Cell Tissue Res. 239: 317–327.

Joly, L. 1960. Fonctions des corpora allata chez *Locusta migratoria* L. Thesis, University of Strasbourg.

Joly, L. 1964. Contrôle du fonctionnement ovarien chez *Locusta migratoria* L. I. Effets de castrations totales et de ligatures unilatérales de l'oviducte. J. Insect. Physiol. 10: 437–442.

Joly, L. and Joly, P. 1970b. Etude biométrique des corpora allata de *Locusta migratoria* ayant subi une destruction des cellules A,B de la pars. C. R. Acad. Sci. Paris 270: 2696–2698.

Joly, L., A. Porte, and A. Girardie. 1967. Caractères ultrastructuraux des corpora allata actifs et inactifs chez *Locusta migratoria*. C. R. Acad. Sci. Paris 265(21): 1633–1635.

Joly, L., P. Joly, A. Porte, and A. Girardie. 1968. Etude physiologique et ultrastructurale des corpora allata de *Locusta migratoria* L. (Orthoptère) en phase grégaire. *Arch. Zool. Exp. Gén.* 109: 703–727.

Joly, L., P. Joly, A. Porte, and A. Girardie. 1969. Analyse ultrastructurale de l'activité des corpora allata de *Locusta migratoria migratorioides* L. et ses conséquences sur la structure que l'on doit attribuer au mécanisme humoral contrôlant la métamorphose. Arch. Zool. Exp. Gén. 109(4): 617–628.

Joly, P. 1945. La fonction ovarienne et son contrôle humoral chez les Dysticides. Arch. Zool. Exp. Gen., 84: 49–164.

Joly, P. 1949. Le système endocrine rétrocérébral chez les Acridiens migrateurs. Ann. Sci. Nat. Zool. [2] 2: 255–262.

Joly, P. 1967. Comparaison du volume et de l'activité physiologique des corpora allata de *Locusta migratoria* L. Ann. Soc. Entomol. Fr. 3(3): 601–608.

Joly, P. 1968. Endocrinologie des Insectes: pp. 1–344. *In:* P. P. Grassé (ed.) "Les grands problèmes de la Biologie" Masson et Cie, Paris.

Joly, P. 1976. Les glandes endocrines des Insectes. Pp. 637–675 *in* P. P. Grassé (ed.), *Traité de zoologie*, Vol. 8. Masson, Paris.

Joly, P. and L. Joly. 1970a. Comparaison des tests cytologiques d'activité des corpora allata des Acridiens. C. R. Acad. Sci. Paris 270: 2560–2562.

Juberthie, C. and P. Cassagnau. 1971. L'évolution du système neurosécréteur chez les Insectes: l'importance des Collemboles et des autres Aptérygotes. Rev. Ecol. Biol. Sol. 8: 59–80.

Judy, K. J., D. A. Schooley, B. J. Hall, B. J. Bergot, and J. B. Siddall. 1973a. Chemical structure and absolute configuration of a juvenile hormone from grasshopper corpora allata *in vitro*. Life Sci. 13: 1511–1516.

Judy, K. J., D. A. Schooley, L. L. Dunham, M. S. Hall, B. J. Bergot, and J. B. Siddall. 1973b. Isolation, structure and absolute configuration of a new natural insect juvenile hormone from *Manduca sexta*. Proc. Natl. Acad. Sci. USA 70: 1509–1513.

Judy, K. J., D. A. Schooley, R. G. Troetschler, R. C. Jennings, B. J. Bergot, and M. S. Hall. 1975. Juvenile hormone production by corpora allata of *Tenebrio molitor* in vitro. Life Sci. 16: 1059–1066.

Kaiser, H. 1980. Licht- und electronenmikroskopische Untersuchung der Corpora allata der Eintagsfliege *Ephemera danica* Müll. (Ephemeroptera: Ephemeridae) während der Metamorphose. Int. J. Insect Morphol. Embryol. 9: 395–403.

Karnovsky, J. J. 1965. A formaldehyde–glutaraldehyde fixative of high osmolarity for use in electron microscopy. J. Cell Biol. 27: 137a.

Kasturi, T. R. and Manithomas, T. 1967. Essential oils of *Ageratum conyzoides:* isolation and structure of two new constituents. Tetrahedron Lett. 1167: 2573–2575.

Khan, M. A. and P. Buma. 1985. Neural control of the corpus allatum in the Colorado potato beetle, *Leptinotarsa decemlineata:* an electron microscopic study utilizing the *in vitro* tannic acid Ringer incubation method. J. Insect. Physiol. 31(8): 639–645.

Khan, M. A., H. M. Romberg-Privee, and H. Schooneveld. 1984. Innervation of the corpus allatum in the Colorado potato beetle as revealed by retrograde diffusion with horseradish peroxidase. Gen. Comp. Endocrinol. 55: 66–73.

Khan, M. A., A. Doderer, A. B. Koopmanschap, and C. A. D. de Kort. 1982. Improved assay conditions for measurement of corpus allatum activity *in vitro* in the adult Colorado potato beetle, *Leptinotarsa decemlineata*. J. Insect Physiol. 28(3): 279–284.

Kerkut, G. A. and L. I. Gilbert (eds.) 1984. *Comprehensive Insect Physiology, Biochemistry and Pharmacology*, Vols. 7 and 8: *Endocrinology*. Pergamon Press, Oxford and Elmsford, New York.

King, R. C., E. A. Koch, and G. A. Cassens. 1961. The effect of temperature upon the ovarian tumors of the *fes* mutant of *Drosophila melanogaster*. Growth 25: 45–65.

King, R. C., S. K. Aggarwal, and D. Bodenstein. 1966a. The comparative submicroscopic morphology of the ring gland of *Drosophila melanogaster* during the second and third larval instars. Z. Zellforschung. Mikrosk. Anat. 73: 272–285.

King, R. C., S. K. Aggarwal, and D. Bodenstein. 1966b. The comparative submicroscopic cytology of the corpus allatum–corpus cardiacum complex of wild type and *fes* adult female *Drosophila melanogaster*. J. Exp. Zool. 161(2): 151–175.

Klug, H. 1959. Histo-physiologische Untersuchungen über die Aktivitäts periodik bei Carabiden. Wiss. Z. Humboldt-Univ. Berlin Math.-Naturwiss. Reihe 8: 405–434.

Kobayashi, M. 1957. Studies on the neurosecretion in the silkworm, *Bombyx mori* L. Bull. Seric. Exp. Stn. (Tokyo) 15(3): 263–273.

Kramer, K. J. and J. H. Law. 1980. Control of juvenile hormone production: the relationship between precursor supply and hormone synthesis in the tobacco hornworm, *Manduca sexta*. Insect Biochem. 10: 569–575.

Kramer, K. J., R. D. Speirs, and C. N. Childs. 1977. Immunochemical evidence for a gastrin-like peptide in the insect neuroendocrine system. Gen. Comp. Endocrinol. 32: 423–426,

Kümmel, G. 1969. Zur Feinstruktur der corpora allata von *Chironomus*. Zool. Anz. (Suppl.) 32: 123–135.

Kuwano, E., R. Takeya, and M. Eto. 1983. Terpenoid imidazoles: new anti–juvenile hormones. *Agric. Biol. Chem.* 47: 921–923.

Kuwano, E., R. Takeya, and M. Eto. 1984. Synthesis and anti–juvenile hormone activity of 1-citronellyl-5-sustituted imidazoles. Agric. Biol. Chem. 48: 3115–3119.

Lafon-Cazal, M. 1976. Radioautographic detection of monoamines in a neuroendocrine gland (corpora cardiaca) of locusts (insects). J. Microsc. Biol. Cell 27(2/3): 257–260.

Lafon-Cazal, M. and M. Arluison. 1976. Localization of monoamines in the corpora cardiaca and the hypocerebral ganglion of locusts. Cell. Tissue Res. 172: 517–527.

Lanzrein, B. 1979. The activity and stability of injected juvenile hormones (JH I, JH II and JH III) in last-instar larvae and adult females of the cockroach *Nauphoeta cinerea*. Gen. Comp. Endocrinol. 39: 69–78.

Lanzrein, B., R. Wilhelm, and J. Buschor. 1981. On the regulation of the corpora allata activity in adult females of the ovoviviparous cockroach *Nauphoeta cinerea*. Pp. 147–160 *in* G. E. Pratt and G. T. Brooks (eds.), *Juvenile Hormone Biochemistry*. Elsevier, Amsterdam and New York.

Lanzrein, B., V. Gentinetta, R. Fehr, and M. Lüscher, 1978. Correlation between haemolymph juvenile hormone titre, corpus allatum volume *in vivo* and *in vitro* activity during oocyte maturation in a cockroach (*Nauphoeta cinerea*). Gen. Comp. Endocrinol. 36: 339–345.

Lanzrein, B., M. Hashimoto, V. Parmakovitch, K. Nakanishi, R. Wilhelm, and M. Lüscher. 1975. Identification and quantification of juvenile hormones from different developmental stages of the cockroach *Nauphoeta cinerea*. Life Sci. 16: 1271–1284.

Laufer, H., D. Borst, F. C. Baker, C. Carrasco, M. Sinkus, C. C. Reuter, L. W. Tsai, and D. A. Schooley. 1986. Identification of a juvenile hormone-L compound in a crustacean. Science (Wash., DC) 235: 202–205.

Lauga-Reyrel, F. 1984a. Etude anatomique et ultrastructurale des organes neurohémaux et des corps allates d'*Hypogastrura tullbergi* (Tullberg) (Collembola: Hypogastruridae). Int. J. Insect Morphol. Embryol. 13(5/6): 399–410.

Lauga-Reyrel, F. 1984b. Modifications ultrastructurales des corps allates au cours du développement et de l'écomorphose chez *Hypogastrura tullbergi* (Tullberg) (Collembola: Hypogastruridae). Int. J. Insect. Morphol. Embryol. 13(5/6): 411–424.

Lea, A. O. 1975. Evidence that the ovaries of *Musca domestica* do not maintain cyclicity by regulating the corpus allatum. J. Insect. Physiol. 21: 1747–1750.

Lea, A. O. and O. Thomsen. 1969. Size independent secretion by the corpus allatum of *Calliphora erythrocephala*. J. Insect Physiol. 15(3): 477–482.

Leclercq-Smekens, M. and J. Naisse. 1986. Relation entre l'aspect cytologique des corps allates du doryphore *Leptinotarsa decemlineata Say* 1880 et la production d'hormone juvénile. Biol. Cell. 57: 12a.

Legay, J. M. 1950. Note sur l'évolution des corpora allata au cours de la vie larvaire de *Bombyx mori*. C. R. Soc. Biol. Paris 144: 512–513.

Lensky, Y., J. C. Baehr, and P. Porcheron, 1978. Dosages radioimmunologiques des ecdysones et des hormones juvéniles au cours du développement post-embryonnaire chez les ouvrières et les reines d'Abeille (*Apis mellifera* L. var. *ligustica*). C. R. Acad. Sci. Paris 287: 821–824.

Lococo, D. J. and S. S. Tobe. 1984. Neuroanatomy of the retrocerebral complex, in particular the pars intercerebralis and partes laterales in the cockroach *Diploptera punctata* Eschscholtz (Dictyoptera: Blaberidae). Int. Insect Morphol. Embryol. 13(1): 65–76.

Lococo, D. J., C. S. Thompson, and S. S. Tobe. 1986. Intercellular communication in a insect endocrine gland. J. Exp. Biol. 121: 407–419.

Loher, N. 1966. Die Steuerung sexueller Verhaltensweisen und der Oocytenentwicklung bei *Gomphocerus rufus*. Z. Vgl. Physiol. 53(3): 277–316.

Loher, W. 1958. An olfactory response of immature adults on the desert locusts. Nature (Lond.) 181: 1280.

Loher, W. 1960. The chemical acceleration of the maturation process and its hormonal control in the male of the desert locust. Proc. R. Soc. Lond. B153: 380–397.

Loher, W. 1961. Die Beschleunigung der Reife durch eine Pheromon des Mannschens der Wustenheuschrecke und die Funktion der Corpora allata. Naturwissenschaft 48: 657–661.

Lukoschus, F. 1955a. Untersuchen zur Metamorphose der Honigbiene *Apis mellifera*. Insectes Soc. 2: 147–162.

Lukoschus, F. 1955b. Die Bedeutung des innersekretorischen Systems für die Ausbildung epidermaler Fastenmerkmale bei der Honigbiene. Insectes soc. 2: 221–236.

Lüscher, M. and B. Lanzrein. 1976. Differential effects of the three juvenile hormones (JH I-II-III) in the cockroach *Nauphoeta cinerea*. Pp. 435–440 *in* Colloq. Int. Cent. Nat. Rech. Sci. No. 251: *Actualité sur les hormones d'invertébrés*. Paris.

Maltête, F. 1962. Contribution à l'étude chronologique de l'embryogenèse chez *Locusta migratoria migratorioides* R. et F.: développement des corps allates et des glandes ventrales de la tête. Thesis of the 3rd cycle, Biologie animale. University of Bordeaux.

Maruthi Ram, G., B. Kishen Rao, and S. S. Thakur. 1982. The volume of corpora allata in relation to the egg maturation of the sweet potato weevil, *Cylas formicarius* F. (Coléoptera: Curculionidae). Curr. Sci. 51(15): 757–758.

Masner, P., S. Dorn, Vogel, M. Kälin, O. Graf, and E. Günthart. 1981. Types of response of insects to a new IGR and to proven standards. Pp. 809–818 *in* F. Sehnal, A. Zabza, J. I. Menn and B. Cymborowski (eds.), *Regulation of Insect Development and Behavior* (Proc. Int. Conf., Karpacz, Poland, June 23–28). Technical University of Wrocław Press, Wrocław, Poland.

Mason, G. 1973. New features of the brain retrocerebral neuroendocrine complex of the locust *Schistocerca vaga* (Scudder). Z. Zellforsch. Mikrosk. Anat. 141: 19–32.

Meinert, [F] 1861. Bidrag til de danske Myrus Naturhistoric. K. Dan. Vidensk. Selsk. Skr. 5: 614–618.

Melnikova, E. J. and A. A. Panov. 1975. Ultrastructure of the larval corpus allatum of *Hypantria cunea* Drury (Insecta, Lepidoptera). Cell Tissue Res. 162: 395–410.

Mendes, M. V. 1948. Histology of the corpora allata of *Melanoplus differentialis* (Orthoptera: Saltoria). Biol. Bull. (Woods Hole) 94: 194–207.

Menn, J. J. and M. Beroza. 1972. *Insect Juvenile Hormones*. Academic Press, Orlando, Florida.

Metzler, M., K. H. Dahm, R. D. Meyer, and H. Roller. 1971. On the biosynthesis of juvenile hormone in the adult *Cecropia* moth. Z. Naturforsch. 26B: 1270–1276.

Metzler, M., D. Meyer, K. H. Dahm, and H. Röller. 1972. Biosynthesis of juvenile hormone from 10-epoxy-7-ethyl-3,11-dimethyl-2,6-tridecadienoic acid in the adult *Cecropia* moth. Z. Naturforsch. 27: 321–322.

Meyer, A. S., E. Hanzmann, and H. A. Schneiderman. 1970. The isolation and identification of the two juvenile hormones from the *Cecropia* silk moth. Arch. Biochem. Biophys. 137: 190–213.

Meyer, A. S., H. A. Schneiderman, E. Hanzmann, and J. A. Ko. 1968. The two juvenile hormones from the *Cecropia* silk moth. Proc. Natl. Acad. Sci. USA 60: 853–860.

Meyer, G. F. and O. Pfugfelder. 1958. Elektronmikroskopische Untersuchungen und der Corpora cardiaca vor *Carausius morosus*. Z. Zellforsch. 48: 556–564.

Monger, D. J., W. A. Lim, F. J. Kezdy, and J. H. Law. 1982. Compactin inhibits insect HMG–CoA reductase and juvenile hormone biosynthesis. Biochem. Biophys. Res. Commun. 105: 1374–1380.

Mordue, W. 1967. The influence of feeding upon the activity of the neuroendocrine system during oocyte growth in *Tenebrio molitor*. Gen. Comp. Endocrinol. 9(3): 406–415.

Morohoshi, S. and J. Shimada. 1975. The control of growth and development in *Bombyx mori*. XXVII. Release of the brain hormone from the corpora allata through nerve axons from brain neurosecretory cells. Proc. Jpn. Acad. 51(10): 743–747.

Morohoshi, S., T. Mori, and S. Sato. 1976a. The control of growth and development in *Bombyx mori*. XXXII. Intercellular microtubules occuring in the corpus allatum. *Proc. Jpn. Acad.* 52(5): 240–243.

Morohoshi, S., T. Mori, and S. Sato. 1976b. The control of growth and development in *Bombyx mori*. 2. Electron-microscopic studies on the activity of corpus allatum in the fifth instar larvae developed from low and high temperature eggs. Proc. Jpn. Acad. 52(8): 196.

Morohoshi, S., T. Mori, H. Akai, T. Oshiki, S. Sato, and J. Shimada. 1975. The control of growth and development in *Bombyx mori*. XXVIII. Light and electron-microscopic studies on elementary neurosecretory granules occuring in nerve axons between the brain and the corpus allatum or the suboesophageal ganglion. *Proc. Jpn. Acad.* 51(10): 747–751.

Müller, H. P. 1965a. Zur Frage der Steuerung des Paarungsverhaltens und der Eireifung beider Feldheuschrecke *Euthystira brachyptera* Ocsk. unter besonderer Berücksichtigung der Rolle der Corpora allata. Z. Vgl. Physiol. 50(5): 447–497.

Müller, H. P. 1965b. Der Einfluss der corpora allata auf paarungsverhalten und an eifnug bei *Euthystira brachyptera* Ocsk. (Acrididae) *Naturwissenschaften* 52(4): 93–94.

Müller, J. 1829. Ueber ein ergentümliches, dem *Nervus sympathicus* analoges Nervensystem der Eingeweide bei den Insekten. Nova Acta Acad. Leopold-Carol. Berlin 14: 227–239.

Müller, P. J., P. Masner, K. H. Trautmann, M. Suchy, and H. K. Wipf. 1974. The isolation and identification of juvenile hormone from cockroach corpora allata *in vitro*. Life Sci. 15: 915–921.

Nabert, A. 1913. Die Corpora allata der Insekten. Z. Wiss. Zool., 104: 181–358.

Nayar, K. K. 1956a. The endocrine organs of the adult wheat-blossom midge, *Sitodiplosis mosellana* Gehin (Cecidomyiidae: Diptera). Proc. Zool. Soc. (Calcutta) 9: 13–18.

Nayar, K. K. 1956b. The structure of the corpus allatum of *Iphita limbata* Hemiptera). Quart. J. Microsc. Sci. 97: 83–93.

Nayar, K. K. 1958. Studies on the neurosecretory system of *Iphita limbata*

Stäl. V. Probable endocrine basis of oviposition in the female insect.
   Proc. Indian Acad. Sci. 67: 233–249.

Nelson, J. A. 1915. *The Embryology of the Honeybee.* Princeton University
   Press, Princeton, New Jersey.

Nickerson, B. 1954. A possible endocrine mechanism controlling locust
   pigmentation. Nature (Lond.) 174: 357–358.

Nijhout, H. F. 1975. Axonal pathways in the brain–retrocerebral neuroendocrine complex of *Manduca sexta* (L.) (Lepidoptera: Sphingidae). Int. J. Insect. Morphol. Embryol. 4: 529–538.

Nishiitsutsuji-Uwo, J. 1960. Fine structure of the neurosecretory system
   in Lepidoptera. Nature (Lond.) 188: 953–954.

Nishiitsutsuji-Uwo, J. 1961. Electron microscopic studies on the neurosecretory system in Lepidoptera. Z. Zellforsch. Mikrosk. Anat.
   54: 613–630.

Normann, T. C. 1965. The neurosecretory system of the adult *Calliphora
   erythrocephala.* I. The fine structure of the corpus cardiacum with
   some observations on adjacent organs. Z. Zellforsch. Mikrosk.
   Anat. 67(4): 461–501.

Norris, M. J. 1952. Reproduction in the desert locust *Schistocerca gregaria*
   Forsk in relation to density and phase. Anti-Locust. Bull. 13: 1–49.

Norris, M. J. 1954. Sexual maturation in the desert locust (*Schistocerca gregaria* Forsk.) with special reference to the effects of grouping. *Anti-
   Locust Bull.* 18(2): 1–44.

Novák, V. J. A. 1951. The metamorphosis hormones and morphogenesis
   in *Oncopeltus fasciatus.* Vestn. Cesk. Spol. Zool. 15: 1–48.

Novák, V. J. A. 1954. Growth of the corpora allata during the post-
   embryonal development in insects. Vestn. Cesk. Spol. Zool. 18: 98–
   133.

Novák, V. J. A. 1966. *Insect Hormones: The Physiology, Morphology and Phylogeny of the Insect Endocrines,* 2nd ed. Methuen, London.

Odhiambo, T. R. 1966a. Morphometric changes and the hormonal activity
   of the corpus allatum in the adult male of the desert locust. J. Insect.
   Physiol. 12: 655–664.

Odhiambo, T. R. 1966b. The fine structure of the corpus allatum of the
   sexually mature male of the desert locust. *J. Insect Physiol.* 12(7): 819–
   828.

Odhiambo, T. R. 1966c. Ultrastructure of the development of the corpus
   allatum in the adult male of the desert locust. J. Insect Physiol. 12(8):
   995–1002.

Odhiambo, T. R. 1966d. Growth and the hormonal control of sexual maturation in the male desert locust, *Schistocerca gregaria* (Forskål).
   Trans. R. Entomol. Soc. Lond. 118(13): 393–412.

Overton, J. 1969. A fibrillar intercellular material between reaggregating
   embryonic chick cells. J. Cell Biol. 40(1): 136–143.

Özbas, S. 1957. Morphological and Histological studies on the corpora al-
    lata and cardiaca in orthoptera. Commun. Fac. Sci. Univ. Ankara,
    Ser. C 8: 19–44.
Palevody, C. 1976. L'ovogenèse chez les Collemboles isotomides:
    cytologie et approche physiologique. Doctoral thesis, University of
    Toulouse.
Palevody, C. and A. Grimal. 1975. Variations cytologiques des corps al-
    lates au cours du cycle reproducteur du Collembole *Folsomia candida*.
    J. Insect Physiol. 22: 63–72.
Panov, A. A. and O. K. Bassurmanova. 1970. Fine structure of the gland
    cells in inactive and active corpus allatum of the bug *Eurygaster integ-
    riceps*. J. Insect. Physiol. 16: 1265–1281.
Papillon, M., S. Lauverjat, and A. M. Cantacuzène. 1972. Influence de la
    température d'élevage sur la fertilité des pontes de *Schistocerca gre-
    garia*. J. Insect Physiol. 18: 2005–2018.
Papillon, M., P. Cassier, J. Girardie, and M. Lafon. 1976. Influence de la
    température d'élevage sur l'activité endocrine de *Schistocerca gre-
    garia*. Arch. Biol. (Bruxelles) 87(1): 103–127.
Pasteels, J. M., and J. Deligne. 1965. Etude du système endocrine au cours
    du vieillissement chez les "reines" de *Microcerotermes parvus*
    (Haviland) et *Cubitermes heghi* (Sjöstedt) (Isoptères: Termitidae).
    Biol. Gabonica 1(4): 325–336.
Paterson, N. F. 1936. Observations on the embryology of *Corynodes pusis*
    (Coleoptera, Chrysomelidae). Q. J. Microsc. Sci. 78: 91–131.
Pener, M. P. 1965. On the influence of corpora allata on maturation and
    sexual behaviour of *Schistocerca gregaria*. J. Zool. (Lond.) 147(2): 119–
    136.
Pener, M. P. 1967. Effects of allatectomy and sectioning of the nerves of
    the corpora allata on oöcyte growth, male sexual behaviour, and
    colour changes in adults of *Schistocerca gregaria*. J. Insect Physiol.
    13(5): 665–684.
Pener, M. P., L. Orshan, and J. de Wilde. 1978. Precocene II causes atro-
    phy of corpora allata in *Locusta migratoria*. Nature (Lond.) 272(5651):
    350–353.
Peter, M. G. and K. H. Dahm. 1975. Biosynthesis of juvenile hormone in
    the *Cecropia* moth: labelling pattern from 1-propionate through deg-
    radation to single carbon atom derivatives. Helv. Chim. Acta 58(4):
    1038–1048.
Peter, M. G., P. D. Shirk, K. H. Dahm, and H. Röller. 1981. On the speci-
    ficity of juvenile hormone biosynthesis in the male *Cecropia* moth. Z.
    Naturforsch. 36C: 579–585.
Pfeiffer, I. W. 1940. Further studies on the function of the corpora allata in
    relation to the ovaries and oviducts of *Melanoplus differentialis*. Anat.
    Rec. 78(Suppl.): 39–40.

Pfeiffer, I. W. 1945. Effect of the corpora allata on the metabolism of adult female grasshoppers. J. Exp. Zool. 99: 183–233.

Pflugfelder, O. 1936. Vergleichend-anatomische, experimentelle und embryologische Untersuchen über das Nervensystem und die Sinnesorgane der Rhynchoten. Zoologica 93: 1–102.

Pflugfelder, O. 1937a. Bau, Entwicklung und funktion der Corpora allata und cardiaca von *Dixippus morosus* Br. Z. Wiss. Zool. 149: 477–512.

Pflugfelder, O. 1937b. Untersuchungen über die Funktion der Corpora allata und cardiaca der Insekten. Zool. Anz. (Suppl.) 10: 121–129.

Pflugfelder, O. 1938. Untersuchungen über die histologischen Veranderungen und das Kernwacstum der Corpora allata von Termiten. Z. Wiss. Zool. 150: 451–467.

Pflugfelder, O. 1948. Volumetrische Untersuchungen an den Corpora allata der Honigbiene *Apis mellifera*. Biol. Zentralbl. 67: 223–241.

Pflugfelder, O. 1958. *Entwicklungs Physiologie der Insekten*. Akademische Verlagsgesellschaft Geest & Portig KG, Leipzig.

Picheral, B. 1968. Les tissus élaborateurs d'hormones steroïdes chez les Amphibiens Urodèles. I. Ultrastructure des cellules du tissu glandulaire du testicule de *Pleurodeles waltli* Michah. J. Microsc. (Paris) 7(1): 115–134.

Piepho, H. 1938a. Wachstum und totale Metamorphose an Hautimplantaten bei der Wachsmotte *Galleria mellonella*. Biol. Zentralb. 58: 356–366.

Piepho, H. 1938b. Ueber die Auslösung der Raupenhätung, Verpuppung und Imaginalentwicklung an Hautimplantaten von Schmetterlingen. Biol. Zentralbl. 58: 481–495.

Piepho, H. 1938c. Nicht-artspezifische Metamorphosehormone bei Schmetterlingen. Naturwissenschaften 26: 383.

Piepho, H. 1938d. Hemmung der Verpuppung durch corpora allata von Junggraupen bei der Wachsmotte *Galleria mellonella*. Naturwissenschaften 27: 675–676.

Pipa, R. L. 1978. Locations and central projections of neurons associated with the retrocerebral neuroendocrine complex of the cockroach *Periplaneta americana* (L.). Cell Tissue Res. 193: 443–455.

Pipa, R. L. 1983. Morphological considerations in the integration of nervous and endocrine systems. Pp. 39–53 *in* R. G. H. Downer and H. Laufer (eds.), *Endocrinology of Insects*. Liss, New York.

Pipa, R. L. and F. J. Novak. 1979. Pathways and fine structure of neurons forming the nervi corporis allati II of the cockroach *Periplaneta americana* (L.). Cell Tissue Res. 201: 227–237.

Polivanova, E. N. and E. V. Bocharova. 1974. Corpora allata in late embryogenesis of *Eurygaster integriceps* Put. (Insecta: Hemiptera). Gen. Comp. Endocrinol. 22: 409–413.

Polivanova, E. N. and E. V. Bocharaova. 1982. The development of the function of corpora allata of *Eurygaster integriceps* embryos under normal conditions and under the effect of juvenile hormone analogue. Zh. Obshch. Biol. 43: 88–95.

Poras, M. 1983. L'innervation protocérébrale des corps allates chez *Locusta migratoria* (Orthoptères). C. R. Acad. Sci. Paris 297: 483–488.

Poras, M., J. Cl. Baehr, and P. Cassier. 1983. Control of corpus allatum activity during the imaginal diapause in females of *Locusta migratoria* L. Int. J. Invertebr. Reprod. 6: 111–122.

Poulson, D. F. 1945. On the origin and nature of the ring gland (Weismann's ring) of the higher Diptera. Trans. Conn. Acad. Arts Sci. 36: 449–487.

Pratt, G. E. and W. S. Bowers. 1977. Precocene II inhibits juvenile hormone biosynthesis by cockroach corpora allata *in vitro*. Nature (Lond.) 265: 548–550.

Pratt, G. E. and K. G. Davey. 1972a. The corpus allatum and oogenesis in *Rhodnius prolixus* Stål. I. The effects of allatectomy. J. Exp. Biol. 56: 201–214.

Pratt, G. E. and K. G. Davey. 1972b. The corpus allatum and oogenesis in *Rhodnius prolixus* Stål. II. The effects of starvation. J. Exp. Biol. 56: 215–221.

Pratt, G. E. and K. G. Davey. 1972c. The corpus allatum and oogenesis in *Rhodnius prolixus* Stål. III. The effect of mating. *J. Exp. Biol.* 56: 223–237.

Pratt, G. E. and S. S. Tobe. 1974a. *In vitro* double-label biosynthesis of juvenile hormones by *Locusta* corpora allata. Gen. Comp. Endocrinol. 22: 402–403.

Pratt, G. E. and S. S. Tobe. 1974b. Juvenile hormones radiobiosynthesised by corpora allata of adult female locusts *in vitro*. Life Sci. 14: 575–586.

Pratt, G. E., S. S. Tobe, and R. J. Weaver. 1974c. Relative oxygenase activities in juvenile hormone biosynthesis of corpora allata of an African locust (*Schistocerca gregaria*) and American cockroach (*Periplaneta americana*). Experientia (Basel) 31(1): 120–122.

Pratt, G. E., R. C. Jennings, A. F. Hamnett, and G. T. Brooks. 1980. Lethal metabolism of precocene. I. To a reactive epoxide by locust corpora allata. Nature (Lond.) 284: 320–323.

Pratt, G. E., S. S. Tobe, R. J. Weaver, and J. R. Finney. 1975. Spontaneous synthesis and release of C_{16} juvenile hormone by isolated corpora allata of female locust *Schistocerca gregaria* and female cockroach *Periplaneta americana*. Gen. Comp. Endocrinol. 26: 478–484.

Quistad, G. B., L. E. Staiger, and D. C. Cerf. 1982. Preparation and biological activity of potential inhibitors of insect juvenile hormone biosynthesis. J. Agric. Food Chem. 30: 1151–1154.

Quistad, G. B., D. C. Cerf, D. A. Schooley, and G. B. Staal. 1981. Fluoromevalonate acts as an inhibitor of insect juvenile hormone biosynthesis. Nature (Lond.) 289: 176–177.

Quistad, G. B., D. C. Cerf, S. J. Kramer, B. J. Bergot, and D. A. Schooley. 1985. Design of novel insect anti–juvenile hormones: allytic alcohol derivatives. J. Agric. Food Chem. 33: 47–50.

Quo Fu. 1965. Etudes sur la reproduction du Criquet migrateur oriental: le rôle des corpora allata. (In Chinese.) *Acta Entomol. Sinica* 14(3): 211–224.

Raabe, M. 1982. Insect Neurohormones. Plenum, New York.

Raina, A. K. and T. K. Borg. 1980. Corpora cardiaca–allata complex of the larvae of the pink bollworm *Pectinophora gossypiella:* an ultrastructural study in relation to diapause. Acta Zool. 61: 65–77.

Rehm, M. 1951. Die zeitliche Folge der Tätigkeitsrhythmen inkretorischer Organe von *Ephestia Kühniella* während der Metamorphose und des Imaginallebens. Wilhelm Roux' Arch. Entwicklungsmech. Org. 145: 205–248.

Riddiford, L. 1980. Interaction of ecydsteroids and juvenile hormone in the regulation of larval growth and metamorphosis of the tobacco hornworm. Pp. 409–439 *in* J. A. Hoffmann (ed.), *Progress in Ecdysone Research*. Elsevier, Amsterdam and New York.

Röller, H. and K. H. Dahm. 1968. The chemistry and biology of juvenile hormone. Recent Prog. Horm. Res. 24: 651–680.

Röller, H. and K. H. Dahm. 1970. The identity of juvenile hormone produced by corpora allata *in vitro*. Naturwissenschaft. 57: 454–455.

Röller, H., K. H. Dahm, C. C. Sweeley, and B. M. Trost. 1967. The structure of juvenile hormone. Angew. Chem. 6: 179–180.

Röseler, P. F. 1977. Juvenile hormone control of oogenesis in bumblebee workers, *Bombus terrestris*. J. Insect Physiol. 23: 985–992.

Röseler, P. F. and I. Röseler. 1978. Studies on the regulation of the juvenile hormone titre in bumblebee workers, *Bombus terrestris*. J. Insect Physiol. 24: 707–713.

Röseler, P. F., I. Röseler, and A. Strambi. 1980. The activity of corpora allata in dominant and subordinated female of the wasp *Polistes gallicus*. Insectes Soc. 27: 97–107.

Rohdendorf, E. 1965. Effects of allatectomy in adult female of *Thermobia domestica* Packard (Lepismatidae: Thysanura). Proc. Symp. Comp. Endocrinol. Invertebra., Jena.

Rohdendorf, E. B. and J. A. L. Watson. 1969. The control of the reproductive cycles in the female firebrat, *Lepismodes inquilinus*. J. Insect Physiol. 15: 2085–2101.

Roonwal, M. L. 1936. Studies on the embryology of the African migratory locust (*Locusta migratoria migratorioïdes* R. et F.). I. The early development with a new theory of multiphased gastrulation among insects. Philos. Trans. R. Soc. Lond. B 226: 391–421.

Ruegg, R. P., S. S. Tobe, and W. Loher. 1986. Juvenile hormone biosynthesis during egg development in the cricket, *Teleogryllus commodus*. J. Insect Physiol. 32(6): 517–521.

Rutz, W., L. Gerig, H. Wille, and M. Lüscher. 1976. The function of juvenile hormone in adult worker honeybees *Apis mellifera*. J. Insect Physiol. 22: 1485–1491.

Saini, R. S. 1966. Neuroendocrine control of oöcyte development in the beetle *Aulacophora foveicollis* Lúc. J. Insect Physiol. 12(8): 1003–1008.

Schaller, F. 1968. Action de la température sur la diapause et le développement de l'embryon d'*Aeschna mixta* (Odonata). J. Insect Physiol. 14(10): 1477–1483.

Scharrer, B. 1961. Functional analysis of the corpus allatum of the insect *Leucophaea maderae* with the electron microscope. Biol. Bull. (Woods Hole) 121(2): 370.

Scharrer, B. 1962a. The fine structure of the neurosecretory system of the insect *Leucophaea maderae*. Pp. 89–97 in *Neurosecretion: Proc. 3rd Int. Symp. on Neurosecretion, Bristol, 1961*.

Scharrer, B. 1962b. The neurosecretory system of *Leucophaea maderae* and its role in neuroendocrine integration. Gen. Comp. Endocrinol. 2(1): 30.

Scharrer, B. 1963. Neurosecretion. XIII. The ultrastructure of the corpus cardiacum of the insect, *Leucophaea maderae* Z. Zellforsch. Mikrosk. Anat. 60: 761–796.

Scharrer, B. 1964a. Histophysiological studies on the corpus allatum of *Leucophaea maderae*. IV. Ultrastructure during normal activity cycle. Z. Zellforsch. Mikrosk. Anat. 62(2): 125–148.

Scharrer, B. 1964b. The fine structure of the blattarian prothoracic glands. Z. Zellforsch. Mikrosk. Anat. 64: 301–326.

Scharrer, B. 1968. Neurosecretion. XIV. Ultrastructural study of sites of release of neurosecretory material in blattarian insects. Z. Zellforsch. Mikrosk. Anat. 89(1): 1–16.

Scharrer, B. 1971. Histophysiological studies on the corpus allatum of *Leucophaea maderae*. V. Ultrastructure of sites of origin and release of a distinctive cellular product. Z. Zellforsch. Mikrosk. Anat. 120: 1–16.

Scharrer, B. and M. von Harnack, 1958. Histophysiological studies on the corpus allatum of *Leucophaea maderae*. I. Normal life cycle in male and female adults. Biol. Bull. (Woods Hole) 115(3): 508–520.

Scharrer, B. and B. von Harnack. 1961. Histophysiological studies on the corpus allatum of *Leucophaea maderae*. III. The effect of castration. Biol. Bull. (Woods Hole) 121(1): 193–208.

Schmialek, P. 1963. Compounds with juvenile-hormone action. Z. Naturforsch. 18B: 516–519.

Schooley, D. A., K. J. Judy, B. J. Bergot, M. S. Hall, and J. B. Siddal. 1973.
Biosynthesis of the juvenile hormones of *Manduca sexta:* labelling
patterns from mevalonate, propionate and acetate. Proc. Natl.
Acad. Sci. USA 70: 2921–2925.

Schooneveld, H. 1970. Structural aspects of neurosecretory and corpus
allatum activity in the adult Colorado beetle, *Leptinotarsa decemlineata* (Say), as a function of day-length. Neth. J. Zool. 20: 151–
237.

Schooneveld, H., S. J. Kramer, H. Privee, and A. van Huis. 1979. Evidence of controlled corpus allatum activity in the adult Colorado potato beetle. J. Insect Physiol. 25: 449–453.

Schrader, K. 1938. Untersuchungen über die Normalentwicklung das
Gehirns und Gehintransplantationen bei Mehlmotte *Ephestia
kühniella,* nebst einigen Bemerkungen über das Corpus allatum.
Biol. Zentralb. 58: 52–90.

Schultz, R. L. 1960. Electron-microscopic observations of the corpora allata and associated nerves in the moth *Celerio lineata.* J. Ultrastruct.
Res. 3: 320–327.

Sedlak, B. J. 1981. An ultrastructural study of neurosecretory fibers within the corpora allata of *Manduca sexta.* Gen. Comp. Endocrinol. 44:
207–218.

Sedlak, B. J. 1982. The ultrastructure of interacting endocrine and target
cells. Pp. 99–103 *in* H. Akai (ed.), *The Ultrastructure and Functioning of
Insect Cells.* Society for Insect Cells, Japan. (Plenum Press, New
York.)

Sedlak, B. J., L. Marchione, B. Devorkin, and R. Davino. 1983. Correlations between endocrine gland ultrastructure and hormone titers in
the fifth larval instar of *Manduca sexta.* Gen. Comp. Endocrinol. 52:
291–310.

Shirk, P. D., G. Bhaskaran, and H. Röller. 1983. Developmental physiology of corpora allata and accessory sex glands in the *Cecropia*
silkmoth. J. Exp. Zool. 227: 69–79.

Shirk, P. D., K. H. Dahm, and H. Röller. 1976. The accessory sex glands as
the repository for juvenile hormone in male *Cecropia* moths. Z.
Naturforsch. 31C: 199–200.

Siew, Y. C. 1965. The endocrine control of adult reproductive diapause in
the Chrysomelid beetle *Galeruca tanaceti* L. *J. Insect. Physiol.* 11: 463–
479.

Slama, K., M. Romanuk, and F. Sorm. 1974. *Insect Hormones and Bio-
analogues.* Springer-Verlag, Berlin and New York.

Smereczynski, S. 1932. Embryologische Untersuchungen über die

Zusammensetzung des Kopfes von *Silpha obscura* L. (Coleopt.) Zool. Jahrb. (Jena) Anat. 55: 233.

Smith, D. S. 1968. *Insect Cells: Their Structure and Functions.* Olivier & Boyd, Edinburgh.

Sparagana, S. P., G. Bhaskaran, K. H. Dahm, and V. Riddle. 1984. Juvenile hormone production, juvenile hormone esterase, and juvenile hormone acid methyltransferase in corpora allata of *Manduca sexta.* J. Exp. Zool. 230: 309–313.

Staal, G. B. 1961. Studies on the physiology of phase induction in *Locusta migratoria migratorioïdes* (R. and F.) Weenmann and Zonen. Ph.D. thesis, University of Wageningen.

Staal, G. B. 1975. Insect growth regulators with juvenile hormone activity. Annu. Rev. Entomol. 20: 417–460.

Staal, G. B. 1977a. Insect control with insect growth regulators based on insect hormones. Pontif. Acad. Sci. Scr. Varia 41: 353–383.

Staal, G. B. 1977b. Essential differences between natural juvenile hormones and juvenile hormone analogs elucidated by use of a substitution assay. Pontif. Acad. Sci. Scr. Varia 41: 333–352.

Staal, G. B. 1986. Anti–juvenile hormone agents. Annu. Rev. Entomol. 31: 391–421.

Stay, B. and S. S. Tobe. 1978. Control of juvenile hormone biosynthesis during the reproductive cycle of a viviparous cockroach. II. Effects of unilateral allatectomy, implantation of supernumerary corpora allata and ovariectomy. Gen. Comp. Endocrinol. 34: 276–286.

Stay, B. and S. S. Tobe. 1981. Control of the corpora allata during a reproductive cycle in a viviparous cockroach. Am. Zool. 21: 663–674.

Steel, C. G. H. 1975. A neuroendocrine feedback mechanism in the insect moulting cycle. Nature (Lond.) 253: 267–269.

Stefano, G. B. and B. Scharrer. 1981. High-affinity binding of an enkephalin analog in the cerebral ganglion of the insect *Leucophaea maderae* (Blattaria). Brain Res. 225: 107–114.

Strambi, C. 1981. Some data obtained by radio-immunoassay of juvenile hormone. Pp. 59–63 *In* G. E. Pratt and G. T. Brooks (eds.), Juvenile Hormone Biochemistry. Elsevier, Amsterdam and New York.

Strangeways-Dixon, J. 1961. The relationships between nutrition, hormones and reproduction in the blowfly *Calliphora erythrocephala* (Meig). II. The effect of removing the ovaries the corpus allatum median neurosecretory cells on selective feeding and the demonstration of the corpus allatum cycle. J. Exp. Biol. 38: 637–646.

Strangeways-Dixon, J. 1962. The relationships between nutrition, hormones and reproduction in the blowfly *Calliphora erythrocephalla* Meig. III. The corpus allatum in relation to nutrition, the ovaries innervation and the corpus cardiacum. J. Exp. Biol. 39: 293–306.

Strindberg, H. 1913. Embryologische Studien and Insekten. Z. Wiss. Zool. 106: 1–227.

Strong, L. 1965a. The relationship between the brain, corpora allata and oocyte growth in the Central American locust, *Schistocerca* sp. I. The cerebral neurosecretory system, the corpora allata and oocyte growth. J. Insect Physiol. 11: 135–146.

Strong, L. 1965b. The relationship between the brain, corpora allata and oocyte growth in the Central American locust, *Schistocerca* sp. II. The innervation of the corpora allata, the lateral neurosecretory complex and oocyte growth. J. Insect. Physiol. 11: 271–280.

Strong, L. and F. P. Amerasinghe. 1977. Allatectomy and sexual receptivity in females of *Schistocerca gregaria*. J. Insect Physiol. 23: 131–135.

Szibbo, C. M. and S. S. Tobe. 1981a. The mechanism of compensation in juvenile hormone synthesis following unilateral allatectomy in *Diploptera punctata*. J. Insect Physiol. 27: 609–613.

Szibbo, C. M. and S. S. Tobe. 1981b. Cellular and volumetric changes in relation to the activity cycle in the corpora allata of *Diploptera punctata*. J. Insect Physiol. 27: 655–665.

Szibbo, C. M., D. Rotin, R. Feyereisen, and S. S. Tobe. 1982. Synthesis and degradation of C_{16} juvenile hormone (JH III) during the final two stadia of the cockroach, *Diploptera punctata*. Gen. Comp. Endocrinol. 48: 25–32.

Takeda, N. 1977. Histophysiological studies on the corpus allatum during prepupal diapause in *Monema flavescens* (Lepidoptera). J. Morphol. 153: 245–262.

Thompson, C. S., D. J. Lococo, and S. S. Tobe. 1987. Anatomy and electrophysiology of neurons terminating in the corpora allata of the cockroach, *Diploptera punctata*. J. Comp. Neurol. 261: 120–129.

Thomsen, E. 1940. Relation between corpora allata and ovaries in adult flies (Muscidae). Nature (Lond.) 145: 28.

Thomsen, E. 1947. The gonadotropic hormones in the Diptera. Bull. Biol. Fr. Belg. (Suppl.) 33: 68–80.

Thomsen, E. and M. Thomsen. 1969. Fine structure of the corpus allatum of the female *Calliphora* with special regard to hormone formation. Gen. Comp. Endocrinol. 13(3): 141–159.

Thomsen, E. and M. Thomsen. 1970. Fine structure of the corpus allatum of the female blow fly, *Calliphora erythrocephala*. Z. Zellforsch. Mikrosk. Anat. 110: 40–60.

Tobe, S. S. 1977. Asymmetry in hormone biosynthesis by insect endocrine glands. Can. J. Zool. 55: 1059–1514.

Tobe, S. S. and G. E. Pratt. 1974. The influence of substrate concentrations of the rate of insect juvenile hormone biosynthesis by corpora allata of the desert locust *in vitro*. Biochem J. 144: 107–113.

Tobe, S. S. and G. E. Pratt. 1974. Dependence of juvenile hormone release from corpus allatum on intraglandular content. Nature (Lond). 252: 474–476.

Tobe, S. S. and G. E. Pratt. 1975a. The synthetic activity and glandular volume of the corpus allatum during ovarian maturation in the desert locust, *Schistocera gregaria*. Life Sci. 17: 417–422.

Tobe, S. S. and G. E. Pratt. 1975b. Corpus allatum activity *in vitro* during ovarian maturation in the desert locust, *Schistocerca gregaria*. J. Exp. Biol. 62: 611–627.

Tobe, S. S. and G. E. Pratt. 1976. Farnesenic acid stimulation of juvenile hormone biosynthesis as an experimental probe in corpus allatum physiology. Pp. 147–163 *in* L. I. Gilbert (ed.), *The Juvenile Hormone*. Plenum Press, New York.

Tobe, S. S. and A. S. M. Saleuddin. 1977. Ultrastructural localization of juvenile hormone biosynthesis by insect corpora allata. Cell Tissue Res. 183: 25–32.

Tobe, S. S. and B. Stay. 1985. Structure and regulation of the corpus allatum. Adv. Insect Physiol. 18: 305–432.

Tobe, S. S., G. E. Pratt, and A. S. M. Saleuddin. 1976. Intracellular visualisation of C_{16} juvenile hormone biosynthesis in corpora allata of adult female desert locust, *Schistocerca gregaria*. Pp. 441–449 *in* Colloq. Int. Cent. Natl. Rech. Sci. No. 251: *Actualités sur les hormones d'Invertébrés*. Lille.

Tobe, S. S., N. Clarke, B. Stay, and R. P. Ruegg. 1984. Changes in cell number and activity of the corpora allata of the cockroach *Diploptera punctata*: a role for mating and the ovary. Can. J. Zool. 62: 2178–2182.

Tobe, S. S., R. P. Ruegg, B. A. Stay, F. C. Baker, C. A. Miller, and D. A. Schooley. 1985. Juvenile hormone titre and regulation in the cockroach *Diploptera punctata*. Experientia (Basel): 41: 1028–1034.

Tombes, A. S. 1972. Scanning electron microscopy of endocrine tissue of *Hypera punctata* (Fabricius) (Coleoptera: Curculionidae). Int. J. Insect Morphol. Embryol. 1: 201–202.

Tombes, A. S. and D. S. Smith. 1970. Ultrastructural studies on the corpora cardiaca–allata complex of the adult alfalfa weevil, *Hypera postica*. J. Morphol. 132(2): 137–147.

Tombes, A. S. and D. S. Smith. 1972. Ultrastructural studies on the corpora cardiaca–allata complex of active and diapausing alfalfa weevil *Hypera postica* (Gyllenhal) adults. Pp. 113–121 *in* V. J. A. Novak and K. Slama (eds.), *Insect Endocrines*, Vol. III. Academia, Prague.

Trautmann, K. H., P. Masner, A. Schuler, M. Suchy, and H. K. Wipf. 1974a. Evidence of the juvenile hormone methyl (2*E*,6*E*)-10,11-epoxy-3,7,11-trimethyl-2,6-dodecadienoate (JH3) in insects of four orders. Z. Naturforsch. 29C: 757–758.

Trautmann, K. H., A. Schuler, M. Suchy, and H. K. Wipf. 1974b. A method for the qualitative and quantitative determination of three natural insect juvenile hormones: Evidence of methyl 10,11-epoxy-3,7,11-trimethyl-2-*trans*-6-*trans* dodecadienoate in *Melolontha melolontha*. *Z. Naturforsch.* 29C: 161–168.

Trautmann, K. H., M. Suchy, P. Masner, H. K. Wipf, and A. Schuler. 1976. Isolation and identification of juvenile hormone by means of a radioactive isotope dilution method: evidence for JH III in eight species from four orders. Pp. 118–129 *in* L. I. Gilbert (ed.), *The Juvenile Hormones*. Plenum Press, New York.

Trujillo-Cenoz, O. 1962. Some aspects of the structural organization of the arthropod ganglia. *Z. Zellforsch. Mikrosk. Anat.* 56: 649–682.

Unnithan, G. C., A. Bern, and K. K. Nayar. 1971. Ultrastructural analysis of the neuroendocrine apparatus of *Oncopeltus fasciatus* (Heteroptera). *Acta Zool.* 52: 117–143.

Varma, R. V. 1979. Corpora allata of *Postelectrotermes nayari* (Isoptera: Kalotermitidae). *Curr. Sci.* 48(15): 699.

Veenstra, J. A., H. M. Romberg-Privee, H. Schooneveld, and J. M. Polak. 1985. Immunocytochemical localization of peptidergic neurons and neurosecretory cells in the neuro-endocrine system of the Colorado potato beetle with antisera to vertebrate regulatory peptides. *Histochemistry* 82: 9–18.

Verhaert, P., J. Geyen, A. de Loof, and F. Vandesande. 1984. Immunoreactive material resembling vertebrate neuropeptides and neurophysins in the brain, suboesophageal ganglion, corpus cardiacum and corpus allatum of the dictyopteran, *Periplaneta americana* L. Cell Tissue Res. 238: 55–59.

Vogt, M. 1942. Weiteres zur Frage der Artspesifital gonadotropen Hormone: Untersuchungen an *Drosophila* Arten. *Wilhelm Roux' Arch. Entwicklungsmech. Org.* 141: 424–454.

Von Harnack, M. 1958. Histophysiological studies on the corpus allatum of *Leucophaea maderae:* the effect of starvation. *Biol. Bull.* (Woods Hole) 115(3): 521–529.

Waku, Y. and L. I. Gilbert. 1964. The corpora allata of the silkmoth, *Hyalophora cecropia:* an ultrastructural study. *J. Morphol.* 115: 69–96.

Warkievi-Granier, S. and J. C. Leonide. 1971. Résultats préliminaires de recherches sur la castration parasitaire et l'existence d'un éventuel relai endocrinien dans la castration des Criquets infestés par des Diptères. *Ann. Zool. Ecol. Anim.* 3: 327–336.

Watson, J. A. L. 1964a. Moulting and reproduction in the adult firebrat, *Thermobia domestica* (Packard) (Thysanura, Lepismidae). I. The moulting cycle and its control. *J. Insect Physiol.* 10: 305–317.

Watson, J. A. L. 1964b. Moulting and reproduction in the adult firebrat, *Thermobia domestica* (Packard) (Thysanura, Lepismidae). II. The reproductives cycles. *J. Insect Physiol.* 10: 399–408.

Weaver, R. J., G. E. Pratt and J. R. Finney. 1974. Cyclic activity of the corpus allatum related to gonotrophic cycles in adult female, *Periplaneta americana*. Experientia (Basel) 31(5): 597–598.

Weed-Pfeiffer, I. 1936a. Removal of corpora allata on egg production on the grasshopper *Melanoplus differentialis*. Proc. Soc. Exp. Biol. Med. 34: 883–885.

Weed-Pfeiffer, I. 1936b. Experimental study of moulting in the grasshopper *Melanoplus*. Proc. Soc. Exp. Biol. Med. 34: 885–886.

Weirich, G. F. and M. G. Culver. 1979. S-Adenosylmethionine–juvenile hormone acid methyltransferase in male accessory reproductive glands of *Hyalophora cecropia* (L.). Arch. Biochem. Biophys. 198: 175–181.

Weismann, R. 1926. Zur Kenntnis der Anatomie und Entwicklungsgeschichte der Stabheuschrecke *Carausius morosus* Br. III. *Entwicklung und Organogenese der Cölomblasen*, pp. 123–328. Fischer, Jena.

Wigglesworth, V. B. 1935. Functions of the corpus allatum of insects. Nature (Lond.) 136: 338–339.

Wigglesworth, V. B. 1936. The function of the corpus allatum in the growth and reproduction of *Rhodnius prolixus*. Q. J. Microsc. Sci. 79: 91–121.

Wigglesworth, V. B. 1948. The functions of the corpus allatum in *Rhodnius prolixus* (Hemiptera). J. Exp. Biol. 25: 1–15.

Wigglesworth, V. B. 1964. The hormonal regulation of growth and reproduction in Insects. Adv. Insect Physiol. 2: 248–335.

Wigglesworth, V. B. 1965. *The Principles of Insect Physiology*, 7th ed. Chapman & Hall, London.

Wilhelm, R. and M. Lüscher. 1974. On the relative importance of juvenile hormone and vitellogenin for oöcyte growth in the cockroach *Nauphoeta cinerea*. J. Insect Physiol. 20: 1887–1894.

Wilkins, J. L. 1969. The endocrine and nutritional control of egg maturation in the fleshfly *Sarcophaga bullata*. J. Insect Physiol. 14: 927–943.

Willey, R. B. 1961. The morphology of the stomodeal nervous system in *Periplaneta americana* (L.) and other Blattaria. J. Morphol. 108: 219–247.

Willey, R. B. and G. B. Chapman. 1960. The ultrastructure of certain components of the corpora cardiaca in orthopteroid insects. J. Ultrastruct. Res. 4: 1–14.

Williams, C. M. 1976. Juvenile hormone. ????

Willis, J. H. 1974. Morphogenetic action of insect hormones. Annu. Rev. Entomol. 19: 97–115.

Wirtz, P. 1973. Differentiation in the honeybee larva: a histological, electron-microscopical and physiological study of caste induction in *Apis mellifera mellifera* L. Meded. Landbouwhogesch. Wageningen 73: 1–155.

Wyatt, G. R. 1972. Insect hormones. Pp. 385–490. *In* G. Litwack (ed.), *Biochemical Action of Hormones*, Vol. 2. Academic Press, New York.

Yin, C. M. and G. M. Chippendale. 1977. Organization of the retrocerebral gland system in lepidopterous larvae of the family Pyralidae. J. Insect Physiol. 23: 755–759.

Yin, C. M. and G. M. Chippendale. 1979a. Diapause of the Southwestern corn borer, *Diatraea grandiosella:* further evidence showing juvenile hormone to be the regulator. J. Insect. Physiol. 25: 513–523.

Yin, C. M. and G. M. Chippendale. 1979b. Ultrastructure characteristics of insect corpora allata in relation to larval diapause. Cell. Tissue Res. 197: 453–461.

Zaretsky, M. and W. Loher. 1983. Anatomy and electrophysiology of individual neurosecretory cells of an insect brain. J. Comp. Neurol. 216: 253–263.

Zurflueh, R. C. 1976. Phenylethers as growth regulators: laboratory and field experiments. Pp. 61–74 *in* L. I. Gilbert (ed.), *The Juvenile Hormones*. Plenum Press, New York.

Morphology, Histology, and Ultrastructure of Cephalic Neurohemal Organs and Their Roles in Morphogenetic Processes in Myriapoda

3

MICHEL DESCAMPS

FRANÇOIS SAHLI

CATHERINE JAMAULT-NAVARRO
and JEANNINE CAPLET

 M. Descamps, F. Sahli, C. Jamault-Navarro, J. Caplet

3.1. Introduction

Cephalic neurohemal organs have been described in all groups of Myriapoda, except in Pauropoda. According to Gersch (1964), cephalic neurohemal organs can be categorized into undifferentiated (i.e., without intrinsic cells) and differentiated (i.e., with intrinsic glandular cells).

Except in Diplopoda, in which there is great diversity, these organs are variously known as cephalic organs, cephalic glands, Gabe organs, or cerebral glands, the latter being mostly confined to Chilopoda.

In Diplopoda, in addition to Gabe organs, there are paraesophageal bodies, connective bodies, periesophageal blood sinus formations, connective bodies, paracommissural plates, and clypeolabral glands.

The experimental demonstration of the involvement of some of these glands in the control of certain morphogenetic processes has been reported for the cerebral glands of Chilopoda; they are thought to regulate molting (moderating effect: Joly, 1961, 1966a; Scheffel, 1965a, 1969) and gametogenesis (moderating action on spermatogenesis: Descamps, 1975; necessary for normal oogenesis: Herbaut, 1976).

In Diplopoda, Gabe organs were found to be involved in the control of molting (Nair, 1974, 1980). Connective bodies are involved in gametogenesis control and are necessary for oogenesis (Nair, 1974).

In this chapter, the location, innervation, histology, and ultrastructure of the various cephalic neurohemal organs or glands of Myriapoda will be presented, with brief accounts of their involvement in various morphogenetic processes. In order to classify, undifferentiated types (= NHOs) will be named "organs," and differentiated type (= NH-EOs) "gland." Other terms for these organs, as used by previous authors, will also be indicated.

3.2. The Cephalic Organ of Symphyla

3.2.1. *Location and Innervation*

The cephalic organs of *Scutigerella pagesi* were studied by Juberthie-Jupeau (1961, 1963), and we will summarize her results here. Also referred to as "cephalic glands" or "cerebral glands," the cephalic neurohemal organs are paired and located lateroventrally behind the posterior edge of the brain (Fig. 3.1). These crescent-shaped organs (50 μm × 15–17 μm), lie close to the excretory gland of the maxillary segment. The nerve, linking each cephalic organ (CO) to the brain, emerges ventrally from the posterior protocerebral lobe and is composed of axons originating from the protocerebral nuchal lobe neurosecretory cells (NSC) (Fig. 3.1).

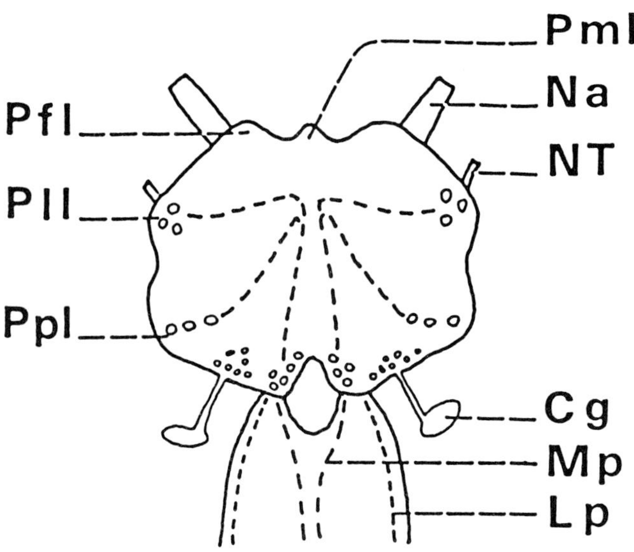

FIGURE 3.1. Dorsal view of the brain of *Scutigerella pagesi* (Symphyla). *Key:* Cg = cerebral gland (= cephalic organ); Lp = laterodorsal pathway of deutocerebral neurosecretory products; Mp = median pathway of protocerebral neurosecretory products; Na = antennal nerve; NT = Tömösvary organ nerve; Pfl = protocerebral frontal lobe; Pll = protocerebral lateral lobe; Pml = protocerebral median lobe; Ppl = protocerebral posterior lobe. (From Joly and Descamps, 1987; courtesy of John Wiley & Sons.)

3.2.2. Histology and Ultrastructure

The ultrastructural study of the COs was conducted by Juberthie-Jupeau and Juberthie (1973a,b). These organs, surrounded by a thin layer of connective tissue (45–80 nm thick), are composed of only glial cells and axonal endings. Two types of glial cells are observed. Peripheral glial cells (1–3) (Figs. 3.2 and 3.3), located in the dorsal part of the organ are characterized by only two cytoplasmic processes running just under the connective sheath (Fig. 3.3). An interstitial glial cell (only one per organ) is found near the entrance of the protocerebral nerve and shows numerous cytoplasmic processes wrapping around the axons or axonal endings. In contrast to nuclei observed in peripheral glial cells, its nucleus is lobate and the chromatin more scattered.

The CO contains three main types of axons: electron-dense globular granules (A type), electron-light granules (B type), and electron-dense, heterogeneous granules (C type). The characteristics of the different types or subtypes are summarized in Table 3.1.

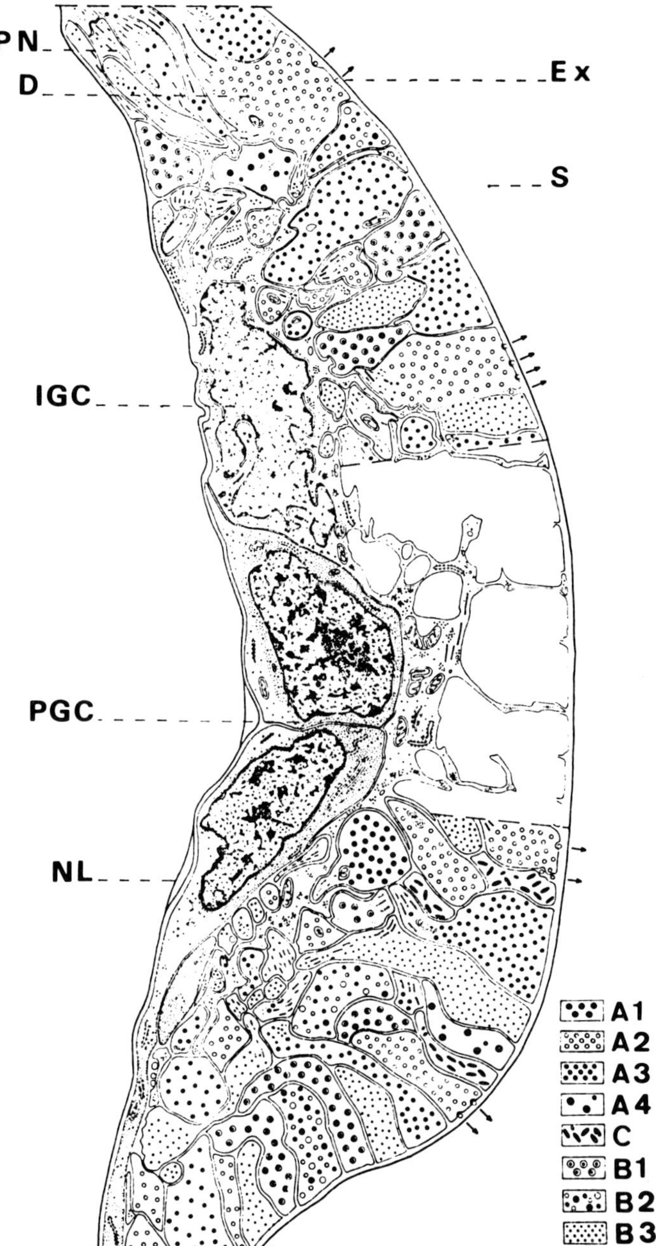

FIGURE 3.2. Schematic drawing of sagittal section of neurohemal cephalic organ of Symphyla. *Key:* D = axonal swelling; Ex = exocytosis; IGC = interstitial glial cell; NL = neural lamellae; PGC = peripheral glial cell; PN = protocerebral nerve; S = blood sinus. (From Juberthie-Jupeau and Juberthie, 1973a; courtesy of the authors and of Gauthier-Villars, Paris.)

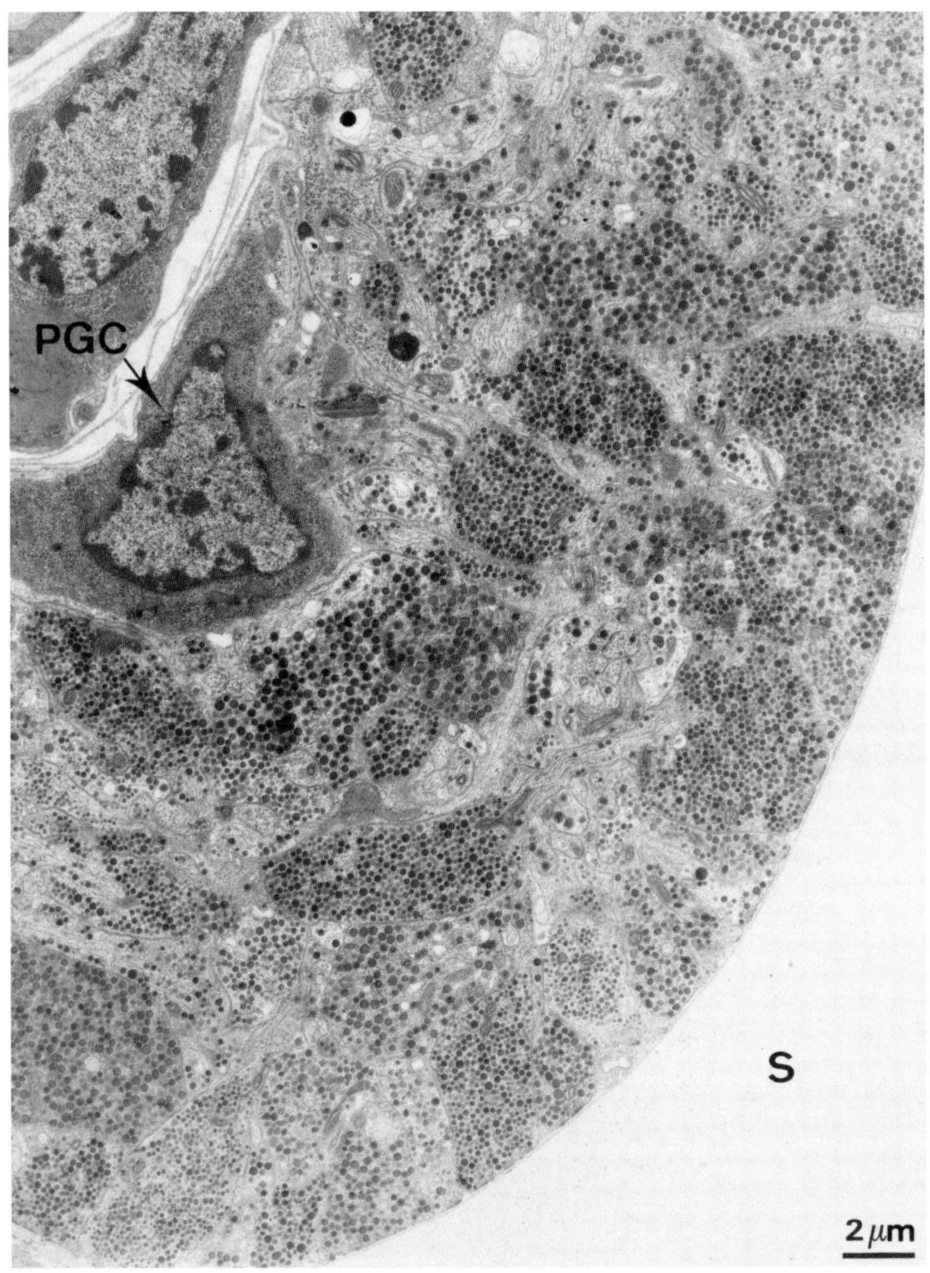

FIGURE 3.3. General view of the cephalic neurohemal organ of *Scutigerella*. *Key:* PGC = peripheral glial cell; S = blood sinus. (From Juberthie-Jupeau and Juberthie, 1973a; courtesy of the authors and of Gauthier-Villars, Paris.)

TABLE 3.1 Characteristics of the Secretory Granules in the Axons of the Cephalic Organ in *Scutigerella*

Axon type	Abundance	Morphology	Electron density	Diameter (nm)
A1	Scarce	Globe-shaped	+++	120–180
A2	+++	Globe-shaped	++	115–160 diameter < 190
A3	++	Globe-shaped	+++	80–140
A4	Scarce	Globe-shaped	+++	75–160 light axoplasm
C		Various (elongated, conical; egg-shaped)	Heterogeneous	
B1		Various (egg-shaped,irregular; globe-shaped)	Dense-cored	115–200
B2			Homogeneous or granular	90–150
B3		subspherical	+	70–115

SOURCE: *Juberthie-Jupeau and Juberthie, 1973a.*

Exocytotic profiles can only be observed in A2 axonal swellings (Juberthie-Jupeau and Juberthie, 1973b), just under the connective sheath or between two axonal endings, of the same type or of different types. These exocytotic profiles are easily observed, almost 100% of A2 axons showing at least one profile.

Another mechanism of release may be involved for axons of other types. A3 or B3 endings show clusters of dense material, just under the plasma membrane, lying by the connective sheath.

3.2.3. *Role in Molting*

There is only indirect evidence of the involvement of COs in the control of molting. Indeed, a secretory cycle related to the molting cycle can be observed in the central nervous system (CNS) of the nuchal lobe, origin of the axons ending in the COs; nevertheless, in the latter it is not possible to show such a cycle (Juberthie-Jupeau, 1963).

3.3. The Cerebral Gland of Chilopoda

As *Lithobius forficatus* is the most studied model in Chilopoda, the results concerning this species will be described first and then compared with the characteristics of other species.

3.3.1. *Location and Innervation*

Cerebral glands (CGs) are paired neurohemal organs of ectodermal origin, linked to the protocerebrum by the nervi glandulae cerebralis I and II (ngc I and ngc II, respectively named N 4 and N 3 by Rilling, 1968). Originally named "frontal organs" by Holmgren (1916), their endocrine nature was shown by Fahlander (1938), who called them "cerebral glands," the name that is now most commonly employed. However, some authors call them "Gabe organs" (Rosenberg, 1976). The axons constituting the ngc I, originate from the protocerebral neurosecretory cells (NSC); the latter, together with ngc I and the cerebral glands, constitute the protocephalic pathway.

In *Lithobius forficatus*, the CGs are located in the subocular area in a lateral fold of the cephalic cuticle (Fig. 3.4). The glandular size is rather small (0.3–0.6 mm long and 0.2–0.3 mm wide) and varies in different animals and sometimes even in the same animal. The ngc I enters the gland at the anterointernal tip. The ngc II arises from the optic nerve and terminates in the gland close to the ngc I.

Differences in other species or groups mainly relate to the gland's location and general aspect (Gabe, 1952, 1966; Joly and Descamps, 1968).

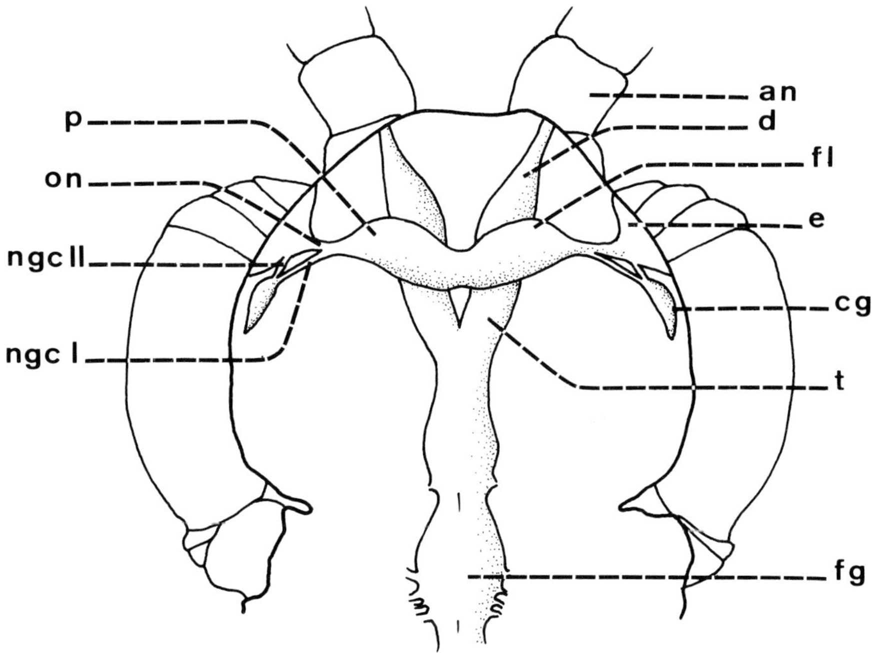

FIGURE 3.4. Dorsal view of the head of *Lithobius forficatus* (Chilopoda: Lithobiomorpha). *Key:* an = antenna; cg = cerebral gland; d = deutocerebrum; e = eye; fg = forcipular ganglion; fl = frontal lobe; ngc I, II = nervus glandulae cerebralis I, II; on = optic nerve; p = protocerebrum; t = tritocerebrum.

Concerning their location, two main positions are recorded: in Geophilomorpha, they are located by the protocerebrum, laterodorsally (Schendylidae) or lateroventrally (Geophilidae, Himantariidae), whereas in Scolopendromorpha they are in the same location as in Lithobiomorpha, i.e., in the subocular area, even in blind species (*Cryptops*). In Scutigeromorpha, the CGs are located ventrally, under the optic lobe (Rosenberg, 1976).

Glandular shapes vary according to the species: in Geophilomorpha, the gland is globular, the outlines being either irregular (*Haplophilus subterraneus, Geophilus carpophagus*), or regular (*Necrophloeophagus longicornis, Scolioplanes acuminatus*). In Scolopendromorpha, the cerebral gland of *Scolopendra cingulata* looks like a bunch of grapes (De Lerma, 1951), whereas in Otostigmini and in Cryptopsidae they are oval (see Fig. 3.7B in Section 3.3.2, below). In Lithobiomorpha, as in Scutigeromorpha, only ovoid shape, with a regular outline was found.

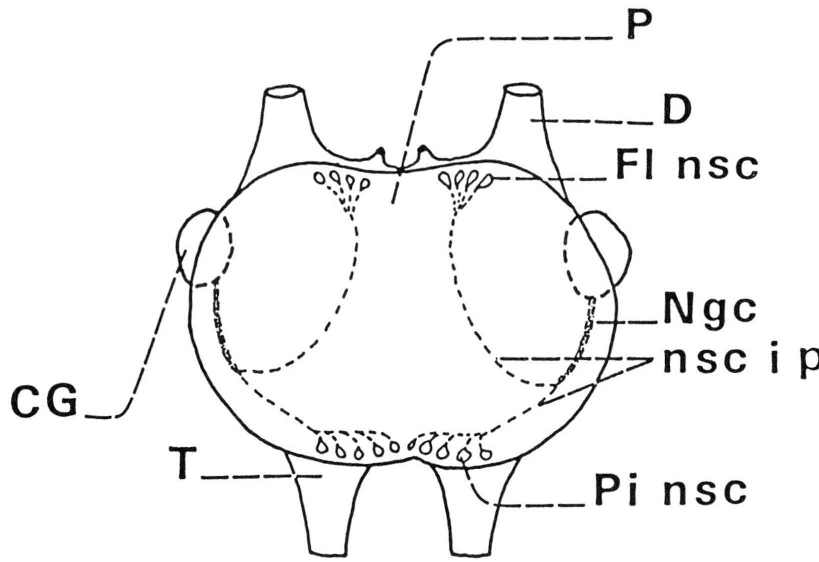

FIGURE 3.5. Dorsal view of the brain of *Geophilus longicornis* (= *Necrophloeo-ephagus longicornis*) (Chilopoda: Geophilomorpha). *Key:* CG = cerebral gland; D = deutocerebrum; Fl nsc = frontal lobe neurosecretory cells; Ngc = nervus glandulae cerebralis; nsc ip = neurosecretory cells intracerebral pathways; P = protocerebrum; Pi nsc = pars intercerebralis neurosecretory cells; T = tritocerebrum. (From Joly and Descamps, 1987; courtesy of John Wiley & Sons.)

The ngc I generally emerges from the protocerebrum at a lateroventral point. However, in Lithobiomorpha, its emergence can be lateral (*Lithobius melanops* and *L. crassipes*) or more axial, near the proto-deutocerebral limit (*L. calcaratus*). The ngc I enters the gland ventrally or laterally in Geophilomorpha, and more or less apically in other orders (Scolopendromorpha, Lithobiomorpha, and Scutigeromorpha). The nerve generally runs transversely, except in some species such as *Necrophloephagus longicornis* (= *Geophilus longicornis*), in which the ngc is parallel to the longitudinal axis (Fig. 3.5).

The ngc II was found only in some Lithobiomorpha. This nerve links the gland to the optic nerve (*L. forficatus*) or links the ngc I to the optic lobe (*L. calcaratus*).

Tracheas can generally be observed in the vicinity of or close to the CGs. In Geophilomorpha, a blood vessel (antennal artery) is also present, and even wrapped in the same connective tissue sheath, compared with the gland in *Geophilus carpophagus* (see Fig. 3.7A in Section 3.3.2, below).

Last, and it is a common character for all the species studied, the CGs lie in a blood sinus.

3.3.2. *Histology and Ultrastructure*

Only few details appear in the gland when observed by light microscopy: cell limits are only seen when the gland is poor in secretion. Generally, the glands show large phloxinophilic areas, and among these are granules stained by chromic hematoxylin. Gabe (1952) suggested that the two types of secretory products were related to different sources; this was later confirmed by ultrastructural studies (Scheffel, 1965b; Joly, 1966b, 1970).

Scheffel (1961) studied these organs in the larvae of *L. forficatus*. The CG is of ectodermal origin and can be found in the first instar of the ana- morphous phase. They are inserted between the lateral lobes of the pro- tocerebrum and the ventral epidermis. At the optical level, cell limits cannot be observed; the diameter of glandular nuclei, smaller than those of the protocerebral nerve cells, is comparable to that of undifferentiated epidermal cells. Related to the gland position, ngc is only observed after the fourth instar. Cerebral gland increases in volume during the animal growth; however, mitoses are seldom observed in newly molted animals.

Phloxinophilic secretion is first observed in the second instar, 2 days after molt, when neurosecretion is not yet stainable in the protocere- brum. According to Scheffel (1961), NSC are too poor in secretory prod- ucts to be stained, but secretions, stored in the gland, are in sufficient amount to be stained. After the third instar, the amount of glandular se- cretion increases, but no secretory cycle can be recorded.

The fine structure of the CGs in adult *L. forficatus* was investigated by Scheffel (1965b) and Joly (1966b). The gland, ensheathed in a layer of con- nective tissue, about 200 nm thick, is constituted of three elements: glan- dular cells, glial cells (scarce), and axonal endings (Fig. 3.6A,B).

Glandular cells are about 15–20 μm by 6–10 μm. Their secretory gran- ules are generally abundant, electron-dense, and have a diameter in the range of 200–300 nm, but it can reach 500 nm in some cases. They are sur- rounded by a membrane less electron-dense than the secretion. In the pe- ripheral glandular cells sometimes vacuoles filled with homogeneous material, originating from several fused secretory granules, are observed. Some cells show cytoplasmic processes with tubules, mitochondria, and numerous secretory granules. Intercellular bridges can sometimes be observed.

Glial cells (Fig. 3.6B) are scarce and often found in the more external part of the gland. Cytoplasmic processes run between axons and/or glan- dular cells; granules are never observed in these cells.

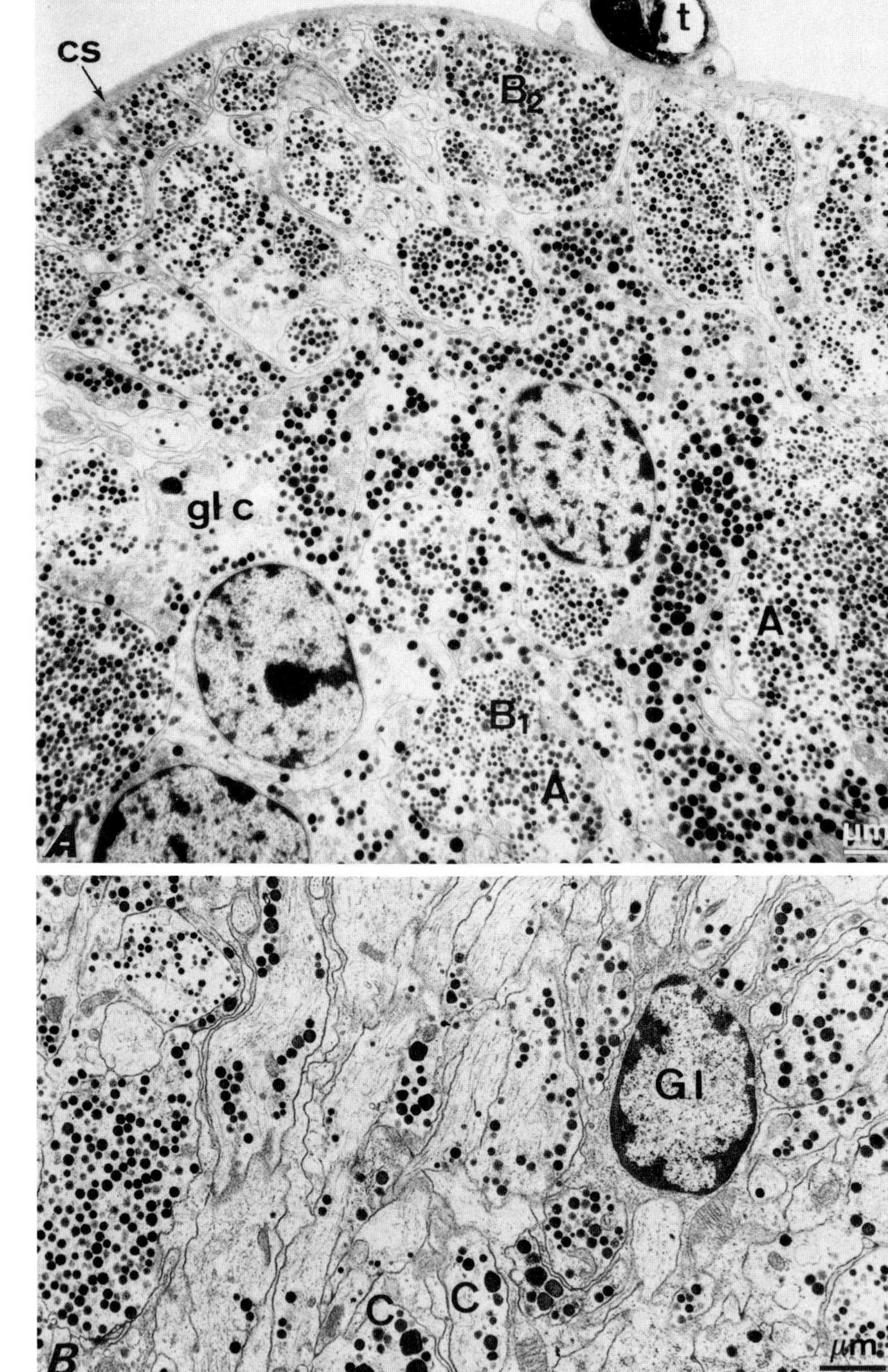

cs
t
B₂
A
gl c
A
B₁
A
µm
A
G.l
C
C
µm
B

Axons represent a high percentage of the gland volume. Most of the axons present secretory granules and a few, elongated and tubular-crested mitochondria. In addition, some axons present peripheral processes with neurotubules. The diameters of axonal granules are smaller than those recorded in glandular cells. Several types of granules are observed, according to their size and electron density; based on the characteristics of protocerebral NSC granules (Jamault-Navarro, 1981), those present in the gland are of A, B1, B2, and (perhaps) C types (frontal lobes NSC) and of B3 type (pars intercerebralis NSC). Note that categorization of granules is not so easy in the ngc as in the gland or the brain.

Some axons, often of small diameter, show either small clear vesicles (diameter: 50–60 nm) or slightly dense granules (50–120 nm); these may or may not be associated with dense granules (diameter more than 120 nm). Up to now, serotonin (5-hydroxytryptamine, 5-HT) and noradrenaline (NA; or, in U.S. parlance, norepinephrine) were found in glandular axons (Descamps et al., 1985).

Thus the secretory products of the CG is exogenous (originating in the brain) or endogenous (synthesized by glandular cells); these glands are, therefore, differentiated neurohemal organs and may be compared to the corpora cardiaca of insects; however these two organs are not of same embryological origin.

Synapses are, for the most part, interaxonal; they are scarce between axons and glandular cells. The release of the endogenous secretory products cannot be easily observed. Nevertheless, in the more external part of the gland, cytoplasmic processes of glandular cells show small vesicles, probably related to release. Released granules are sometimes found in the internal part of the connective sheath, near a cluster of small vesicles. These granules are without membranes, and it is not possible to follow their migration through the connective sheath. Cytoplasmic lobes, filled with endogenous or exogenous secretory granules, can be released all at one time. Such glandular parts are wrapped in a thin layer of connective tissue. Pictures of this release are more abundant and varied in axons: granular fragmentation, small electron-lucent vesicles, granular interconnections, and exocytosis profiles (Joly and Descamps, 1977).

The control of glandular activity by protocerebral NSC has been evidenced by experiments of microcauterization of these NSC (either of the frontal lobes or of the pars intercerebralis area) or by sectioning the ngc:

FIGURE 3.6. Ultrastructural aspects of the cerebral gland in *L. forficatus*: (A) general view; (B) aspect of the gland near the entrance point of the *nervus glandulae cerebralis*. *Key:* A, B1, B2, C = axon types; cs = connective sheath; Gl = glial cell; gl c = glandular cell; t = trachea.

both lead to injuries in the axonal endings and, as a consequence, to dysfunctioning of glandular activity (Joly, 1976; Joly and Jamault-Navarro, 1978). Electrical stimulation of protocerebrum (frontal lobes or pars intercerebralis) leads to glandular hyperactivity (Joly and Descamps, 1977). Somehow, unilateral removal of the cerebral gland leads to compensatory hypersecretion in the remaining gland, with an increase in axonal diameter.

External factors also act on the secretory activity of the gland. In starved animals, there is first an increase in glandular syntheses, with increased release; then, injuries appear in glandular cells and the secretion decreases. Feeding after starvation leads to an increase in glandular activity, but only if injury is not severe. The structure of axonal endings is not changed in experiments involving external factors: it is, therefore, concluded that the latter act directly on CGs (Joly, 1977).

In the other orders of Chilopoda, study by light microscopy revealed only differences in the connective septa inside the gland of some Geophilomorpha (Fig. 3.7A); such structures are never seen in other orders (Lithobiomorpha, Scolopendromorpha, Scutigeromorpha).

Ultrastructural studies of glands in *Geophilus longicornis* (Geophilomorpha) (Ernst, 1971), *Scutigera coleoptrata* (Scutigeromorpha) (Rosenberg, 1976), and in three Scolopendromorpha (*Scolopendra cingulata, Cryptops savignyi,* and *C. hortensis*) (Descamps and Joly, 1985) have shown that in these species also, the CGs are differentiated neurohemal organs.

In *Scutigera coleoptrata*, in the middle of the molting cycle, two types of glandular cells are found, based on the electron density of their cytoplasm. The ngc is made of about 200 axons, and in the cerebral gland five axon types are observed, one of them being characterized, in the middle of the molting cycle, by numerous small electron-lucent vesicles. In *G. longicornis*, four axon types are present in the gland, two of which are aminergic and modulate the activity of glandular cells. A third type is filled with granules of same aspect and size as those observed in pars intercerebralis (pi) NSC; it should be kept in mind that, in *G. longicornis*, a part of pi secretion goes into the cerebral gland (Fig. 3.5).

FIGURE 3.7. (A) Cerebral gland (GC) in *Geophilus carpophagus* (Chilopoda: Geophilomorpha). *Key:* bv = blood vessel; t = trachea. (B,C) Cerebral gland (GC) aspects in *Cryptops savignyi* (Chilopoda: Scolopendromorpha). *Key:* ax = axonal section; gl = glial cell; m = muscle; ngc = nervus glandulae cerebralis. Note that the nervous and glandular parts are well separated, and not mixed as they are in *Lithobius* (see Fig. 3.6). The dashed line (in C) emphasizes the separation between the two compartments of the gland.

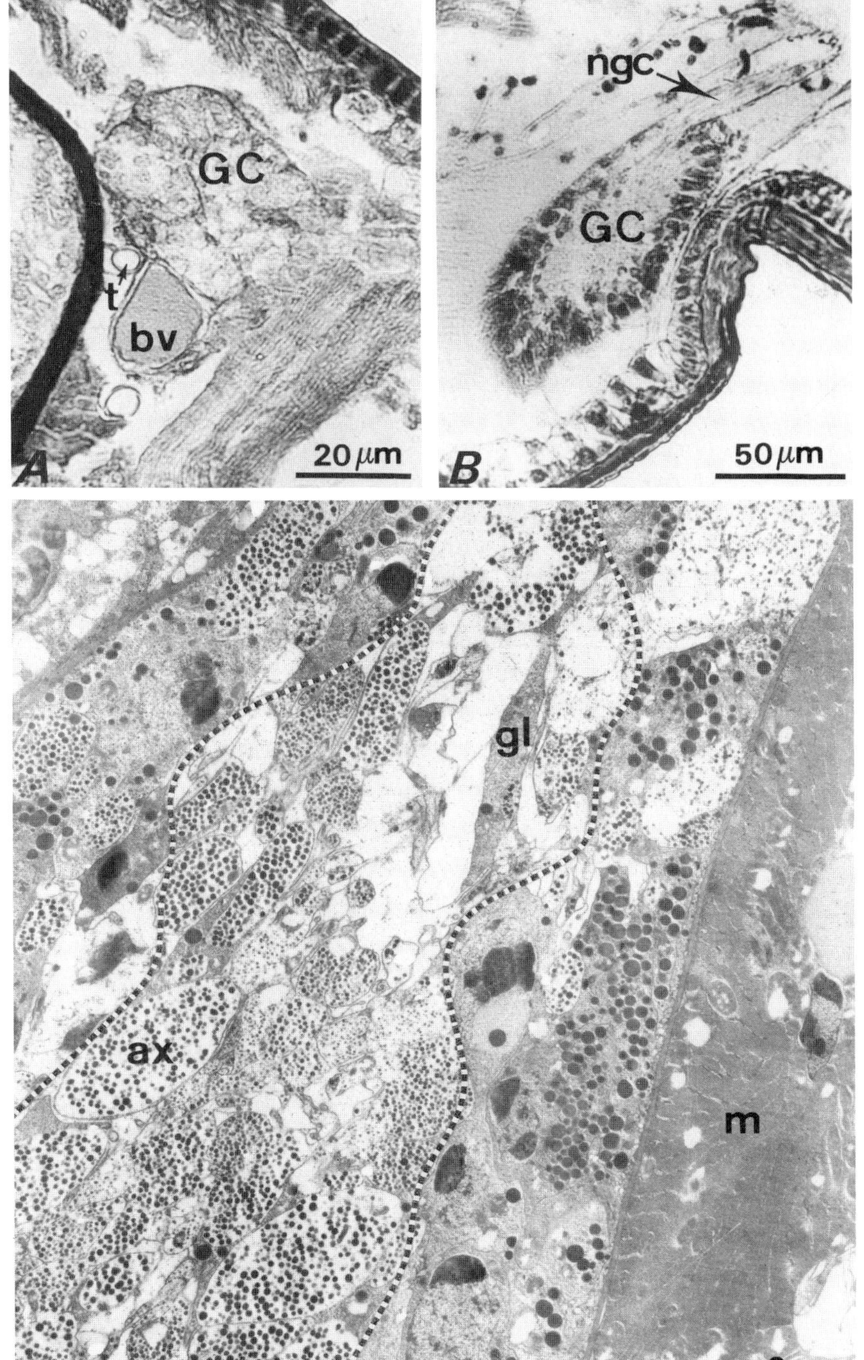

GC
t
bv
A
20 μm
ngc
GC
B
50 μm
gl
ax
m
C
2 μm

In Scolopendromorpha, large differences are recorded according to the genus studied. Indeed, if the structure remains classic in *Scolopendra* (glandular cells and axons are mixed inside the gland), in *Cryptops*, glandular cells are only peripheral, the central compartment being occupied by axonal endings, glial cells, and interaxonal lacunae. A comparable partition is observed in the corpora cardiaca of *Locusta migratoria* (Insecta: Orthoptera), where pars nervosa and pars glandularis are well separated. In addition, one glial cell type and two glandular cell types are found in *Scolopendra*, whereas two and three are respectively observed in *Cryptops*.

The chemical nature of the secretion was investigated by histo- and cytochemical methods by Joly and Devauchelle (1970) in adult *Lithobius forficatus*. Polysaccharides or lipids are not components of the secretory granules. Cytochemical procedures and action of proteolytic enzymes (pepsin, trypsin) showed that secretory granules, both endogenous and exogenous, are made of non-histone proteins, with cystine and cysteine residues. In addition, there are arginine and lysine in endogenous granules.

Enzymatic digestion in two neighboring axons filled with granules of similar appearance and size showed, respectively, more or less peptic digestion; this finding agrees with the concept of chemical plurality for exogenous secretory products.

3.3.3. Role in Regulation of Molting and Gametogenesis

A secretory cycle, related to the molting cycle, can be shown in the glandular cells (Joly, 1970). The secretory activity is observed at the end of one molting cycle and at the beginning of the next, and is characterized by a well-developed rough endoplasmic reticulum (RER), dictyosomes, and numerous ribosomes. Release of granules occurs during this period. Then a rest period ensues (middle of the molting cycle), during which the dictyosomes do not secrete, the ribosomes are less numerous, and RER less developed. Generally, the glandular cells secrete synchronously. Nevertheless, "dark" and "light" cells are sometimes observed in the same gland, evidence of asynchronism. "Dark" cells show numerous ribosomes; these active glandular cells often show large secretory granules, more or less damaged, which are synthesized during the previous cycle but not released.

Experimental demonstration of the involvement of CGs in control of molting was performed in larvae (Scheffel, 1965a, 1969) and in adults (Joly, 1961, 1966a, 1980) of *L. forficatus*, and in adults of *Scolopendra cingulata* (Joly, 1962). CGs exert a moderating action; the removal of the glands induce an increase in the percentage of molting and a shortening of the duration of the molting cycle. Implantations of CGs in animals deprived of their own can inhibit molting when implantations (four glands at a time) occur every 20 days.

The moderating action of the hormone released by the CGs is more likely exerted at the epidermal level but could, in addition, control the activity of the ecdysial glands.

The CGs also play a moderating role on the course of the spermatogenetic cycle. Their removal leads to precocious spermiogenesis (Descamps, 1975) and to increased syntheses in the spermatocytes (Descamps, 1978). The CGs are necessary for normal oogenesis; removal of the glands lead to degeneration of most of the oocytes (Herbaut, 1976). In control of molting, as well as in the control of gametogenesis, a quantitative effect of the CGs was shown experimentally by unilateral removal.

3.4.   The Cephalic Organs of Diplopoda*

3.4.1.   *The Gabe Organs (GOs)*

Neurohemal organs in Diplopoda were first observed by Gabe (1954). However, his descriptions of these organs were so vague that, in fact, they needed to be "rediscovered" (e.g., see Sahli, 1958, 1966). According to Sahli, Gabe never found the true neurohemal organs in Julidae. He probably mistook these organs for paraesophageal bodies, just as he confused the intracerebral course of the GO nerves with PW4 pathway and Holmgren's globuli I with globuli II. Sahli (1963) proposed to rename the "cerebral glands" as "Gabe organs" in his honor.

3.4.1.1.   LOCATION AND INNERVATION

The GOs are paired organs symmetrically located in the head with respect to the sagittal plane (Fig. 3.8). Their position, however, is variable (Fig. 3.9). In Juliformia (Spirostreptida, Spirobolida, Julida; see Fig. 3.9F,G,H), GOs are placed in the ventral and posterior part of the head, on either side of the subesophageal nervous mass (Sahli, 1966, 1974, 1977b, 1979, 1985a; Sahli and Petit, 1978). Tripathi (1976) erroneously showed GOs near the brain in the Harpagophoridae (belonging to spirostreptid Juliformia). In Polydesmida (Fig. 3.9A,B), GOs are located behind the brain, and more dorsally and rather well in front of those in

---

*Written by F. Sahli and J. Caplet.

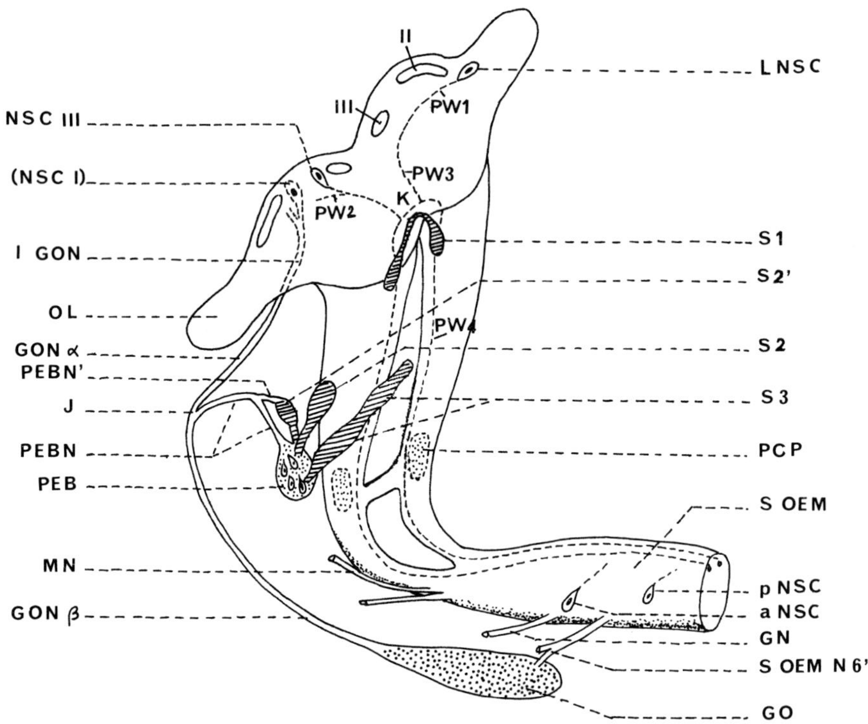

FIGURE 3.8. Posterior view of the neurohemal system in a Julidae. The Gabe organ has been laterally moved to the lower part and only the left neurohemal complex drawn. *Key:* a NSC = anterior neurosecretory cells of the subesophageal nerve mass; GN = nerve of the gnathochilarium; GO = Gabe organ; GON α = α part of the Gabe organ nerve; GON β = β part of the Gabe organ nerve; I GON = intracerebral course of the GO nerve; J = junction of the GO and paraesophageal body nerves; K = K commissure; L NSC = lateral neurosecretory cells; MN = mandibular nerve; (NSC I) = NSC of the globulus I (situated anterior side); NSC III = NSC of the globulus III; OL = optic lobe; PCP = paracommissural plate; PEB = paraesophageal body; PEBN = PEB nerve; PEBN' = branch of the PEB nerve; p NSC = posterior NSC of the subesophageal nerve mass; PW1, PW2, PW3, PW4 = neuropilar pathways of neurosecretory products; S1, S2, S2', S3 = periesophageal blood sinus formations 1, 2, 2', 3; S OEM = branch of nerve No. 6 of the subesophageal nerve mass; II = globulus II; III = globulus III. (Modified after Juberthie-Jupeau, 1983.)

Juliformia (Gabe, 1954; Gersch, 1958; Glaser, 1958; Prabhu, 1959, 1961, 1962; Sahli, 1962, 1966, 1974, 1977a, 1979, 1985a and b). In Callipodida (Sahli, 1977c) and Chordeumatida (Gabe, 1954; Palm, 1955; Sahli, 1966, 1977c; Seifert, 1971), both constituting the Coelocheta (cf. Hoffman, 1979), GOs lie behind the brain, laterally and relatively close to the integument of the cephalic capsule (Fig. 3.9C,D). In Glomerida (Gabe, 1954; Juberthie-Jupeau, 1967), the GOs are, among those of Diplopoda, located closest to the brain, being almost united with it (Fig. 3.9E). In Penicillata (Polyxenidae) (Nguyen Duy-Jacquemin, 1971, 1973; Seifert and El-Hifnawi, 1972), GOs have a posterior location near the tubular coils of the maxillary nephridia, which lie dorsal and caudal. Although often far away from the brain (F. Sahli, personal observations), GOs here are much closer to the brain than in Juliformia.

In all cases GOs are innervated by a paired nerve, always containing, at least, axons arising from protocerebral NSC. Depending on the location of the GOs, the GO nerve is long in Juliformia, shorter in Polyxenidae, relatively short in Callipodida and Chordeumatida, still shorter in Polydesmida, and shortest in Glomerida.

GOs can have various other nervous connections: (a) They can be directly (Glomeridae, Glomeridellidae, Callipodidae, Paradoxomatidae) or indirectly (Juliformia) connected to the paraesophageal bodies (PEBs) and/or the periesophageal blood sinus formations (SFs); such nerves are called PEB nerves. (b) In some cases (at least in Julidae and Polyxenidae), GOs are respectively connected to the subesophageal nervous mass by one or two nerves (Fig. 3.8). (c) In *Craspedosoma*, GOs are connected to the periesophageal ring by a tritocerebral nerve (Seifert, 1971).

3.4.1.2. HISTOLOGY AND ULTRASTRUCTURE

After staining with Gabe's PAF (paraldehyde fuchsin) picroindigocarmine and Gomori's CH phloxine, GOs show drops and droplets of secretory products, polymorphic nuclei, some axons, some tracheae, and sometimes blood cells. Cellular limits are not visible.

Prabhu (1961) was the first who emphasized that the only nuclei occurring in GOs are those of glial cells ("Schwann cells"), which do not appear to be secretory. Contrary to what Gabe described (1954, 1966), GOs contain no secretory cells of their own and thus no intrinsic material. In Julidae, Gabe probably mistook the "cerebral gland" for PEBs (see Gabe, 1966), which obviously contain secretory cells. He probably generalized these observations to all other aforementioned Diplopoda, and also because he was misled by the presence of two main types of neurosecretory products in the GOs: acidophilic and basophilic.

In Julidae, which possess PEBs (these bodies dispatch their secretion

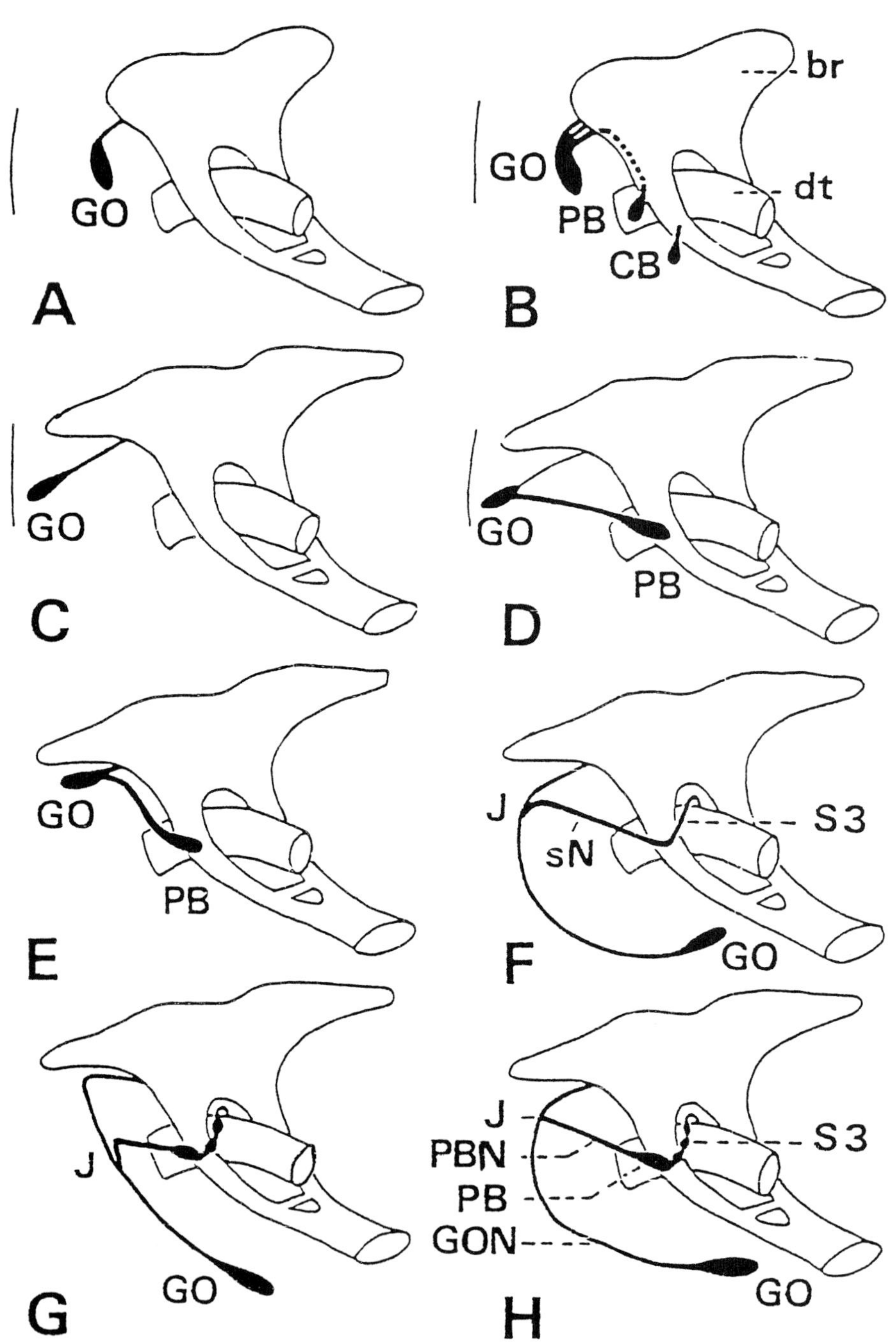
GO
A
br
GO
PB
CB
dt
B
GO
C
GO
PB
D
GO
PB
E
J
sN
S3
GO
F
J
GO
G
J
PBN
PB
GON
S3
GO
H

material towards the GOs, via PEB nerves; cf. Section 3.4.2), one can observe the same types of droplets (acidophilic or both acidophilic and basophilic) in the PEB nerves and in the β part of the GO nerve (Sahli, 1966; see his Fig. 101). The β part of the GO nerve and GO exhibit the same secretion material, which, therefore, has an extrinsic origin.

In Diplopoda without PEBs, basophilic secretion always originates from the protocerebral NSC via GO nerves. On the other hand, the origin of the other kinds of droplets (e.g., in Polydesmidae, Platyrhacidae, Craspedosomatidae, Chordeumatidae, and Polyxenidae) is unknown.

Modification or maturation of the secretion products in the GO nerves and in the GOs has been suggested (Sahli, 1966). Gabe's (1954, 1966) suggestion that the acidophilic product originates from the intrinsic secretory cells seems to be incorrect, because numerous other structural and ultrastructural investigations have shown the absence of intrinsic secretory cells. The organs located in the clypeolabrum and described under the name "GOs" by El-Hifnawi and Seifert (1972) in *Ommatoiulus sabulosus* (= *Schizophyllum subulosum*) are not GOs (see Section 3.4.6).

All ultrastructural investigations of GOs have shown the absence of secretory cells and, thus, their strictly neurohemal nature (Craspedosomatidae: Seifert, 1971; Polyxenidae: Seifert and El-Hifnawi, 1972; Polydesmidae and Julidae: Sahli and Petit, 1972; Glomeridae: Juberthie and Juberthie-Jupeau, 1974; Paradoxosomatidae: F. Sahli and J. Caplet, unpublished data). Enveloped by a thin neural lamina, GOs appear to be

FIGURE 3.9. Diagram of the neurosecretory systems in Diplopoda (solid black). Only the left side is shown. NSC and PCPs are not represented. (A) Polydesmida Polydesmidae and Platyrhacidae. (Based on data from Sahli, 1977a, 1979, and other unpublished work.) (B) Polydesmida, Paradoxosomatidae. (Based on data from Prabhu, 1962; Sahli, 1966, 1980; Sahli and Petit, 1974, 1975b.) (C) Coelocheta Chordeumatida. (Based on data from Sahli, unpublished work.) (D) Coelocheta Callipodida. (Based on data from Sahli, 1977c.) (E) Pentazonia Glomerida. (Based on data from Juberthie-Jupeau, 1967; Sahli, unpublished work.) (F) Juliformia Spirostreptida. (Based on data from Sahli, 1977b, and unpublished work.) (G) Juliformia Spirobolida. (Based on data from Sahli and Petit, 1978.) (H) Juliformia Julida. (Based on data from Sahli, 1966.) *Key:* br = brain; CB = connective body; dt = digestive tract; GO = Gabe organ; GON = GABE organ nerve; J = junction of the GO and PEB nerves; PB = paraesophageal body; PBN = paraesophageal body nerve; S3 = one of the four periesophageal blood sinus formations; sN = sinus nerve taking the place of the paraesophageal body nerve. (From Sahli, 1985a; courtesy Verlag Paul Parey.)

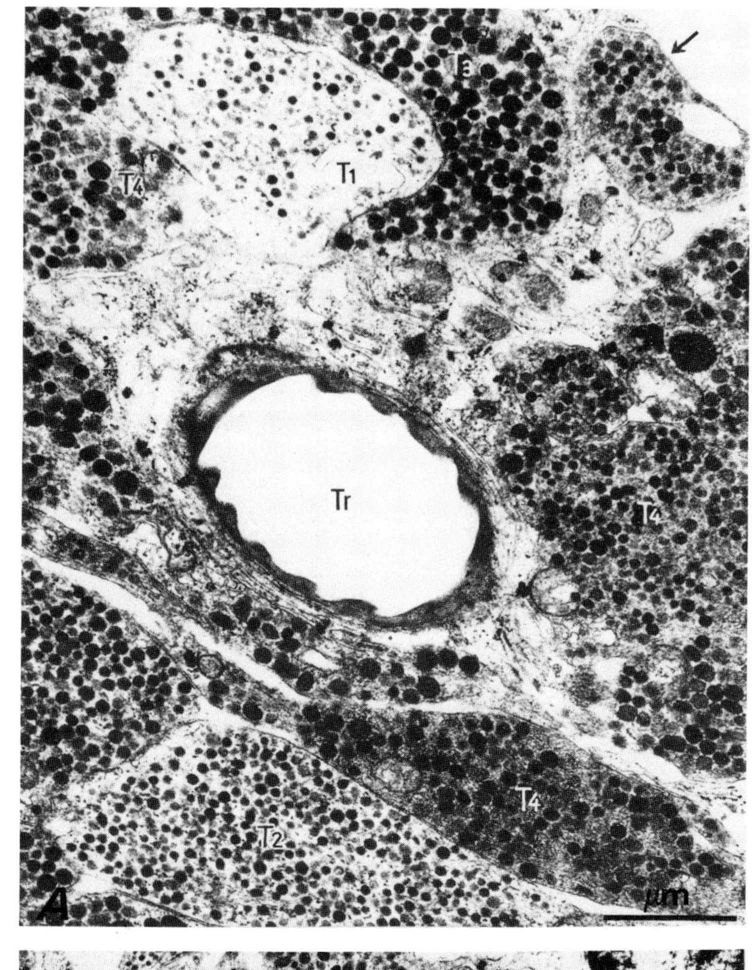

T3
T4
T1
Tr
T4
T4
T2
T4
µm
A
T4
T3
T3
µm
B

made up of (a) numerous axonal endings with elementary secretory granules, (b) nerve fibers, (c) a few glial cells (except in *Craspedosoma rawlinsii*: Seifert, 1971), (d) some small tracheae, and occasionally (e) some blood cells.

Three or four types of axonal endings can be found, depending on the size, shape, and electron density of the granules and of the background (hyaloplasm). Three types are described in Polyxenidae (Seifert and El-Hifnawi, 1972) and in Julidae (Petit and Sahli, 1975). Four types are found in Craspedosomatidae (Seifert, 1971: four T-types, Fig. 3.10) and Glomeridae (Juberthie and Juberthie-Jupeau, 1974: three A- and one B-type).

The julid *O. sabulosus* is the only diplopod in which NSC of globuli I (= NSC I, whose axons take part in the constitution of the GO nerve), the two parts (α and β) of the GO nerve, the GOs, the PEBs, and the PEB nerves have all been studied by means of sections of heads fixed and embedded *in toto* (Sahli and Petit, 1972, 1973; Petit and Sahli, 1973, 1975, 1977). Two types of NSC I have been observed: one with small secretory granules (ax 1), and another with middle-size granules (ax 3) (Petit and Sahli, 1977).

The GO nerve contains the two aforementioned types of axons in its α part (i.e., before its connection with the PEB nerve). In its β part (i.e., after the connection with the PEB nerve), it contains the two previous axon types and a third one (ax 2, which is also observed in PEBs and in PEB nerves). The axons of the three types are observed in the GOs (Fig. 3.11).

3.4.1.3.   ROLE IN MOLTING

In *Jonespeltis splendidus*, bilateral ablation of GOs resulted in an acceleration of molting (Nair, 1974). Moreover, the influence of these glands was assayed in an insect, *Dysdercus cingulatus* (Heteroptera). Implantation of GOs (two pairs) in fourth- or fifth-instar nymphs, newly molted, or in 1-day-old fifth instar postponed molting, while in 2-day-old fifth instar the same implantation is without effect (Nair, 1980).

3.4.2.   *The Paraesophageal Bodies (PEBs)*

PEBs were first described in the Julidae and called "hypocerebral formations" (Sahli, 1961). The same year, Prabhu named them "visceral

FIGURE 3.10. Ultrastructure of the Gabe organ of *Craspedosoma rawlinsii* (Coelocheta Craspedosomatidae): (A) T1, T2, T3, T4 = the four granule types; Tr = trachea; arrow = exocytosis site; (B) peripheral section of the Gabe organ in *C. rawlinsii* showing two granule types, T3 and T4. (From Seifert, 1971; courtesy of the author and of Springer Verlag.)

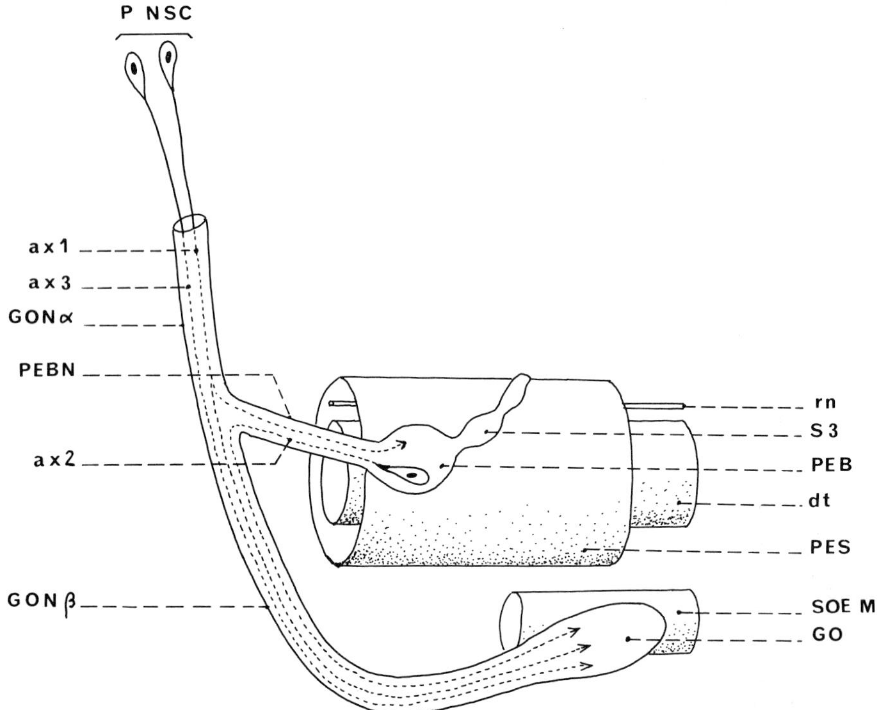

FIGURE 3.11. Location of the Gabe organ (GO), paraesophageal body (PEB), and blood sinus formation S3 in Julidae. Arrows indicate the supposed migration direction of the NS products. *Key:* ax1, ax2, ax3 = axon types 1, 2, 3; dt = digestive tract; GON α = α part of the GO nerve; GON β = β part of the GO nerve; PEBN = PEB nerve; PES = periesophageal blood sinus; P NSC = protocerebral neurosecretory cells; rn = recurrent nerve. (From Sahli and Caplet, unpublished work.)

ganglia" in *Anoplodesmus* (= *Jonespeltis*) because he thought they corresponded to the visceral ganglia mentioned by Newport (1843), which in fact probably correspond to the posterior salivary gland. Sahli (1963) renamed the hypocerebral formations as PEBs. The organs called PEBs by El-Hifnawi and Seifert (1972) in *O. sabulosus* are not PEBs.

PEBs exist in all Juliformia, except in Spirostreptida (Sahli, 1966; Sahli, 1977b; Sahli and Petit, 1978), in Glomerida (Juberthie-Jupeau, 1967), in Callipodida (Sahli, 1977c), and in Paradoxosomatidae (Prabhu, 1961, 1962; Sahli, 1962, 1963, Sahli and Petit, 1978). PEBs do not exist in some important families.

PEBs are paired organs located on each side of the foregut, under the brain and behind the periesophageal collar. Their location is constant in all Diplopoda (Figs. 3.8 and 3.9). Typically the PEBs contain (a) intrinsic nervous secretory cells, (b) occasionally some glial cells, (c) secretory material, (d) nerve fibers, (e) some tracheae, and (f) sometimes one or several blood cells (Prabhu, 1961, 1962; Sahli, 1962, 1966; Juberthie-Jupeau, 1973).

The role or putative role of PEBs has never been experimentally investigated.

3.4.3. *The Periesophageal Blood Sinus Formations (SFs)*

Periesophageal blood sinus formations were found for the first time in Julidae by Sahli (1966). Some of these formations also occur in Spirobolidae (Sahli and Petit, 1978), Spirostreptidae (Sahli, 1977b), Callipodidae (Sahli, 1977c), and Glomeridae (Sahli, unpublished data). The role of these structure remains unknown.

In Julidae, SFs consist of four paired groups (S1, S2, S3, S4). They are all closely associated with the periesophageal blood sinus, which is a forward extension of the cephalic aorta. S1, S2, and S3 have been found in Spirobolida (Sahli and Petit, 1978), but only S2 and S3 in Spirostreptidae (Sahli, 1977b). S2s also occur in Callipodidae. Lastly, S3 tracts, without swellings, exist in Glomeridae (Sahli, unpublished data). The structure of the four SFs is uniform in Julidae. The components of all the SFs are the same as those of PEBs (see Sahli, 1966, p. 203).

Only the ultrastructures of the S2s of two Julidae, *Tachypodoiulus niger* (= *albipes*) and *Ommatoiulus sabulosus*, have been published (Sahli and Petit, 1979). The intrinsic secretory cells along with the two axonal types ax 1 (small granules) and ax 2 (large granules) are similar to those of the PEBs.

3.4.4. *The Connective Bodies (CBs)*

CBs were found for the first time by Prabhu (1959, 1961) in *Anaplodesmus* (= *Jonespeltis*) *splendidus* and by Sahli (1962) in *Oxidus* (= *Orthomorpha*) *gracilis*. Both are Polydesmida and belong to the family Paradoxosomatidae. Tripathi's (1976) account of CBs in *Gonoplectus*, an harpagophorid belonging to the Spirostreptida, seems to be erroneous.

3.4.4.1.   LOCATION AND INNERVATION

The paired CBs are located behind and in the vicinity of each connective of the circumesophageal ring. Their location is not strictly constant: in *Oxidus*, they are near the subesophageal commissure (see Fig. 3.14 in Section 3.4.7, below); in *Anoplodesmus*, they lie much more dorsally, near the junction of the connective with the brain.

Each CB is typically innervated by a short nerve emerging from the connective. In some cases (*Anoplodesmus:* Prabhu, 1961), CBs are sessile.

3.4.4.2. HISTOLOGY AND ULTRASTRUCTURE

The CBs consist of enlarged nerve fiber terminations (Prabhu, 1959, 1961, 1962). Each CB is enveloped by a neural lamella. Semithin sections stained with Azure blue B, show two types of material: large light colloidal masses and dark droplets (Sahli and Petit, 1975b). CBs are often filled with secretory products; they show sometimes one or several blood cells.

CBs are undifferentiated neurohemal organs without intrinsic secretory cells. Secretions originate from cerebral NSC (Sahli, 1966; Sahli and Petit, 1975b); the products migrate along the connective via a pathway called PW 4 (= TR 4: Sahli, 1966). At the junction between the brain and the connectives exists an anteroventral area called "commissure K," which belongs to the pathway PW 4. The commissure K is a chiasma and site of axonal swellings, which, in all Diplopoda, present no NSC, contrary to what was described by Prabhu (1962), who probably mistook NSC (C type) for infiltrated blood cells.

Ultrastructural studies confirm that CBs are undifferentiated neurohemal organs: they are made of some glial cells and axonal terminations of three morphological types (Fig. 3.12), called T 1, T 2, and T 3 (Sahli and Petit, 1974). Terminations T 1 show elementary granules of small diameter (max. 100 nm) and of various electron density. In the animals investigated, only few terminations T 1 could be observed. Terminations T 2, which are numerous, are made of spherical granules whose size ranges from 100 to 160 nm and which are very electron-dense. Terminations T 3 are darker than terminations T 2; they are filled with numerous granules, variable in shape and with a maximum diameter of about 200 nm.

3.4.4.3. ROLE IN OOGENESIS

In *Jonespeltis splendidus* (Nair, 1974), bilateral removal of CBs leads to a blocking of oogenesis and to oocyte degeneration. Reimplantation of CBs restores an oogenetic cycle comparable to that observed in controls. So, CBs appear to be the source of a gonadotropic factor, necessary for oocyte growth and vitellogenesis.

3.4.5. *The Paracommissural Plates (PCPs)*

PCPs were found by Sahli (1966) in Julidae and first called "trainées Tr nv" (i.e., trails of neurosecretory products in the neurilemma of the circumesophageal ring). After ultrastructural studies (Sahli and Petit, 1975a), they were renamed "paracommissural plates." PCPs have been observed

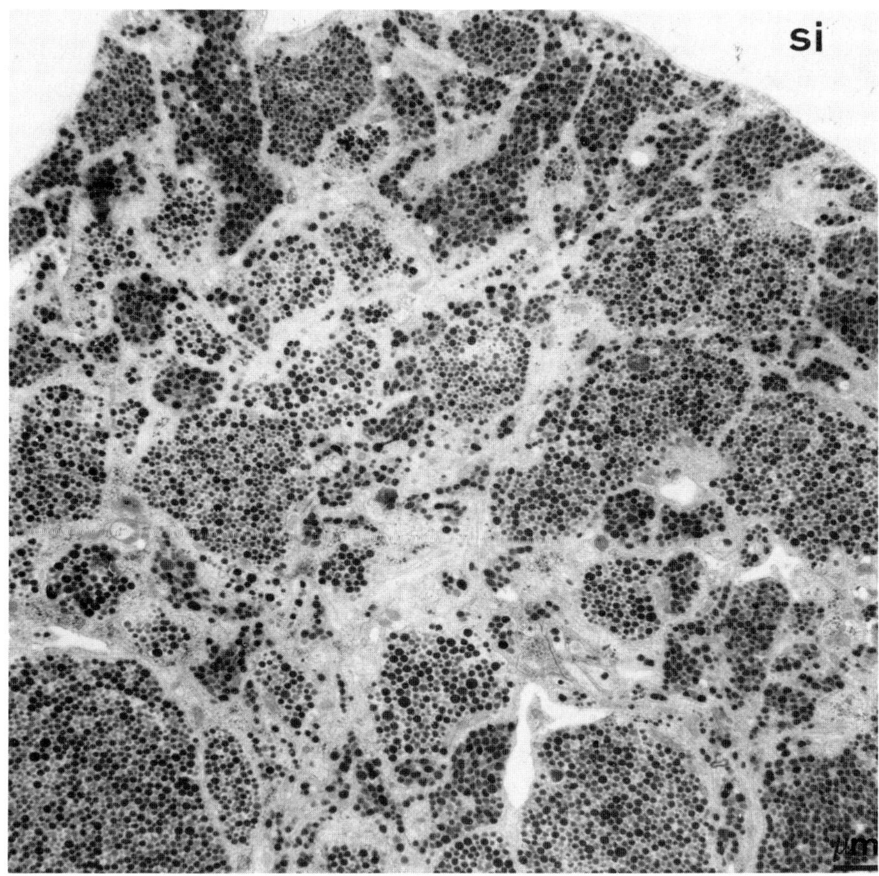

FIGURE 3.12. Ultrastructure of a connective body in *Oxidus gracilis* (Polydesmida: Paradoxosomatidae). General view showing several neurosecretory axon types and absence of secretory cells. *Key:* si = blood sinus. (From Sahli and Caplet, unpublished work.)

in Julidae, Polydesmidae, and Spirobolida (Sahli, unpublished data). Their role is unknown.

PCPs are anatomically not distinctly separated from the connectives. Owing to their cerebral incorporation, PCPs are not innervated by an extracerebral nerve. They represent the axonal endings of NSC, which are located in the ganglia of the subesophageal commissure in Julidae (Petit and Sahli, 1978).

Ultrastructural studies show that in Julidae (Petit and Sahli, 1978) PCPs are composed of at least two distinct types of axonal endings, glial cells, tracheae, and a neural lamella facing a blood sinus. Two types of axonal terminations are present in PCPs whose secretion products correspond strictly and respectively to those of the NSC (namely A and B types in *O. sabulosus;* cells with light and dark backgrounds in *T. niger*) (Petit and Sahli, 1978).

3.4.6.   The Clypeolabral Glands (CLGs)

In *Ommatoiulus sabulosus,* El-Hifnawi and Seifert (1972) incorrectly described Gabe organs. Sahli and Petit (1972) called them "clypeolabral glands" (= "clypeolabral formations": Juberthie-Jupeau, 1983). They have been found in Julidae, Polydesmidae, Callipodidae, and Glomeridae (Sahli, unpublished data). CLGs role remains unknown.

CLGs are paired organs located in the clypeolabrum in front of the stomatogastric bridge, lying on the course of the nervus labri medialis. According to El-Hifnawi and Seifert (1972), in electron microscopy, each CLG appears to be delimited from the surrounding hemocoel by a thick laminated stroma. Intercellular spaces are also filled with stroma. CLGs are made up of (a) axonal profiles, considered as "extrinsic axonal terminations" by the aforementioned authors; (b) one type of intrinsic secretory cells ("parenchyma cells" of the authors); (c) axon-like profiles regarded as the processes of the intrinsic secretory cells; (d) glial-like cells and their processes, which, with exceptions, envelop both the secretory cells and the "extrinsic axons"; and (e) tracheae.

3.4.7.   Evolution of Cephalic Organs of Diplopoda

Sahli (1985b) assumes that in primitive Myriapoda, GOs may have been located as they are in modern Juliformia, in Penicillata, and Symphyla, i.e., far behind the brain (= extended type). Primitively, GOs might have been innervated by three metameral nerves: one mandibular (cf. modern Penicillata, Nguyen Duy-Jacquemin, 1973) and two maxillar nerves (only one maxillar metamere—that of the gnathochilarium—seems to remain in modern Diplopoda; cf. Dohle, 1974). In Julidae, GOs are in connection with one maxillar nerve (Fig. 3.8: S OEM N 6' [= Nsoe 6' in Sahli, 1966]).

Chaudonneret (1978) put forward a hypothesis according to which Arthropoda inherited two primitive lateral nerve cords (PLCs) from an annelid-like ancestor. He suggested that the two lateral cords have given rise to many NHOs and NH-EOs in Arthropoda. According to this hypothesis, GOs might be relics of PLCs and belong to two or three postoral cephalic metameres. If so, GOs might be considered partly homologous with the corpora allata of Insecta (which belong to two postoral metameres; Chaudonneret, 1978).

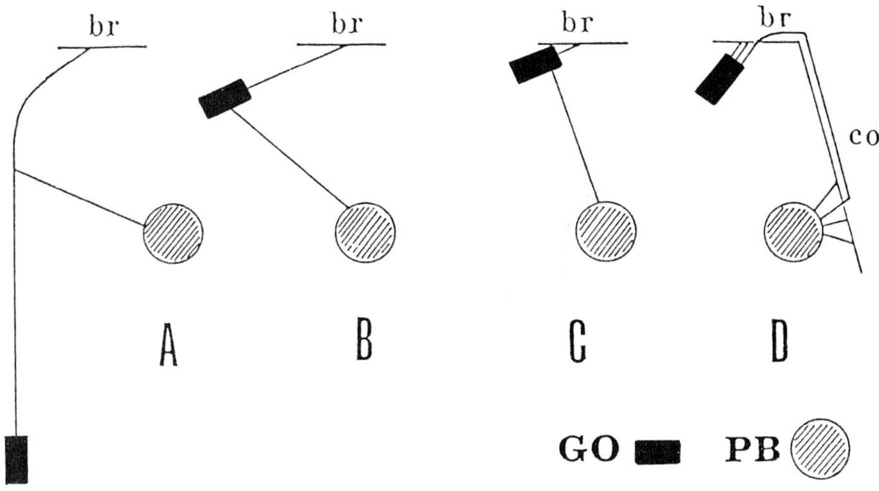

FIGURE 3.13. Gabe organ (GO) and paraesophageal body (PB) systems: (A) Gabe organ, extended type in *Tachypodoiulus*; (B,C,D) condensed type (after the GO has migrated upward) in *Callipus*, *Glomeris* and *Oxidus*, respectively. br = brain; co = connective. (From Sahli, 1985b; courtesy of Artis Bibliotheek.)

In Diplopoda, we can tentatively suggest the scenario illustrated in Fig. 3.13. GOs are progressively brought nearer to the brain, while PEBs always occupy the same position. This gradual trend of the GOs toward the brain ("cephalization") led to a condensed type in which GOs and PEBs are directly connected. A closer approach led nearly to an incorporation of the PEB nerves into the brain (Figs. 3.13 and 3.14). According to Sahli (1985b), not only PEBs, but perhaps also CBs and PCPs, might have their origin in PLCs. If so, PEBs are not the result of a fusion of the first four lateral ganglia of PLCs (as suggested by Sahli, 1985b) but represent only the first ganglion or possibly a fusion of the two first ganglia. Owing to an incorporation process, from extracerebral organs, PCPs might have become intracerebral ones (Fig. 3.15). PCPs might belong to the tritocerebrum (i.e., the third metamere) or to the tetrocerebrum (i.e., a transient metamere being posterior to the tritocerebrum and anterior to the mandibular ganglia; Chaudonneret, 1987). Sahli (1985b) argues that the cerebral glands in Chilopoda may be regarded as GOs that belong to the condensed type and have acquired additional secretory cells during embryonic development (cf. the data of Heymons, 1901, in Scheffel, 1961). Nevertheless, according to J. Chaudonneret (personal communication), cerebral glands of Chilopoda are GOs derived from the ganglia of

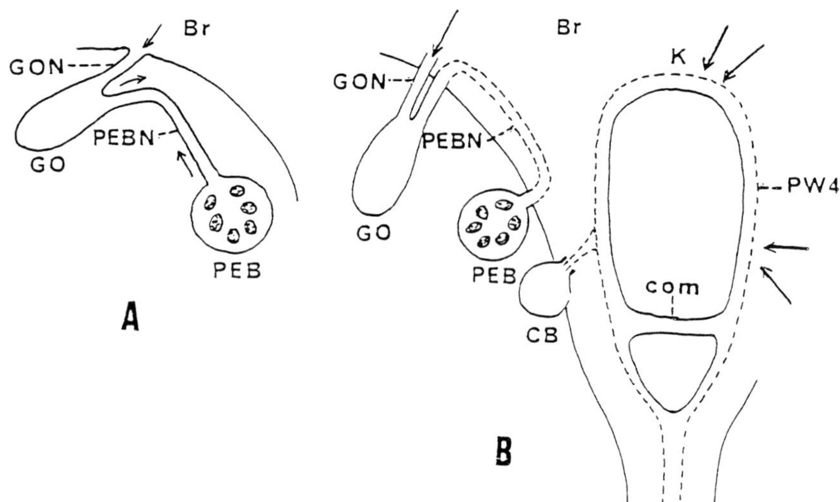

FIGURE 3.14. Diagram of the Gabe organ (GO), paraesophageal body (PEB) and connective body (CB) showing the evolution from a Glomeridae-type (A) to a Paradoxosomatidae-type (B) and the incorporation of the bulk of the paraesophageal body nerve (PEBN) into the brain (in B). Only the left side is represented. Arrows indicate the arrival of neurosecretory material. [Based on data from Juberthie-Jupeau, 1973 (A) and from Sahli and Petit, 1975b (B).] *Key:* Br = brain; com = subesophageal commissure; GON = Gabe organ nerve; K = K commissure. (Modified from Sahli, 1974.)

the PLCs, in which the original neurons have evolved into intrinsic secretory cells, instead of being lost as in Diplopoda and Symphyla.

In short, Myriapoda possess several systems:

(a) A typical myriapodal system (MYR-S) consisting of the GOs (or equivalents), present in all investigated Myriapoda. The MYR-S enters in connection with the PEBs in some Diplopoda.

(b) Modern lateral paraconnective organs (MLPC Os) which might have derived from PLCs. These organs, which exist only in some Diplopoda, are located on each side of the circumesophageal connectives. There are one NH-EO (i.e., the PEBs) and two NHOs: the CBs in connection with the neuropil neurosecretory pathway PW 4 (a typical continuous symphylan–diplopodan Myriapoda system, stretching from the brain to the end of the ventral nerve cord) and the PCPs.

(c) The cephalic blood sinus system made up in Julidae, of four SFs, which might correspond more or less to the aortal neurohemal system in Insecta. But in most Juliformia this system is made up of NH-EOs. SFs

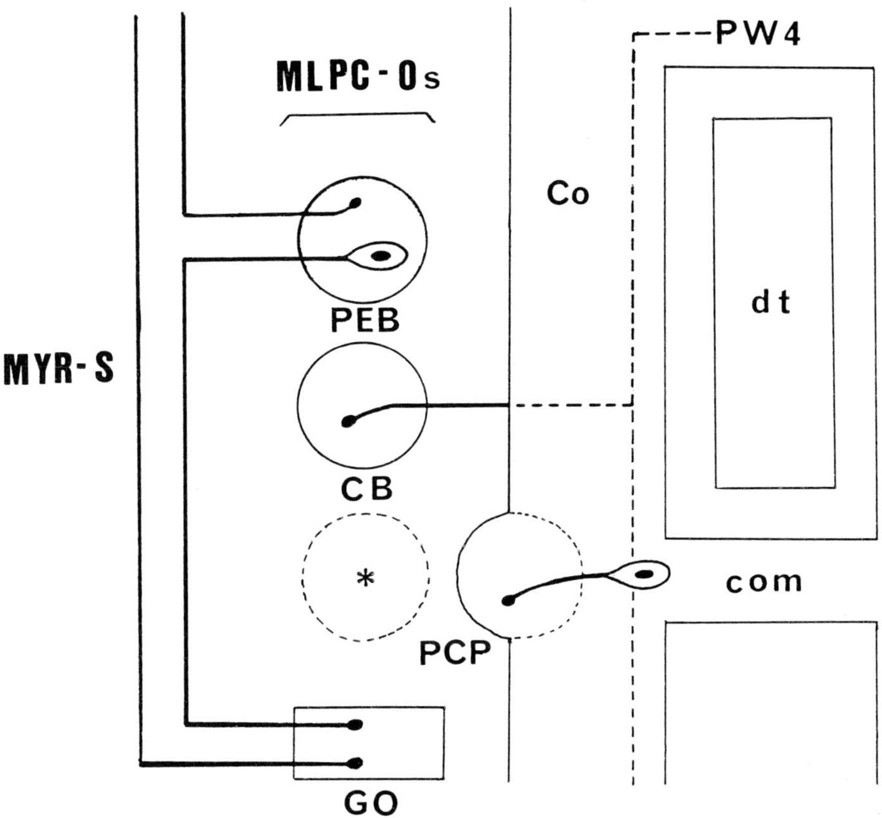

FIGURE 3.15. The myriapodan system (MYR-S) and the three modern lateral paraconnective organs (MLPC-Os: circles). Connective bodies (CB) exist only in Paradoxosomatidae. The dotted circle with an asterisk might represent the primitive location of the paracommissural plates (PCP). *Key:* Co = circumesophageal connective; com = subesophageal commissure; dt = digestive tract; GO = Gabe organ; PEB = paraesophageal body; PW4 = neuropil neurosecretory pathway No. 4. (From Sahli, unpublished work.)

and PEBs are closely associated in Juliformia, as the aortal NH system is with the corpora cardiaca in Insecta.

(d) A clypeolabral system which is in connection with the stomatogastric bridge.

3.5. Summary

Myriapoda present different types of cephalic neurohemal organs. In all classes studied (Symphyla, Chilopoda, Diplopoda), either Gabe organs

(GOs) or cerebral glands (CGs) (or a homologous organ) can be found. In Symphyla and Diplopoda, this neurohemal organ is of undifferentiated type (= NHO) and made of axonal fibers, axonal swellings, and glial cells, surrounded by a connective sheath. In Chilopoda, cerebral glands are of the differentiated type (= NH-EO), i.e., with, in addition to elements previously cited, intrinsic secretory cells. Major variations concern organ location and shape, and internal components such as number of glial cell types (or glandular cell types, if any), axon types, or the presence (or not) of connective septa. In Chilopoda, CG of *Cryptops* (Scolopendromorpha) is the only one in which nervous elements are not mixed with glandular cells.

In Symphyla the involvement of the cephalic organ in molting control is only putative, but in Diplopoda and Chilopoda GOs or CGs have a certain moderating influence on molting. Chilopod's CG is also involved in gametogenesis control, having a moderating effect on spermatogenesis and being necessary for normal oogenesis.

From other kinds of neurohemal organs (only described in Diplopoda), only the role of connective bodies has been experimentally investigated; these NHOs, located in the vicinity of the periesophageal ring, consist of axon swellings and glial cells enveloped by a neural lamella. CBs are necessary for normal oogenesis.

Paraesophageal bodies (NH-EO), periesophageal blood sinus formations (NHO), paracommissural plates (NHO), and clypeolabral glands (NH-EO) have no reported morphogenetic role.

Acknowledgments

We sincerely thank Professor J. Chaudonneret (University of Dijon, France) for his many comments. We also thank Dr. L. Juberthie-Jupeau (Laboratoire souterrain CNRS, Moulis, France) and Professor G. Seifert (University of Giessen, Federal Republic of Germany) for sending original micrographs.

References

Chaudonneret, J. 1978. La phylogenèse du système nerveux annélido-arthropodien. Bull. Soc. Zool. Fr. 103: 69–95.

Chaudonneret, J. 1987. Evolution of insect brain, with special reference to the so-called tritocerebrum. Pp. 3–26 *in* A. P. Gupta (ed.), *Arthropod Brain*. Wiley, New York.

De Lerma, B. 1951. Note originali e critiche sulla morfologia comparata degli organi frontali degli arthropodi. Ann. Ist. Mus. Zool. Univ. Napoli 3: 1–22.

Descamps, M. 1975. Etude du contrôle endocrinien du cycle spermatogénétique chez *Lithobius forficatus* L. (Myriapode Chilopode): rôle du complexe "cellules neurosécrétrices des lobes frontaux du protocérébron–glandes cérébrales." Gen. Comp. Endocrinol. 25: 346–357.

Descamps, M. 1978. Rôle des centres endocrines céphaliques dans la régulation de la spermatogenèse chez *Lithobius forficatus* L. (Myriapode Chilopode). Bull. Soc. Zool. Fr. 103: 367–373.

Descamps, M. and R. Joly. 1985. Ultrastructure of the cerebral glands in *Scolopendra cingulata* Latr., *Cryptops savignyi* Leach and *C. hortensis* Leach (Myriapoda: Scolopendromorpha). Int. J. Insect. Morphol. Embryol. 14: 105–113.

Descamps, M., R. Joly, and C. Jamault-Navarro. 1985. Autoradiographic localization of 5-hydroxytrypamine and noradrenaline in the central nervous system of *Lithobius forficatus* L. (Myriapoda Chilopoda). Bijdr. Dierkd. 55: 47–54.

Dohle, W. 1974. The segmentation of the germ band of Diplopoda compared with other classes of arthropods. Symp. Zool. Soc. Lond. 32: 143–160.

El-Hifnawi, E. and G. Seifert. 1972. Die Ultrastruktur des Gabeschen Organs von *Schizophyllum sabulosum* L. (Diplopoda, Juliformia). Z. Zellforsch. 132: 273–285.

Ernst, A. 1971. Licht- und electronenmikroskopische Untersuchungen zur Neurosekretion bei *Geophilus longicornis* Leach unter besonderer Berücksichtigung der Neurohämalorgane. Z. Wiss. Zool. 182: 62–130.

Fahlander, K. 1938. Beiträge zur Anatomie und systematischen Einteilung der Chilopoden. Zool. Bidr. Uppsala 17: 1–148.

Gabe, M. 1952. Sur l'emplacement et les connexions des cellules neurosécrétrices dans les ganglions cérébroïdes de quelques Chilopodes. C. R. Acad. Sci. Paris 235: 1430–1432.

Gabe, M. 1954. Emplacement et connexions des cellules neurosécrétrices chez quelques Diplopodes. C. R. Acad. Sci. Paris 239: 828–830.

Gabe, M. 1966. *Neurosecretion.* Pergamon Press, Oxford and Elmsford, New York.

Gersch, M. 1958. Neurohormone bei wirbellosen Tieren. Verh. Dtsch. Zool. Ges. 22: 40–76.

Gersch, M. 1964. *Vergleichende Endokrinologie der wirbellosen Tiere.* Akademische Verlagsges., Leipzig.

Glaser, R. 1958. Histologische Untersuchungen über das neurosekretorische System bei *Polydesmus testaceus* Ltz. (Diplopoda). Cited by Gersch (1958), above.

Heymons, R. 1901. Die Entwicklungsgeschichte der Scolopender. Zoologica (Stuttg.) 13: 1–224.

Herbaut, C. 1976. Etude expérimentale de la régulation endocrinienne de l'ovogenèse chez *Lithobius forficatus* L. (Myriapode Chilopode): rôle du complexe "cellules neurosécrétrices protocérébrales–glandes cérébrales." Gen. Comp. Endocrinol. 28: 264–276.

Hoffman, R. L. 1979. *Classification of Diplopoda*. Museum d'Histoire naturelle, Geneva.

Holmgren, N. 1916. Zur vergleichenden Anatomie des Gehirns von Polychaeten, Onychophoren, Xiphosuren, Arachniden, Crustaceen, Myriapoden und Insekten. K. Sven. Vetenskapsakad. Handl. 56: 1–103.

Jamault-Navarro, C. 1981. Cellules neurosécrétrices et trajets axonaux protocérébraux chez *Lithobius forficatus* L. (Myriapode Chilopode): étude ultrastructurale. Arch. Biol. (Bruxelles) 92: 203–218.

Joly, R. 1961. Déclenchement expérimental de la mue chez *Lithobius forficatus* L. (Myriapode Chilopode). C. R. Acad. Sci. Paris 252: 1673–1675.

Joly, R. 1962. Les glandes cérébrales, organes inhibiteurs de la mue chez les Myriapodes Chilopodes. C. R. Acad. Sci. Paris 254: 1679–1681.

Joly, R. 1966a. Contribution à l'étude du cycle de mue et de son déterminisme chez les Myriapodes Chilopodes. Bull. Biol. Fr. Belg. 3: 379–480.

Joly, R. 1966b. Sur l'ultrastructure de la glande cérébrale de *Lithobius forficatus* L. (Myriapode Chilopode). C. R. Acad. Sci. Paris 263: 374–377.

Joly, R. 1970. Evolution cyclique des glandes cérébrales au cours de l'intermue chez *Lithobius forficatus* L. (Myriapode Chilopode). Z. Zellforsch. Mikrosk. Anat. 110: 85–96.

Joly, R. 1976. Influence de quelques interventions expérimentales sur l'activité sécrétoire des glandes cérébrales chez *Lithobius forficatus* L. (Myriapode Chilopode). Gen. Comp. Endocrinol. 30: 301–312.

Joly, R. 1977. Influence de quelques facteurs externes sur l'activité sécrétoire des glandes cérébrales chez *Lithobius forficatus* L. (Myriapode Chilopode): étude en microscopie photonique et électronique. Gen. Comp. Endocrinol. 32: 167–178.

Joly, R. 1980. Evolution ultrastructurale et action de greffons de glandes cérébrales sur le cycle de mue chez *Lithobius forficatus* L. (Myriapode Chilopode). Bull. Soc. Zool. Fr. 105: 49–56.

Joly, R. and M. Descamps. 1968. Etude comparative du complexe endocrine céphalique chez les Myriapodes Chilopodes. Gen. Comp. Endocrinol. 10: 364–375.

Joly, R. and M. Descamps. 1977. Influence de l'électrostimulation cérébrale sur l'histologie ultrastructurale et le rôle physiologique des glandes cérébrales chez *Lithobius forficatus* L. (Myriapode Chilopode). Arch. Biol. (Bruxelles) 88: 333–347.

Joly, R. and M. Descamps. 1987. Histology and ultrastructure of the myriapod brain. Pp. 135–157 *in* A. P. Gupta (ed.), *Arthropod Brain*. Wiley, New York.

Joly, R. and G. Devauchelle. 1970. Etude cytochimique de la glande cérébrale de *Lithobius forficatus* L. (Myriapode Chilopode), nature des sécrétions. J. Microsc. (Paris) 9: 631–642.

Joly, R. and C. Jamault-Navarro. 1978. Rôle de la *pars intercerebralis* sur l'activité sécrétoire des glandes cérébrales chez *Lithobius forficatus* L. (Myriapode Chilopode): étude ultrastructurale. Arch. Zool. Exp. Gén. 119: 487–496.

Juberthie, C. and L. Juberthie-Jupeau. 1974. Etude ultrastructurale de l'organe neurohemal cérébral de *Spelaeoglomeris doderoi* Silvestri, Myriapode Diplopode cavernicole. Symp. Zool. Soc. Lond. 32: 199–210.

Juberthie-Jupeau, L. 1961. Données sur la neurosécrétion protocérébrale et mise en évidence de glandes céphaliques chez *Scutigerella pagesi* Jupeau (Myriapode, Symphyle). C. R. Acad. Sci. Paris 253: 3081–3083.

Juberthie-Jupeau, L. 1963. Recherches sur la reproduction et la mue chez les Symphyles. Arch. Zool. Exp. Gén. 102: 1–172.

Juberthie-Jupeau, L. 1967. Données sur le système endocrinien de quelques Diplopodes Oniscomorphes (Myriapodes). C. R. Acad. Sci. Paris 265: 1525–1529.

Juberthie-Jupeau, L. 1973. Etude ultrastructurale des corps para-oesophagiens chez un Diplopode Oniscomorphe *Loboglomeris pyrenaica* Latz. C. R. Acad. Sci. Paris 276: 169–172.

Juberthie-Jupeau, L. 1983. Neurosecretory systems and neurohemal organs of Myriapoda. Pp. 204–278 *in* A. P. Gupta (ed.), *Neurohemal Organs of Arthropods*. Thomas, Springfield, Illinois.

Juberthie-Jupeau, L. and C. Juberthie. 1973a. Etude ultrastructurale de l'organe neurohémal céphalique chez un Symphyle *Scutigerella silvatica* (Myriapode). C. R. Acad. Sci. Paris 276: 1577–1581.

Juberthie-Jupeau, L. and C. Juberthie. 1973b. Décharge par exocytose d'une catégorie de granules de neurosécrétion dans l'organe neurohémal d'un Symphyle (Myriapode). C. R. Acad. Sci. Paris 277: 1357–1360.

Nair, V. S. K. 1974. Studies on the probable neurosecretory control of moulting and vitellogenesis in the millipede *Jonespeltis splendidus* Verhoeff (Myriapoda Diplopoda). Thesis, University of Kerala, India.

Nair, V. S. K. 1980. Moult inhibition in the insect *Dysdercus cingulatus* (Insecta: Heteroptera) by the cerebral glands of the millipede *Jonespeltis splendidus* (Myriapoda: Diplopoda). Experientia (Basel) 36: 607–608.

Newport, G. 1843. On the structure, relations and development of the nervous and circulatory system and on the existence of the complete circulation of the blood in vessels, in Myriapoda and macrurous Arachnida. Philos. Trans. R. Soc. Lond. 2: 243–302.

Nguyen Duy-Jacquemin, M. 1971. Mise en évidence de glandes cérébrales chez le Diplopode Penicillate *Polyxenus lagurus* (Myriapodes). C. R. Acad. Sci. Paris 272: 2195–2196.

Nguyen Duy-Jacquemin, M. 1973. Contribution à la connaissance de l'anatomie céphalique, des formations endocrines et du développement postembryonnaire de *Polyxenus lagurus* (Diplopodes Penicillates). Thesis Sci., University of Paris, No. AO 8186.

Palm, N. B. 1955. Neurosecretory cells and associated structures in *Lithobius forficatus* L. Ark. Zool. K. Sven. Vetenskap. 9: 115–129.

Petit, J. and F. Sahli. 1973. Etude cytochimique ultrastructurale des corps paraoesophagiens chez le Diplopode Iulide, *Schizophyllum sabulosum* (L.). J. Microsc. (Paris) 17: 86a.

Petit, J. and F. Sahli. 1975. Cytochemical and electron-microscopic study of the paraoesophageal bodies and related nerves in *Schizophyllum sabulosum* (L.), Diplopoda Julidae. Cell Tissue Res. 162: 367–375.

Petit, J. and F. Sahli. 1977. Etude ultrastructurale des cellules neurosécrétrices des globuli I chez *Schizophyllum sabulosum* (Diplopoda: Julidae). Bull. Soc. Zool. Fr. 102: 431–437.

Petit, J. and F. Sahli. 1978. Etude ultrastructurale des plages paracommissurales (organes neurohémaux céphaliques) des Diplopodes Iulidae. Abh. Verh. Naturwiss. Ver. Hamb. 21/22: 295–309.

Prabhu, V. K. K. 1959. Note on the cerebral gland and a hitherto unknown connective body in *Jonespeltis splendidus* Verhoeff (Myriapoda, Diplopoda). Curr. Sci. (Bangalore) 28: 330–331.

Prabhu, V. K. K. 1961. The structure of the cerebral glands and connective bodies of *Jonespeltis splendidus* Verhoeff (Myriapoda: Diplopoda). Z. Zellforsch. Mikrosk. Anat. 54: 717–733.

Prabhu, V. K. K. 1962. Neurosecretory system of *Jonespeltis splendidus* (Verhoeff) (Myriapoda, Diplopoda). Pp. 417–420 *in* H. Heller and R. B. Clark (eds.), *Neurosecretion*, Mem. Soc. Endocrinol., Vol. 12. Academic Press, Orlando, Florida.

Rilling, G. 1968. *Lithobius forficatus*, Grosses Zool. Prakt., Vol. 13b. Fischer Verlag, Stuttgart.

Rosenberg, J. 1976. Die Ultrastruktur der Gabeschen Organe ("Cerebraldrüsen") von *Scutigera coleoptrata* (Chilopoda, Notostigmophora). Zool. Beitr. 22: 281–305.

Sahli, F. 1958. Quelques données sur la neurosécrétion chez le Diplopode *Tachypodoiulus albipes* C. L. Koch. C. R. Acad. Sci. Paris 246: 470–472.

Sahli, F. 1961. Sur une formation hypocérébrale chez les Diplopodes Iulides. C. R. Acad. Sci. Paris 252: 2443–2444.

Sahli, F. 1962. Sur le système neurosécréteur du Polydesmoïde *Orthomorpha gracilis* C. L. Koch (Myriapoda Diplopoda). C. R. Acad. Sci. Paris 254: 1498–1500.

Sahli, F. 1963. Sur le système neurosécréteur des Myriapodes Diplopodes. Proc. 16th Int. Congr. Zool., Washington, D.C. Vol. 2, p. 149.

Sahli, F. 1966. Contribution à l'étude de la périodomorphose et du système neurosécréteur des Diplopodes Iulides. Thesis Sci., University of Dijon, No. 94.

Sahli, F. 1974. Sur les organes neurohémaux et endocrines des Myriapodes Diplopodes. Symp. Zool. Soc. Lond. 32: 217–230.

Sahli, F. 1977a. Sur les cellules neurosécrétrices des globuli I et sur la voie neurosécrétrice protocéphalique des Myriapodes Diplopodes: premières observations chez les Polydesmida Platyrrhacidae, les Nematophora Lysiopetalidae, les Spirostreptida et les Spirobolida. C. R. Acad. Sci. Paris 284: 815–817.

Sahli, F. 1977b. Sur le système neurosécréteur des Spirostreptidae (Diplopoda, Spirostreptida). C. R. Acad. Sci. Paris 285: 1315–1318.

Sahli, F. 1977c. Présence de corps paraoesophagiens chez les Nematophora Lysiopetalidae (Myriapoda, Diplopoda). C. R. Acad. Sci. Paris 284: 211–213.

Sahli, F. 1979. Different types of neurosecretory system in Diplopoda. Pp. 279–285 *in* M. Camatini (ed.), *Myriapod Biology.* Academic Press, Orlando, Florida.

Sahli, F. 1980. On the nerves of the Gabe organs and paraoesophageal bodies in *Oxidus gracilis* (C.L.K.) (Myriapoda Diplopoda) and on the utmost importance of thorough anatomical knowledge. J. Adv. Zool. 1: 24–27.

Sahli, F. 1985a. Aspects of the neurosecretory system relevant to the classification of the juliform millipedes. Z. Zool. Syst. Evolutionsforsch. 23: 10–15.

Sahli, F. 1985b. Neurohemal organs in Myriapoda and phylogeny. Bijdr. Dierkd. 55: 193–201.

Sahli, F. and J. Petit. 1972. Observations sur l'ultrastructure des organes de Gabe des Polydesmidae et des Iulidae (Diplopoda). C. R. Acad. Sci. Paris 275: 2017–2020.

Sahli, F. and J. Petit. 1973. Observations sur l'ultrastructure des corps paraoesophagiens des Diplopodes Iulides. C. R. Acad. Sci. Paris 276: 2019–2022.

Sahli, F. and J. Petit. 1974. Observations sur l'ultrastructure des corps connectifs (organes neurohémaux céphaliques) d'*Orthomorpha gracilis* (C.L.K.) (Diplopoda, Polydesmoidea). C. R. Acad. Sci. Paris 279: 2055–2058.

Sahli, F. and J. Petit. 1975a. Les plages paracommissurales (formations neurohémales céphaliques) des Diplopodes. C. R. Acad. Sci. Paris 280: 2001–2004.

Sahli, F. and J. Petit. 1975b. Sur les voies de neurosécrétion ("TR 4") dans le collier perioesophagien des Diplopodes et plus particulièrement du Strongilosomidae *Orthomorpha gracilis* (C.L.K.). Bull. Soc. Zool. Fr. 100: 521–529.

Sahli, F. and J. Petit. 1978. Sur le système neurosécréteur céphalique de quelques Spirobolida (Diplopoda). C. R. Acad. Sci. Paris 286: 473–476.

Sahli, F. and J. Petit. 1979. The latero-oesophageal complex in Iulidae (Diplopoda). Pp. 308–313 *in* M. Camatini (ed.), *Myriapod Biology.* Academic Press, Orlando, Florida.

Scheffel, H. 1961. Untersuchungen zur Neurosekretion bei *Lithobius forficatus* L. (Chilopoda). Zool. Jahrb. Anat. 79: 529–556.

Scheffel, H. 1965a. Über die Wirkung implantierter Cerebraldrüsen auf die Larvenhäutungen von *Lithobius forficatus* L. (Chilopoda). Zool. Anz. 174: 173–178.

Scheffel, H. 1965b. Elektronenmikroskopische Untersuchungen über den Bau der Cerebraldrüse der Chilopoden. Zool. Jahrb. Physiol. 71: 624–640.

Scheffel, H. 1969. Untersuchungen über die hormonale Regulation von Häutung und Anamorphose von *Lithobius forficatus* L. (Myriapoda, Chilopoda). Zool. Jahrb. Physiol. 74: 436–505.

Seifert, G. 1971. Ein bisher unbekanntes Neurohämalorgan von *Craspedosoma rawlinsii* Leach (Diplopoda, Nematophora). Z. Morphol. Tiere 70: 128–140.

Seifert, G. and E. El-Hifnawi. 1972. Die Ultrastruktur des Neurohämalorgans am Nervus protocerebralis von *Polyxenus lagurus* (L.) (Diplopoda, Penicillata). Z. Morphol. ökol. Tiere 71: 116–127.

Tripathi, S. P. 1976. Nervous and neurosecretory system of the millipede *Gonoplectus malayus* (Carl.). Ph.D. Thesis, University of Gorakhpur, India.

Morphology, Histology, and Ultrastructure of the Ecdysial Gland (Y-Organ) in Crustacea

4

EUGENE SPAZIANI

4.1. Introduction

The Y-organs are glands in crustaceans that are specialized for the synthesis and secretion of one or more steroid hormones (ecdysteroids). Functional homologues of the Y-organs occur in insects: the prothoracic glands. In arthropods generally, the secreted ecdysteroids, but especially their 20-hydroxy metabolites, regulate growth and development, including molting and regeneration. Collectively, then, the glands (generically termed "ecdysial") play a central role in the life cycles of these abundant, economically important animals. This role is sufficient as rationale to attempt understanding the physiology of these glands. However, they are under intense study also because experiments in recent years make it apparent that they are useful as models in the study of (a) steroid hormone biosynthesis generally in animals and (b) regulation of that synthesis by peptide hormones. For example, in common with their vertebrate analogues (ovary, testis, and adrenal cortex as endocrine organs), ecdysial glands utilize cholesterol as hormone precursor, apparently employ the same second messenger systems (although with potentially revealing variations), and are controlled by peptide hormones of cephalic origin (see Watson and Spaziani, 1985b; Mattson and Spaziani, 1987). For reviews of Y-organ physiology, readers are directed to recent surveys of crustacean endocrinology by Skinner (1985), Mattson (1986), Quackenbush (1986), Fingerman (1987), and Watson et al. (1989). That by Watson et al. (1989) compares ecdysial gland mechanisms in crustaceans and insects. The present review focuses on Y-organ structure. Coverage is intended to be comprehensive and to relate, where possible, structure with function.

Where structural changes are described in relation to the stages of the molting cycle, stages and substages are designated according to the scheme of Drach (1939; cf. Skinner, 1985). Progressively from the molt itself (ecdysis: stage E), there occur early and late postmolt (metecdysis: stages A and B, respectively), intermolt (anecdysis: C_{1-4}), and premolt (proecdysis: D_{0-4}).

4.2. Gross Anatomy

Y-organs have been studied morphologically in representative adults of most macrocrustaceans, including the Peracarida (amphipods and isopods) and in developmental stages of several species of decapod larvae. The Hoplocarida (stomatopods) and Cirripedia (barnacles) have unaccountably been neglected, but in other major taxonomic groups discrete glands have been described and accurately identified as the homologues of the classical Y-organs of brachyuran crabs. However, readers of the earlier literature should be aware that tissues of glandular appearance often were termed Y-organs, molting glands, or ecdysial glands, but have since been shown to be patches of lymphatic cells, antennary glands, or the mandibular organs. These errors entered the literature beginning with the papers of M. Gabe (1953, 1956), who discovered the Y-organs of brachyurans, and persisted to as late as 1975 (Hinsch and Hajj, 1975; cf. Hinsch, 1977). The problem arose apparently because of difficulties in assigning a segmental origin to the putative Y-organs, in part due to the lack of traceable innervation or vasculature. Gabe (1953, 1956), for example, examined the "Y-organs" in over 100 species of malacostracans, but did not provide sufficient written details or drawings to enable subsequent investigators to distinguish between Y-organs and other tissues with similar histology. This vagueness included the brachyuran Y-organ, but through the efforts of Gabe's student, Echalier (1954–1959), probably in collaboration with Gabe (cf. Sochasky et al., 1972), the Y-organs in *Carcinus maenas* were clearly described and their location recorded by drawings. Moreover, Echalier (1954–1959) provided the original experimental proof of Gabe's hypothesis that the Y-organs are the source of a molt-controlling hormone, a functional homologue of that of insect prothoracic glands. In view of this history, it is my opinion that the glands traditionally referred to as the Y-organs of Gabe should, instead, be referred to as the Y-organs of Gabe and Echalier. Subsequent publications by others, in which morphological or experimental data were incorrectly attributed to Y-organs, or in which the identity of the Y-organs is inadequately documented, are not included in the present review. However, they are cited and critically analyzed by Sochasky and colleagues (1972, 1974). Descriptive studies that distinguish Y-organs from other glands in the cephalothorax, especially the mandibular organs, are those of Sochasky et al. (1972), Aoto et al. (1974), Burghause (1975), Demeusy (1975), Bazin (1976), Buchholz and Adelung (1980) (see also Hinsch, Chapter 10, in this volume); Le Roux (1974, 1977) distinguished those tissues in 27 species encompassing three species of Euphausiacea and four infraorders of Decapoda (Caridea, Astacidea, Anomura and Brachyura). Additional discussions of misidentity of tissues appear in Demeusy (1975), Vernet (1976), Vernet et al. (1978), McConaugha (1980), Spindler et al. (1980), and Skinner (1985).

Y-organs are most conspicuous as discrete glands in brachyuran crabs, usually being 1–2 mm in diameter, but are as much as 3 mm in diameter and weigh in excess of 10 mg in adults of the larger species that have been examined. Within a species, the glands may be ovoid to round in shape, yellowish to opalescent white in color, and are often embedded in yellowish to brown tissues of fatty appearance (e.g., *Cancer* sp.: Simione and

Hoffman, 1975). In older individuals, this periglandular tissue may penetrate the gland with the enveloping sheath (Hinsch et al., 1980). Viewed from the dorsal aspect, location of the glands in brachyurans is invariably "ventral to the insertion of the lateral portion of the mandibular external adductor muscle and just above the junction of the branchiostegite with the lateral body cuticle at the anterior end of the branchial chamber" (Passano, 1960). On each side, they lie against the epidermis (hypodermis) of the ventral anterior face of the cephalothorax, lateral to the eyestalks. The glands are epidermal derivatives (see Section 4.3, below) but in brachyurans are not attached to the epidermis, except in occasional species by a residual strand (e.g., *Pachygrapsus:* Vernet et al., 1978). I consider the most useful illustrations of the glands' position to be the line drawings by Echalier (1959), one of which was reprinted by Passano (1960, his Fig. 4) and by Skinner (1985, her Fig. 8B). The accessibility of the glands permits their surgical removal, for which purpose Passano (1961, his Fig. 1) provided a photograph of a gland *in situ* through a window cut in the exoskeleton (of *Sesarma*). Johnson (1980, her Fig. 1) shows the position clearly in *Callinectes* with a marked photograph of the whole animal. The segmental position of the Y-organs appears generally to be maxillary, but Le Roux (1974) examined frontal sections and other anatomic evidence and concluded that adults of the crabs *Inachus* and *Macropodia* are among exceptions in that the Y-organs are in the maxillulary segment.

In the infraorders Caridea (prawns, shrimps) and Astacidea (crayfishes, lobsters), Y-organs are more difficult to locate, relative to those in brachyurans, owing to less prominent profiles and to variation in shape and, to some extent, position. The Y-organs of the prawns *Palaemon* and *Pandalus* and the crayfish *Procambarus* are in the junction of the prebranchial and branchial chambers, just beneath the inner wall of the branchiostegite (Aoto et al., 1974); an important additional characteristic is that the glands are just under the lateral insertion of the posterior dorsoventral muscle. Aoto et al. (1974) provided detailed written descriptions and drawings of whole-animal cross sections and frontal sections of *Palaemon paucidens* that pinpoint the location of the Y-organs relative to other organs, including the mandibular gland. Lucid drawings of lateral views of the crayfish *Orconectes* (= *Cambarus*) by Burghause (1975) and of the lobster *Homarus* by Sochasky et al. (1972) make it clear that the position of the Y-organs is identical or very similar in all the genera of Caridea and Astacidea so far mentioned. Le Roux (1974, 1977) would add to this list the adults of *Palaemon serratus* and *Palaemonetes* and of the anomuran *Pagarus*, and Bourquet et al. (1977) the shrimp *Penaeus*. Except for the larvae of some species (see below), the position of the Y-organs is generally in the maxillary segment, as in the Brachyura. The glands in these groups have been described as "translucent" (Aoto et al., 1974), but differ from brachyuran glands in at least two respects: (a) they are always closely applied to the epidermis, often appearing to be locally hypertrophied portions of the ventral epidermis projecting into a cavity, and (b) they are elongate, flattened, or tubular strips of tissue. Le Roux (1974) observed cross-striations on the glands of *Homarus*. They range from 2 to 6 mm long and from 0.2 to 0.3 mm wide in mature adults. *Orconectes* is an exception, with glands that are flattened, ovoid disks (Burghause, 1975).

Among the Peracarida, Ducruet (1986) describes the position of the Y-organs of an amphipod, *Gammarus pulex*, as ventrolateral in the back part of the head. Each is a cellular mass of the epidermis that juts out into the hemocoel. The Y-organs are immediately behind local cuticular thickenings that are easily seen from the outside. Their position in the maxillary segment was established early in studies of the isopods *Porcellio* (Maissiat and Legrand, 1970), *Ligia* (Maissiat and Maissiat, 1976), and *Sphaeroma* and *Cymadoce* (Charmantier and Trilles, 1973, 1975, 1979). Their identity as Y-organs was demonstrated unequivocally through experiments in which molting could be prevented by cautery ablation and restored by implanting glands or injecting ecdysteroids (Maissiat, 1970a,b; Maissiat and Legrand, 1970; Maissiat and Graf, 1973; Blanchet, 1974). It may be characteristic of isopods (except *Ligia:* see Maissiat et al., 1979) that the Y-organs degenerate and disappear in males within 2–4 months after the pubertal molt (Charmantier and Trilles, 1973) whereas they remain in females throughout life (Charmantier and Trilles, 1975). Similarly, in brachyurans, among the Oxyrhncha (decorator and spider crabs), the last molt in males occurs on attaining sexual maturity; this is followed in *Maja* (Carlisle, 1957) and *Acanthonyx* (Chaix et al., 1976) with degeneration of the Y-organ within 2 weeks. In contrast, the portunid crab *Carcinus* undergoes the pubertal molt early when it is quite small; it retains the Y-organs and continues growing with several subsequent molts (Carlisle, 1957).

Careful microscopic examinations in the modern era permit the conclusion that Y-organs are ductless glands that are not innervated and are not penetrated by systemic blood vessels. Echalier (1959) claimed to see both nervous and capillary penetrations of Y-organs in *Carcinus*, but these could not be confirmed (Passano, 1960; Knowles, 1965). Passano (1960) observed nerves approaching from the subesophageal ganglion, but these bypassed the glands. Johnson (1980) found the same in *Callinectes*. Echalier (1959) noted that cutting these nerves in *Carcinus* did not affect Y-organ function. Searches for nerve connections were similarly negative for carideans and astacideans: *Pandalus* and *Procambarus* (Aoto et al., 1974), *Orconectes* (Burghause, 1975), and *Palaemon* (Aoto et al., 1974; Le Roux, 1977). Generally in crustaceans, the Y-organs are positioned in—or sufficiently near—a hemocoel so that nutrient, metabolite, hormone, and gas exchanges occur through superfusion of hemolymph over the surface of the glands (see Passano, 1960; Aoto et al., 1974). Circulation within the glands occurs readily, facilitated by sinusoids that are sufficiently large to permit transport of hemocytes (see Section 4.4, below).

In concluding this section on gross morphology, a remarkable structure should be mentioned that appears to be unique to the isopods. This is the lateral cephalic nerve plexus (LCNP) located adjacent to the Y-organs in all seven genera that have thus far been studied in detail. Isopods have two neurohemal organs in the head. One is the classical sinus gland of crustaceans generally, which is located on the surface of an optic lobe and is made up of neurosecretory axon terminals whose cell bodies are mostly in the eyestalks. The other is the LCNP, which is a collection of axon terminals of neurosecretory cells in several parts of the central nervous system, including one type that originates in the plexus itself. It is near or closely applied to the anterior end of the Y-organ, but is not structurally integrated with the gland (see Fig. 4.3A,B). [*Note:* All figures are grouped at the end of the text, before the references.] Its function is not known, but is widely believed to control Y-organ activity in concert with neurosecretions from the sinus gland. Maissiat and colleagues (1979) present evidence to suggest that it is responsible for degeneration of Y-organs after the pubertal molt. With extracellular electrodes placed near the Y-organs, Chiang (1987) has recorded long duration, large-amplitude action potentials originating from the LCNP that may influence Y-organ secretion. The literature on this neurohemal organ is summarized in the research papers just mentioned, and in that of Martin et al. (1983), which includes a diagram of all the relevant cephalic neural-endocrine elements and their interconnections in *Ligia*.

4.3.   Development

After hatching, crustaceans briefly exist (for some hours) in embryonic form as a prezoea, which molts to the first zoeal larval stage. There are two to six zoeal molt stages, depending on species (five, for example, in *Cancer* sp.: see Anderson, 1978, for a full description of development). These are similar in appearance, but there follows a molt and modified metamorphosis to the megalopa stage. These stages occur in the plankton and occupy one to several months, after which the larvae settle to the substratum and molt to the first stage in which the young resemble adults.

The literature on Y-organ development is sparse, and virtually nothing is known of the roles that the developing endocrine glands play in larval molting. McConaugha (1980) followed Y-organ development through the larval stages of *Cancer*. The gland appears in the first zoeal stage as a cord of 6–10 epidermal cells with sparse cytoplasm and progressively undergoes hyperplasia, then folding and mixing of the cellular cords as development proceeds through the megalopa stage. By the first crab stage molt, the glands resemble those in adults: they are spindle-shaped and subdivided into lobules with hemocoelic sinuses. McConaugha found that only a single cell type was evident, but the cells showed changes with the 7-day molt cycle. Starting on day 4 after the previous molt (apparently molt stage C_{3-4}), cytoplasmic volume increases greatly, especially in cells near the hemocoel. During premolt, extensive cytoplasmic vacuolization was seen, suggesting the presence of a secretory product. Similar observations were made by Buchholz (1984), who studied Y-organ development in a mud crab, the xanthid *Rithropanopeus harrisii*. From the first zoeal stage to the megalopa, cell number and volume increased four- to fivefold; from first zoea to the first juvenile stage, gland size increased 11- to 13-fold. The glands are quite discrete in form by the fourth zoeal stage; Fig. 4.1 shows a three-dimensional reconstruction of such a gland in *R. harrissi*.

Y-organ differentiation in *Palaemon* was seen as early as the prezoeal stage, arising from the lateral epidermis at the level of the maxillule (Le Roux, 1984). The single line of cells first increased in height, then the cytoplasm became vacuolized on the side of each cell next to the hemocoel. As in *Cancer* (above), this suggests an influence by an agent in the hemolymph. The cells then multiply and by 48 h after the first sign of differentiation have assumed the shape of the definitive organ. In Le Roux's (1974) landmark paper in which the Y-organs and mandibular organs were distinguished in 27 representative crustaceans, he noted that the position of the Y-organs relative to appendages vary with developmental stage and species. Thus, the glands occur in the maxillulary segment in the larvae and young of Euphausiacea, adults of *Inachus* and *Macropodia*, and the zoeae of *Clibanarius*, *Pagarus*, *Crangon*, and *Palaemonetes*. In contrast, the glands are maxillary in adult *Pagarus* and larvae of *Homarus*, *Inachus*, and *Macropodia*. Generally, in Euphausiacea, the larvae of decapod swimmers, and anomurans the relative position of Y-organs and legs are least modified by cephalization.

4.4.   Histology and General Cytology

In all crustaceans, the Y-organs are made up of anastomosing cords of cells. The cells are of a single epithelial type, but within a gland in a given stage the cells may differ from one another in cytostructure, apparently reflecting degrees of biosynthetic activity or capacity. All glands are lobulated to some extent and are generally encased in a membranous sheath of varying thickness. Lobulation is most pronounced in the Brachyura, in which the sheath penetrates the gland, defines the lobules, and is continuous with the lining of hemal sinuses within the gland (Johnson, 1980). In some Brachyura (e.g., *Pachygrapsus, Uca, Eriocheir*), the sheath is a thin lamina or is virtually absent (Bressac, 1973; Spindler et al., 1980). Among the astacideans and carideans, in which the Y-organs are closely applied to, or appear to be integrated with, the epidermis, the lobules tend to be cone-shaped (Fig. 4.3C) and the cells to be spindle-like, with long processes extending to the basal lamina toward the cuticle (e.g., *Astacus:* Birkenbeil and Gersch, 1979; see especially *Orconectes:* Burghause, 1975). Aoto et al. (1974) found the appearance of *Pandalus* Y-organs to differ considerably from those of *Palaemon* and *Procambarus;* the glands were multiple infoldings of the epidermis enclosing many lumena. Cellular processes also appear in brachyuran Y-organs, but these are more extensions of rounder cells, often restricted to cells in the periphery (e.g., *Cancer:* Hinsch et al., 1980) and directed to the hemocoelic surface of the gland (*Pachygrapsus:* Herman, 1967; Bressac, 1973; *Hemigrapsus* and *Carcinus:* Buchholz and Adelung, 1980; *Cancer:* Hinsch et al., 1980). The effect appears to be one of increasing the surface area for secretion and absorption (Fig. 4.4).

All Y-organs include hemocoelic sinuses or lacunae, in which hemocytes can be seen (Knowles, 1965; Chassard-Bouchaud and Hubert, 1975b; Diener et al., 1975; Bazin, 1976; Bressac, 1976; Birkenbeil and Gersch, 1979; Buchholz and Adelung, 1980), described as "granulocytes" in *Carcinus* by Chassard-Bouchaud and Hubert (1975b) and as "granular lymphocytes" in *Cancer* by Simione and Hoffman (1975). These last authors also reported the presence of "capillaries" (see also Aoto et al., 1974), and Johnson (1980) of "arterioles," within the sinuses. Since the Y-organs are supplied only by an "open" circulation that bathes the surface of the gland (see Section 4.2, above), the internal sinuses clearly function to facilitate exchanges between the hemolymph and cells in deeper layers of the gland.

4.4.1. *Intermolt*

Y-organs at this stage are in their unstimulated state, with ecdysteroid output at a minimum or absent. More precisely, the glands are presumed to be under the tonic inhibitory influence of the molt-inhibiting hormone (MIH) from the X-organ, sinus gland complex of the eyestalks. A degree of uncertainty exists on this point since no one has yet measured MIH titers in the blood. Y-organ cells in the Brachyura are compacted and tend to be cuboidal, but in other crustaceans are less crowded and assume a more cylindrical, high columnar shape (Vernet et al., 1978). Accordingly, the latter cells contain more cytoplasm; however, generally in crustaceans, the amount of cytoplasm relative to nuclear volume in the "resting" gland is always described as sparse (Figs. 4.5A and 4.6) (Herman, 1967; Aoto et al., 1974; Le Roux, 1974, 1977; Zerbib et al., 1975; Bressac, 1976; Bourquet et al., 1977; Birkenbeil and Gersch, 1979; Buchholz and Adelung, 1980; Hinsch et al., 1980; Johnson, 1980). The cytoplasm usually appears homogeneous in texture (e.g., *Ligia:* Maissiat and Maissiat, 1976), containing few organelles (*Astacus:* Birkenbeil and Gersch, 1979) or very fine granules (*Callinectes:* Johnson, 1980), resulting in even staining (*Cancer:* Simione and Hoffman, 1975). However, vacuolization is occasionally seen in cells of part of the gland (Simione and Hoffman, 1975).

The absence of mitotic activity is quite characteristic of this stage (e.g., Simione and Hoffman, 1975; Zerbib et al., 1975; Bressac, 1976, 1978; Le Roux, 1977) and of postmolt (Le Roux, 1977; Bressac, 1978; Johnson, 1980), although some species did not show mitoses in any stage of the cycle (*Palaemon:* Aoto et al., 1974; *Orconectes:* Burghause, 1975; *Penaeus:* Bourquet et al., 1977). Nuclei are often irregularly shaped and of different sizes (Diener et al., 1975; Simione and Hoffman, 1975) but are generally circular or ovoid. One or two nucleoli are seen, surrounded by heterochromatin that is also condensed in masses next to the nuclear envelope (Figs. 4.5 and 4.6) (Diener et al., 1975; Simione and Hoffman, 1975; Maissiat and Maissiat, 1976; Hinsch et al., 1980; Johnson, 1980).

4.4.2. *Premolt–Postmolt*

Typically, in representatives of all taxa, the Y-organs hypertrophy during premolt, involving increases in cytoplasm relative to nuclear volume, but also absolute increases in the volumes of both; the glands appear obviously larger (Aoto et al., 1974; Burghause, 1975; Maissiat and Maissiat, 1976; Le Roux, 1977; Birkenbeil and Gersch, 1979; Hinsch et al., 1980; Johnson, 1980; Girard and Maissiat, 1983). Most of the hypertrophy occurs during the D_0–D_1 stage and peaks from D_1 to D_4, depending on species, at a volume at least twofold greater than the original volume (Aoto et al., 1974; Maissiat and Maissiat, 1976). Immediately after the molt, volumes sharply decrease back to those of stage C (Aoto et al., 1974; Burghause, 1975) or temporarily to less than in stage C (Johnson, 1980). Nuclei, which were round and full in stage D, become small and dense in stages A and B (Johnson, 1980).

The incidence of mitoses appears to be a sensitive index of glandular activity. In brachyurans, Bressac (1978) and Johnson (1980) saw increases in mitoses, starting as early as late C_4, with the largest number in D_1 (see also Le Roux, 1977), thereafter dropping off to an absence of mitoses in postmolt. Bressac (1978) shows the pattern clearly with a graph of mitoses against cycle stage. More revealingly, in *Ligia*, Girard and Maissiat (1983) measured both mitotic incidence and hemolymph ecdysteroid titers and found coincident peaks during C_2 and D_0–D_1. While the change in C_2 is not typical and not readily explained, the early increases in mitoses in C_4 to D_1 clearly reflect increased biosynthesis of ecdysteroids, coinciding also with large increases in the uptake of the precursor, cholesterol, by the Y-organs (Spaziani and Kater, 1973). Turnover of sterol is quite rapid (Spaziani and Kater, 1973; Vensel et al., 1984), and both biochemical and structural evidence suggests that the glands normally do not accumulate secretion product (Karlson and Skinner, 1960; Aoto et al., 1974; Mattson and Spaziani, 1985; Watson and Spaziani, 1985b). It is presumed that the titer of MIH at these times are very low to absent.

A convenient means of inducing the acute premolt condition is to excise the eyestalks (remove MIH). Within 24 h, the amount of cholesterol taken up by the Y-organs vastly increases, both *in vivo* (Vensel et al., 1984) and *in vitro* (Watson and Spaziani, 1985a). Within 3–8 days after excision, increases are seen in Y-organ RNA synthesis, cytoplasmic and nuclear volumes (Simione and Hoffman, 1975), and number of mitoses (Bressac, 1976).

4.5.   Ultrastructure

The first electron microscopic study of Y-organs was conducted by Knowles (1965) in the brachyuran *Carcinus*. The study was primarily a search for innervation to the gland, but no such connections could be found. However, he noted great variability in fine structure between individuals "as might be expected of an organ so intimately concerned in the cyclical activities of growth, differentiation and moulting." The cytoplasm of many cells was filled with mitochondria and microsomes, but no organized endoplasmic reticulum could be seen. Other observations suggest the presence of artifacts. Fixation of invertebrate tissue for ultrastructural work is always difficult, and the Y-organs present special problems. Contradictory results for Y-organs in separate studies of the same or related species may be due in part to uneven or incomplete fixation, leading variously to such artifacts as swelling and the appearance of vacuoles, perinuclear spaces, or localized degeneration. Another source of problems is subtle species differences in fixability of the glands by a given procedure. For example, Buchholz and Adelung (1980) reported that a method which they found effective in fixing *Hemigrapsus* Y-organs gave unacceptable results in the glands of *Carcinus*. Generally, working solutions of high intrinsic osmolarity appear to be necessary. The present author finds the following to be routinely effective in *Cancer* species (modifications of a method of Buchholz and Adelung, 1980): prefixing in 3% glutaraldehyde in 0.125 *M* Sörensen phosphate buffer (pH 7.6), containing sucrose at 0.45 *M* (15.4%), rinsing in the buffer with 0.45 *M* sucrose, and post-fixing in 2% osmium in the 0.125 *M* buffer.

The summaries of Y-organs' ultrastructure, below, are distilled from the extant electron microscopic studies as follows:

(a) Brachyura: *Pachygrapsus* (King, cited in Herman, 1967; Bressac, 1973), *Callinectes* (Diener et al., 1975), *Carcinus* (Knowles, 1965; Chassard-Bouchaud and Hubert, 1975a,c; Zerbib et al., 1975; Bazin, 1976; Buchholz and Adelung, 1980), *Hemigrapsus* (Buchholz and Adelung, 1980), *Cancer* (Hinsch et al., 1980), *Portunas* (Taketomi and Hyodo, 1986).

(b) Caridea and Astacidea: *Palaemon*, *Pandalus*, and *Procambarus* (Aoto et al., 1974), *Astacus* (Birkenbeil and Gersch, 1979), *Orconectes* (Birkenbeil and Eckert, 1983).

(c) Peracarida: *Ligia* (Maissiat and Maissiat, 1976; Maissiat et al., 1979), *Sphaeroma* (Charmantier and Trilles, 1979).

Cells in the periphery of Y-organs generally exhibit slender pseudopod-like processes extending to the basal lamina (e.g., see Herman, 1967; Bressac, 1973; Bazin, 1976); they contain electron-dense tips resembling hemidesmosomes (Hinsch et al., 1980). The arrangement of processes suggest that they function to increase the surface area for absorption and secretion.

4.5.1. *Intermolt*

4.5.1.1. NUCLEUS

In crabs, nuclei are generally ovoid. Heterochromatin surrounds one or two nucleoli but characteristically also condenses in masses near the nuclear membranes (Figs. 4.5 and 4.6) (Bressac 1973; Bazin, 1976; Chassard-Bouchaud and Hubert, 1975c; Zerbib et al., 1975; Buchholz and Adelung, 1980; Hinsch et al., 1980). Blebbing of the outer nuclear membrane was occasionally seen (Hinsch et al., 1980). The nuclei of shrimps and crayfishes (Aoto et al., 1974) and isopods (Maissiat and Maissiat, 1976) appear similar to those of crabs.

4.5.1.2. MITOCHONDRIA

Numerous mitochondria occur throughout the cytoplasm of all Y-organs thus far studied. Virtually all contain tubular cristae. Exceptions appear among isopods in which cristae are described as lamellar, arranged at 90°

to the long axis (Maissiat and Maissiat, 1976), and as long and anastomosing (Charmantier and Trilles, 1979). Mitochondria are polymorphic, usually mixed populations of small, ovoid and elongated, rod-like types. In
Carcinus, the former contain a dense matrix, some with vacuoles. Groups
of the small type often appear associated with the smooth endoplasmic
reticulum near the nucleus (Bazin, 1976). The "macromitochondria" are
fewer in number, are more packed with cristae, and show a granular matrix (Chassard-Bouchaud and Hubert, 1975c; Zerbib et al., 1975). In
Astacus, the "giant" mitochondria show longitudinally oriented double
membranes through most of the central axis of the organelle (Birkenbeil
and Gersch, 1979).

4.5.1.3. ENDOPLASMIC RETICULUM (ER)
AND RIBOSOMES

In all crustaceans studied, there is great abundance of free ribosomes in
the cytoplasm of Y-organ cells, and their numbers generally do not
change during the cycle. Only rarely are ribosomes observed to be attached to membranous structures; i.e., rough (granular) ER generally is
poorly developed or absent. However, there is some disagreement on
this point in *Carcinus*. None was seen by Buchholz and Adelung (1980),
nor was any mentioned by Chassard-Bouchaud and Hubert (1975a,c),
whereas Zerbib et al. (1975) found the granular ER to be well developed;
Bazin (1976) reported only granular masses associated with or near the
smooth ER. Moderate amounts of rough ER appear in the isopod *Ligia* as
flattened saccules (Maissiat and Maissiat, 1976).

The smooth ER has been of special interest, since its presence in abundance is a characteristic of steroid-secreting cells in vertebrates. Reports of
its occurrence in brachyuran Y-organs present a very mixed picture. Little
to no smooth ER could be found in any stage of *Cancer* (Hinsch et al., 1980)
or *Hemigrapsus* (Buchholz and Adelung, 1980), or during intermolt (the
only stage examined) of *Pachygrapsus* (Herman, 1967). Among several individuals of *Carcinus maenas* in stages C and D, Buchholz and Adelung
(1980) could find only part of a gland in one crab (in stage D_1; see Section
4.5.2, below) exhibiting a significant amount of smooth ER. This was of
the dramatic lamellar type typical of vertebrate glands. However, in examining *Carcinus* in stage C_4, when crab Y-organs show subtle metabolic
and cytological indications of earliest premolt (see Section 4.4.2, above),
smooth ER was reported to be abundant, mostly in vesicular or tubular
form in the cellular processes at the periphery of the gland (Chassard-
Bouchaud and Hubert, 1975a; Zerbib et al., 1975). Apparently, the same
was seen in *Pachygrapsus marmoratus* by Bressac (1973). On the other hand,
the Y-organs of shrimps and crayfishes show abundant smooth ER as

cisternae and vesicles throughout intermolt (Aoto et al., 1974; Birkenbeil and Gersch, 1979). The same was reported for *Ligia* (Maissiat and Maissiat, 1976), but in *Sphaeroma* it is sparse, in the form of small vesicles (Charmantier and Trilles, 1979).

4.5.1.4. GOLGI APPARATUS

With rare exceptions, the Golgi structures in Y-organs cells have been described as absent or inconspicuous and unaffected by the molting cycle. The Golgi apparatus was described as well developed only in *Carcinus mediterraneus* (Zerbib et al., 1975) and as usually inconspicuous but more pronounced during D stages in *Astacus* (Fig. 4.7A; Birkenbeil and Gersch, 1979). Taken together with the observations on the rough ER (above), the impression emerges that the cells do not produce proteins for export.

4.5.1.5. OTHER ORGANELLES

All groups of crustaceans studied exhibit lysosomal bodies in the cytoplasm of Y-organ cells, some among several types in *Pachygrapsus crassipes* having been identified by positive staining for acid phosphatase (Herman, 1967). Some cells contain autophagic vacuoles (Bressac, 1973; Chassard-Bouchaud and Hubert, 1975a; Charmantier and Trilles, 1979). Frequently seen also are aggregations of microvesicles and microtubules and small granules mostly in the periphery (Herman, 1967; Chassard-Bouchaud and Hubert, 1975a; Bazin, 1976; Buchholz and Adelung, 1980). The microvesicles observed by Buchholz and Adelung (1980) appear to be transported from cell to cell and into the hemolymph lacunae. With special staining, glycogen "rosettes" have been identified (Maissiat and Maissiat, 1976; Buchholz and Adelung, 1980), which appear to increase in early premolt. Aoto et al. (1974) delineated several types of inclusion: one is a vesicle with an electron-dense "cap" resembling a structure found in testicular Leydig's cells; another contains a concentric membrane system and may be a type of lysosome; a third comprises electron-dense spheres in the cell processes of the gland periphery that are most abundant during stages C and D_0–D_1. These last may be the same as the peculiar inclusions that were found in the same location in *Ligia* cells and proved to contain proteinaceous material in crystalline state (Maissiat et al., 1979). The functions of most of these several inclusions are not known. In only one case has an inclusion been positively associated with a secretory product (see Section 4.5.3., below).

4.5.2. *Premolt to Postmolt*

Starting in late stage C_4, a frequent but by no means universal observation is that the amount of smooth ER increases during premolt (e.g., *Carcinus:* Bazin, 1976; Buchholz and Adelung, 1980; *Astacus:* Birkenbeil and Gersch,

1979) (Fig. 4.7A). In crayfishes and shrimps, usually the nature rather than the quantity of the ER changes during premolt; it is much vesiculated in C stages and D_0, but is undulating tubular during D_1–D_4 and postmolt; occasionally in D_2, the ER consists of lamellar cisternae with anastomoses (Aoto et al., 1974; see also Maissiat and Maissiat, 1976). Similarly, ribosomes (Birkenbeil and Gersch, 1979) and mitochondria (Hinsch et al., 1980) appear to increase during premolt, but in other cases the organelles change in character rather than in numbers; for example, in *Palaemon*, mitochondria, large and packed with tubular cristae in intermolt, are reduced as much as 15-fold in size and contain few cristae in premolt and postmolt (Aoto et al., 1974).

The quite remarkable observations of Buchholz and Adelung (1980) on the Y-organs of *Hemigrapsus* and *Carcinus* may explain some apparently contradictory results between laboratories. A Y-organ from only one *Hemigrapsus nudus* in D_1 (among several animals of both species in this and other stages) showed cells packed with mitochondria but no smooth ER, while other cells or cell processes were full of smooth ER to the point of virtually excluding other organelles (Fig. 4.2). The ER cells seemed to be segregated to specific lobules of the gland and were separated from the other cell type by lamina-lined sinusoids. Buchholz and Adelung's publication includes three-dimensional reconstructive drawings of tissue sections that clearly show the cell types (cf. Fig. 4.2). Cellular diversity also was highlighted in the study of isopod Y-organs by Charmantier and Trilles (1979).

4.5.3. *Induced Premolt: Structural and Cytochemical Effects of De-eyestalking*

About a day after eyestalk excision, Y-organ cells of *Cancer* showed increased blebbing of the nuclear membrane, and increased rough ER and small vesicles in the peripheral cellular processes. From 24 to 45 days afterward, the cells exhibited increased numbers of mitochondria and electron-dense particles in the cellular processes in contact with the external lamina (Hinsch et al., 1980). Six days after eyestalk excision in *Carcinus*, the glands were somewhat hypertrophied and the Golgi apparatus was more pronounced (Bazin, 1976). Generally, these effects are consistent with observations in natural premolt, but results have been less marked than might have been expected.

In contrast, quite revealing results have been obtained with de-eyestalked preparations in the first successful attempt to localize a Y-organ secretory product immunocytochemically. Birkenbeil and Eckert (1983) immunostained sections of activated *Orconectes* glands and found ecdysone-positive reactions localized in the tubular cristae of mitochondria

(Fig. 4.7B,C). Not all mitochondria were stained, but the numbers stained increased with length of time after de-eyestalking, presumably reflecting increased biosynthetic activity. Also seen occasionally were stained granular structures and caveolae of the plasma membrane, apparently indicating secretory granules and sites of exocytosis, respectively (Fig. 4.7D). Also remarkable: the smooth ER was rarely stained. This is in marked contrast with insect prothoracic glands in which the same technique mostly stained the smooth ER, never the mitochondria (Birkenbeil, 1983).

4.6.   General Discussion

The structural features of Y-organ cells conform extensively with those of the steroid-hormone-secreting models in mammals, namely, testicular Leydig's cells, fascicular and glomerular zones of the adrenal cortex, cells of the ovarian theca interna, interstitium and corpora lutea, and the placenta. These features principally and generally include smooth ER in much greater relative abundance than granular ER, numerous free ribosomes, polymorphic mitochondria with tubular cristae, relative inconspicuousness of organelles involved with protein synthesis, nuclei with clumped heterochromatin in the periphery, and a relative lack of stored secretion product. Apparent variations in the occurrence or appearance of organelles between and within crustacean species and with physiological state are not inconsistent with the vertebrate models. For example, work has been reviewed that shows that the smooth ER exhibits extremes in variation between and within species and even within a single gland, during any given stage of the molt cycle. Thus, virtually no smooth ER was seen in the crab *Cancer* during intermolt, whereas it was regularly abundant in crayfishes and shrimps at that stage. In the crab *Carcinus*, one laboratory reported some Y-organ cells in early premolt packed with smooth ER, while cells virtually adjacent showed none. The mammalian adrenal cortex exhibits within-gland variability in that the smooth ER is generally most developed in the zona fasciculata and least in the glomerulosa. For a given zone, there are marked species differences, varying from very little smooth ER in the guinea pig to great quantities in the human (cf. Fawcett et al., 1969). The smooth ER also varies in form in Y-organs, seen as empty vesicles, random tubes, tubular or laminar sheets, or flattened cisternae. All these forms have been observed in mammalian cells (Christensen and Gillim, 1969), occasionally two different forms of the smooth ER in a given cell (mouse Leydig's cells: Christensen, 1965). The vesicular form is considered likely to be an artifact of fixation (the smooth ER is particularly prone to this problem), but the forms otherwise occur as functions of species, organ, developmental or degenerative stage, and presence or absence of stimulation by tropic hormones (Christensen, 1965; Fawcett et al., 1969). One or more of these factors could account for the observation in *Carcinus* (Buchholz and Adelung, 1980) of smooth ER appearing in only segregated cells in a Y-organ. Similarly, in both crustacean and mammalian steroidal cells, mitochondria occur with tubular or lamellar cristae. A peculiar variation seen in *Astacus* mitochondria, parallel membranes in the long axis (Birkenbeil and Gersch, 1979), has also been observed in the rat (Christensen and Gillim, 1969). The unusual inclusion seen in *Ligia*, a large body with crystalline latticework of protein, appears to be the same as the "crystal of Reinke" found in human Leydig's cells (cf. Fawcett et al., 1969).

In further consideration of comparative cellular structure between phyla, the effects of regulatory peptide hormones should be considered. Virtually all of the instances of sparse or absent smooth ER observed in Crustacea were in intermolt animals. Cells at this stage also showed a sparse cytoplasm with few organelles. Crustaceans are peculiar and probably unique in that intermolt apparently results from *tonic suppression* of Y-organ function by the eyestalk peptide, MIH. A remarkably similar cytological picture is presented in the vertebrate glands after hypophysectomy; the loss of smooth ER is particularly striking, but also cytoplasmic volume is decreased and mitochondria are affected in numbers and internal form (see Fawcett et al., 1969). Activation of Y-organs starting in very late intermolt and progressing through premolt results from at least the *absence* of suppression by MIH and features cytoplasmic and nuclear hypertrophy, and increased quantities of smooth ER and mitochondria. As would be expected, similar changes are seen after premolt is induced by eyestalk removal. These Y-organs more closely resemble the normal vertebrate glands, which are under low tonic stimulation by the pituitary tropins, or the glands recovering from hypophysectomy with replacement hormone therapy. Also, it should be noted that the absolute amount of smooth ER in Y-organs of any stage generally is lower than is typical for mammalian glands. Of all the studies on Y-organ structure reviewed, only one reported cells full of smooth ER, as is common in, for example, mammalian Leydig's cells. Crustaceans cannot synthesize the nuclear ring structure of steroids and thus must obtain cholesterol, the steroid hormone precursor, from the diet directly or produce it by converting phytosterols (see Spaziani and Kater, 1973; Watson and Spaziani, 1985b). The vertebrate glands, in contrast, can synthesize cholesterol facultatively, in varying degrees, utilizing enzyme systems in the smooth ER and cytosol (Tamaoki, 1973). Christensen (1965) marshaled evidence from comparative work on Leydig's cells to permit the conclusion that the amount of smooth ER occurs in direct proportion to the cells' capacity to synthesize cholesterol. Thus, it is reasonable to surmise that the relative lack of smooth ER in Y-organs is related to the inability to synthesize the precursor of the ecdysteroid secretion product.

In both crustaceans and vertebrate steroidogenic cells, the Golgi apparatus and the granular ER are much less pronounced than in other secretory cell types, although among steroidogenic cells these organelles appear generally more well developed in vertebrates. The organelles' function in these cells is not understood. However, it may be expected that they assemble maintenance and structural proteins. Perhaps more to the point, steroidogenesis in the stimulated cells is known to depend in part upon induced protein synthesis. Thus, ecdysteroid production in Y-organs is inhibited by MIH or cycloheximide, which also inhibit protein synthesis (Mattson and Spaziani, 1986). Similarly, stimulation of steroidogenesis in vertebrate glands by pituitary tropins is blocked by specific inhibitors of protein synthesis (e.g., see Garren et al., 1965; Janszen et al., 1977; Landefeld et al., 1979). Instances were reviewed in which both Golgi structures and granular ER became more pronounced in the activated cells.

Extensive biochemical and structural research in vertebrate steroidogenic cells has provided broad understanding of the function of the organelles in steroid hormone biosynthesis. Cholesterol is synthesized by the cells or preferentially enters from the circulation by active endocytosis while bound to lipoprotein. Freed by lysosomal digestion of the complex, cholesterol enters mitochondria, apparently by a form of transport. Side-chain cleavage occurs in the inner mitochondrial membrane through the action of a cytochrome P-450 respiratory enzyme system. The product, pregnenolone, then enters biosynthetic pathways that produce progesterone, glucocorticoids, mineralocorticoids, estrogens, or androgens, depending upon the gland. For production of estrogens, androgens, and most glucocorticoids, the necessary hydroxylations, aromatizations, removal of remnant side chain, etc. requires that pregnenolone leaves the mitochondria for the smooth ER which contains the appropriate enzymes, in most cases different P-450 systems for each step. However, C-11 hydroxylation (cortisol) or C-18 hydroxylation (aldosterone) requires that a C-11 deoxy metabolite return to the mitochondrial inner membrane (Tamaoki, 1973; Hall, 1985). The division of labor for organelles in Y-organ cells is not known. However, two experiments have been done which begin to address the question. Segments of activated Y-organs from crabs were incubated with [^{14}C]cholesterol, after which the tissue was homogenized and subjected to subcellular fractionation by ultracentrifugation. Of the total radioactivity present, over half was in the cytosol, approximately 20% each in the microsomes (smooth ER) and heavy (nuclear) fraction, and only 1% in the mitochondria (Watson and Spaziani, 1985a). Birkenbeil and Eckert (1983) subjected sections of activated Y-organs of crayfish to immunocytochemical procedures to localize ecdysteroids. Many of the mitochondria showed the positive reaction, but not the smooth ER. These experiments give contradictory impressions of the site of major activity in ecdysteroid biosynthesis, results which may reflect fundamental species differences. However, neither is conclusive, as both are open to a variety of interpretations. Clearly, much more work is needed. Since ecdysteroid production does not require side-chain cleavage, and features hydroxylations and other reactions at nuclear and side-chain sites different from those of the vertebrate hormones, the results of studies that relate structure with function in Y-organs should be awaited with considerable interest by comparative physiologists and evolutionists.

4.7.   Summary

The Y-organs of Gabe and Echalier are discrete epithelioid glands in crustaceans that are specialized for the synthesis and secretion of steroid hormones controlling growth and differentiation. They are homologous with insect prothoracic glands in structure and function and analogous to steroid hormone-secreting glands of vertebrates. They are of a single cell type, derivative of the epidermis, and are located in the cephalothorax, the precise location and form differing with infraorder and developmental stage. Y-organs are not innervated, and their circulation consists of open superfusion of hemolymph and percolation through internal sinusoids. They are extensively lobulated in internal organization. The cytostructure is highly variable with individuals, species, and molt-cycle stage, but it generally features numerous mitochondria with tubular cristae and much more smooth endoplasmic reticulum relative to the granular type. Large numbers of free ribosomes also are characteristic. Under apparent suppression by the eyestalk peptide, molt-inhibiting hormone (MIH), during postmolt and intermolt, the cytoplasm and its organelles and inclusions tend to be sparse. The picture changes to hypertrophy and more prominent cytostructures in premolt, when the influence of MIH is absent. At this time, ecdysteroids are actively being produced. This review has compared the cellular structure of Y-organs with those in vertebrate glands in the presence and absence of stimulation by tropic hormones. Unlike the vertebrate glands, Y-organs cannot synthesize the steroid precursor, cholesterol, and structurally most resemble those mammalian glands with the least capacity for *de novo* synthesis of the steroid nucleus. The division of labor by the organelles in steroid production is discussed.

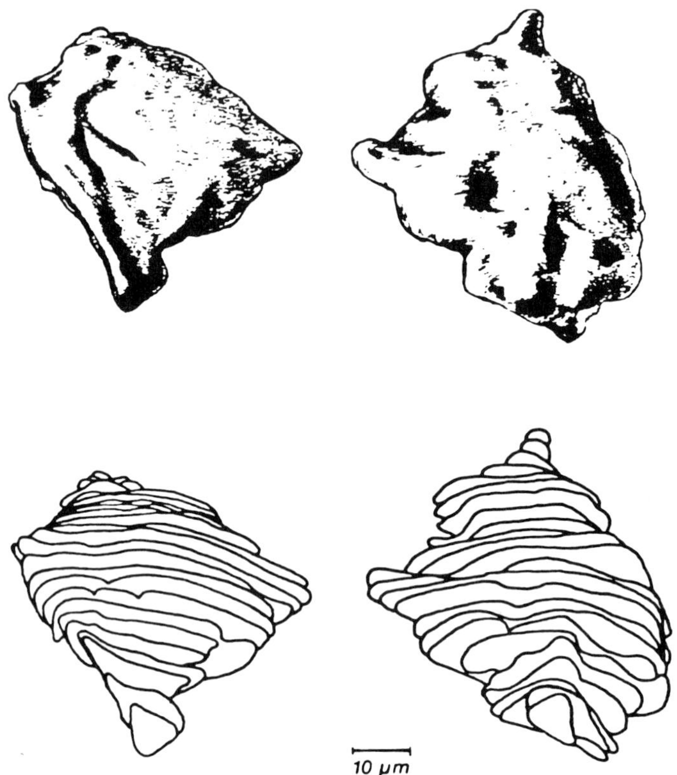

FIGURE 4.1. Three-dimensional reconstructions of a Y-organ from a fourth-stage zoeal larva of the crab *Rithropanopeus harrisii*. Transverse serial sections were cut (1 μm) of which every second section was drawn. On the left are drawings of the whole gland and its reconstruction showing the surface facing the hemocoel. On the right are drawings of the gland turned over to the right, exposing the underside, which is associated with the cuticular surface. ×1000. (From Buchholz, 1984.)

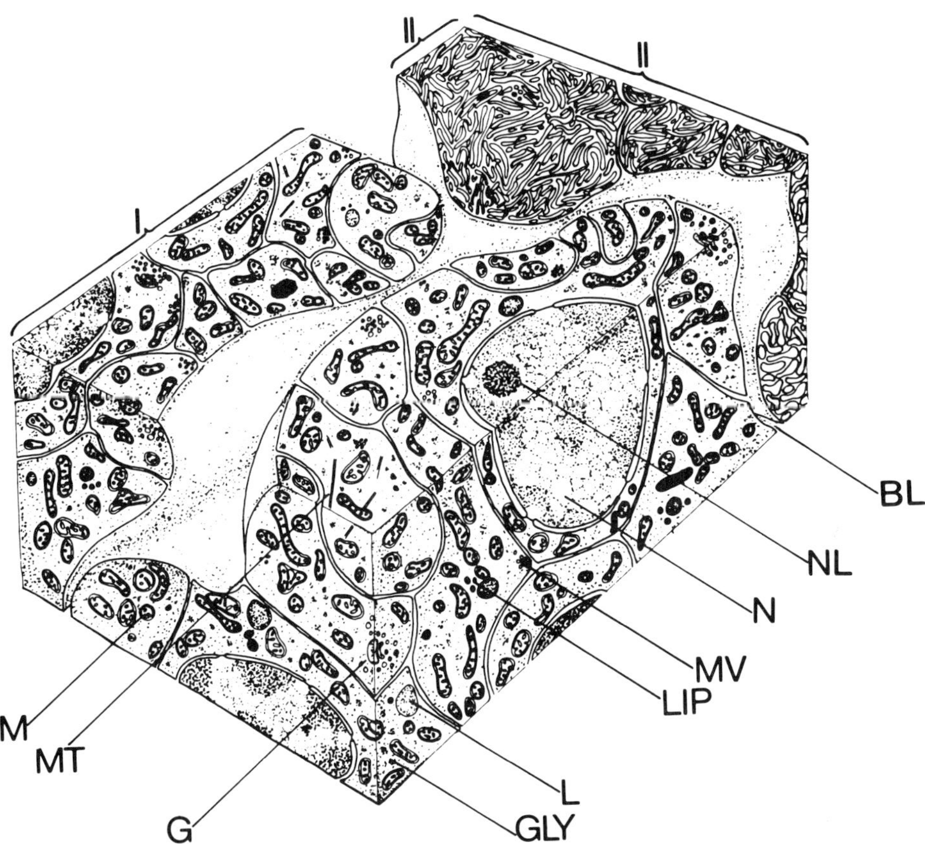

FIGURE 4.2. Three-dimensional reconstructive drawing of a portion of a Y-organ from the crab *Hemigrapsus nudus*. Composite cellular features are shown from an individual in premolt stage D_1. Hemocytes omitted. *Key:* I = lobule with the most frequent cell type; II = lobule with abundant endoplasmic reticulum appearing to have displaced all other oganelles; BL = basal lamina; G = Golgi element; GLY = glycogen; L = lysosome; LIP = liposome; M = mitochondrion; MT = microtubule; MV = microvesicle; N = nucleus; NL = nucleolus. (From Buchholz and Adelung, 1980.)

FIGURE 4.3. (A&B) Light micrographs of the lateral cephalic nerve plexus (LCNP) of a prepubertal isopod, *Sphaeroma serratum:* (A) Neurosecretory products (arrows). ×3750. (B) Y-organ and LCNP. ×12,000. (C) Light micrograph of longitudinal section through the Y-organ of an adult *Homarus americanus*. *Key:* GC = glial cell; MG = molting gland (Y-organ); NE = neurilemma; NSG = neurosecretory granules; PC = plexus cell. (A and B from Maissiat et al., 1979; C from Le Roux, 1974.)

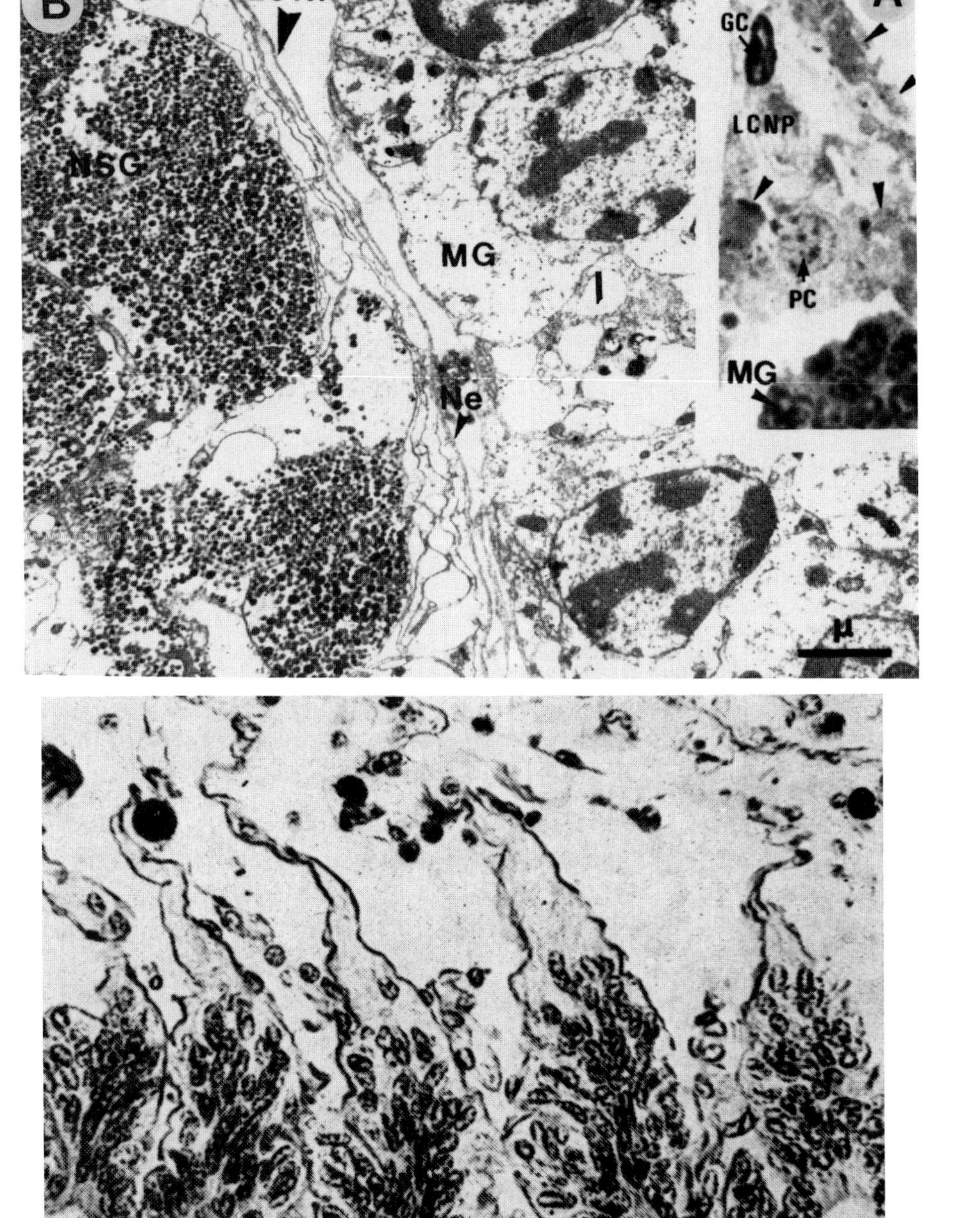

B
LCNP
NSG
MG
Ne
I
A
GC
LCNP
PC
MG
M
C
25 µm

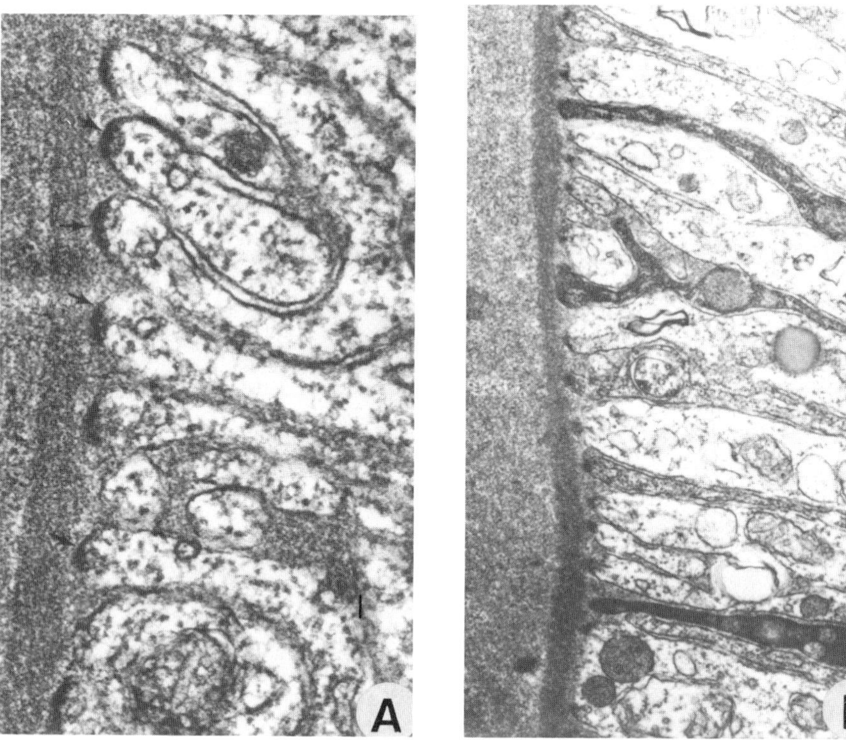

FIGURE 4.4. Ultrastructure of Y-organs of *Cancer antennarius:* (A) Numerous cellular processes adjacent to basal lamina in gland from an intermolt crab. Note the dense thickenings of each process (arrows) adjacent to basal lamina. ×57,950. (B) Cytoplasmic processes adjacent to basal lamina in premolt crabs; numerous vesicles and electron-dense particles are seen. The cytoplasm varies in electron density from cell process to cell process. ×16,150. (From Hinsch et al., 1980.)

FIGURE 4.5. Ultrastructure of Y-organs of the crab *Cancer antennarius:* (A) Intermolt; epithelial nature of tissues is evident. The sparse cytoplasm contains vesicles, few mitochondria, and numerous free ribosomes. Aggregates of heterochromatin are adjacent to the nuclear envelope. ×4275. (B) Premolt. Compared with A, mitochondria are more numerous and heterochromatin is reduced. ×9120. From Hinsch et al., 1980.

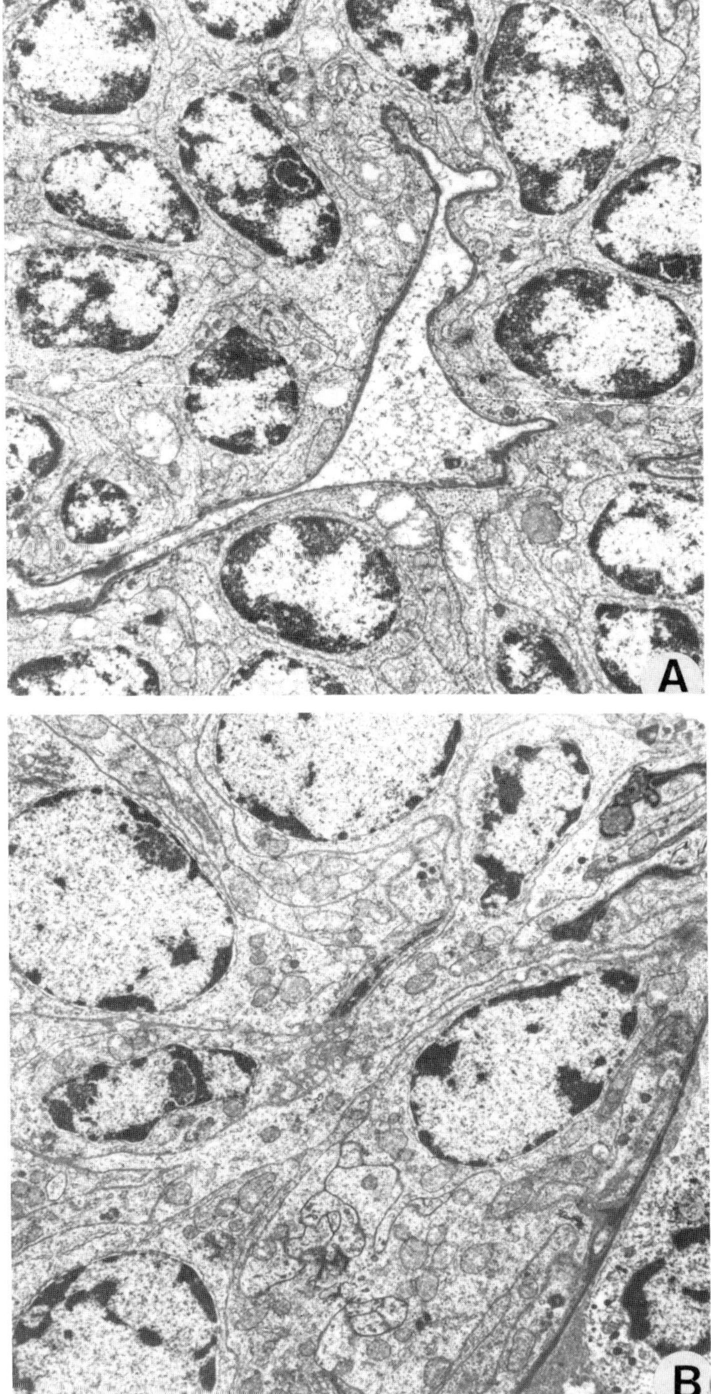

FIGURE 4.6. Ultrastructure of the Y-organ of the crayfish *Astacus astacus* in the intermolt stage. Note the small amount of cytoplasm and scarcity of organelles. (A) ×3800; (B) ×12,000. *Key:* M = mitochondrion; N = nucleus. (From Birkenbeil and Gersch, 1979.)

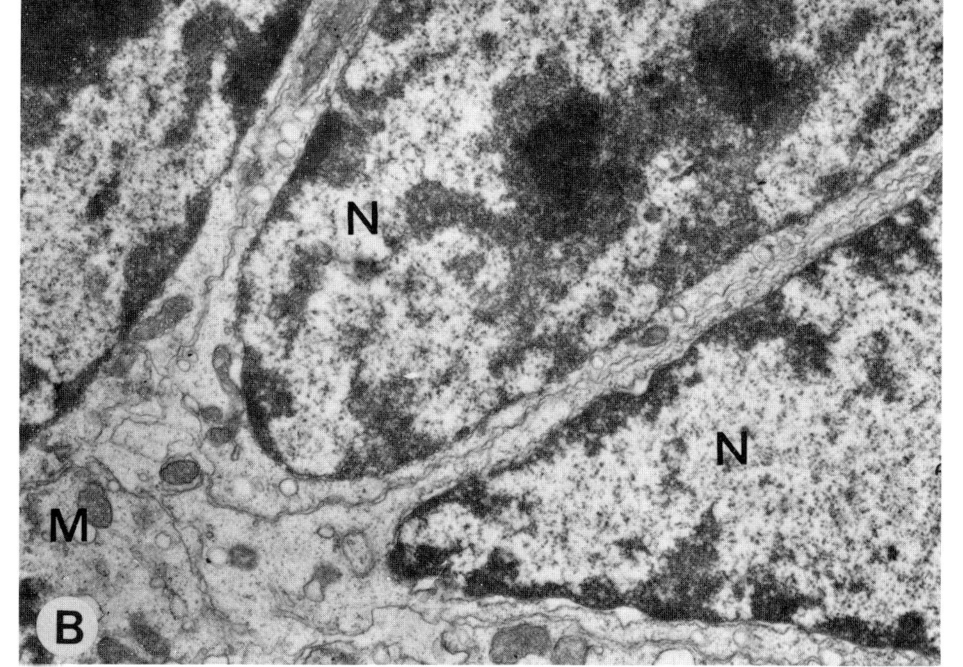

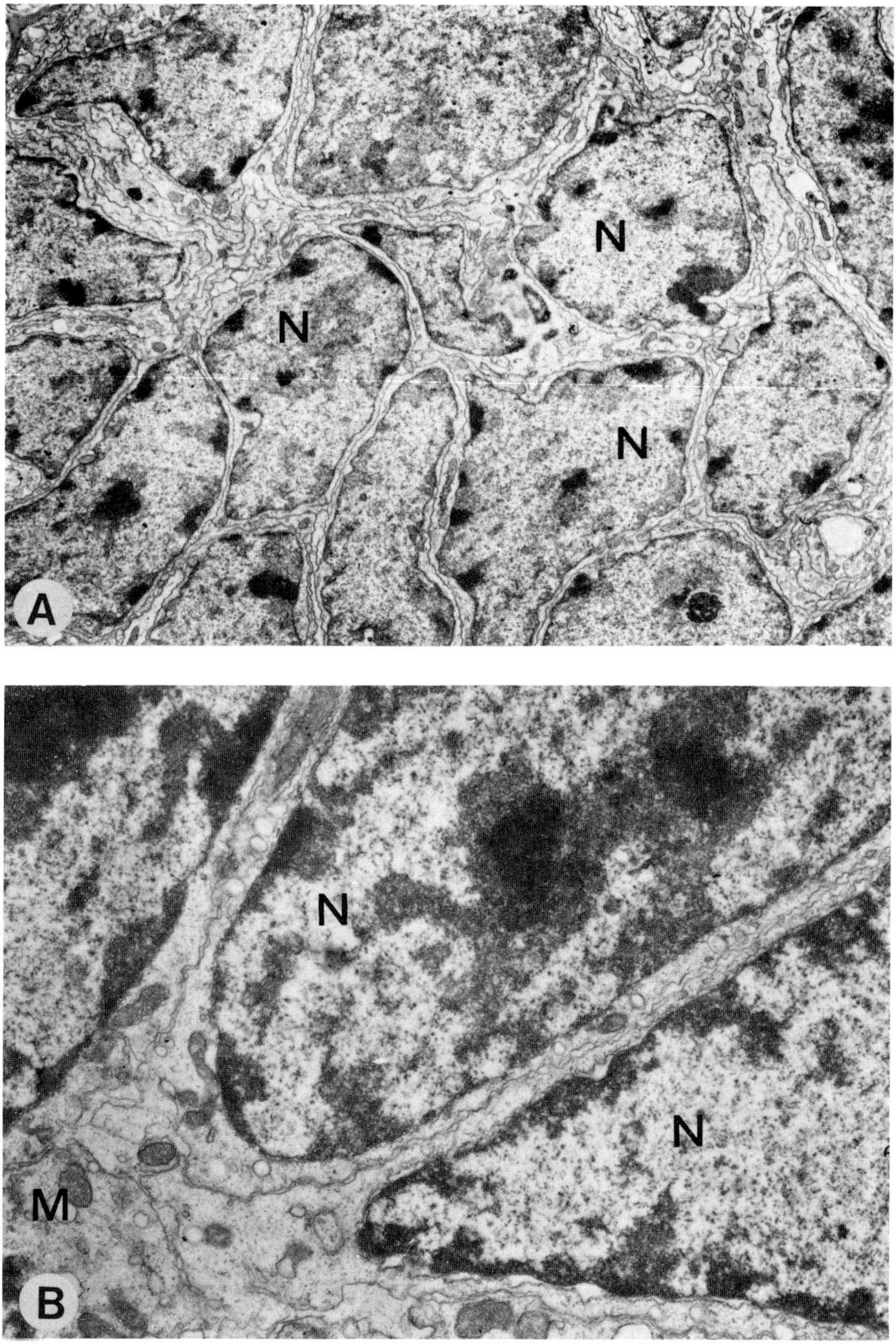

 E. Spaziani

FIGURE 4.7. Ultrastructure of crayfish Y-organs: (A) *Astacus astacus* in molting stage D (premolt). Cytoplasm well developed. Majority of endoplasmic reticulum of the agranular type, in close spatial relationship with the Golgi apparatus. ×24,600. (B–D) *Orconectes limosus* after immunostaining for ecdysone. ×36,000. (A) Only one mitochondrial profile exhibits ecdysone immunostaining. Hyaloplasm and smooth endoplasmic reticulum are unstained. (B) Mitochondrial profile showing ecdysone immunoreactivity in and around tubular cristae (arrows). (C) Pit of plasma membrane containing extracellular immunoreactive material suggesting ecdysone release site. *Key:* ER = endoplasmic reticulum; GA = Golgi apparatus; M = mitochondrion; N = nucleus. (A from Birkenbeil and Gersch, 1979; B–D from Birkenbeil and Eckert, 1983.)

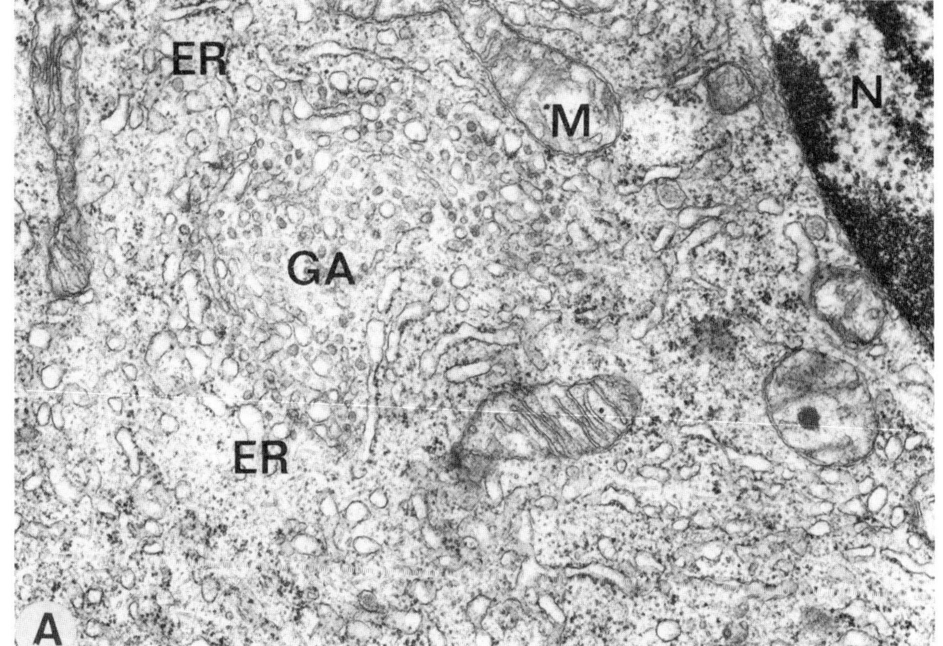
ER
M
N
GA
ER
A

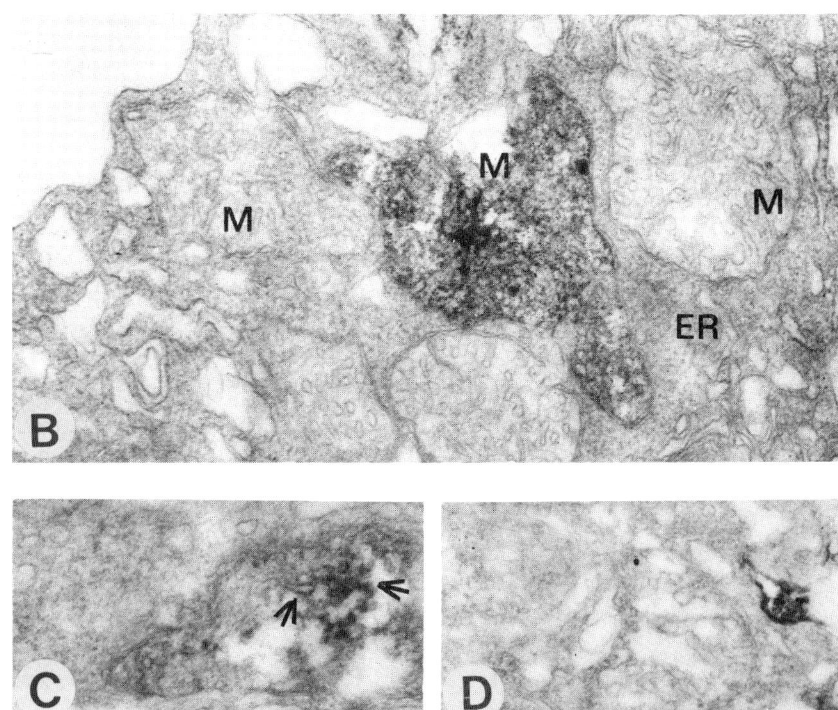
M
M
M
ER
B
C
D

Acknowledgments

Investigations from my laboratory that are reported in this review were supported by grants from the National Science Foundation (U.S.A.) and institutional grants from the National Institutes of Health to the University of Iowa.

References

Anderson, W. R. 1978. A description of laboratory-reared larvae of the yellow crab, *Cancer anthonyi* Rathbun (Decapoda, Brachyura), and comparisons with larvae of *Cancer magister* Dana and *Cancer productus* Randall. Crustaceana (Leiden) 34: 55–68.

Aoto, T., Y. Kamiguchi, and S. Hisano. 1974. Histological and ultrastructural studies on the Y-organ and the mandibular organ of the freshwater prawn, *Palaemon paucidens*, with special reference to their relation with the molting cycle. J. Fac. Sci. Hokkaido Univ. 19: 295–308.

Bazin, F. 1976. Mise en évidence des caractères cytologiques des glandes stéroïdogènes dans les glandes mandibulaires et les glandes Y du Crabe *Carcinus maenas* (L.) normal et épédonculé. C. R. Acad. Sci. Paris 282D: 739–741.

Birkenbeil, H. 1983. Ultrastructural and immunocytochemical investigation of ecdysteroid secretion by the prothoracic glands of the waxmoth *Galleria mellonella*. Cell Tissue Res. 229: 433–441.

Birkenbeil, H. and M. Eckert. 1983. Immuno-electron-microscopic evidence of ecdysteroids in Y-organ of *Orconectes limosus* Crustacea, Decapoda. Cell Tissue Res. 233: 233–236.

Birkenbeil, H. and M. Gersch. 1979. Ultrastructure of the Y-organ of *Astacus astacus* (L.) (Crustacea) in relation to the moult cycle. Cell Tissue Res. 196: 519–524.

Blanchet, M. F. 1974. Etude du contrôle hormonal du cycle d'intermue et de l'exuviation chez *Orchestia gammarella* par microcautérisation des organes Y suivie d'introduction d'ecystérone. C. R. Acad. Sci. Paris 278D: 509–512.

Bourquet, J.-P., J.-M. Exbrayat, J.-P. Trilles, and G. Vernet. 1977. Mise en évidence et description de l'organe Y chez *Penaeus japonicus* (Bate, 1881) (Crustacea, Decapoda, Natantia). C. R. Acad. Sci. Paris 285D: 977–980.

Bressac, C. 1973. Données sur l'ultrastructure de la glande de mue (organe Y) du crabe *Pachygrapsus marmoratus*. C. R. Acad. Sci. paris 277D: 1165–1167.

Bressac, C. 1976. Effets de l'ablation des pédoncles oculaires sur les organes Y du crabe *Pachygrapsus marmoratus*. C. R. Acad. Sci. Paris 282D: 1873–1875.

Bressac, C. 1978. L'activité mitotique au niveau de l'organe Y du crabe *Pachygrapsus marmoratus* (Fabricius). C. R. Acad. Sci. Paris 286D: 289–291.

Buchholz, C. 1984. Die Häutungsdrüse larvaler decapoder Crustaceen: Untersuchungen der Struktur und Funktion am Beispiel der Schlammkrabbe *Rithropanopeus harrisii* (Gould). Ph.D. thesis, Institut für Meereskunde, Christina-Albrechts-Universität, Kiel, West Germany.

Buchholz, C. and D. Adelung. 1980. The ultrastructural basis of steroid production in the Y-organ and the mandibular organ of the crabs *Hemigrapsus nudus* (Dana) and *Carcinus maenas* L. Cell Tissue Res. 206: 83–94.

Burghause, F. 1975. Das Y-organ von *Orconectes limosus* (Malacostraca, Astracura). Z. Morphol. Tiere 80: 41–57.

Carlisle, D. B. 1957. On the hormonal inhibition of moulting in decapod crustacea. II. The terminal anecdysis in crabs. J. Mar. Biol. Assoc. U.K. 36: 291–307.

Chaix, J.-C., J.-P. Trilles, and G. Vernet. 1976. Dégénérescence de l'organe Y chez les mâles pubères d'*Acanthonyx lunulatus* (Risso) (Crustacea, Decapoda, Oxyrhyncha). C. R. Acad. Sci. Paris 283D: 523–525.

Charmantier, G. and J.-P. Trilles. 1973. Dégénérescence de la glande de mue chez les mâles pubères se *Sphaeroma serratum* (Crustacé, Isopode). C. R. Acad. Sci. Paris 276D: 581–583.

Charmatier, G. and J.-P. Trilles. 1975. Evolution des organes Y chez certains Sphaeromatidae (Crustacea, Isopoda, Flabellifera). C. R. Acad. Sci. Paris 281D: 1109–1110.

Charmantier, G. and J.-P. Trilles. 1979. La dégénérescence de l'organe Y chez *Sphaeroma serratum* (Fabricius, 1787) (Isopoda, Flabellifera): étude ultrastructurale. Crustaceana (Leiden) 36: 29–38.

Chassard-Bouchaud, C. and M. Hubert. 1975a. Sur l'existence de vésicules de réticulum endoplasmique lise dans l'organe Y de *Carcinus maenas* L. (Crustacé, Décapode). C. R. Acad. Sci. Paris 281D: 707–709.

Chassard-Bouchaud, C. and M. Hubert. 1975b. Etude ultrastructurale des hémocytes présents dans l'organe Y de *Carcinus maenas* L. (Crustacé Decapode). C. R. Acad. Sci. Paris 281D: 807–810.

Chassard-Bouchaud, C. and M. Hubert. 1975c. Etude infrastructurale de l'organe Y de *Carcinus maenas* L.: comparaison entre des animaux sains et des animaux parsités par *Sacculina carcini* (Thompson). C. R. Acad. Sci. Paris 281D: 893–895.

Chiang, R. G. 1987. Neurosecretory action potentials recorded extracellularly from a neurohemal region associated with the Y-organ in the terrestrial isopod, *Oniscus asellus*. Experientia (Basel) 43: 1220–1221.

Christensen, A. K. 1965. The fine structure of testicular interstitial cells in guinea pigs. J. Cell Biol. 26: 911–935.

Christensen, A. K. and S. W. Gillim. 1969. The correlation of fine structure and function in steroid-secreting cells, with emphasis on those of the gonads. Pp. 475–488 in K. W. McKerns (ed.), *The Gonads*. Appleton-Century-Crofts, New York.

Demeusy, N. 1975. Observations sur le fonctionnement des glandes mandibulaires du Décapode Brachyoure *Carcinus maenas* L.: animaux témoins et animaux sans pédoncles oculaires. C. R. Acad. Sci. Paris 281D: 1887–1889.

Diener, R. A., A. I. Yudin, and W. H. Clark, Jr. 1975. The ecdysial gland of the blue crab, *Callinectes sapidus* Rathbun. Am. Zool. 15: 786 (abstr.).

Drach, P. 1939. Mue et cycle d'intermue chez les crustacés Décapodes. Ann. Inst. Océanogr. 19: 103–391.

Ducruet, J. 1986. Localisation de l'orange Y chez le crustacé amphipode *Gammarus pulex* (L., 1758). Crustaceana (Leiden) 51: 231–234.

Echalier, G. 1954. Recherches expérimentales sur le rôle de "l'organe Y" dans le mue de *Carcinus maenas* (L.) Crustacé Dècapode. C. R. Acad. Sci. Paris 238D: 523–525.

Echalier, G. 1955. Rôle de l'organe Y dans le déterminisme de la mu de *Carcinides (Carcinus) maenas* (L.) (Crustacés Dècapodes): expériences d'implantation. C. R. Acad. Sci. Paris 240D: 1581–1583.

Echalier, G. 1959. L'organe Y et le déterminisme de la croissance et de la mue chez *Carcinus maenas* (L.) Crustacé Décapode. Ann. Sci. Nat. Zool. Biol. Anim. [12] 1: 1–59.

Fawcett, D. W., J. A. Long, and A. L. Jones. 1969. The ultrastructure of endocrine glands. Recent Progr. Horm. Res. 25: 315–380.

Fingerman, M. 1987. The endocrine mechanisms of crustaceans. J. Crustacean Biol. 7: 1–24.

Gabe, M. 1953. Sur l'existence chez quelques Crustacés Malacostracés d'un organe comparable à la glande de mue des Insectes. C. R. Acad. Sci. Paris 237D: 1111–1113.

Gabe, M. 1956. Histologie comparée de la glande de mue (organe Y) des Crustacés Malacostracés. Ann. Sci. Nat. Zool. Biol. Anim. [11] 18: 145–152.

Garren, L. D., R. L. Ney, and W. W. Davis. 1965. Studies on the role of protein synthesis in the regulation of corticosteroid production by adrenocorticotropic hormone *in vivo*. Proc. Natl. Acad. Sci. U.S.A. 53: 1443–1450.

Girard, P. and R. Maissiat. 1983. Variations du taux des ecdystéroïdes hémolymphatiques chez le mâle de *Ligia oceanica* (L.) (Crustacea, Isopoda, Oniscoidea) en fonction du cycle de mue et des modifications structurales de l'organe Y. Can. J. Zool. 61: 534–538.

Hall, P. F. 1985. Role of cytochromes P-450 in the biosynthesis of steroid hormones. Vitam. Horm. 42: 315–368.

Herman, W. S. 1967. The ecydsial glands of arthropods. Int. Rev. Cytol. 22: 269–347.

Hinsch, G. W. 1977. Fine structural changes in the mandibular gland of the male spider crab, *Libinia emarginata* (L.) following eyestalk ablation. J. Morphol. 154: 307–315.

Hinsch, G. W. and H. A. Hajj. 1975. The ecdysial gland of the spider crab, *Libinja emarginata* (L.). I. Ultrastructure of the gland in the male. J. Morphol. 145: 179–187.

Hinsch, G. W., E. Spaziani, and W. H. Vensel. 1980. Ultrastructure of Y-organs of *Cancer antennarius* in normal and de-eyestalked crabs. J. Morphol. 163: 167–174.

Janszen, F. H. A., B. A. Cooke, and H. L. van der Molen. 1977. Specific protein synthesis in isolated rat testis Leydig cells: influence of luteinizing hormone and cycloheximide. Biochem. J. 162: 341–346.

Johnson, P. T. 1980. *Histology of the Blue Crab, Callinectes sapidus.* Praeger, New York.

Karlson, P. and D. M. Skinner. 1960. Attempted extraction of crustacean moulting hormone from isolated Y-organs. Nature (Lond.) 185: 543–544.

Knowles, F. G. W. 1965. Neuroendocrine correlation at the level of ultrastructure. Arch. Anat. Microsc. Morphol. Exp. 54: 343–358.

Landefeld, T. D., K. L. Campbell and A. R. Midgley. 1979. Rapid changes in the synthesis of specific ovarian granulosa cell proteins induced by human choriogonadotropin. Proc. Natl. Acad. Sci. U.S.A. 76: 5153–5157.

Le Roux, A. 1974. Mise au point à propos de la distinction entre l'organe Y et l'organe mandibulaire chez les Crustacés eucarides. C. R. Acad. Sci. Paris 278D: 1261–1264.

Le Roux, A. 1977. L'organe Y de *Palaemon serratus* (Pennant) (Décapode, Natantia): localisation et aspects histologiques. Cah. Biol. Mar. 18: 413–425.

Le Roux, A. 1984. Histogenèse de l'organe Y (glande de mue) chez l'embryon de la crevette *Palaemon serratus.* (Pennant) Boll. Zool. 51: 62 (abstr.).

Maissiat, J. 1970a. Etude expérimentale du rôle de l'organe Y dans le déterminisme endocrine de la mue chez isopode Oniscoïde *Porcellio dilatatus* Brandt. C. R. Acad. Sci. Paris 270D: 2573–2574.

Maissiat, J. 1970b. Anecydsis experimentale provoquée chez l'Oniscoïde *Ligia oceanica* L. et rétablissement de la mue par injection d'ecdysone ou réimplantation de glande maxillaire. C. R. Soc. Biol. 164: 1607–1609.

Maissiat, J. and F. Graf. 1973. Action de l'ecdystérone sur l'apolysis et l'ecdysis de divers Crustacés isopodes. J. Insect. Physiol. 19: 1265–1276.

Maissiat, J. and J. J. Legrand. 1970. Contribution a l'étude expérimentale du contrôle hormonal du cycle de mue chez les Oniscoïdes *Porcellio dilatatus* et *Ligia oceanica*. C. R. Soc. Biol. 164: 359–364.

Maissiat, R. and J. Maissiat. 1976. Structure et ultrastructure de la glande de mue et synthèse des ecdysones en fonction des étapes du cycle de la mue chez *Ligia oceanica* (Crustacé Isopode Oniscoïde). Bull. Soc. Zool. Fr. 101: 545–558.

Maissiat, R., G. Martin, J. Maissiat, and P. Juchault. 1979. Ultrastructural development of the neurohemal organ joined to the ecdysial gland after imaginal moulting in the male isopod *Sphaeroma serratum* FAB. (Crustacea Flabellifera). Cell Tissue Res. 203: 403–414.

Martin, G., R. Maissiat, and P. Girard. 1983. Ultrastructure of the sinus gland and lateral cephalic nerve plexus in the isopod *Ligia oceanica* (Crustacea Oniscoidea). Gen. Comp. Endocrinol. 52: 38–50.

Mattson, M. P. 1986. New insights into regulation of the crustacean molt cycle. Zool. Sci. 3: 733–744.

Mattson, M. P. and E. Spaziani. 1985. Characterization of molt-inhibiting hormone (MIH) action on crustacean Y-organ segments and dispersed cells in culture and a bioassay for MIH activity. J. Exp. Zool. 236: 93–101.

Mattson, M. P. and E. Spaziani. 1986. Regulation of Y-organ ecdysteroidogenesis by molt-inhibiting hormone in crabs: involvement of cyclic AMP–mediated protein synthesis. Gen. Comp. Endocrinol. 63: 414–423.

Mattson, M. P. and E. Spaziani. 1987. Demonstration of protein kinase C activity in crustacean Y-organs, and partial definition of its role in regulation of ecdysteroidogenesis. Mol. Cell. Endocrinol. 49: 159–172.

McConaugha, J. R. 1980. Identification of the Y-organ in the larval stages of the crab, *Cancer anthoni* Rathbun. J. Morphol. 164: 83–88.

Passano, L. M. 1960. Molting and its control. Pp. 473–536 *in* T. H. Waterman (ed.), *The Physiology of Crustacea*, Vol. 1. Academic Press, Orlando, Florida.

Passano, L. M. 1961. The regulation of crustacean metamorphosis. Am. Zool. 1: 89–95.

Quackenbush, L. S. 1986. Crustacean endocrinology, a review. Can. J. Fish. Aquat. Sci. 43: 2271–2282.

Simione, F. P., Jr. and D. L. Hoffman. 1975. Some effects of eyestalk removal on the Y-organs of *Cancer irroratus* Say. Biol. Bull. (Woods Hole) 148: 440–447.

Skinner, D. M. 1985. Molting and regeneration. Pp. 44–146 *in* D. E. Bliss and L. H. Mantel (eds.), *The Biology of Crustacea*, Vol. 9. Academic Press, Orlando, Florida.

Sochasky, J. B. and D. E. Aiken. 1974. The Y organ–molting gland problem in decapod Crustacea: the "organ of Madhyastha" and "organ of Carlisle," two putative Y organs. Can. J. Zool. 52: 1251–1257.

Sochasky, J. B., D. E. Aiken, and N. H. F. Watson. 1972. Y organ, molting gland, and mandibular organ: a problem in decapod Crustacea. Can. J. Zool. 50: 993–997.

Spaziani, E. and S. B. Kater. 1973. Uptake and turnover of cholesterol-^{14}C in Y-organs of the crab *Hemigrapsus nudus* as a function of the molt cycle. Gen. Comp. Endocrino. 20: 534–549.

Spindler, K.-D., R. Keller, and J. D. O'Connor. 1980. The role of ecdysteroids in the crustacean molting cycle. Pp. 247–280 *in* J. A. Hoffmann (ed.), *Progress in Ecdysone Research*. Elsevier/North-Holland Publ., Amsterdam and New York.

Taketomi, Y. and M. Hyodo. 1986. The Y-organ of the crab, *Portunas trituberculatus:* effects of ecdysterone on the ultrastructure. Cell Biol. Int. Rep. 10: 367–374.

Tamaoki, B.-I. 1973. Steroidogenesis and cell structure. J. Steroid Biochem. 4: 89–118.

Vensel, W. H., E. Spaziani, and L. S. Ostedgaard. 1984. Cholesterol turnover and ecdysone content in tissues of normal and de-eyestalked crabs (*Cancer antennarius*). J. Exp. Zool. 229: 383–392.

Vernet, G. 1976. Données actuelles sur le déterminisme de la mue chez les Crustacés. Ann. Biol. 15: 155–188.

Vernet, G., C. Bressac, and J.-P. Trilles. 1978. Quelques données récentes sur l'organe Y (glande de mue) des Crustacés décapodes. Arch. Zool. Exp. Gén. 119: 201–225.

Watson, R. D. and E. Spaziani. 1985a. Effects of eyestalk removal on cholesterol uptake and ecdysone secretion by crab (*Cancer antennarius*) Y-organs *in vitro*. Gen. Comp. Endocrinol. 57: 360–370.

Watson, R. D. and E. Spaziani. 1985b. Biosynthesis of ecdysteroids from cholesterol by crab Y-organs, and eyestalk suppression of cholesterol uptake and secretory activity, *in vitro*. Gen. Comp. Endocrinol. 59: 140–148.

Watson, R. D., E. Spaziani, and W. E. Bollenbacher. 1989. Regulation of ecdysone biosynthesis in insects and crustaceans: a comparison. Pp. 188–203 *in* J. Koolman (ed.), *Ecdysone*. Thieme-Verlag, Stuttgart.

Zerbib, C., N. Andrieux, and J. Berreur-Bonnenfant. 1975. Données préliminaires sur l'ultrastructure de la glande de mue (organe Y) chez le crabe *Carcinus meditteraneus* sain et parasité par *Sacculina carcini*. C. R. Acad. Sci. Paris 281D: 1167–1169.

Ecdysial Glands and Ecdysteroids in Terrestrial Chelicerata

5

CHRISTIAN JUBERTHIE and J. C. BONARIC

5.1.   Introduction

It is mainly during the past 10 years that experimental and structural studies of the endocrinology of some arthropod groups, such as Chelicerata, have been conducted. Molting hormones (ecdysteroids) have been detected in Merostomata, Arachnida, and Pycnogonida. Certain cells or organs that were thought to synthesize and secrete molting hormone have been described. The neuroendocrine complex, which regulates growth, molting, reproduction, and diapause, has also been described in nearly all the groups of Arachnida; however, very few experiments have been performed, because of the difficulties in working with these species.

Various books and reviews, such as those of Gabe (1966) on neurosecretion, Herman (1967) on ecdysial glands, Tombes (1970, 1979) and Gupta (1983) on neurohemal organs, and Barth (1985) on neurobiology, as well as the comparative studies of Highnam and Hill (1977), contain a wealth of recent information on these topics. In addition, several general books have been published on particular groups of Arachnida, such as spiders (Foelix, 1982; Nentwig, 1987) and ticks (Binnington, 1986).

5.2.   Morphology and Histology

5.2.1.   *Ecdysial or Molting Glands*

5.2.1.1.   OPILIONIDA

In the immature stages of four species of Opilionida (Phalangiidae)—*Leiobunum rotundum, Phalangium opilio, Oligolophus tridens,* and *Odiellus gallicus*)—Herlant-Meewis and Naisse (1957) and Naisse (1959) described an endocrine tissue behind the syncerebrum, surrounded by the fat body, that shows periodical changes corresponding to the molting cycle. Gersch (1964) called the endocrine cells described by Naisse in Phalangiidae the "molting gland," a term since used by Gabe (1966), Streble (1966), and Herman (1967). It is in the form of a distinct dorsoventral stripe divided into two branches from the level of the esophagus up to the endosternite. It persists in the adult, but then has the appearance of an interstitial tissue. However, no ultrastructural studies or experiments have been performed to confirm that interpretation.

In a troglophil species, *Ischyropsalis luteipes* (Ischyropsalidae), such a structure could not be found in the area described by Naisse (C. Juberthie, unpublished data).

5.2.1.2.   SCORPIONIDA

Prior to the detection of ecdysteroids in Scorpionida, several authors proposed putative molting glands from histological studies. Habibulla (1961), in the young forms of the Indian scorpion *Heterometrus swammerdami*, described paired, symmetrically arranged, blind end-organs situated at the end of symmetrical nerves which arise from the pedipalpal ganglia. They are composed of two types of cells, one large, another much smaller and bipolar, and would possess a delimiting wall exclusively during the postmolt stage. The organs are bathed in the hemolymph, and hemocytes are present near the cells of the organs in the premolt stages. Habibulla suggested that they play a role in the molting process. However, according to Gabe (1971), the end-organs of Habibulla do not possess any of the histological characteristics of a molting gland and cannot be homologized with Legendre's (1958) "anterior organ." Habibulla's figures give good arguments for Gabe's position. Gabe proposed another organ as the molting gland, in four species: *Buthus occitanus* and *Centruroides margaritatus* (Buthidae), *Diplocentrus scaber* (Diplocentridae), and *Euscorpius flavicaudis* (Chactidae); it is the so-called glomus (preglomerular mass) of the coxal gland—an approximatively spherical mass of cells located at the point where the artery of the coxal gland reaches the saccule and ensheathed by a basal lamina. Its general features are those of gland cells, and there is no evidence of a functional link with the other part of the coxal gland. The histological characteristics led Gabe to consider a possible analogy with the molting gland. However, he indicated that the position of the preglomerular mass is different from the supposed prosomatic molting gland known at that time (Legendre, 1958, 1959).

Stockmann (1984) also did not support Habibulla (1961), because he did not find the end-organ in juveniles and adults of *Euscorpius flavicaudis* and *E. carpathicus*. He also disagreed with Gabe's interpretation regarding the preglomerular mass, in which he found a lumen and muscle fibers, and proposed instead that "anterior endocrine cells" at the base of the pedipalps, in a position similar to the anterior organ of spiders, formed the presumed molting gland, although Streble (1966) had earlier reported the absence of molting glands in the prosoma of scorpions.

In conclusion, the putative molting gland in scorpions remains unknown, since neither experimental data nor ultrastructural studies nor radioimmunoassay (RIA) evidence of such organs in tissue culture have been reported to date.

5.2.1.3.   ARANEAE

Millot (1926, 1930) first described clusters of prosomatic cells in several species of Labidognatha spiders. He called these groups of cells "tissue endocrine," based on the presence of numerous vacuoles, which probably liberated their product directly into the blood. According to Millot, these endocrine cells originated from reticulated cephalothoracic tissue, which, according to Legendre (1959), derived from the mesoderm; this is in contrast to the prothoracic glands of insects, which are derived from the ectoderm. Millot suggested that most spiders possess this endocrine tissue and demonstrated that its position and function differed in various families. However, during the early postembryonic instars (nymphs according to the terminology of Vachon, 1953), Millot (1926, 1930) did not observe differentiated endocrine cells. On the other hand, Browning (1942) did observe such cells and reported that they divided by mitosis and could transform into nephrocytes.

Legendre (1958, 1959) more clearly described the location of this endocrine tissue and distinguished a more or less dispersed anterior organ. The cytological changes in this organ during postembryonic development led him to conclude that it played a role in the molting process, and he homologized it to the prothoracic glands of insects. Streble (1966) studied the endocrine tissue described by Millot (1926, 1930) and distinguished anterior, median, and posterior organs in the prosoma. He confirmed the presence of mitosis and described cytological changes related to the intermolt cycle; the secretory activity of the system during the intermolt cycle could suggest a role in the molting phenomenon. More recently, Bonaric (1980) regarded the endocrine tissue of Millot, as a "molting gland," based on histological and ultrastructural characteristics described in *Pisaura mirabilis* (Pisauridae).

Molting glands have recently been described in many other spiders: in Mygalomorpha (Legendre, 1981); in a primitive spider, *Heptathela kimurai* (Liphiistiidae) (Legendre, 1981); in a relict troglobite, *Telema tenella* (Bonaric and Juberthie, 1983), and in other Labidognatha such as *Micrommata ligurinum* (Sparassidae) and *Filistata insidiatrix* (Filistatidae) (Bonaric et al., 1984); and in the Bolas spider, *Mastophora cornigera* (Lopez et al., 1985).

The location of the molting gland differs in Araneomorpha, Mygalomorpha, and Liphistiomorpha: in the first group, it is mainly prosomatic; in the second, mainly opisthosomatic; and in the third, both in the prosoma and opisthosoma. Bonaric (1980) has provided the most comprehensive description of the molting gland in this group, based on histological studies of several hundred young and adult spiders, representatives of a majority of species. The molting gland consists of numerous prosomatic masses of cells, distributed symmetrically and metamerically, and generally located in the ventral region of the prosoma (Fig. 5.1). The gland is divided into lateral and posterior groups of endocrine cells.

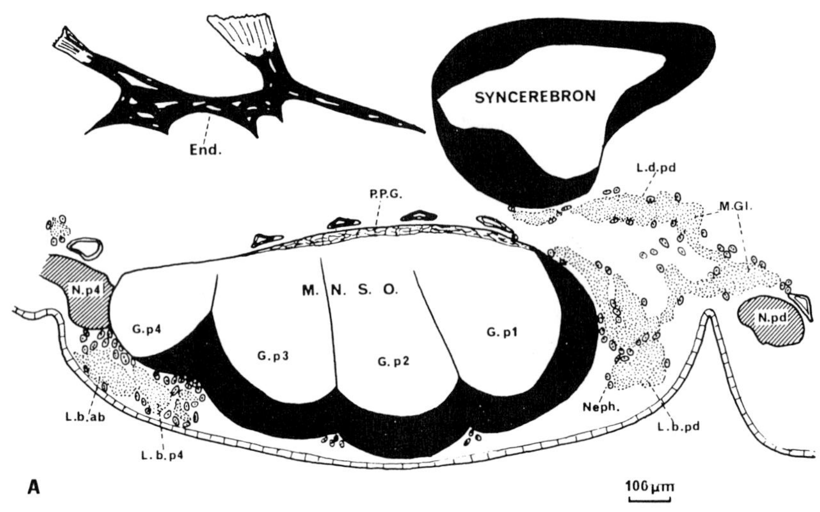

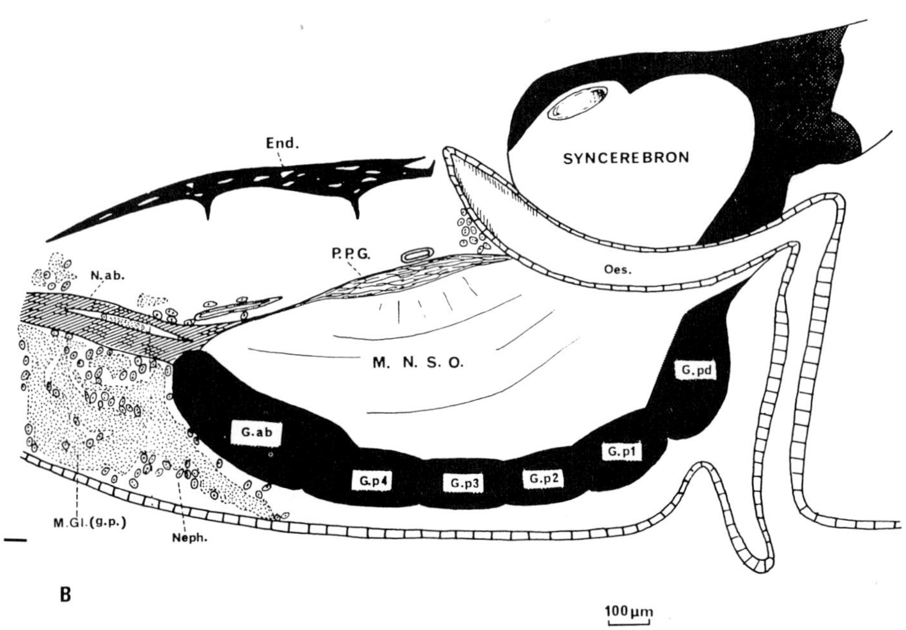

FIGURE 5.1. Ecdysial gland (M.Gl.) in *Pisaura mirabilis* (Pisauridae). *Key:* End. = Endosternite; G.pd, G.p1, G.p2, G.p3, G.p4, G.ab = nerve ganglia of pedipalp, leg 1, 2, 3, 4, abdominal; N.ab, N.p4, N.ped = abdominal, leg 4, pedipalp nerv; Neph. = nephrocyte; Oes. = oesophagus; P.P.G. = paraganglionic plates. (A) Lateral view. (B) Sagittal view. (From Bonaric, 1980.)

Lateral Groups of Endocrine Cells

These groups of cells are located on each side near the nerves that arise from the pedipalp and leg ganglia of the subesophageal nervous mass. They include the following:

Pedipalp groups, or anterior organs of Legendre.   These are large, paired islets of endocrine cells, isolated from each other in the frontal area by a space filled with nephrocytes, in the region of the blood vessel and the nerve of the pedipalps. Each of them consists of: a dorsal band (d.pd.b), which enters the part of the coxa differentiated in the maxillar lobe; a posterodorsal mass, between the syncerebrum and the endosternite, and also the chelicera and the pedipalp ganglia; and a voluminous laterobasal mass (l.b.pd.m.), which extends to the inferior face of the pedipalp ganglion with lateral branches.

In the young inside the cocoon (*larva* in Vachon's terminology), the two lateral groups of pedipalp endocrine cells fuse in the medioanterior region of the prosoma, which exactly corresponds to the "anterior organ" of Legendre (1958). As early as in the first free instar (*nymph 1* in Vachon's terminology), the two groups are separated. For this reason, Bonaric (1980) and Bonaric and Juberthie (1980) included that group with the other groups of the appendages under the term "lateral groups."

Leg (L1, L2, L3, L4) groups.   Each fused prosomatic metamere with legs possesses a paired, symmetrical mass of endocrine cells located dorsally and posterior to the blood vessel and the nerve of the corresponding leg. Each mass of cells consists of: a dorsal and lateral band running on the dorsal face of the nerve, then entering the coxa, and ending near the distal limit of the coxa; and a laterobasal band located in the space between the nerves of two successive legs.

Posterior Group

This group consists of a voluminous mass of cells surrounding the abdominal nerves near the cauda equina. Most of the cells are sternal, and the remaining ones are divided into several cell bands above the abdominal nerves and the subesophageal nerve mass. The posterior and laterobasal groups of leg 4 are connected.

Individual differences and changes during embryonic and postembryonnic development.   In *Pisaura mirabilis*, Bonaric (1960) observed individual differences and postembryonic changes.

He found that in the last postembryonic instars (*nymphs 9 and 10* in Vachon's terminology), the posterior and each lateral group are isolated. In contrast, in some individuals of sixth and seventh instars, all the lateral and posterior groups are fused in two continuous bands; the other specimens showed the general features described previously. In the first and second postembryonic free instars, each group is reduced to a few cells. The endocrine cells form bands only in the third instar.

According to Legendre (1958), in *Dolomedes fimbriatus*, the endocrine cells can be observed very soon after the end of embryonic development, marked by the first apolysis, which produces a pharate instar in spiders and in others Chelicerata (Juberthie, 1988). The first endocrine cells can be observed prior to that apolysis, during the differentiation of neuromeres when the coelomic vesicles are clearly distinct. Later, some of the cells migrate to the pedipalp segment and others form jackets surrounding each segmentary nerve (days 14 and 15 of embryonic development). According to the description by Legendre (1958), we can place the first apolysis near day 15.

During the embryonic concentration of the nervous mass, the presumptive ganglia migrate to the prosoma with their embryonic endocrine cells, which come together in the posterior region of the abdominal ganglia.

All authors have noted some differences among species or members of the same family in other Araneomorpha.

According to Millot (1930), the endocrine tissue is reduced to lateral prosomatic groups in Dysderidae and to peripharyngial clusters in Salticidae. In *Telema tenella* (Telemidae), the endocrine cells are concentrated in two prosomatic laterodorsal stripes that are more developed in the posterior area of the prosoma, with metamere-like constrictions, and are anteriorly tapered ending near the base of the pedipalps (Fig. 5.1 and 5.2). Streble confirmed the presence of a typical endocrine tissue in Sicariidae (*Scytodes thoracica*). According to Lopez et al. (1985), the Bolas spider, *Matophora cornigera*, has some groups of endocrine cells in the opisthosoma, an exceptional feature in Araneomorpha.

Uncertainties about the presence and location of the molting gland in Mygalomorpha have been resolved by the investigations of Legendre (1981) in three species: (*Nemesia incerta*) (Ctenizidae), (*Scodra calceata*) (Theraphodidae), and (*Legendrella pauliani*) (Migidae). In all these species, he found that the gland was mainly localized in the opisthosoma and formed masses of cells surrounding and running along the two main branches of the abdominal nerve for the spinnerets; some of the cells penetrated the blood spaces, between the phyllotracheae, testes, and the midgut diverticules. In *Scodra calceata*, the prosoma is practically devoid of molting gland cells, except for the scarce endocrine cells near the pedipalp nerve. In *Legendrella pauliani*, two small groups of endocrine cells are localized in the rostral area of the prosoma and another small mass near the cauda equina.

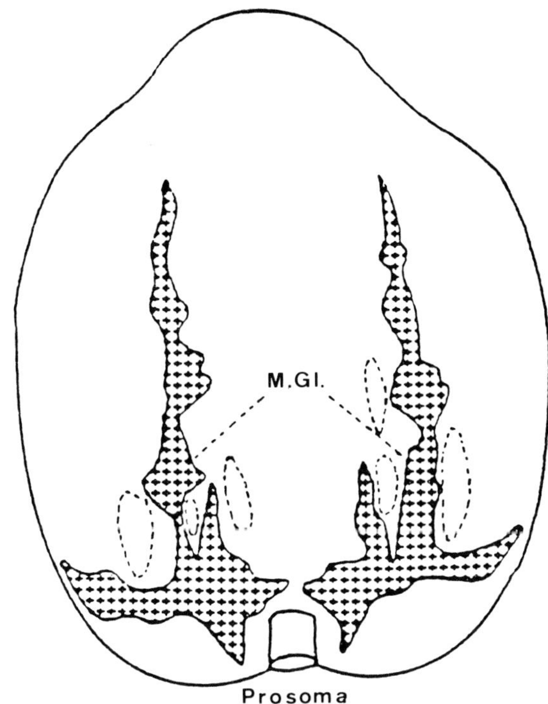

FIGURE 5.2. Location of the ecdysial gland (M.Gl.) in the prosoma of *Telema tenella* (Telemidae); horizontal view. (From Bonaric and Juberthie, 1983.)

In *Nemesia incerta*, in addition to the two previous groups in the rostral area, in front of the syncerebrum is a small medioventral group near the pedipalp nerves surrounding blood vessels and running medially below the subesophageal nervous mass.

Streble (1966) described in three neotropical species, *Pamphobeteus roseus*, *P. tetracanthus*, and *Grammostola actaeon* (Theraphosidae), voluminous masses of endocrine cells (50–250 μm) surrounding the blood vessels in the prosoma; however, the opisthosoma of these species should be reexamined, as well as the assumption by Millot, who did not find endocrine cells in the prosoma of *Nemesia caementaria* and *N. carminani* (Ctenizidae) and in *Atypus affinis* (Atypidae); according to Legendre's description (1981), these species could be of the *Scodra* type.

In Liphistiomorpha, e.g., *Heptathela kimurai*, the most primitive spider, the molting gland is as well developed in the opisthosoma (anterior part) as in the prosoma (Legendre and Lopez, 1981). In all the known cases, the endocrine cells of the molting gland are bathed in the hemolymph of the

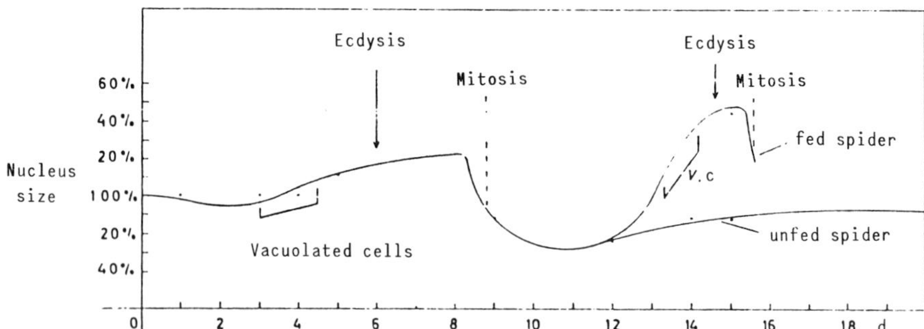

FIGURE 5.3. Mitotic cycle and size of the nucleus during the intermolt cycle in *Pardosa amentata*. (Modified from Streble, 1966.)

sinus, partly surrounded with blood cells, and often mixed with nephrocytes.

Histological structure of the molting gland. Although described earlier as a syncytium, the endocrine tissue consists (Fig. 5.4A) of individual cells (Legendre, 1958; Streble, 1966; Bonaric and Juberthie, 1980; Bonaric, 1980). No glial cells were observed. A mitotic phase was observed during the C period of the intermolt cycle (Figs. 5.3 and 5.4B); in *Pisaura mirabilis*, the mitotic phase appears from days 7 to 12 in an intermolt cycle of 22 days for the eighth postembryonic free instar. (For histology see Section 5.3.)

5.2.1.4.   ACARI

In mites, several organs or groups of cells compete for the ecdysial gland title; however, the molting gland remains hypothetical. According to Streble (1966), groups of cells located in the hemolymphatic sinus in *Trombidium holosericeum* could represent endocrine cells. They are active in females prior to the imaginal ecdysis in this species and during the juvenile instars in *Argas reflexus*. Their cytological modifications seem to be related to changes in the tegumentary cells during molting. In the cattle tick, *Boophilus microplus*, according to Binnington (1986), the "lateral

FIGURE 5.4. Eighth nymph of *Pisaura mirabilis:* (A) Light micrograph showing the morphology of the ecdysial gland (Mgl), nephrocyte (neph), hemocytes (H) (×600). (B) Mitoses (Mit) in the ecdysial gland during the C period of the intermolt cycle (×550). (C) Electron micrograph of ecdysial gland in A period, 20 h after ecdysis, showing SER, Golgi apparatus (G), and tubular body (Ft) (×34,500). (D) Ecdysial gland at the beginning of C period, 6 days after ecdysis showing RER, mitochondria with tubular cristae (T cr), and polysomes (p) (×12,000).

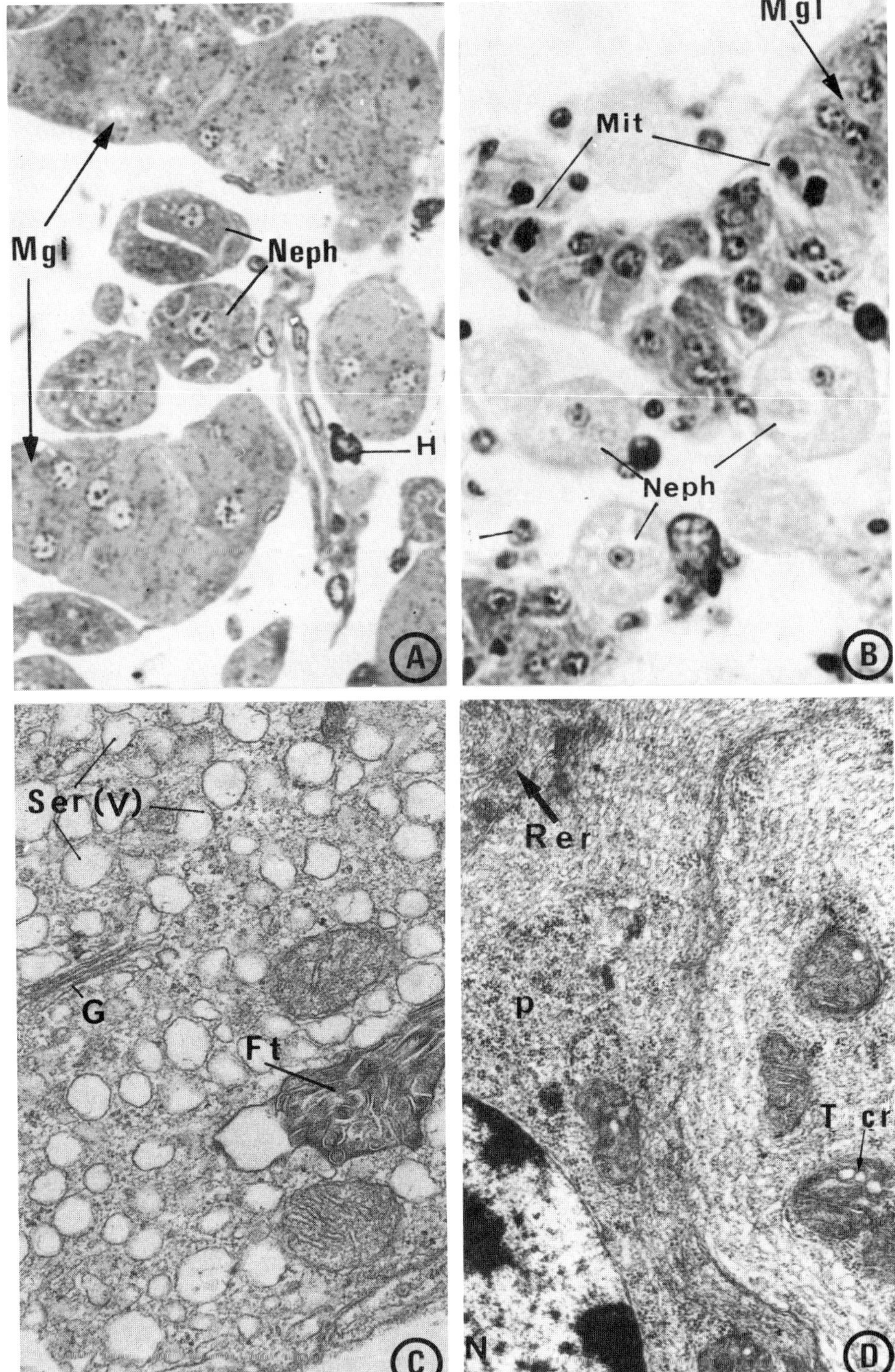
Mgl
Neph
H
Mit
Mgl
Neph
Ser(V)
G
Ft
Rer
p
N
T cr
A
B
C
D

organs" (in view of their ultrastructure) could represent the site of synthesis of the molting hormone.

Another potential molting gland could be the "peritracheal gland" described in *Argas arboreus* (Roshdy and Marzouk, 1982), since Ellis and Obenchain (1984) noted that, in contrast to other organs, cells associated with the tracheal trunks secreted high amounts of immunoreactive material. The gland is a discrete cell mass associated with tracheal branchings and/or adhering to the hypodermis near the central nerve mass of each postembryonic developmental stage of *A. arboreus*. These cells bud from ectoderm during embryogenesis, as does the prothoracic gland of insects. According to the authors, the histology of these glandular cells supports homology to the insect peritracheal "prothoracic" gland responsible for secretion of ecdysones. However, the secretory granules of these cells are stained blue with chrome–hematoxylin–phyloxine staining technique of Gomori that is used to demonstrate neurosecretory material. These data would seem to indicate that these cells have a neurosecretory function and should not be regarded as constituting an ecdysial gland.

5.2.2. Oenocytes as Sources of Ecdysteroids

The prothoracic gland of insects represents the main organ that synthesizes insect molting hormones. In addition, insect oenocytes have been shown to synthesize ecdysones (Romer et al., 1974; Ruhland and Romer, 1977). In various species of harvestmen, *Phalangium opilio, Leiobunum limbatum, Opilio parietinus,* and *Opilio ravennae* (Phalangiidae), Romer and Gnatzy (1981) described cells measuring up to 130 μm in the proximal segments of the femora of all four pairs of walking legs. These cells exhibit all the fine structural characteristics of insect oenocytes, in particular, the conspicuous agranular smooth endoplasmic reticulum (SER). Radioimmunoassay after *in vitro* incubation of these cells has demonstrated the synthesis of α- and β-ecdysones. The similarity between the fine structure of these cells in harvestmen and that of oenocytes of insects, as well as the results of ecdysone analysis, induced the authors to call these large cells in Opilionida "arachnid oenocytes."

5.2.3. Other Ecdysteroid-Producing Organs

In *Opilio ravennae,* one of the species of harvestmen discussed in the previous subsection, Romer and Gnatzy (1981) found by means of RIA a relatively large amount of RIA-positive material in the ovaries of the females collected in September and October 1979. By contrast, practically no ecdysteroids were found in the ovaries and the testes of animals collected in the middle of November in Frankfurt-Eschersheim; the females entered into dormancy. The last measure was performed after the last egg laying, and presumably no more than 1 or 2 months before the animals death.

In contrast, a high level of ecdysone persists in the tergite of the opisthosoma of males and females collected in November. According to the aforementioned authors, another center of synthesis is present in the region of the tergite of the opisthosoma; however, the exact site is unknown.

In conclusion, other centers of ecdysone synthesis have been demonstrated in harvestmen (tergites, ovaries, and femoral oenocytes), but it is unknown whether or not they play a role with the putative molting gland of Naisse and control the molting cycle.

In spiders and other Arachnid groups, no RIAs have been performed with dissected organs and in tissue culture.

5.3. Cytological and Ultrastructural Changes in the Ecdysial Glands

Since V. B. Wigglesworth (1952) first demonstrated the cycle of secretory activity in the prothoracic glands (PG) of two nymphal instars of *Rhodnius*, cytological studies have been performed in several orders of insects. The cyclical changes in the PG cells have been reinvestigated by means of electron microscopy in Heteroptera by Gras and Beaulaton (1979), Beaulaton and Gras (1980), and Beaulaton et al. (1984) (see also Chapter 7 in this volume). According to these authors, the main cyclical changes in the molting gland can be summarized as follows:

During the first period of the molting cycle (A, B, and part of C), when the ecdysteroid level is very low, the PG cells are flat, with a reduced amount of cytoplasm; the nucleoli present a compact structure. The first phase of the secretory cycle is characterized by an increase in the amount of cytoplasm and in organelle populations; the mitochondria are rod-shaped with a dense matrix and flat lamellar cristae. Large numbers of ribosomes are accumulated in cytoplasm, and the rough endoplasmic reticulum (RER) is limited to a few scattered cisternae. At the beginning of the critical period, the RER and ribosomes increase and elongated cisternae are plentiful; these characteristics were reported in several other insects, such as *Locusta migratoria* by Fain-Maurel and Cassier (1968) and Joly et al. (1969). The Golgi sacks become more numerous and increase in size, and the Golgi complexes containing numerous vesicles are randomly distributed in the cytoplasm.

During the period from the critical point to the peak of ecdysteroids, the PG cells display changes, such as deep infolding of the nucleus and enlargement of the RER compartment. One of the most striking changes is the gradual development of the SER related to ecdysteroid production.

At the beginning of the increase in ecdysteroid concentration, the first smooth-surfaced cisternae appear; at the periphery, the SER predominates and is intermingled with ribosomes and glycogen particles, whereas RER predominates in the inner cytoplasm. At the peak of ecdysteroid concentration, the RER reaches its maximal development and the secretory cells attain their maximal size. Numerous dense bodies and autophagic vacuoles, sequestering a mitochondrion or a cytoplasmic islet, are present in greater number than previously.

From the peak to ecdysis, the concentration of ecdysteroid is characterized by a sharp decrease; the size of PG cells declines, giving a high nucleus–cytoplasm ratio, and the number of dense bodies and autophagic vacuoles increases; the SER disappears, and the RER as well as the other organelles decrease. According to Birkenbeil (1983), in the wax moth, *Galleria mellonella*, the SER forms the secretory granules in which ecdysone has been shown immunocytochemically; the Golgi apparatus does not seem to be directly involved in ecdysone secretion.

5.3.1.   *During Intermolt Cycle in Juvenile Instars*

5.3.1.1.   ARANEAE

According to Streble (1966), in the subadult instar of *Pardosa amentata* (Fig. 5.3), the endocrine cells present a rhythmic activity. In the egg sac, 2 days before the ecdysis of the last pharate instar, the size of the nuclei of the ecdysial cells increases and rises to its maximum 1 day after ecdysis; then a series of mitoses occurs; the new nuclei are small and remain of the same size as long as the first free instar does not feed. When it catches and feeds on prey, the size of nuclei increases and a new cycle begins. In the integument, mitosis occurs 3–5 days past the mitoses in the ecdysial gland.

According to Bonaric (1980) and Bonaric and Juberthie (1980), in *Pisaura mirabilis* (Figs. 5.4B,C,D, 5.5A,D, and 5.6), a mitotic phase develops from days 7–13 of the eighth intermolt, which is 22 days long; the size of nuclei

FIGURE 5.5. Eighth nymph of *Pisaura mirabilis*. (A) Beginning of C period, 6–8 days after ecdysis, showing long cisternae of SER and mitochondria (m) (×24,000). (B&C) C period, 11 days after ecdysis showing vesiculated SER, Golgi (G), and multivesicular bodies (mc) (×28,500; ×34,500). (D) End of the C period, 14–17 days after ecdysis; pynocytotic vesicles (PV), exocytosis (E), polysomes (P), smooth endoplasmic reticulum (S), plasma membrane (Pm) (×40,500).

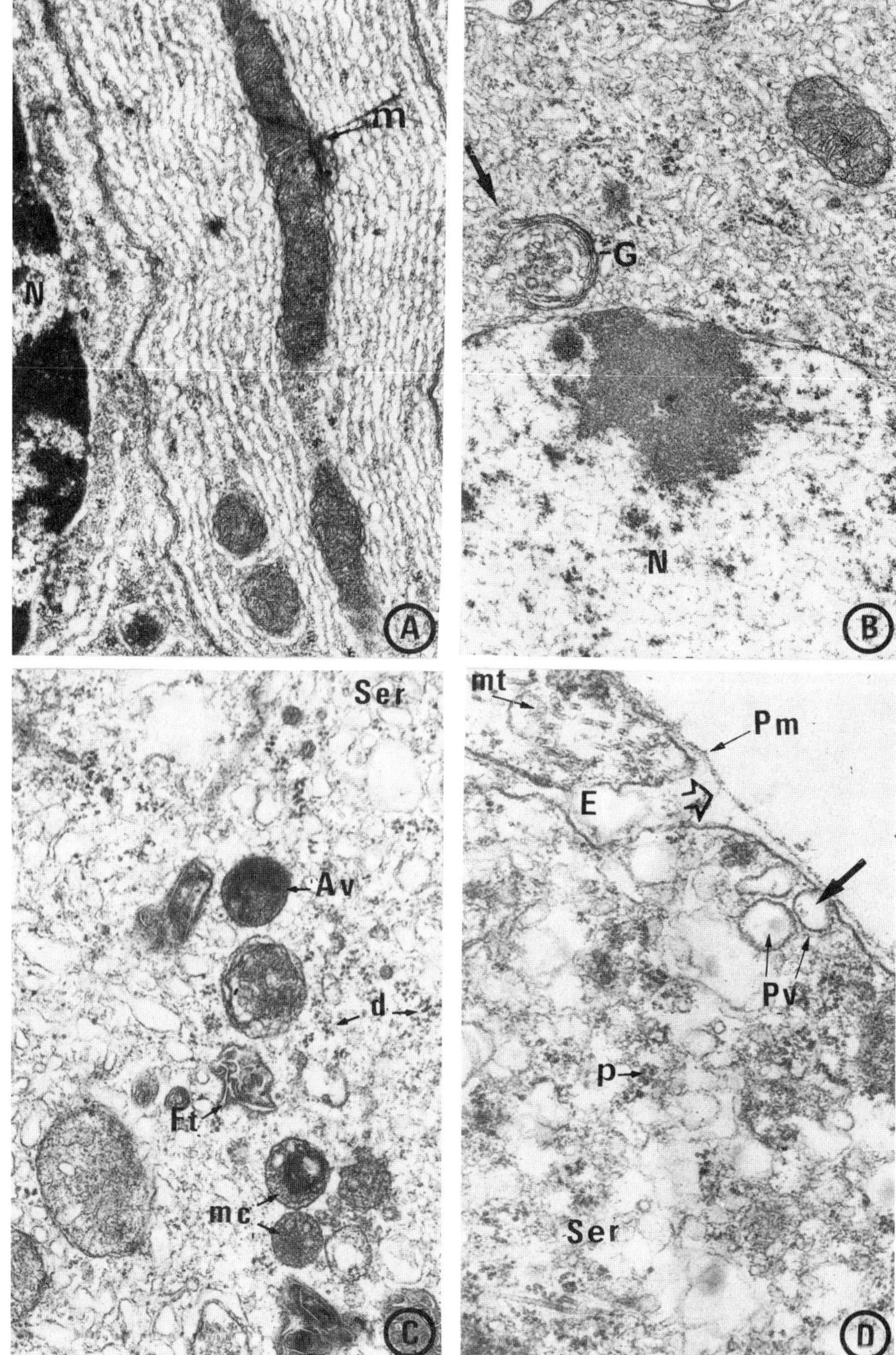

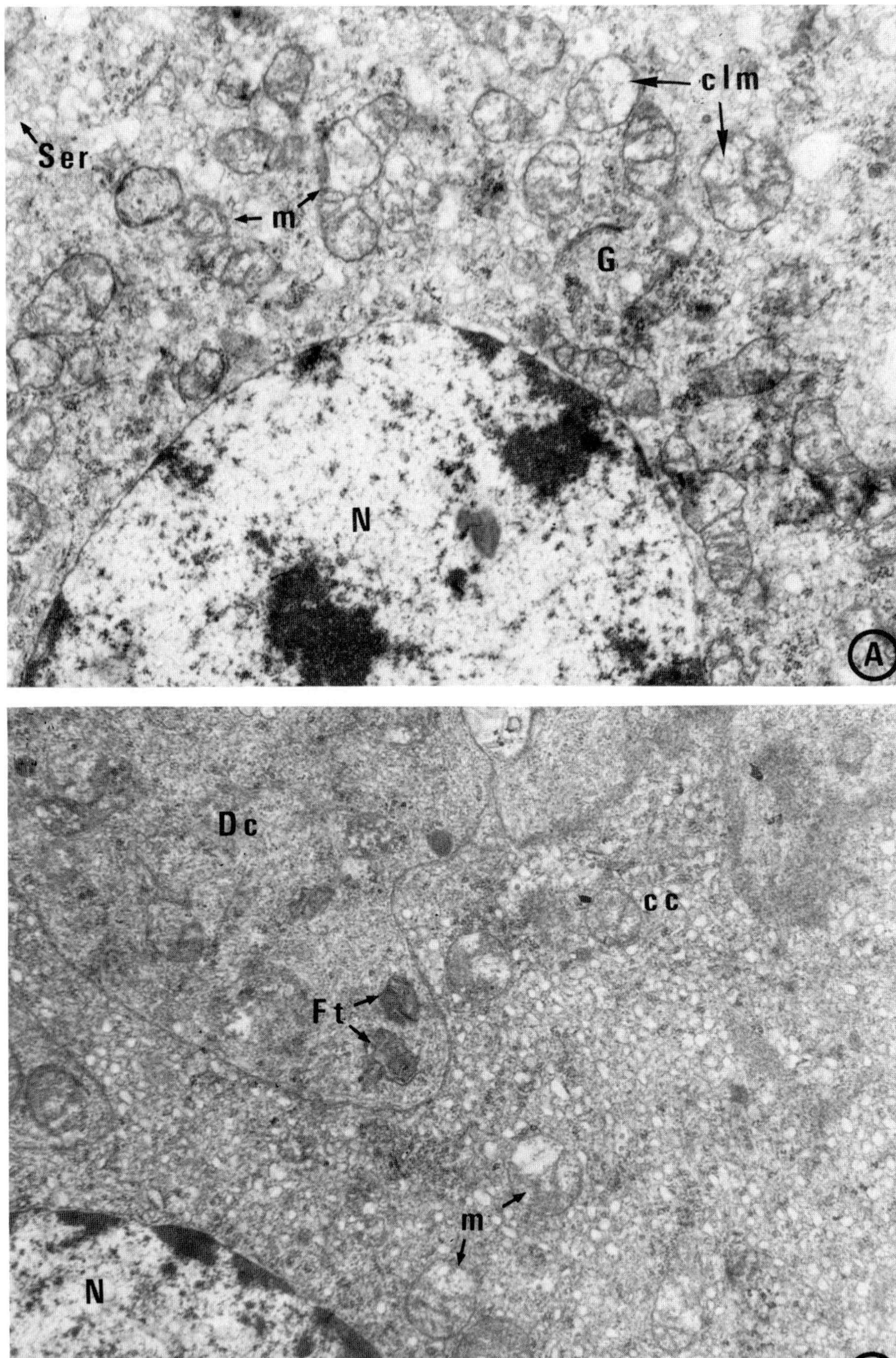

Ser
m
clm
G
N
Dc
cc
Ft
m
N
A
B

increases to the maximum at the end of the A period and at the beginning of the C period (from 4 to 7 days); the several small nucleoli of the A and B periods increase in number and become larger. The cells exhibit a slight asynchronism in their secretory cycle. The data in *Pardosa* and *Pisaura* are similar.

Among Arachnida, only the presumed molting gland of spiders has been examined by electron microscopy. Streble (1966) was the first to publish a micrograph of the "endocrine tissue" in *Meta menardi*. Later, Bonaric (1980) and Bonaric and Juberthie (1980) provided detailed data on the ultrastructural features of the molting gland in the juveniles and adults of *Pisaura mirabilis*.

The endocrine tissue in this animal consists of a basal lamina and endocrine cells. The basal lamina, 20–40 nm thick, surrounds the islets of endocrine cells; it is composed of an amorphous material that is finely granular. The gland cells present changes during the intermolt cycle. The authors described these changes in the free eighth postembryonic instar (*nymph 8* in Vachon's terminology). The eighth intermolt stage is 22 days long; on average, the A, B, C, and D periods are 1, 4, 13, and 4 days long, respectively.

At the beginning of the cycle (A and B periods), when ecdysteroid concentration is low (200 pmol/g), free ribosomes and polysomes are gathered in perinuclear areas; the RER is limited to short cisternae. At the beginning of the C period, the SER appears and greatly increases. At first, it is composed of a network of elongated tubules, then transforms in vesiculate cisternae (40–100 nm in diameter), which reach their maximal development (with cisternae of 150–200 nm in diameter) at the end of the C period concomitantly with the peak of ecdysteroid (1300 pmol/g) concentration. Polysomes are dispersed or gathered in small areas; ribosomes are dispersed. From the peak of ecdysteroid concentration to ecdysis, characterized by a fast decrease of ecdysteroid concentration, the vesiculation of SER regresses and progressively disappears after ecdysis during the A and B periods of the following intermolt cycle.

During the A period, the mitochondria are small, round, and have a dense matrix. During the B period, the mitochondria develop tubulovesicular cristae. In the first part of the C period, the mitochondria expand and may be rounded, elongated, curved, and with diverse cristae and relatively dense matrices. In the second part of the C period and during the increase of ecdysteroid concentration, vacuolized mitochondria

FIGURE 5.6. Eighth nymph of *Pisaura mirabilis:* (A) Beginning of the D period, 17 days after ecdysis. Light endocrine cell showing mitochondria (m) with clear matrix ($\times$18,000). (B) Diapausing seventh nymph, ecdysial gland showing light (C) and dark cells (D) ($\times$14,700).

with reduced cristae appear, and mitochondria increase in number until the peak of ecdysteroid concentration is reached. After the peak, they become reduced in size, and some again become rounded with dense matrices and are similar to the mitochondria of the A period. During the second half of the intermolt cycle, autophagic vacuoles sequestering cytoplasmic islets or mitochondria develop and numerous multivesicular bodies or heterogeneous and angular inclusions appear, more or less dense and composed of aggregated tubular structures and masses of dense material. Near the peak of ecdysteroid production, the plasma membrane is covered with numerous pinocytotic vesicles and some clear exocytotic vesicles below the basal lamina.

Similar ultrastructural features, particularly the very high amount of SER, have been observed in other spiders; the cave dweller species, *Telema tenella* (Bonaric and Juberthie, 1983), and the mygalomorph from Chile, *Tryssothele pissii* (J. C. Bonaric, unpublished data).

In conclusion, the endocrine cells of spiders present the ultrastructural features and changes, particularly SER and mitochondria with tubular cristae, described in the prothoracic glands of pterygote insects and in cells which synthesize steroids.

5.3.1.2. OPILIONIDA

In *Phalangium opilio* and *Leiobunum rotundum* (Phalangiidae), Naisse (1959) described cyclical modifications in the presumptive molting gland.

During A and B periods the cells possess small nuclei, surrounded with dense cytoplasm, and the gland is compact. During the last third of the cycle, vacuoles appear, grow, and fuse into large intracytoplasmic spaces. The nuclei enlarge and the nucleoli are distended. The mitotic phase appears in the middle of the cycle, and at the end of the cycle the secretory products pass into the hemolymph and the number and size of the vacuoles are reduced. At the time of ecdysis, the nuclei are small in size, and the cytoplasm is again dense and the gland compact.

5.3.2. *During Winter Diapause*

The molting gland of the seventh instar that is in winter diapause (Fig. 5.6B) presents characteristics similar to those mentioned during the B period and at the beginning of C period of the intermolt cycle of active and normal individuals and of clear and dark cells. The nuclei are round before the mitotic event.

5.3.3. *Ecdysial Glands in Adults*

In spiders with postpubertal ecdysis in females, preliminary observations have been performed with light microscopy in male of *Scodra calceata*

(Mygalomorpha: Theraphosidae) (Legendre, 1981), male of *Legendrella pauliani* (Migidae) (Legendre, 1981), female of *Tryssothele pissii* (Dipluridae) (Bonaric, unpublished data), female of *Nemesia caementaria* (Ctenizidae) (Bonaric, unpublished data), male of *Nemesia incerta* (Legendre, 1981), and female of *Filistata insidiatrix* (Araneomorpha: Filistatidae) (Bonaric et al., 1984).

The molting gland is present in the adults and seems active in the female with postpubertal and annual ecdysis. In *Filistata insidiatrix* (Bonaric et al., 1984), the molting gland is in a stage similar to that observed after the mitotic phase in the active juvenile instar of *Pisaura*. In the adult males of spiders (Mygalomorpha and Araneomorpha), which have no ecdysis, the molting gland still persists. In *Scodra calceata* (Legendre, 1981), the presence of several aggregated endocrine cells suggests that these parts of the gland begin to regress. In the female of Araneomorpha, which has no postpubertal ecdysis, the molting gland persists without regressive changes in *Pisaura mirabilis* (Bonaric, unpublished data); however, it is more fragmented than in normal juvenile instars. In female of *Pardosa amentata* (Streble, 1966), the cells have active secretory phases characterized by vacuole peaks, concomitant with cocoon carrying and a new cycle of ovogenesis; the gland degenerates just before death. In the male of *Pardosa amentata*, the gland persists until just before the death.

5.4. Physiological Roles of Ecdysial Glands and Ecdysteroids

The importance of ecdysteroids in the molting process is now well known in Insecta and Crustacea. The ecdysial glands in insects, cultured *in vitro*, synthesize ecdysones from cholesterol present in the food. In addition, other tissues (such as those of ovaries; also oenocytes) synthesize ecdysteroids. 20-Hydroxyecdysone (20-HE), or ecdysterone, has been demonstrated to be the main molting hormone.

Research in certain groups of arthropods, such as Chelicerata, began only in the late 1960s, and relatively few contributions have been published in this group. Krishnakumaran and Schneiderman (1968) were pioneers in noting the effect of exogenous ecdysteroids on the molting phenomena in spiders. It took 10 years for the ecdysteroid hormones to be detected in Arachnida, for the first time by Bonaric and De Reggi (1977) in a spider. Since then research in Acari and, to a lesser degree, in Opilionida and Scorpionida has progressed.

5.4.1   Existence of Molting Hormone in Arachnida

5.4.1.1.   ARANEAE

Using the protocol recommended by De Reggi et al. (1975), Bonaric and De Reggi were the first to demonstrate the presence of ecdysteroids in *Pisaura mirabilis*. The concentration of endogenous ecdysteroids was estimated by RIA, carried out on total extracts. The combination of thin-layer chromatography (TLC) with the RIA technique, carried out on each fraction and separated by means of chromatography, proved the existence of an RIA-positive product in the zone corresponding to the α- and β-ecdysones, the classical molting hormones in insects.

5.4.1.2.   OPILIONIDA

Romer and Gnatzy (1981) demonstrated the presence of ecdysteroids in tissues other than the molting gland (oenocytes, epidermis of tergites, ovaries) in *Opilio ravennae*.

5.4.1.3.   SCORPIONIDA

Recently, El Bakary (1986) and El Bakary et al. (1987) isolated ecdysteroids in Scorpionida from hemolymph and total extracts of eggs and young larvae of *Leiurus quinquestriatus* (Buthidae). Ecdysteroids were separated from the different extracts by means of high-pressure liquid chromatography (HPLC), then identified and measured by RIA, using the protocol described by Porcheron (1979); α - and β-ecdysones and polar and less polar ecdysteroids were detected.

5.4.1.4.   ACARI

Ecdysone and ecdysterone have been detected in nymphs of the tick *Amblyomma hebraeum* by Delbecque et al. (1978). The presence of RIA-positive material has been confirmed in the same species (Diehl et al., 1982) and in another ixodid, *Boophilus microplus* (K. P. Wigglesworth et al., 1985). During deposition of the new epicuticle, the highest ecdysteroid concentration was measured in the hemolymph and in the total extracts in the fifth-instar nymph of the argasid tick, *Ornithodoros moubata* (Germond et al., 1982). Recently, Connat (1987) confirmed the presence and nature of ecdysteroids in the three aforementioned species and in another ixodid, *Amblyomma variegatum*. In addition, he detected ecdysone, ecdysterone, and esters of these hormones in the eggs.

5.4.1.5.   OTHER CHELICERATA

Among Chelicerata, ecdysones have been recorded in *Limulus polyphemus* (Merostomata) by means of bioassays (Jegla, 1974; Winget and Herman, 1979), and by RIA (Jegla and Costlow, 1979; Jegla, 1982). More recently,

Behrens and Büchmann (1983) have extracted and identified 20-HE and detected an ecdysone-like hormone by bioassay and RIA in the pycnogonid *Pycnogonum littorale;* the ecdysteroid levels in this animal are very high. Ecdysteroids, well known in numerous groups of insects (see the review by Porcheron, 1979) and Crustacea (see the review by Spindler et al., 1980), have also been isolated in Myriapoda: first in the centipede *Hanseniella* then in *Lithobius forficatus* (Chilopoda) (Joly et al., 1979).

In conclusion, the ecdysones, particularly the active hormone 20-HE previously described in insects, are present in all major groups of Arthropoda, especially in Chelicerata.

5.4.2. Changes in the Levels of Endogenous Ecdysteroids

5.4.2.1. ARANEAE

The concentration of ecdysteroids change during the molting cycle in a manner similar to that in insects (Bonaric and De Reggi, 1977; Bonaric, 1980). In *Pisaura mirabilis*, during the eighth instar, which is 22 days long, the variation of the ecdysteroids is as shown in Fig. 5.7. During ecdysis (E_1), in individuals sampled exactly at ecdysis, the level of ecdysones is relatively low [75 ng ecdysterone-equivalents (EE)/g of fresh weight]. From ecdysis to day 5 (A and B periods of the intermolting cycle), the hormone level increases slowly up to some 100–150 ng/g. From days 5–17 (intermolting period C), the level of ecdysones remains stable and low ($\sim$50 ng/g). From day 17 to day 22 (ends of the C period and the D period, respectively), a main peak of ecdysone appears, showing an average of 400 ng/g; the individual peak can rise to 750 ng/g. Individual variability in duration of the intermolt cycle produces a shift of the main peak. Before the next ecdysis (E_2) and at the beginning of the next instar, the values recorded show that the level of ecdysone drops drastically.

The combination of TLC with RIA for each of the fractions separated by chromatography reveals that β-ecdysone is the major contributor. Histological studies of the successive events in the integument during the intermolting cycle show that the increase in the concentration of ecdysones could be correlated with apolysis, and the peak with the secretion of new epicuticle and exocuticle. Significant quantities of ecdysone were revealed in males and in adult female spiders (250 ng/g) and are probably related to the maturation of ovocytes. The organ of synthesis is still unknown.

Fluctuations in ecdysteroid levels during the winter diapause in *Pisaura mirabilis* were followed in a population from the south of France (Bonaric, 1980). The diapause period is characterized by the persistence of

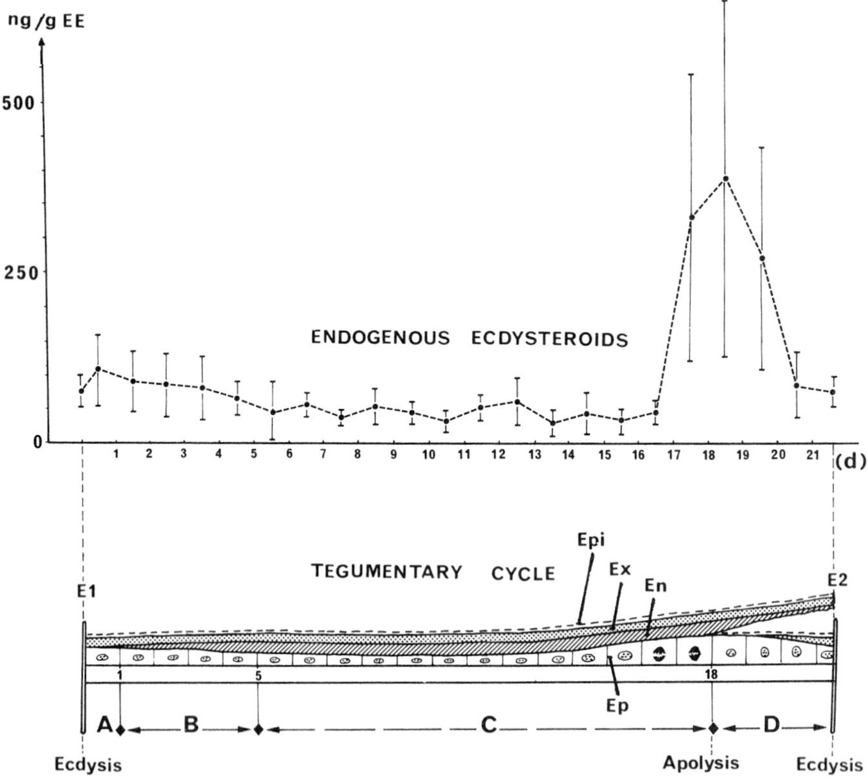

FIGURE 5.7. Ecdysteroid levels and intermolt cycle in the spider *Pisaura mirabilis*. Endogenous ecdysteroid levels expressed as nanograms ecdysterone-equivalents (= EE) per gram of fresh weight (+ SD). Tegumentary cycle of nymphal stage 8 during the intermolt from period A to period D. Integument layers: Ep, epidermis; En, endocuticle; Ex, exocuticle; Epi, epicuticle. (From Bonaric, 1980.)

a low hormone level (150–250 ng/g) for 3 months, from December to February; this level is close to that which characterizes the C period in the active *Pisaura* juveniles (Fig. 5.7). At the end of the diapause, the ecdysteroid concentration suddenly increases to reach average values ranging from 1000 to 1500 ng/g. This increase is linked to a renewal of development, marked by a stimulation of metabolism and induction of molting process. The decrease of ecdysteroid level and the persistence of a low level in diapausing seventh instar of *Pisaura mirabilis* agree with results obtained during the diapause of the butterfly *Pieris brassicae* (Mauchamps et al., 1977), *Hyalophora cecropia* (MacDaniel, 1979), *Pimpla in-*

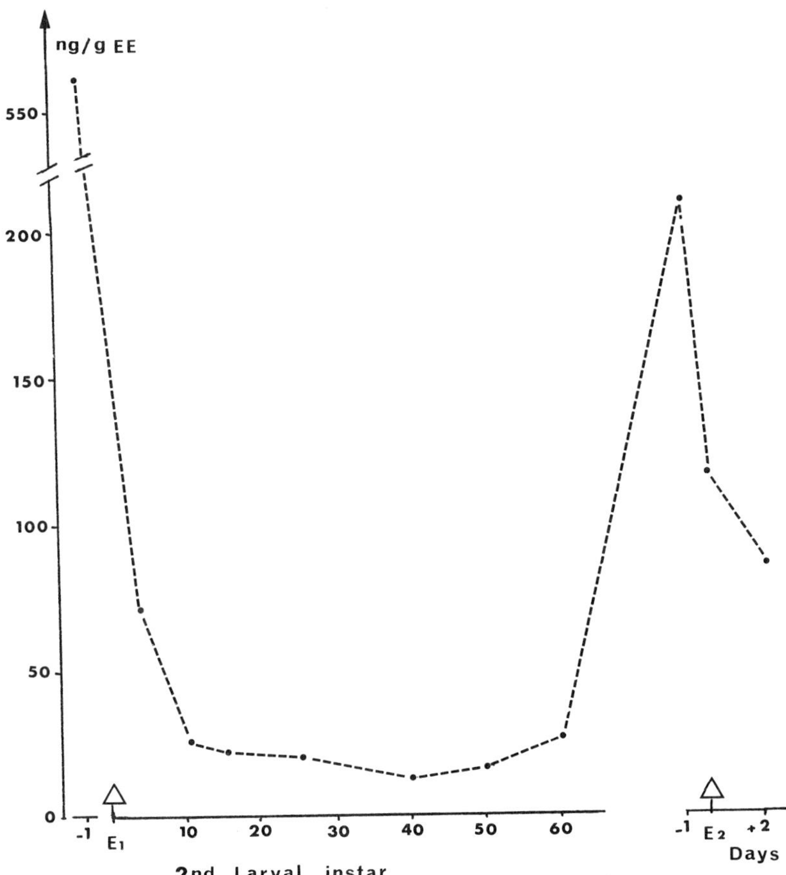

FIGURE 5.8. Ecdysteroid levels in the scorpion *Leiurus quinquestriatus* expressed as nanograms ecdysterone-equivalents (= EE) per gram of fresh weight (+ SEM). *Key:* E1, E2 = first and second molts. (From El Bakary et al., 1987).

vestigator (Claret et al., 1978), and the cricket *Grillus campestris* (Ismail et al., 1979).

5.4.2.2. SCORPIONIDA

Endogenous ecdysteroid levels were recently estimated in the scorpion *Leiurus quinquestriatus* (El Bakary, 1986; El Bakary et al., 1987) (Fig. 5.8).

During the last day of the first larval instar, which lives 3 days, the concentration of ecdysteroid is very high (569 ± 50 ng/g). The ecdysteroid level decreases at the following ecdysis (E$_1$) to 40 ± 5 ng/g and remains stable during most of the intermolting cycle of the second larval instar. During the preexuvial period, it reaches a maximum of 210 ± 30 ng/g, and it remains relatively high during the ecdysis, which produces the third instar (119 ± 4 ng/g).

The ecdysteroid level is very low in adult males and females (1.5 ± 0.5 ng/g).

Ecysteroids are present in eggs and embryos, and their levels fluctuate from 6 to 75 pg/mg; however, since the developmental stage of the eggs sampled was not known, the fluctuations could not be related to embryonic events. RIA applied in fractions separated by HPLC demonstrated that ecdysone is present at a low level and the main components are more polar than the 20-HE in larvae, eggs, and embryos. In contrast, at the beginning of the second larval instar the relatively high level of ecdysteroid is due to an increase of 20-HE.

5.4.2.3. ACARI

The variation of ecdysteroid levels measured by means of RIA were related to the events of the cuticular molting cycle in the ixodids *Amblyomma hebraeum* (Diehl et al., 1982), *Amblyomma variegatum* (Connat, 1987), and *Ornithodoros moubata* (Argasidae) (Germond et al., 1982). The ecdysteroid levels during the last nymphal instar in *Amblyomma hebraeum* is representative of results in Acari (Fig. 5.9). Each nymphal stage, as in other tick species, requires a blood meal to complete a molting cycle. During the 16 days after the blood meal, the ecdysteroid level in total extracts is very low (0.5–1.0 ng EE/tick). The ecdysteroid level begins to rise after day 17, and reaches a maximum (14 ng EE/tick) near day 23.

From days 23–25, the hormone concentration decreases until ecdysis (days 31–34) and reaches a minimum of 0.5 ng EE/tick. The variations are similar in *Ornithodoros moubata* and *Amblyomma variegatum*; however, Connat (1987) described a second moderate peak a few days after the main peak. The presence of ecdysone and ecdysterone were demonstrated by HPLC/RIA during the molting cycle, but—as in insects—ecdysterone is the dominant hormone.

Recent investigations have focused on the endogenous levels of ecdysteroids (ES) and on the metabolism of these hormones in female ticks. Since these animals do not molt during the adult stage, it was supposed that ES are most likely used in the reproductive process. In the hard tick *Boophilus microplus*, according to Connat (1987), a peak of free ecdysteroid (E) occurred just prior to complete engorgement and detachment. (Afterward, the levels of E decreased to a very low value, since the free hormone was conjugated to esterase-labile apolar conjugates.) These apolar products synthesized by the female were found in newly laid eggs of *B. microplus* (K. P. Wigglesworth et al., 1985). This result is similar to that

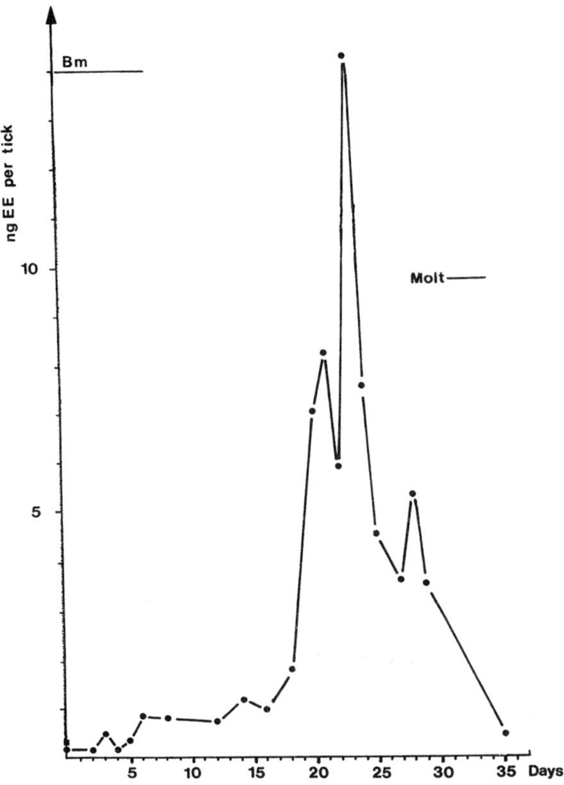

FIGURE 5.9. Ecdysteroid levels during development from nymphs to adults in the tick *Amblyomma hebraeum*. Ecdysteroid titer is expressed as nanograms ecdysterone-equivalents (= EE) per tick as a function of number of days after beginning of feeding. *Key:* Bm = bloodmeal; Molt = ecdysis between days 31–34. (From Diehl et al., 1982.)

previously observed in the female of the soft tick *Ornithodoros moubata* (Connat et al., 1984).

A quite different situation has been observed in the female hard tick *Amblyomma hebraeum*, in which only low amounts of conjugated ecdysteroids were found; their level, measured by RIA, increased progressively during the single gonotrophic cycle. About 350 ng 20-E equivalents were measured in a female prior to oviposition; there ES corresponded to free 20-E and E and were incorporated into the eggs (Connat et al., 1985). According to Dees et al. (1984), a similar situation seems to be present in the female of *Dermacentor variabilis*. RIA on crude extracts showed a great increase in immunoreactive products during engorgement; then the ES level rose during vitellogenesis to reach 19 ng 20-E equivalent per tick. RIA, coupled with HPLC analysis showed that part of the RIA-positive material comigrated with 20-E. Lower amounts of ES than those found in *A. hebraeum* were incorporated into the eggs.

In conclusion, two different strategies are used in Acarina: some species incorporate apolar conjugates in the eggs, and others free hormones. As was demonstrated in Insecta and Crustacea, the ecdysteroids play the role of molting hormone during the postembryonic development and another role during the reproductive period of the adult female.

5.4.3. *Effects of Exogenous Ecdysteroids on Biology of Arachnida*

5.4.3.1. ARANEAE

Krishnakumaran and Schneiderman (1968, 1970) were the first to test the effect of the molting hormones (α- and β-ecdysone) as well as their synthetic analogues on Mandibulata (Arachnida and Merostomata). According to these authors, injected ecdysterone dissolved in insect Ringer solution induces molting in the two araneids *Araneus cornutus* and *Dugesiella hentzi*, as well as in the horseshoe crab *Limulus polyphemus*. These early experiments were improved upon by specialists of each arachnida group.

Bonaric (1976, 1977a,b) used *Pisaura mirabilis*, in which the knowledge of the life cycle and the number and duration of the postembryonic instars and intermolt stages has been a model for physiological experiments. Inoculation of a concentrated alcoholic solution of ecdysterone (50 μg OH-E/g of fresh weight) induces early molting in the free eighth juvenile instar. Doses above 100 μg/g may prematurely trigger phenomena preparatory to molting, such as secretion of new epicuticle and exocuticle, but ecdysis is not normal. The effect of hormonal supply varies, depending on the moment of molting cycle. Any injection of an ecdysterone solution with an adjusted concentration (50 μg/g) during the A and B periods of the molting cycle is manifested by a lengthening of the intermolt cycle. The same dose inoculated during the C period induces early molting phenomena, apolysis, secretion of epicuticle and exocuticle, and ecdysis itself; early ecdysis, caused by the ecdysterone treatment, occurs in a synchronic manner in the same experimental batch of juveniles. The same ecdysterone supply during the D period, when the endogenous hormone concentration is high, only induces a slight reduction in intermolt. This suggests that normal ecdysis can occur only if the hormone level remains low; ecdysis is postponed or abnormal when the ecdysterone concentration is maintained high experimentally.

According to Bonaric (1977c, 1980), injections of ecdysterone (from 50 to 100 μg/g) causes a break in the winter diapause and a renewal of the growth and molting process in the diapausing seventh instar of *Pisaura mirabilis*. The results show a significant reduction in the response time of spiders to the same ecdysterone treatment from the beginning to the end of diapause. These differences in sensitivity to ecdysterone of diapausing *Pisaura* reflect a gradation in the intensity of winter diapause. These results are comparable to those that Bodnaryk (1977) obtained on the diapausing pupae of the butterfly *Mamestra configurata*.

5.4.3.2.   SCORPIONIDA

Kumar (1980) observed that a synthetic form of ecdysone found most active in tests on *Calliphora*, when injected into batches of the scorpion *Palammaeus bengalensis* of different ages, could not bring the scorpions to molt and failed to produce any change related to molting. These experiments should be reexamined in order to confirm or refute this unusual result in Arachnida. Moreover, endogenous ecdysteroids were recently found in a scorpion in all its postembryonic instars (El Bakary, 1986).

5.4.3.3.   ACARI

The effect of exogenous ecdysteroids was tested in numerous species of Acari. According to Mango et al. (1976), in Argasidae, ecdysone and ponasterone A, a phytoecdysone, induced supernumerary molting and perfectly viable "super-ticks"; ecdysone in *Ornithodoros porcinus* (Obenchain and Mango, 1980) produced similar results. Ingestion of 22,25-dideoxyecdysone during the blood meal provoked supermolting in all females of *Ornithodoros moubata* given doses as low as 35 ng/ml blood approximately 10 days after feeding, which corresponds to the normal amount of time required for a nymphal molt; other ingested ecdysteroids such as ecdysone, 20-HE, makisterone A, and ponasterone A also induced supermolting, but never in more than 50% of the females, and only with levels of hormone 500 times greater than normal. However, the supermolting females had malformations of distal parts of the legs and the mouthparts. Topical application was ineffective, except for makisterone A, which induced slight effects (Connat et al., 1983). In contrast to ingested ecdysone, injected hormone in the hemocoel of *O. moubata* from 100 to 250 ng induced integument molting at the lower dose, and death at the higher dose (Connat, 1987). Pound et al. (1984) also demonstrated that *Ornithodoros parkeri* was more sensitive to injected 20-E, which induced supermolting with 1 ng/ml blood, than to ingested 20-E, inducing supermolting only with 5 μg/ml blood. These differences in sensitivities between ingested and injected hormones are related to the efficient power of detoxification of the gut of females, which are well protected from exogenous classical ecdysteroids.

Great differences in the sensitivities to supermolt by exogenous ES has been recorded in several argasid species. Some are very sensitive, e.g., female *Argas arboreus,* in which topical application of only 1 µg 20-E, inokosterone, or cyasterone provoked supermolting (Ahmed and Bassal, 1982). *Argas persicus* and *Ornithodoros tholozani* were much less sensitive to ingested ES than *O. porcinus* (Mango, 1979). *A. japonicus* failed to supermolt when ES was applied by dipping, continuous substrate contact, or incorporation in the blood meal (Kitaoka, 1972).

By contrast, in Ixodidae, topical applications or injected ES in engorged females *Amblyomma hebraeum, A. variegatum,* and the cattle tick, *Boophilus microplus,* did not induce integument molting (Connat, 1987). In addition, ES postponed vitellogenesis in *A. hebraeum* and caused a decrease in the number of eggs produced, and inhibited the oviposition in *B. microplus* (Connat, 1987) and in *Hyalomma dromedarii* (Khalil et al.,1984).

In Argasidae, inhibited vitellogenesis was observed in *O. moubata* (Kitaoka, 1972; Mango et al., 1976; Connat et al., 1983) and in *B. microplus* (Mansingh and Rawlins, 1977; Connat, 1987). These authors concluded that high ecdysteroid levels are incompatible with vitellogenesis and oocyte maturation.

MacDaniel and Oliver (1978) reported that topical application of β-ecdysone on *Dermacentor variabilis* induced no effect on spermatogenesis and testis development. In Ixodidae, topical applications of 20-E in *D. variabilis* nymphs did not change the duration of nymphal development but caused an increase in deathrate at ecdysis (Dees et al., 1984). By contrast, topical applications of 1–5 ng ES, 3 days after the attachment of nymphs, induced an early ecdysis in *H. dromedarii;* outside of this period from 10 to 20 µg/tick were necessary to obtain a similar result (Khalil et al., 1984).

Wright (1969) in *D. variabilis,* and Sannasi and Subramonian (1972) in *Rhipicephalus sanguineus,* by means of topical applications of E and 20-E— or 20,25-DDE (22,25-dideoxyecdysone) in the first species—produced a break in the larval diapause.

5.4.4.   *Metabolism and Detoxification of Ingested Ecdysteroids*

Since spiders exclusively ingest fluids from arthropod prey, which often contain high amounts of ecdysteroids, physiological problems of detoxification must be overcome by these predators. For example, pupae of *Tenebrio molitor* used to feed spiders in some laboratories contain 200 µg of 20-E in the whole body (Delbecque, 1976) without causing any developmental problems for the spiders. This suggests that ingested ecdysteroids are retained in the gut and/or efficiently detoxified in order to prevent action on normal development. Indeed, spiders have development efficient detoxification mechanisms. According to Connat et al. (1986), in addition to the polar pathway, which probably acts by hydroxylation and conjugation with phosphates as generally found in insects (see the review by Lafont and Koolmann, 1984), the spiders could be using an apolar pathway similar to that found in ticks, and now also found in insects, which acts by conjugation of ecdysteroids with long-chain fatty acids. (Diehl et al., 1985; Connat and Diehl, 1986; Crosby et al., 1986; Kubo et al., 1987). The metabolic study of Connat et al. (1986) demonstrates unambiguously that spiders (*Araneus sclopetarius*, *Pholcus phalangioides*, *Tegeneria atrica*, *Zygiella-x-notata*, and *Pardosa* sp.) detoxify the ingested ecdysteroids mainly by converting them to apolar conjugates, which are excreted in the feces. The aforementioned authors reported similar results in two scorpions, *Androctonus australis* and *Buthus occitanus*, which converted tritiated injected [^{3}H]ecdysone and [^{3}H]20-E into polar and apolar metabolites; the apolar pathway was more efficient in *Buthus occitanus*, soft ticks *Onithodoros parkeri* and *O. moubata*, and two species of hard ticks, *A. hebraeum* and *B. microplus* (Ixodidae), which converted [^{3}H]20-E into apolar products.

In all cases, hydrolysis of the apolar metabolites (esterase-labile conjugates) with esterase liberated the free injected hormone. However, the role of these compounds is not well established; besides the previous role, they may be a form of hormone storage in eggs for future use in the development of the embryos and in the intra-chorion and intra-cocoon ecdysis.

5.5. Summary

The ecdysial gland is well known in spiders, while many putative organs compete for the title in other Arachnida. By means of RIA, several other tissues have been demonstrated as ecdysteroid-secreting centers (oenocytes, integument, ovaries) in harvestmen, Acari, or scorpions. In spiders, the molting gland presents features and changes during the intermolt cycle similar to those described in the prothoracic glands of pterygote insects; in particular, it possesses a dominant smooth endoplasmic reticulum and mitochondria with tubular cristae characteristic of cells that synthesize steroids. The ultrastructure of these organs provides good criteria by which we can homologize "endocrine cells" of spiders with a molting gland. The embryogenesis of the endocrine cells is known in *Dolomedes fimbriatus*. The cells appear very early in the embryo and are present before the first apolysis, which produces a pharate instar in the chorion (*prelarva* in Vachon's terminology).

By contrast to insects, cytoimmunochemical detection of ecdysteroids in putative ecdysial glands and gland culture have not been performed in spiders to demonstrate the synthesize of ecdysteroid. Similarities between spiders and insects, according to ultrastructural features, cycle level, and relation with integument cycle, are suitable to show that these organs are ecdysial glands.

We can speculate as to an evolutionary trend for the molting gland of spiders. From a system present in the most primitive spiders (Liphistiomorpha, Liphistiidae), characterized by endocrine cells present in both the opisthosoma and prosoma, two divergent lines developed: (1) one in Mygalomorpha, characterized by a molt gland mainly concentrated in the opisthosoma; (2) another in Araenomorpha, characterized by a molt gland concentrated almost exclusively in the prosoma.

The presence of ecdysteroids in ovaries of harvestmen and in ovaries and embryo of Acari, and the presence of endocrine cells in embryo with vitellus before the first apolysis in spiders, suggest that the so-called embryonic molt in the chorion or in the chorion partly opened (which gives rise to incompletely developed instars that are not mobile and live in the cocoon) is induced by ecdysteroids. These ecdysteroids can have two origins: (1) possibly a maternal origin from the ovaries, with transformation into β-ecdysone to induce the first apolysis as well as the others; or (2) possibly the conversion is accomplished by the embryonic endocrine cells in spiders by a synthesis in those cells. Experiments and dosage by means of RIA on fractions separated by mass spectrometry are necessary to examine the postulate of a first role of the embryonic endocrine cells (conversion of maternal ecdysteroids) and later a second (direct) synthesis of ecdysone.

In spiders and in groups where the ecdysial gland is not well known (Scorpionida, harvestmen, Acari), RIA procedure on hemolymph, or crude extracts, have demonstrated the presence of ecdysteroids (α- and β-ecdysone; polar conjugates). In Arachnida, a change of the ecdysteroid level during the molting cycle with a peak of β-ecdysone about the apolysis, identical to the variations observed in insects, has been described herein. These changes are related to integument events (apolysis; deposition of layers of the new integument) in spiders and Acari.

Ingested or injected ecdysteroid experiments in spiders and Acari have confirmed that arachnid ecdysteroids play a similar role as the insect molting hormone. Spiders have efficient detoxification mechanisms for ingested or injected ecdysteroids, and presumably so do the other carnivorous Arachnida, which feed on prey with high levels of ecdysteroids. The detoxification pathways are similar to those in insects.

References

Ahmed, S. H. and T. T. M. Bassal. 1982. Introduction of adult tick molting by ecdysteroids. J. Egypt. Soc. Parasitol. 12: 383–387.

Barth, F. G. 1985. *Neurobiology of Arachnids.* Springer-Verlag, Berlin and New York.

Beaulaton, J. and R. Gras. 1980. Influence on aldehyde fixatives on the ultrastructure of the prothoracic gland cells in *Rhodnius prolixus* (Insecta, Heteroptera) with special reference to their osmotic effects. Microsc. Acta 82: 351–366.

Beaulaton, J., P. Porcheron, R. Gras, and P. Cassier. 1984. Cytophysiological correlation between prothoracic gland activity and hemolymph ecdysteroid concentration in *Rhodnius prolixus* during the fifth larval instar: further studies in normal and decapited larvae. Gen. Comp. Endocrinol. 53: 1–16.

Behrens, W. and D. Büchmann. 1983. Ecdysteroids of the pycnogonid *Pycnogonum litorale* (Strom) (Arthropoda, Pantopoda). Gen. Comp. Endocrinol. 51: 8–14.

Binnington, K. C. 1986. Ultrastructure of the tick neuroendocrine system. Pp. 152–162 in J. R. Sauer and J. A. Hair (eds.), *Morphology, Physiology and Behavioral Biology of Ticks.* Ellis Horwood, Chichester, England.

Birkenbeil, H. 1983. Ultrastructural and immunocytochemical investigation of ecdysteroid secretion by the prothoracic gland of the waxmoth *Galleria mellonella.* Cell Tissue Res. 229: 433–441.

Bodnaryk, R. P. 1977. Stages of diapause development in the pupa of *Mamestra configurata* based on the β=ecdysone sensitivity index. J. Insect. Physiol. 23: 537–542.

Bonaric, J. C. 1976. Effects of ecdysterone on the molting mechanisms and duration of the intermolt period in *Pisaura mirabilis* Cl. Gen. Comp. Endocrinol. 30: 267–272.

Bonaric, J. C. 1977a. Action de l'ecdystérone sur la réduction de durée de la période d'intermue chez *Pisaura mirabilis* Cl. (Araneae, Pisauridae): détermination de l'effet optimal par inoculation de concentrations croissantes. Arch. Zool. Exp. Gen. 118: 43–51.

Bonaric, J. C. 1977b. Action de l'ecdystérone sur le cycle de mue de l'araignée *Pisaura mirabilis* Cl.: variations des réponses en fonction de la date d'intervention au cours de l'intermue nymphal. Arch. Zool. Exp. Gen. 118: 409–421.

Bonaric, J. C. 1977c. Existence d'une diapause hivernale révélée par l'action de traitements hormonaux chez l'Araignée *Pisaura mirabilis* Cl. (Pisauridae). C. R. Acad. Sci. Paris 284: 1297–1300.

Bonaric, J. C. 1980. Contribution à l'étude de la biologie du développement chez l'Araignée *Pisaura mirabilis* (Clerck, 1758): approche physiologique des phénomènes de mue et de diapause hivernale. Doctoral thesis, University of Montpellier, France.

Bonaric, J. C. and M. De Reggi. 1977. Changes in ecdysone levels in the Spider *Pisaura mirabilis* nymphs (Araneae, Pisauridae). Experientia (Basel) 33: 1664–1665.

Bonaric, J. C. and C. Juberthie. 1980. La glande de mue des araignées: étude structurale et ultrastructurale de cette formation endocrine chez *Pisaura mirabilis* Cl. (Araneae, Pisauridae). C. R. Vème Colloq. Eur. Arachnol. (1979) 9: 21–30.

Bonaric, J. C. and C. Juberthie. 1983. Le système neuroendocrine rétro-cérébral et la "glande de mue" de l'araignée cavernicole *Telema tenella* (Telemidae). Mém. Biospéol. 10: 427–434.

Bonaric, J. C., M. Emerit, and R. Legendre. 1984. Le complexe neuroen-docrine rétrocérébral et la "glande de mue" de *Filistata insidiatrix* Forsköl (Araneae, Filistatidae). Rev. Arachnol. 5: 301–310.

Browning, H. C. 1942. The integument and moult cycle of *Tegenaria atrica* (Araneae). Proc. R. Soc. Lond. 131: 65–86.

Claret, J., P. Porcheron, and F. Dray. 1978. La teneur en ecdysones cir-culantes au cours du dernier stade larvaire de l'hyménoptère endo-parasite *Pimpla instigator* et l'entrée en diapause. C. R. Acad. Sci. Paris 286: 639–641.

Connat, J. L. 1987. Aspects endocrinologiques de la physiologie du dével-oppement et de la reproduction chez les tiques. Doctoral thesis, University of Dijon, France.

Connat, J. L. and P. A. Diehl. 1986. Probable occurrence of ecdysteroids fatty acid esters in different classes of arthropods. Insect Biochem. 16: 91–97.

Connat, J. L., P. A. Diehl, and M. Morici. 1984. Metabolism of ec-dysteroids during the vitellogenesis of the tick *Ornithodoros moubata* (Ixodoidea, Argasidae): accumulation of apolar metabolism in the eggs. Gen. Comp. Endocrinol. 56: 100–110.

Connat, J. L., P. A. Diehl, and M. Thompson. 1986. Possible inactivation of ingested ecdysteroids by conjugation with long chain fatty acids in the female tick *ornithodoros moubata* (Acarina, Argasidae). Arch. Insect Biochem. Physiol. 3: 235–252.

Connat, J. L., E. M. Dotson and P. A. Diehl. 1987. Metabolism of ec-dysteroids in the female tick *Amblyomma hebraeum* (Ixodoidae, Ix-odidae): accumulation of free ecdysone and 20-hydroxyecdysone in the eggs. J. Comp. Physiol. 157(5): 689–699.

Connat, J. L., P. A. Diehl, N. Dumont, S. Carminati, and M. J. Thompson. 1983. Effects of exogenous ecdysteroids of the female tick *Ornithodoros moubata:* induction of supermolting and influence on oogenesis. Z. Angew. Entomol. 96(5): 520–530.

Connat, J. L., P.A. Diehl, H. Gfeller, and M. Morici. 1985. Ecdysteroids in females and eggs of the ixodid tick *Amblyomma hebraeum*. Int. J. Invertebr. Reprod. Dev. 8: 103–116.

Crosby, T., R. P. Evershed, D. Lewis, K. P. Wigglesworth, and H. H. Rees. 1986. Identification of ecdysone 22–long chain fatty acyl esters in newly laid eggs of the cattle tick, *Boophilus microplus*. Biochem. J. 240(1): 131–138.

Dees, W. H., D. E. Sonenshine, and E. Breidling. 1984. Ecdysteroids in the American dog tick *Dermacentor variabilis* (Acari: Ixodidae) during different periods of tick development. J. Med. Entomol. 21: 514–523.

De Reggi, M., M. Hirn. and M. Delaage. 1975. Radioimmunoassay of ecdysone an application to *Drosophila* larvae and pupae. Biochem. Biophys. Res. Commun. 66: 1307–1315.

Delbecque, J. P. 1976. Taux d'ecdysones et cycle de mue au cours de la métamorphose de *Tenebrio molitor* L. Thesis, third cycle, University of Dijon, France.

Delbecque, J. P., P. A. Diehl, and J. D. O'Connor. 1978. Presence of ecdysone and ecdysterone in the tick *Amblyomma hebraeum* Koch. Experientia (Basel) 34: 1379-1380.

Diel, P. A., J. E. Germond, and M. Morici. 1982. Correlation between ecdysteriod titers and integument structure in nymphs of the tick *Amblyomma hebraeum* Koch. (Acari: Ixodidae). Rev. Suisse Zool. 89: 859-868.

Diehl, P. A., J. L. Connat, J. P. Girault, and R. Lafont. 1985. A new class of apolar ecdysteroid conjugates: esters of 20-hydroxyecdysone with long chain fatty acids in ticks. Int. J. Invertebra. Reprod. Deve. 8: 1–13.

El Bakary, Z. 1986. Contribution à l'étude biologique du Scorpion *Leiurus quinquestriatus* (Buthidae). Doctoral thesis, University of Orsay, France.

El Bakary, Z., P. Porcheron, M. Moriniere, and S. Fuzeau-Braesch. 1987. Mise en évidence et dosage d'ecdystéroïdes chez le Scorpion *Leiurus quinquestriatus*. C. R. Acad. Sci. Paris 304: 453–456.

Ellis, B. J. and F. D. Obenchain. 1984. In vivo and in vitro production of ecdysteroids by nymphal *Amblyomma variegatum* tick. Pp. 400-404 *in* D. A. Griffiths and C. E. Bowman (eds.), *Acarology VI*, Vol. 1. Ellis Horwood, Chichester, England.

Fain-Maurel, M. A. and P. Cassier. 1968. Etude infrastructurale des glandes de mue de *Locusta migratoria migratorioides*. R. et F. I. Evolution cyclique au cours des stades larvaires et genèse du produit de sécrétion, l'ecdysone. Arch. Zool. Exp. Gen. 109: 445–476.

Foelix, R. 1982. *Biology of Spiders*. Harvard University Press, Cambridge, Massachusetts.

Gabe, M. 1954. La neurosécrétion chez les Invertébrés. Ann. Biol. 30: 5–62.

Gabe, M. 1966. *Neurosécrétion*. Pergamon Press, Oxford and Elmsford, New York.

Gabe, M. 1971. Données histologiques sur le glomus coxal (massif préglomérulaire), glande de mue possible des scorpions. Ann. Sci. Nat. Zool. Paris 13: 609–622.

Germond, J. E., P. A. Diehl, and M. Morici. 1982. Correlation between integument structure and ecdysteroid titers in fifth-stage nymphs of the tick, *Ornithodoros moubata* (Murray, 1877; *sensu* Walton, 1962). Gen. Comp. Endocrinol. 46: 255–266.

Gersch, M. 1964. *Vergleichende Endokrinologie der wirbellosen Tieren*. Akademishe Verlagsges. Geest & Portig, Leipzig.

Gras, R. and J. Beaulaton. 1979. Developpement cyclique du réticulum endoplasmique agranulaire dans les glandes thoraciques (glandes ecdysiales) de *Rhodnius prolixus* Stäl (Insecte-Hétéroptère) aux deux derniers stades larvaires. Experientia (Basel) 35: 386–388.

Gupta, A. P. 1983. *Neurohemal Organs of Arthropods: Their Development, Evolution, Structures, and Functions*. Thomas, Springfield, Illinois.

Habibulla, M. 1961. Secretory structures associated with the neurosecretory system of the immature scorpion *Heterometrus swammerdami*. Q. J. Microsc. Sci. 102: 475–479.

Herman, W. S. 1967. The ecdysial glands of arthropods. Int. Rev. Cytol. 22: 269–347.

Herlant-Meewis, H. and J. Naisse. 1957. Phénomènes neurosécrétoires et glandes endocrines chez les Opilions. C. R. Acad. Sci. Paris 245: 858–860.

Highnam, K. C. and L. Hill. 1977. *The Comparative Endocrinology of Invertebrates*. Arnold, London.

Ismail, S., P. Porcheron and S. Fuzeau-Braesh. 1979. L'hormone de mue et la diapause chez *Gryllus campestris* L. (Orthoptères): dosage, injections et métabolisme. C. R. Soc. Biol. 107: 37–40.

Jegla, T. C. 1974. Ecdysone activity in *Limulus polyphemus*. Am. Zool. 14: 1288.

Jegla, T. C. 1982. A review of the molting physiology of the trilobite larva of *Limulus*. Pp. 83–101 *in* R. L. Alan (ed.), *Physiology and Biology of Horseshoe Crabs: Studies on Normal and Environmentally Stressed animal*. Plenum Press, New York.

Jegla, T. C. and J. D. Costlow. 1979. Ecdysteroids in *Limulus* larvae. Experientia (Basel) 35: 554–555.

Joly, L., P. Joly, and A. Porte. 1969. Remarques sur l'ultrastructure de la glande ventrale de *Locusta migratoria* L. (Orthoptères) en population dense. C. R. Acad. Sci. Paris 269: 917–920.

Joly, R., P. Porcheron, and F. Dray. 1979. Etude des variations du taux d'ecdystéroïdes au cours du cycle de mue dans l'hémolymphe de

Lithobius forficatus (Myriapode: Chilopode). C. R. Acad. Sci. Paris 288: 243–246.

Juberthie, C. 1988. Critères de fin de développement embryonnaire et début du développement postembryonnaire: approche endocrine. C. R. Xème Colloq. Eur. Arachnol. (in press).

Juberthie-Jupeau, L. A. Strambi, M. De Reggi, and C. Juberthie. 1979. The presence of ecdysteroids and the variation of their level during the first adult stage of the myriapod *Hanseniella ivorensis* Juberthie-Jupeau and Kehe (Symphyla). Experientia (Basel) 35: 1405–1407.

Kahlil, G. M., A. A. Shaarawy, D. E. Sonenshine, and S. M. Gad. 1984. Ecdysone effects on the camel tick, *Hyalomma dromedarii* (Acari: Ixodidae). J. Med. Entomol. 21: 188–193.

Kitaoka, S. 1972. Effects of ecdysone in ticks, especially on *Ornithodoros moubata* (Acarina–Argasidae). Proc. 14th Int. Congr. Entomol., Australia p. 272.

Krishnakumaran, A. and H. A. Schneiderman. 1968. Chemical control of molting in arthropods. Nature (Lond.) 220: 601–603.

Krishnakumaran, A. and H. A. Schneiderman. 1970. Control of molting in mandibulate and chelicerate arthropods by ecdysones. Biol. Bull. (Woods Hole) 139: 520–538.

Kubo, I., S. Komatsu, Y. Asaka, and G. De Boer. 1987, Isolation and identification of apolar metabolites of ingested 20-hydroxyecdysone in frass *Heliothis virescens* larvae. J. Chem. Ecol., 13 (4): 785–794.

Kumar, R. 1980. Effect of administration of synthetic ecdysone on the moulting of *Palamnaeus bengalensis*. Experientia (Basel) 36: 411–412.

Lafont, R. and J. Koolman. 1984. Ecdysone metabolism. Pp. 196–226 *in* J. Hoffmann and M. Porchet (eds.), *Biosynthesis, Metabolism and Mode of Action of Invertebrate Hormones*. Springer-Verlag, Berlin and New York.

Legendre, R. 1958. Contribution à l'étude du système nerveux des Aranéides. Ann. Biol. 34: 193–223.

Legendre, R. 1959. Contribution à l'étude du système nerveux des Aranéides. Ann. Sci. Nat. Zool. 1: 339–473.

Legendre, A. 1981. Les éléments endocrines de la glande de mue chez les araignées mygalomorphes. Rev. Arachnol. 3: 133–138.

Legendre, R. and A. Lopez. 1981. Observations histologiques complémentaires chez l'Araignée Liphistiomorphe *Heptathela kimurai* Kishida, 1923 (Liphistiidae). Atti Soc. Toscana Sci. Nat. 88: 34–44.

Lopez, A., M. K. Stowe, and J. C. Bonaric. 1985. Anatomie interne de l'Araignée à Bolas nord-américaine *Mastophora cornigera* (Hentz, 1850) (Araneae: Araneidae) après sa sortie du cocon. Publ. Sci. Acc. 8: 1–9.

MacDaniel, C. N. 1979. Haemolymph ecdysone concentrations in

Hyalophora cecropia pupae, dauer pupae and adults. J. Insect. Physiol. 25: 143–145.

MacDaniel, R. S. and J. H. Oliver, Jr. 1978. Effects of two juvenile hormone analogs and β-ecdysone on nymphal development, spermatogenesis and embryogenesis in *Dermacentor variabilis* (Say) (Acari: Ixodidae). J. Parasitol. 64: 571–573.

Mango, C. K. A. 1979. The effects of beta-ecdysone and ponasterone A on the soft tick *Ornithodoros porcinus* Walton, 1962. Ph.D. thesis, University of Nairobi, Kenya.

Mango, C., T. R. Odhiambo, and R. Galun. 1976. Ecdysone and the super tick. Nature (Lond.) 260: 318–319.

Mansingh, A and S. C. Rawlins. 1977. Antigonadotropic action of insect hormone analogues on the cattle tick *Boophilus microplus*. Naturwissenschaften 64: 41.

Mauchamps, B., J. L. Pennetier, P. Tarroux, and R. Lafont. 1977. Metabolisme de la diapause hivernale chez *Pieris brassicae:* aspects biochimiques et corrélations endocrines. 4ème Séminaire Physiologie Insectes, Bordeaux.

Millot, J. 1926. Contribution à l'histophysiologie des Aranéides. Bull. Biol. Fr. Belg. 8: 1–238.

Millot, J. 1930. Le tissu réticulé du céphalothorax des Aranéides et ses dérivés: néphrocytes et cellules endocrines. Arch. Anat. Microsc. 26: 43–81.

Naisse, J. 1959. Neurosécrétion et glandes endocrines chez les Opilions. Arch. Biol. 70: 217–264.

Nentwig, W. 1987. *Ecophysiology of Spiders.* Springer-Verlag, Berlin and New York.

Obenchain, F. D. and C. K. A. Mango. 1980. Effects of exogenous ecdysteroids and juvenile hormones on female reproductive development in *Ornithodoros procinus.* Am. Zool. 20: 936.

Porcheron, P. 1979. L'hormone de mue des arthropodes: dosage radioimmunologique, production, divers aspects de son rôle physiologique. Doctoral thesis, University of Paris.

Pound, J. M., J. H. Oliver, and R. H. Andrews. 1984. Induction of apolysis and cuticle formation in female *Ornithodoros parkeri* (Acari: Argasidae) by hemocoelic injections of beta-ecdysone. J. Med. Entomol. 21: 612–614.

Romer, R. and W. Gnatzy, 1981. Arachnid oenocytes: ecdysone synthesis in the legs of harvestmen (Opilionidae). Cell Tissue Res. 216: 449–453.

Romer, F., H. Emmerich, and A. Nowock. 1974. Biosynthesis of molting hormones in isolated prothoracic glands and oenocytes of *Tenebrio molitor* in vitro. J. Insect Physiol. 20: 1975–1987.

Roshdy, A. and A. S. Marzouk. 1982. The subgenus *Persicargas* (Ixodoidea: Argasidae: Argas). 35. The lateral segmental organs and peritracheal gland in immature and adult *A. (P.) arboreus*. Z. Parasitenkd. 66: 335–343.

Ruhland, U. and F. Romer. 1977. Nachweis von häutungsaktiven Stoffen in isolierten Prothorakaldrüsen und Oenocyten bei *Bombyx mori* während des 5 Larvenstadiums. Wilhelm. Roux's Arch. Dev. Biol. 181: 123–134.

Sannasi, A. and T. Subramonian. 1972. Hormonal rupture of larval diapause in tick *Rhipicephalus sanguineus* (Latr.) Experientia (Basel) 28: 666–667.

Spindler, K. D., R. Keller, and J. D. O'Connor. 1980. The role of ecdysteroids in the crustacean molting cycle. Pp. 247–280 *in* J. A. Hoffmann (ed.), *Progress in Ecdysone Research*. Elsevier/North-Holland Publ., Amsterdam and New York.

Stockmann, R. 1984. Description du système endocrine chez les scorpions du genre *Euscorpius* (Chactidae). Rev. Arachnol. 5: 368–369.

Streble, H. 1966. Untersuchungen über das hormonale System der Spinnentiere (Chelicerata) unter besonderer Berücksichtigung des "endokrinen Gewebes" der Spinnen (Araneae). Zool. Jahrb. Physiol. 72: 157–234.

Tombes, A. S. 1970. *An Introduction to Invertebrate Endocrinology*. Academic Press, Orlando, Florida.

Tombes, A. S. 1979. Comparison of arthropod neuroendocrine structures and their evolutionary significance. Pp. 645–667 *in* A. P. Gupta (ed.), *Arthropod Phylogeny*. Van Nostrand Reinhold, New York.

Vachon, M. 1953. Commentaires à propos de la distinction des stades et des phases du développement post-embryonnaire chez les Aranéides. Bull. Mus. Natl. Hist. Nat. 25: 294–297.

Wigglesworth, K. P., Lewis, D. and H. H. Rees. 1985. Ecdysteroid titre and metabolism to novel apolar derivatives in the adult female *Boophilus microplus* (Ixodidae). Arch. Insect. Biochem. Physiol. 2: 39–54.

Wigglesworth, V. B. 1952. The thoracic gland in *Rhodnius prolixus* (Hemiptera) and its role in moulting. J. Exp. Biol. 29: 561–570.

Winget, R. R. and W. S. Herman. 1979. Influence of molt cycle and beta-ecdysone on protein synthesis in the Chelicerate Arthropod, *Limulus polyphemus*. Comp. Biochem. Physiol. 62: 119–122.

Wright, J. E. 1969. Hormonal termination of larval diapause in *Dermacentor albipictus*. Sciences (NY) 163: 390–391.

Morphology, Histology, and Ultrastructure of the Ecdysial Glands in Myriapoda

6

GERHARD SEIFERT

6.1. Introduction

As in other arthropods, the postembryonic development of Myriapoda is characterized by the presence of several juvenile stages, the last of which molts into the adult.

As in most insects, the definitive number of metameres of the scolopendromorph and geophilomorph chilopods are formed during embryonic development (epimorphic). Diplopoda, Pauropoda, Symphyla, and lithobiomorph and scutigeromorph Chilopoda, however, hatch as larvae with fewer trunk metameres, reaching their definite number of segments by a gradual teloblastic growth from stage to stage (anamorphic). Within the anamorphic chilopods, the definitive number is reached after a few molts; following these anamorphic stages, few epimorphic juvenile stages occur prior to molting into the adult. Many myriapod species continue to molt after reaching the adult stage.

The interest in the regulation of molting in Myriapoda, the hormones involved, and their sites of synthesis arose after these processes became known in insects and decapod crustaceans. This report surveys our present knowledge about the organs that seem to produce the specific ecdysial hormones.

6.2. Ecdysial Glands in Chilopoda

Since Joly (1966a) observed that implants of integument into *Lithobius forficatus* synchronously molted with its host, it has been known that ecdysis in chilopods also is regulated by hormones. The significantly higher rate of molting in adults after ecdysone injection (Joly, 1964) showed further that the ecdysial hormone is an ecdysteroid. This is confirmed by the results of Joly et al. (1979) and Leubert et al. (1979). There is experimental evidence that the activity of the ecdysial hormone is under neuroendocrine control of the lobus frontalis–cerebral gland complex (Joly, 1961, 1962, 1966b; Joly and Descamps, 1968; Scheffel, 1961, 1963, 1965a, 1969, 1987).

The first experiments to localize the "molting center" were performed on *Lithobius forficatus* (Scheffel, 1963, 1965b). Ligation experiments on second-stage anamorphic larvae showed that the active center is located in the region of the posterior head and maxilliped metameres. These results were supported by investigations on the effect of destruction of tissues from this region, using ultraviolet (UV) irradiation (Scheffel, 1969), especially that of "lymphatic strands" surrounding the salivary glands, and the glandulae maxillares secundae, both of which were presumed to be ecdysteroid-producing organs. In an electron microscopical investigation, Rosenberg and Seifert (1975) identified the latter to be an exocrine gland. The "lymphatic strands," however, were shown to be active metabolic organs composed of a high number of podocytes similar to the pericardial cells or nephrocytes in many groups of arthropods (Seifert and Rosenberg, 1974). There is also a high micromorphological correspondence with the prothoracic gland of insects. Ultrastructural changes during the molting cycle further confirm the ecdysial function. Leubert (1982, 1986) demonstrated *in vitro* secretion of ecdysone and 20-hydroxyecdysone from isolated "lymphatic strand"–salivary gland complexes of adult *Lithobius forficatus* during proecdysis. Recently, ecdysteroid-binding sites were discovered by immunohistochemical methods (Seifert and Bidmon, 1988). Hence, the glandula ecdysalis is most probably the ecdysial gland in *Lithobius*.

Ultrastructure studies of only the head gland (glandula capitis) of *Scutigera coleoptrata* (Rosenberg, 1974), the morphologically identical "lymphatic tissue" of *Cryptops hortensis*, and some geophilomorphic species (Rosenberg, 1978, 1979) are available to date.

6.2.1. Topography, Histology, and Ultrastructure of Ecdysial Glands

6.2.1.1. ECDYSIAL GLAND (GLANDULA ECDYSALIS) OF LITHOBIOMORPHA

The ecdysial glands in some species of the genera *Lithobius* and *Lamyctes* surround large parts of the salivary glands (mandibular glands), the foregut, the dorsal tracheal stems, and the maxillary nephridia. Concomitant with the growth of the salivary glands during postembryonic development, the relative size of the ecdysial glands also increases (Fig. 6.1). In the first anamorphic larval stage (=fetus according to Verhoeff, 1905), they are restricted to the posterior part of the head up to the level of the maxillary nephridia; in the second stage (=larva 1 according to Verhoeff), they reach up to the posterior part of the maxilliped segment; and in the third stage (=larva 2 according to Verhoeff), they reach far into the second trunk metamere and have their largest (relatively) longitudinal extension. Hence, Scheffel (1969), in his localization experiments on the second larvae, always made his ligature caudally to the first pair of legs to separate them as safely as possible from the rest of the trunk.

The growth of the salivary glands particularly narrows the hemocoel of this region and pushes the ecdysial glands toward the visceral fat body, the musculature of the foregut, the dorsal longitudinal muscles, the dorsal vessel, and—on the other side—against the ventral nerve cord (Fig. 6.2A).

The ecdysial glands occur as lobes of podocytes loosely attached to

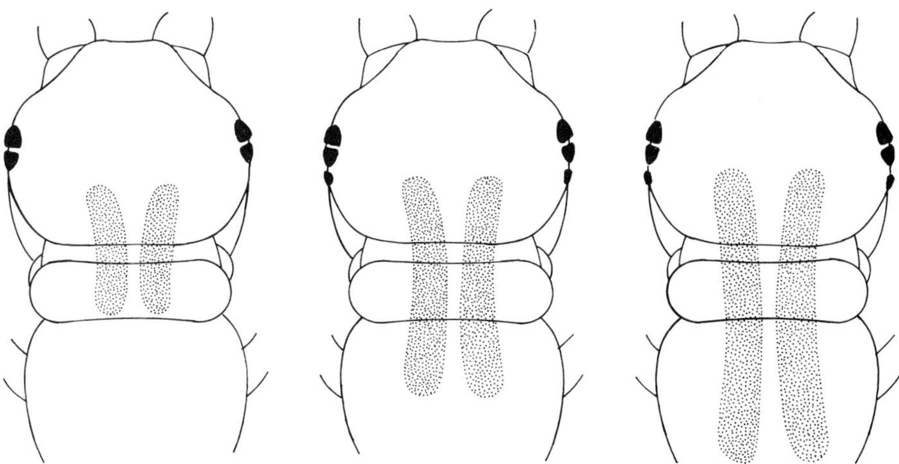

FIGURE 6.1. Schematic drawing showing progressive increase in the proportions and relative lengths of the ecdysial glands in the head and anterior trunk of the postembryonic instars 1–3 of *Lithobius forficatus*.

each other. Their periphery (cortex) is characterized by large spaces, as a consequence of deep invaginations (Fig. 6.2A,C). More centrally, there are cytosomes, which in some cases may reach the size of nuclei in adults. Their content binds to specific antibodies against ecdysteroids (Fig. 6.2B), indicating the occurrence of molting hormone (Seifert and Bidmon, 1988). The ecdysteroid-binding sites are restricted to the cytosomes (see the inset of Fig. 6.2B). Ultrastructural investigations by Seifert and Rosenberg (1974) showed that the lobes of the podocytes are enclosed and separated from neighboring lobes by a basal lamina (Fig. 6.3). Numerous deep invaginations reach from the basal cell membrane proximally, producing a typical cortical labyrinth. Occasionally, these invaginations are peripherally joined to one another by desmosomes (Fig. 6.3A,B). But often the surface of the podocytes is formed by rows of pedicells, which are fingershaped projections, mostly oriented parallel to the cell surface. They support each other by porous diaphragms resembling desmosomes and keep the cortical labyrinth permanently open toward the hemocoel (Fig. 6.3A–C). Thus, the hemolymph is filtered through the basal lamina and the diaphragms, and an exchange of molecules and ions into and from the labyrinth takes place. Electron microscopy shows that alongside the extremely enlarged surfaces predominantly receptor-coupled cytoses take place, as indicated by coated pits (arrowheads) and, later on, coated vesicles with an obvious clathrin sheath. Within the cortical region of the

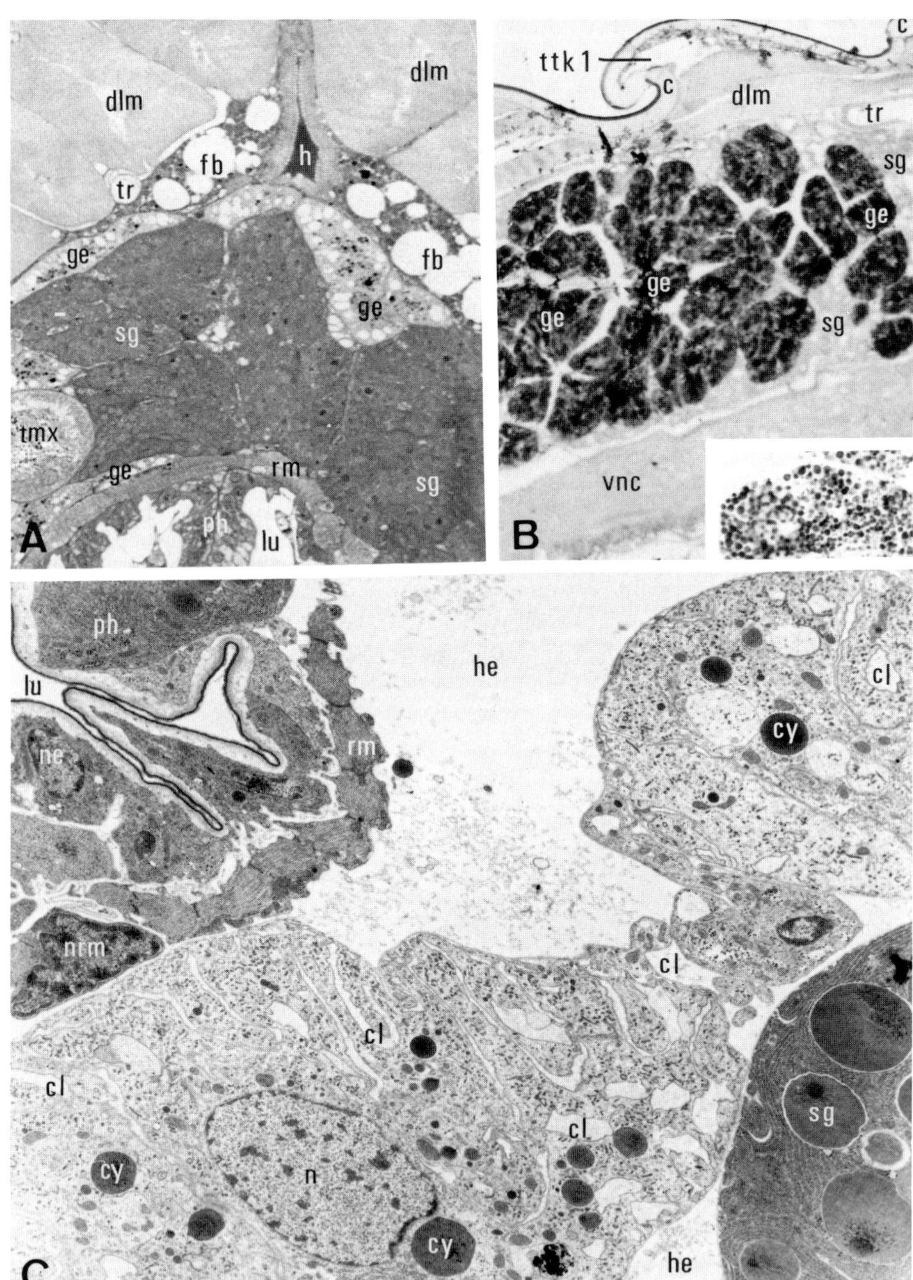
dlm
dlm
fb
h
tr
ge
fb
ge
sg
tmx
ge
rm
ph
lu
sg
A
ttk 1
c
c
dlm
tr
sg
ge
ge
ge
sg
vnc
B
ph
he
cl
lu
cy
rm
ne
cl
nrm
cl
cl
cy
cl
sg
n
cy
he
C

cytoplasm, there are also large translucent vacuoles that sometimes contain membrane fragments. Their enveloping membrane is, as in coated vesicles, internally covered by structures that are regarded as receptor-coupled ligands. A clathrin coat on the outside (toward the cytoplasm) is lacking. It is presumed that numerous coated vesicles, after losing their clathrin sheaths, fuse to produce the translucent vacuoles, thus becoming endosomes. Tubular structures (Fig. 6.3A, arrows) whose contents gradually become more electron-dense, are observed to develop in the endosomes and have been found to detach from them. The cortical cytoplasm is filled with these tubular structures. It is believed that receptors and ligands separate within the endosomes. The ligands remain in the vesicles, whereas the receptors are recycled by the tubular "vehicles" and exocytotically reincorporated into the cytoplasm membrane of the labyrinth.

There are some mitochondria and very few cisternae of endoplasmic reticulum in the cortical areas surrounding the labyrinth. Dictyosomes are completely lacking (Fig. 6.3A–C). Occasionally, there are microtubules that may transport the vesicles and tubular structures described above.

In the more central regions of the podocytes, there are vacuoles with different electron densities, which are called "cytosomes," their function being as yet unclear (Fig. 6.3A,D). They are presumed to be mostly fused lysosomes and endosomes (=phagolysosomes) in which the ecdysteroids may be synthesized from cholesterol or its metabolites taken up from the hemolymph; there may also be primary lysosomes among them.

FIGURE 6.2. Topography, histology, and biochemical features of the ecdysial glands of *L. forficatus:* (A) Cross section through the first trunk segment of an adult animal. (B) Sagittal and paramedian freeze section through the anterior trunk region showing reactions with antibodies against ecdysteroids. DAB (3,3-diaminobenzidinetetrahydrochloride) staining is restricted to the lobes of the ecdysial glands, and most intensive within the cytosomes (inset). (C) Electron micrograph of a cross section through the first trunk segment of an anamorphic instar. *Key:* c = cuticle; cl = spaces of the cortical labyrinth; cy = cytosomes; dlm = dorsal longitudinal muscles; fb = fat body lobe; ge = ecdysial gland; h = heart; he = hemocoel; lu = lumen of the pharynx; n = nucleus of a podocyte; ne = nucleus of a foregut epithelium cell; nrm = nucleus of a circular muscle cell; ph = pharynx; rm = circular muscles of the pharynx; sg = salivary gland; ttk 1 = tergum of the first trunk segment; tmx = tubule of maxillary nephridium; tr = trachea; vnc = ventral nerve cord. A, ×320; B, ×110; inset, ×450; C, ×3900.

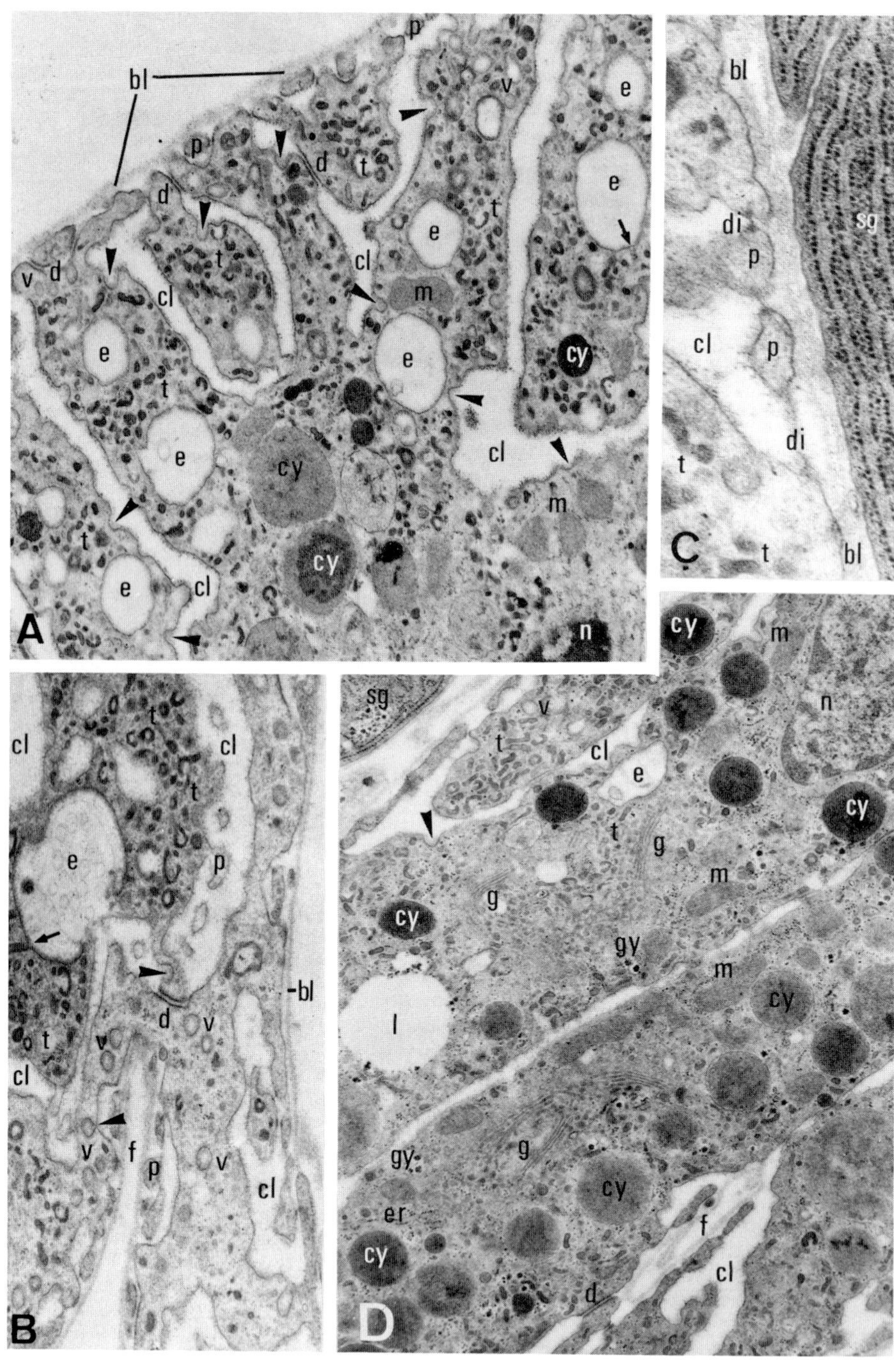

The stained particles shown in the inset of Fig. 6.2B can only be these cytosomes, as indicated by their localization, number, and size. Figure 6.4 schematically summarize the hypothesis of the function of the cortical areas of the podocytes.

A high rate of metabolism of the podocytes' central regions is indicated by the presence of numerous dictyosomes, relatively frequent mitochondria, endoplasmic reticulum with rough and smooth portions, single lipid droplets, and glycogen supply (Figs. 6.3D and 6.5.). The scattered distribution of chromatin within the nucleus and a large nucleolus suggest a high rate of transcription.

The content of podocytes changes during the molting cycle. Lipid droplets have never been observed before the critical phase when the subsequent molting becomes irreversible, whereas lipid droplets are frequent afterward. At that time, the increase in glycogen depots becomes obvious. Furthermore, mitochondria, dictyosomes, and especially the cytosomes seem to have increased in number. Whether this condition changes later, up to the start of ecdysis, is unknown. Investigations of the ecdysial glands, 12–24 h after ecdysis, however, showed remarkable changes: within the podocyte lobes are large areas of lysed cells containing membrane fragments (Fig. 6.6A). Some areas of these destroyed cells show cytoplasm with intact tubular structures and mitochondria. The lysed regions are always enclosed in native cells, often containing coated vesicles. The endosomes are relatively small at the time when the fusion of a new generation of coated vesicles starts. Glycogen depots and lipid droplets are absent immediately after ecdysis. The composition of the cytosomes also changes. Osmiophilic vesicles are fewer and smaller, whereas large vesicles are often translucent or show a heterogeneous content: patches of electron-dense material are enclosed in a matrix of fine granular material. There is a gradual change from translucent to

FIGURE 6.3. Electron micrographs from sections through the cortical region of podocytes of the ecdysial gland of *L. forficatus:* (A & B) The labyrinth and typical structures of receptor-coupled endocytosis. (C) Peripheral region with a row of pedicells interconnected by diaphragms. (D) A more central region of the podocyte. *Key:* bl = basal lamina; cl = cortical labyrinth; cy = cytosome; d = desmosome; di = diaphragm; e = endosome; er = endoplasmic reticulum; f = fold; g = Golgi complex; gy = glycogen; l = lipid droplet; m = mitochondria; n = profile of a nucleus; p = pedicells; sg = salivary gland; t = tubular structures; v = coated vesicles; arrows point to developing tubular structures at endosomes; arrowheads, to coated pits along the membrane of the cortical labyrinth. A, B, & D, ×14,100; C, ×34,800.

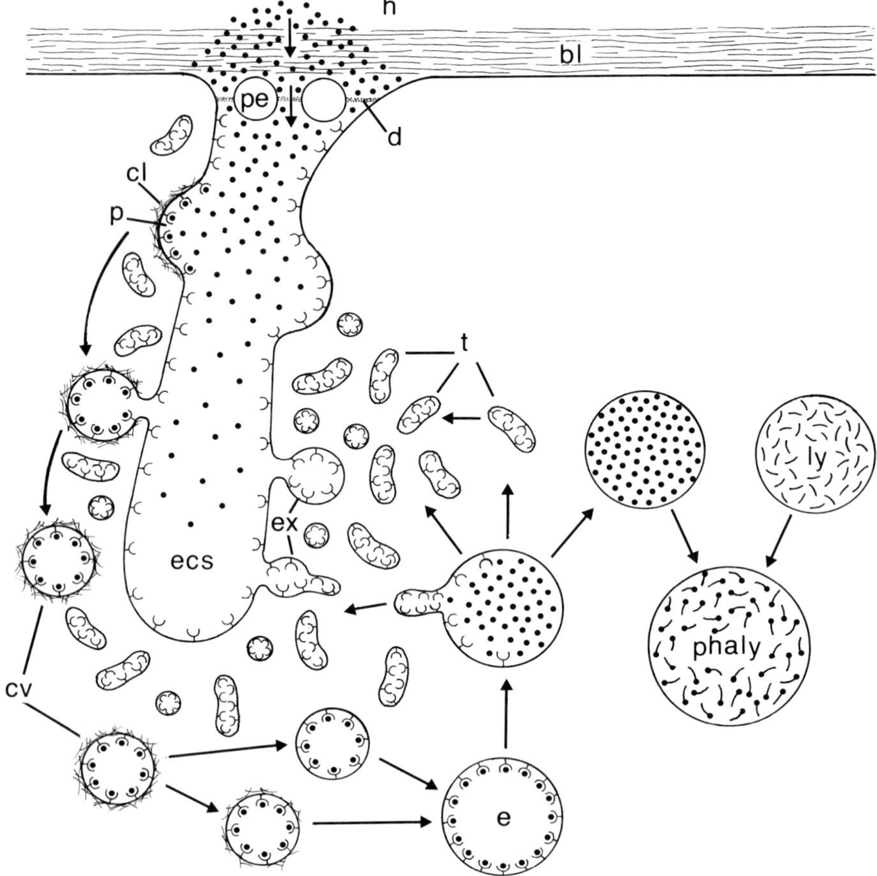

FIGURE 6.4. Diagram showing the structures and processes involved in receptor-coupled endocytosis (see text). *Key:* bl = basal lamina; cl = clathrine sheath; cv = coated vesicles; d = diaphragm; e = endosome; ecs = extracellular space of the cortical labyrinth; ex = exocytotic structures; h = hemocoel, ly = lysosome; p = coated pit; pe = pedicell; phaly = phagolysosome.

FIGURE 6.5. Electron micrographs from sections through central regions of podocytes of the ecdysial gland of *L. forficatus:* (A) Area where glycogen and a lipid droplet are stored. (B) Prominently developed endoplasmic reticulum and a cytolysosome. *Key:* cl = cortical labyrinth; cy = cytosome; cyl = cytolysosome; e = endosome; er = endoplasmic reticulum; f = fold; gy = glycogen; l = lipid droplet; m = mitochondria; n = nucleus; nc = nucleolus; sg = salivary gland; v = coated vesicles. ×14,100.

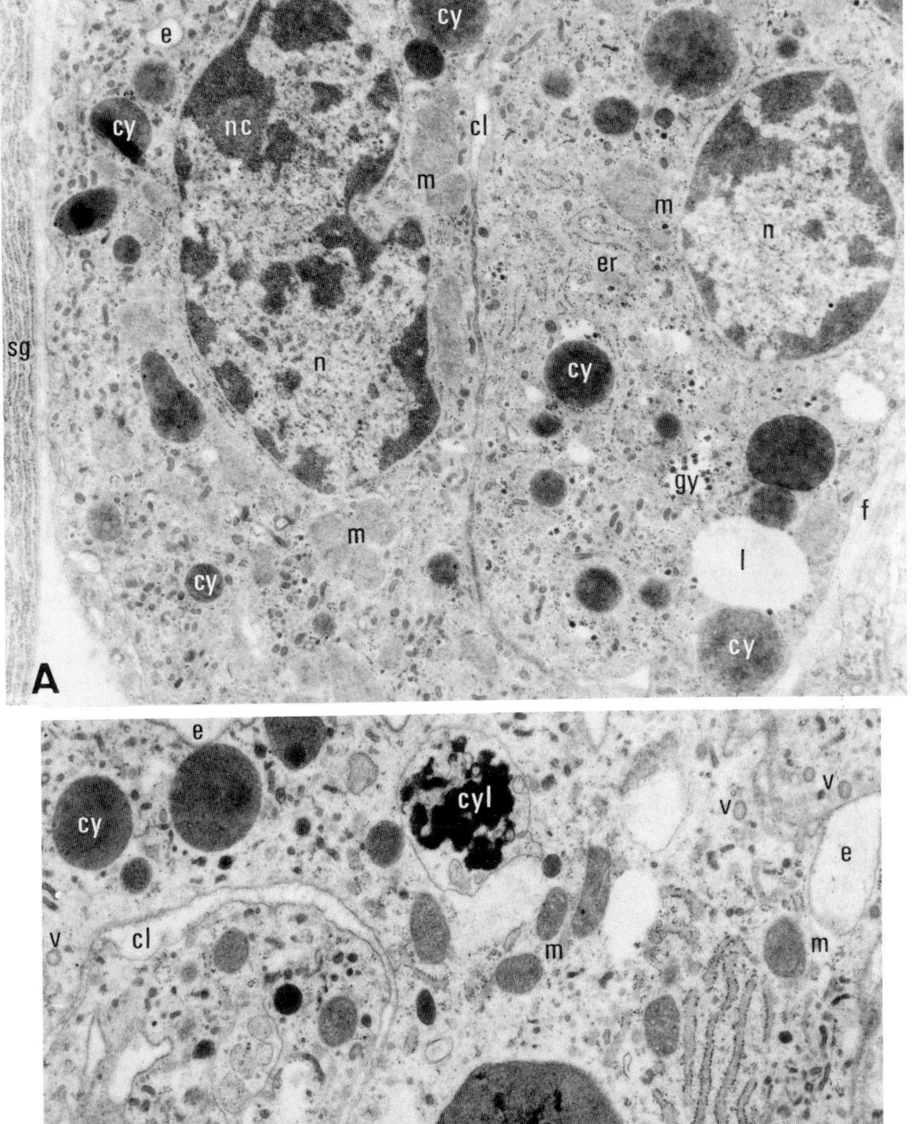

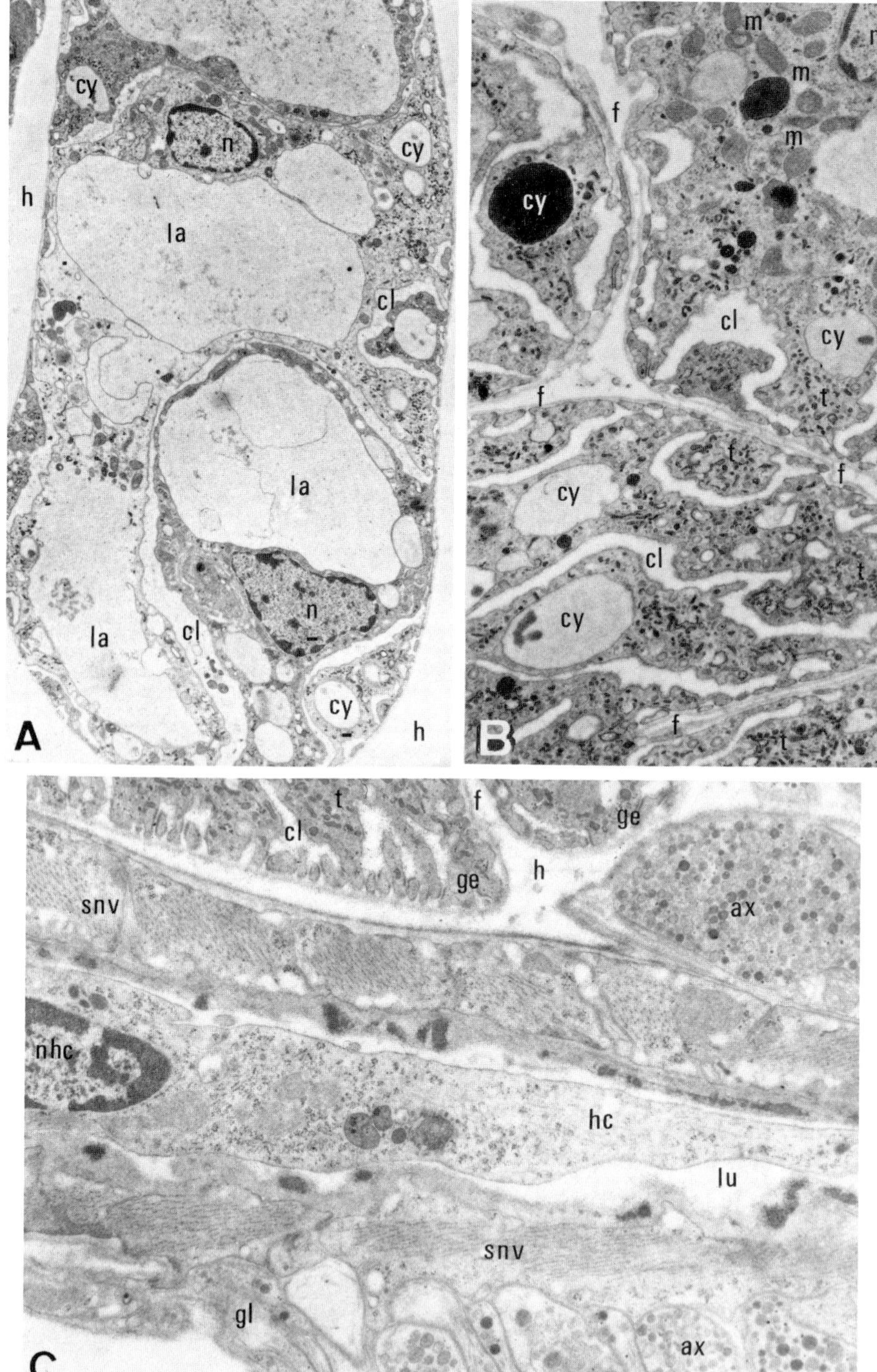

dense cytosomes. All cytosomes, however, show an irregularly folded membrane, indicating a lower interior pressure. Presumably this is due to the release of ecdysteroids into the hemocoel; this fact is surprising, as Joly et al. (1979) observed an extraordinary low titer of ecdysteroids within the hemolymph at the time of ecdysis. Further investigations on the morphological changes and their physiology are necessary for a comprehensive understanding.

Little is known about the innervation of the ecdysial gland (Richter, 1986). There are thin nerves or scattered axons containing neurosecretory droplets, ventral to the foregut; these axons directly contact the podocytes (Fig. 6.6C). They could be branches of a nerve that innervates the supraneural vessel. Neither synaptoid structures nor indications of changes during the molting cycle could be found within these axons. Their origin is also unknown.

6.2.1.2.   HEAD GLAND (GLANDULA CAPITIS) OF *Scutigera coleoptrata*

Rosenberg (1974) described a paired head gland in *Scutigera coleoptrata*, located paramedially in the posterior head region and caudally from the brain. This gland is attached to a hemolymph vessel reaching dorsally from the pharynx to the vertex, penetrating between the pharyngeal dilator muscles and fat body lobes (Fig. 6.7 and 6.8). In contrast to the ecdysial gland of *L. forficatus*, this organ is compact and smaller in extension. The occurrence of ecdysteroids has not yet been proved. Thus, the presumption that the head gland is the ecdysial gland of *S. coleoptrata* is based only on their ultrastructure, their localization in the anterior part of the body (as expected in all anamorphic taxa, showing teloblastic segment proliferation during postembryonic development), and the concentration of podocytes in two compact organs.

Histologically, the head gland is nearly identical with the ecdysial

FIGURE 6.6. Electron micrographs from sections through the ecdysial gland of *L. forficatus* third epimorphic instar a few hours after ecdysis: (A) Whole lobe showing vacuolized areas. (B) Cortical region at higher magnification. (C) The periphery of podocytes and the attached hemolymph vessel with its nerve and a neurosecretory axon between vessel and ecdysial gland. *Key:* ax = axon; cl = cortical labyrinth; cy = cytosome; f = fold of podocyte; ge = surface of podocytes; gl = glial envelope of the vessel nerve; h = hemocoel; hc = hemocyte; la = vacuolized area; lu = lumen of the supraneural vessel; m = mitochondria; n = nucleus; nhc = nucleus of a hemocyte; snv = wall of the supraneural vessel; t = tubular structures. A, ×4500; B, ×7400; C, ×14,100.

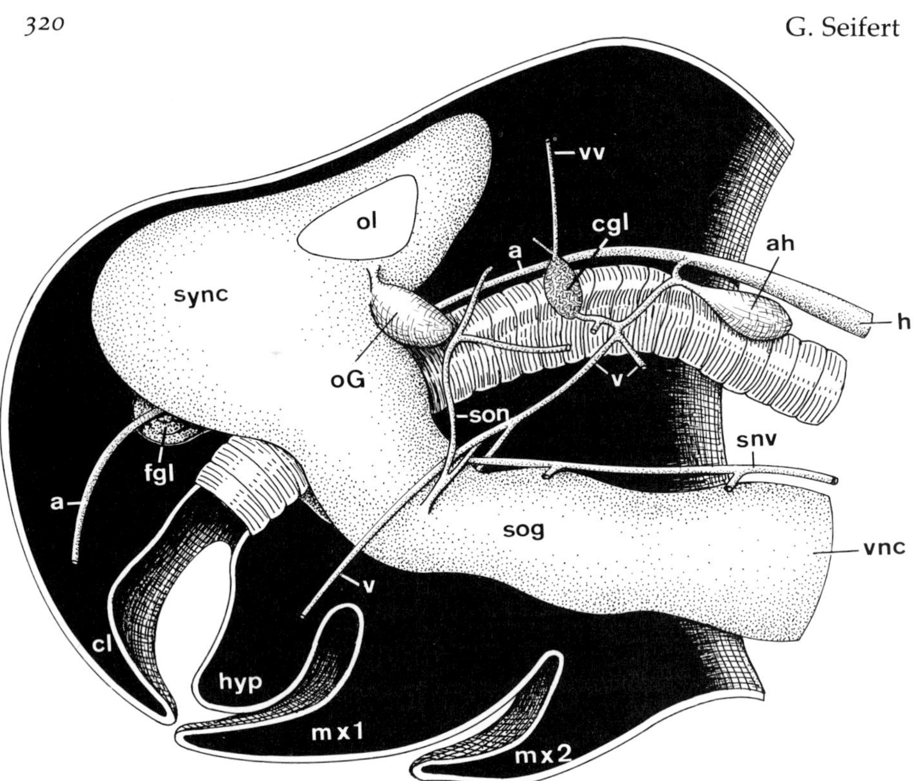

FIGURE 6.7. A schematic drawing of a laterally opened head capsule of *Scutigera coleoptrata* showing the topography of the head gland. To avoid confusion, some organs were not drawn. *Key:* a = aorta; ah = accessory heart; cgl = head gland; cl = clypeus; fgl = frontal ganglion; h = heart; hyp = hypopharynx; mx1, 2 = first and second maxilla; oG = Gabe organ; ol = optic lobe; snv = supraneural vessel; sog = subesophageal ganglion; son = subesophageal nerve; sync = syncerebrum; v = hymolymph vessel; vnc = ventral nerve cord; vv = vertical hemolymph vessel.

FIGURE 6.8. Paramedian sagittal sections through the region of the head gland of *S. coleoptrata*: (A&B) Light micrographs of semithin sections. (C) Electron micrograph. *Key:* b = posterior margin of the protocerebrum; c = cuticle of the vertex region; cl = cortical labyrinth; cy = cytosome; ep = epidermis; f = fold of podocyte; fb = fat body; gc = head gland; h = hemocoel; hc = hemocyte; m = mitochondria; mu = pharynx dilatator muscles; nm = nucleus of the vessel wall; np = nucleus of a podocyte; ve = hemolymph vessel. A, ×450; B, ×1150; C, ×4500.

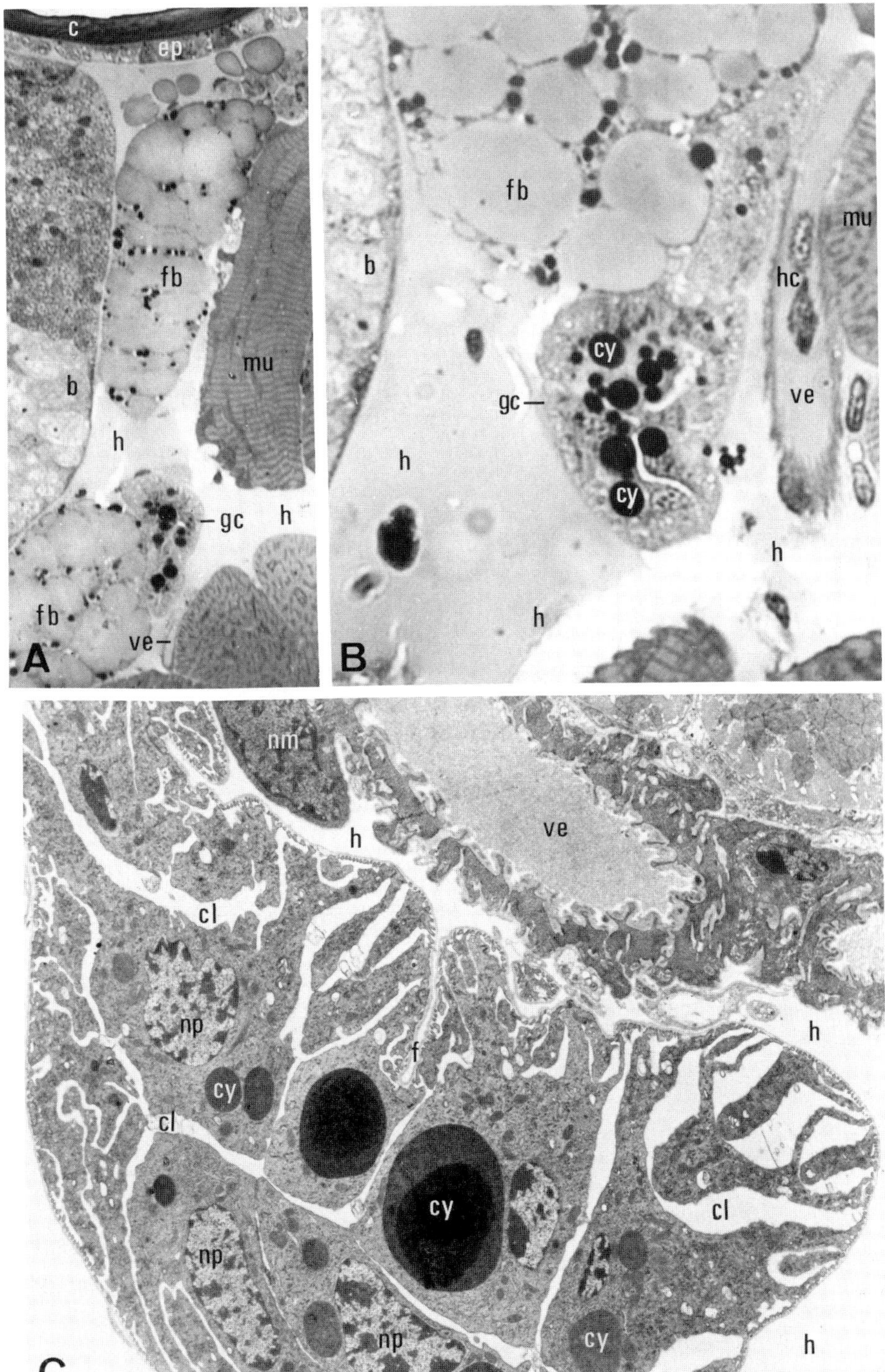

gland of *Lithobius* (Fig. 6.8). Higher magnification shows peripherally located large intercellular spaces between the cells. Centrally, refractile cytosomes and nuclei are the most characteristic features.

Their fine structure also corresponds to that of the ecdysial gland of *Lithobius*. It is also composed of podocytes surrounded by a common basal lamina. The single podocytes are separated by deep intercellular spaces. The surface of each cell, directed toward the hemocoel, is characterized by large and often branched invaginations, forming an extracellular cortical labyrinth and a strongly increased cell surface.

Extensive parts of the superficial border of each podocyte arise proximally from the basal lamina ranks of pedicells (Fig. 6.9). Diaphragms between neighboring pedicells keep filtration clefts open, which lead into the cortical labyrinth. The membrane alongside the labyrinth shows numerous coated pits (arrowheads), indicating the vesiculation of coated vesicles into the cytoplasm. Between these vesicles are typical tubular transport structures for receptors and also endosomes, where the former are produced. Except for a few mitochondria, there are no other organelles within the cell cortex.

Organelles are more frequent within the central cell regions proximal to the cortical labyrinth (Figs. 6.9A and 6.10A). Nuclei, numerous cytosomes, dictyosomes, and mitochondria occur in these regions. The mostly rough endoplasmic reticulum is well elaborated in some areas. Glycogen deposits are found throughout these regions.

The extraordinary elaboration of dictyosomes is a prerequisite for a high membrane turnover, caused by receptor-coupled endocytoses. However, dictyosomes also seem to play an important role in the development of lysosomes, as these organelles are occasionally seen to be formed within concentrically arranged Golgi cisternae (Fig. 6.10A, asterisk).

Thin nerve processes reach close to the head gland in *S. coleoptrata*. Single axons, most of which contain only little neurosecretory products, are observed within the hemocoel between the podocyte lobes (Fig. 6.10B). Similar axons are found between the nearby hemolymph vessel

FIGURE 6.9. Electron micrographs from sections through podocytes of the head gland of *S. coleoptrata:* (A) A survey of peripheral and central regions. (B) Cell cortex with typical structures of receptor-coupled endocytosis. *Key:* bl = basal lamina; cy = cytosome; di = diaphragm; e = endosome; er = endoplasmic reticulum; g = Golgi complex; m = mitochondria; n = profile of a nucleus; p = pedicells; t = tubular structures; v = coated vesicles; arrowheads point to coated pits along the membrane of the cortical labyrinth, arrows to regions in endosomes separating the tubular receptor structure from the vesicular ligand structure. A, ×4100; B, ×42,000.

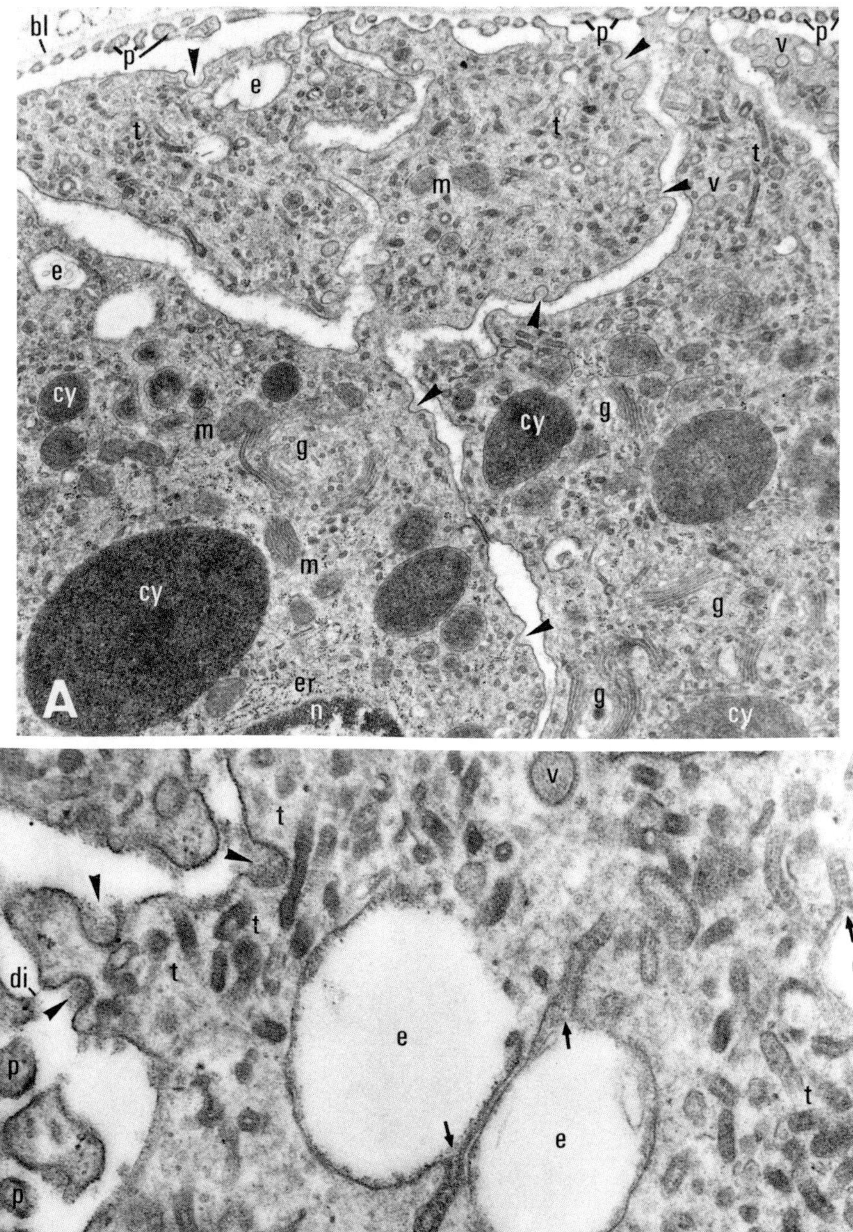

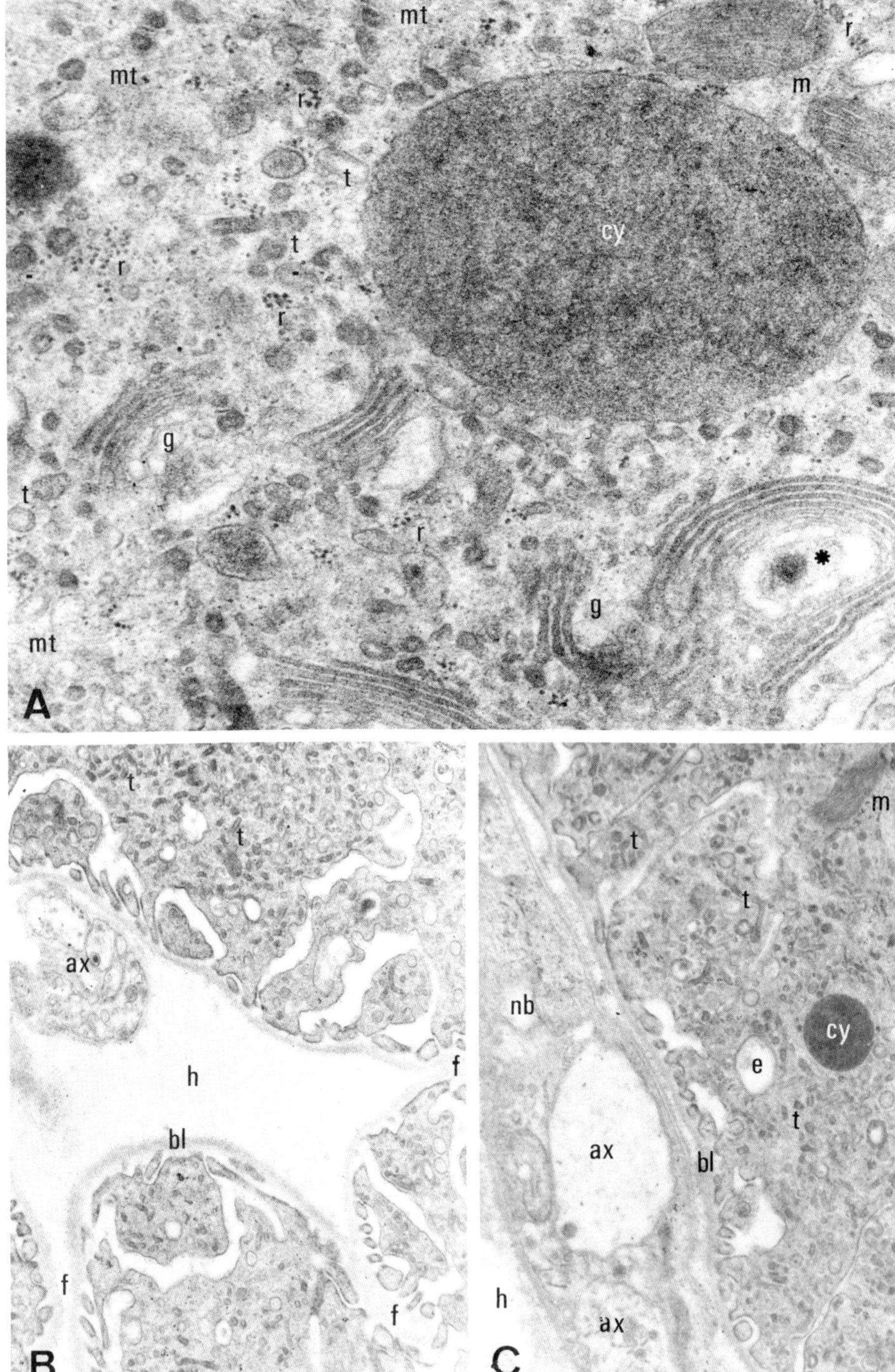
mt
mt
r
t
t
r
r
cy
m
r
t
g
g
r
mt
*
A
t
t
ax
f
h
bl
f
f
B
m
t
t
nb
cy
e
ax
t
bl
h
ax
C

and the head gland (Fig. 6.10C). Synaptoid structures, however, have not been observed in *Scutigera;* thus, the origin and functions of these axons remain uncertain.

6.2.1.3. ECDYSIAL GLANDS IN SCOLOPENDROMORPHA AND GEOPHILOMORPHA

Within the epimorphic chilopods, ecdysial glands corresponding to those in *Lithobius* and *Scutigera,* in terms of their location and structure, have not been found. In epimorphic animals, however, ecdysial glands are thought to be present in different parts of the body; and there are groups of podocytes that show high similarity to the ecdysial glands described above in other body regions. Immunological tests on ecdysteroids have not yet been performed.

Kowalevsky (1895) described a "lymphatic tissue" surrounding the Malpighian tubules of *Scolopendra cingulata.* As these cells were observed to accumulate stains, they were later called nephrocytes (Duboscq, 1898; Palm, 1954). Corresponding cells have also been found in *Cryptops hortensis,* a central European species. Their topography and ultrastructure have been described by Rosenberg (1979): a monolayered "lymphatic tissue" consisting of podocytes surrounds the Malpighian tubules over their whole length. It is sometimes supported by tracheae, hemocytes, and parts of the fat body. At various intervals, they attach to hemolymph vessels with small pores in their walls. The ultrastructure of the podocytes corresponds to that of the ecdysial glands of *Lithobius* and *Scutigera* in respect of all parts of the cells: there is a common basal lamina, pedicells with diaphragms, a cortical labyrinth, coated pits and coated vesicles, endosomes, tubular structures for recycling of receptors, and—more centrally within the cells—numerous cytosomes closely related to Golgi complexes, mitochondria, and a weakly elaborated endoplasmic reticulum. Nearly all cells characteristically contain a huge concrement vacuole with patchy, moderately electron-dense material in a translucent matrix. Its occurrence and proximity to vessels and hemocytes indicate that the

FIGURE 6.10. Electron micrographs from sections through the head gland of *S. coleoptrata:* (A) A cell region with Golgi-complexes. (B&C) The surface of podocytes shown attached to small nerves. *Key:* ax = axons; bl = basal labyrinth; cy = cytosome; e = endosome; f = folds of podocytes; g = dictyosomes surrounded by many vesicles; h = hemocoel; m = mitochondria; mt = microtubules; nb = nerve branch attached to the surface of podocytes; r = free ribosomes; t = tubular structures; the asterisk marks the center of a dictyosome probably including a developing lysosome. A, ×38,400; B&C, ×17,400.

"lymphatic tissue" is involved in excretion. Future investigations will have to show whether these cells are also capable of synthesizing ecdysteroids.

In Geophilomorpha, nephrocytes are distributed in the perivisceral sinus of nearly all trunk regions (Rosenberg, 1978). They predominantly produce intersegmentally arranged strands of cells along both sides of the gut; these strands are about as long as the connectives of the ventral nerve cord between the two ganglia. Within one strand, the cells may be compact or loosely arranged, depending on species. All strands are surrounded by a common basal lamina, which is often stiff and has a diameter of about 150 nm. They are attached to the gut by additional connective tissue and tracheae. The ultrastructure is very similar to that of the "lymphatic tissue" of the scolopendromorphs and, therefore, corresponds to the ecdysial glands of *Lithobius*. Only the size of many cytosomes (referred to as excretory or concrement vacuoles by Rosenberg, 1978) indicates a probably excretory function. As for the "lymphatic tissue" of scolopendromorphs, future investigations will have to show whether the nephrocyte strands, or only parts of them, are the sites of synthesis of ecdysteroid hormones.

6.3. Ecdysial Glands in Diplopoda

Compared with Chilopoda there is even less information on the ecdysial glands and ecdysis regulation in Diplopoda. The function of ecdysteroid hormones in molting is still unproved, and no experiments or determinations of ecdysteroid titers have yet been made. However, Bidmon et al. (1987) showed recently that there are ecdysteroid-binding sites within the nuclei of various tissues in *Polyxenus lagurus*.

The parthenogenetic females of this species (belonging to the subclass Penicillata) regularly molt before oviposition (Schömann, 1956; Seifert, 1960). Some 15–20 deposited eggs are protected against moisture and predators by setae, projecting like lancets from the bundle of the trunk end of their mother, preventing them from direct contact with the surroundings. Obviously, the molting, which occurs before each egg-laying, is necessary to ensure an intact integument and thus a reasonable number of these setae. As shown by Seifert (1966), removal of the setae induces molting, and thus a regeneration of setae. Molting occurs less regularly after amputation of antennae or cold shock (Seifert, 1966; Nguyen Duy–Jacquemin, 1972).

Because diplopods are anamorphic animals, their ecdysial organs are expected to be located at the anterior region of the body. In *Polyxenus lagurus, Craspedosoma rawlinsi,* and *Schizophyllum sabulosum,* all putative glands in the head and anterior trunk segments have been investigated (El-Hifnawi, 1972). Only in *P. lagurus* was found an endocrine organ that is morphologically identical with the ecdysial glands of the chilopods (Seifert and El-Hifnawi, 1972). This organ also showed changes during the molting cycle, indicating a function in ecdysis regulation. Current

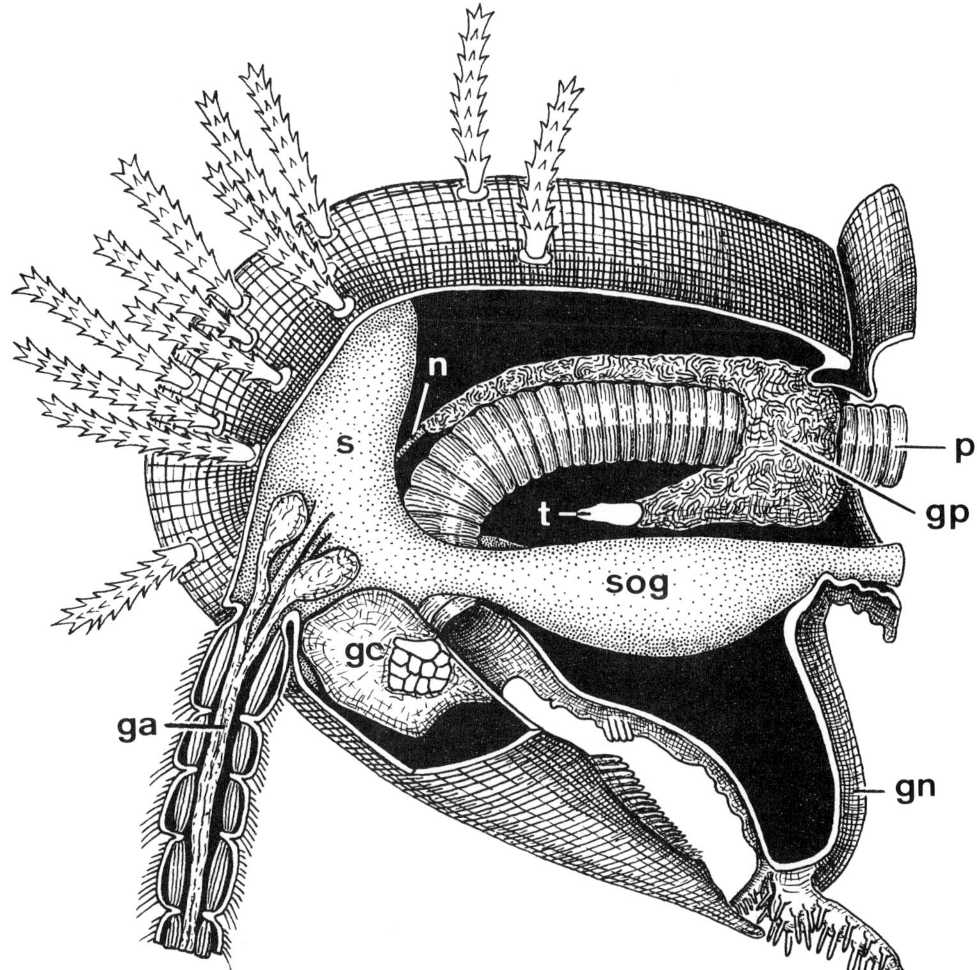

FIGURE 6.11. Schematic drawing of a laterally opened head capsule of *Polyxenus lagurus*, illustrating the topography of the collar gland. *Key:* ga = antennary gland; gc = cerebral gland; gn = gnathochilarium; gp = glandula perioesophagealis (collar gland); n = nervus glandulae perioesophagealis; p = pharynx; s = syncerebrum; sog = subesophageal ganglion.

immunohistochemical investigations will show whether there are ecdysteroid-binding sites within the cells of this "collar gland." For species
of the subclass Chilognatha, there are as yet no indications for a corresponding ecdysone-producing organ.

6.3.1. Topography, Histology, and Ultrastructure of the Periesophageal or Collar Gland (Glandula Perioesophagealis) of Polyxenus lagarus

The unpaired collar gland (glandula perioesophagealis) is most likely the
ecdysial organ in *Polyxenus lagurus* (Fig. 6.11, gp). It runs alongside the
unpaired median nervus glandulae perioesophagealis (originating on
the ventral part of the primary syncerebrum) between the brain and the
anterior foregut, and surrounds the esophagus like a collar in the posteri-
or head region. Ventrally to the foregut, it is attached to the tentorium.
The collar region of the gland has a rather large volume. As in the ecdysial
glands of chilopods, light microscopy reveals that the organ is characterized by a loose structure and numerous inclusions (Fig. 6.12A). Histologically, the collar gland represents a typical podocyte tissue, the lobes of
which overlaps at some parts (Fig. 6.12B). The whole organ is surrounded
by a basal lamina. The cell cortex is characterized by deep clefts, forming a
labyrinth with an enlarged surface. Thus, large extracellular compartments are produced where the endocytotic uptake of filtered components
into the cytoplasm may take place. The exterior parts of the clefts are kept
open by rows of pedicells with interconnected diaphragms.

Most of the uptake is receptor coupled, as indicated by numerous
coated pits at the cell membrane and coated vesicles within the cortical
cytoplasm. Tubular structures responsible for the recycling of receptors,
as well as high amounts of glycogen, which serve as an energy reserve,
are also characteristic. Centrally, endosomes and cytosomes are the dominant structures (Fig. 6.13). Some endosomes just separate the tubular receptor region from the vesicular ligand region (arrows). In the center of

FIGURE 6.12. Micrographs of sections through the collar gland of *P.
lagurus:* (A) Light micrograph of a median sagittal semithin section. (B)
Electron micrograph of a cross section through the posterior head region.
Key: b = brain; cl = cortical labyrinth; cy = cytosome; dilf = frontal, dilt =
tentorial, and dilv = vertical dilatator muscles of the foregut; f = folds of
podocytes; fg = frontal ganglion; gp = glandula perioesophagealis (collar
gland); h = hemocoel; lu = lumen of the pharynx; n = nuclei of
podocytes; nf = nucleus of a pharynx epithelium cell; ngp = nervus glandulae perioesophagealis; p = pharynx; rm = circular muscles of the pharynx; sog = subesophageal ganglion; t = tentorium; tr = trachea. A, ×450;
B, ×4500.

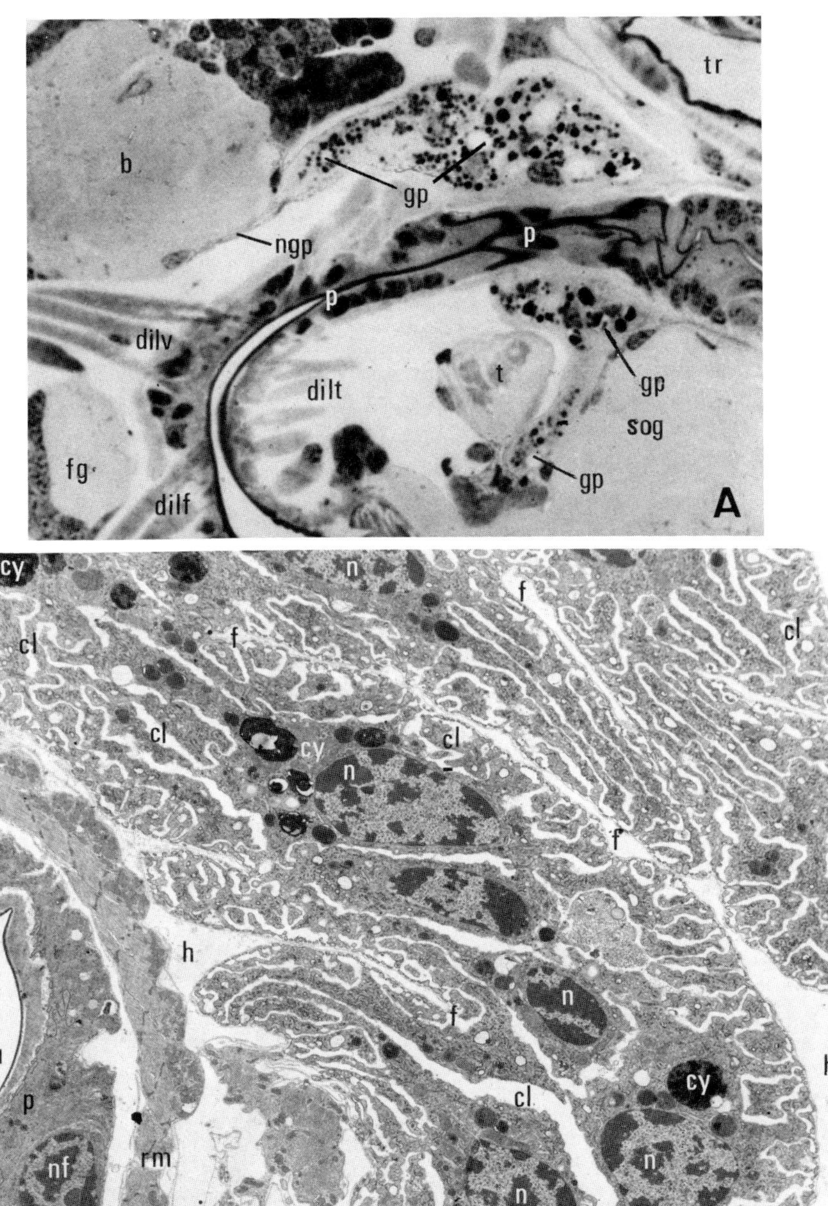

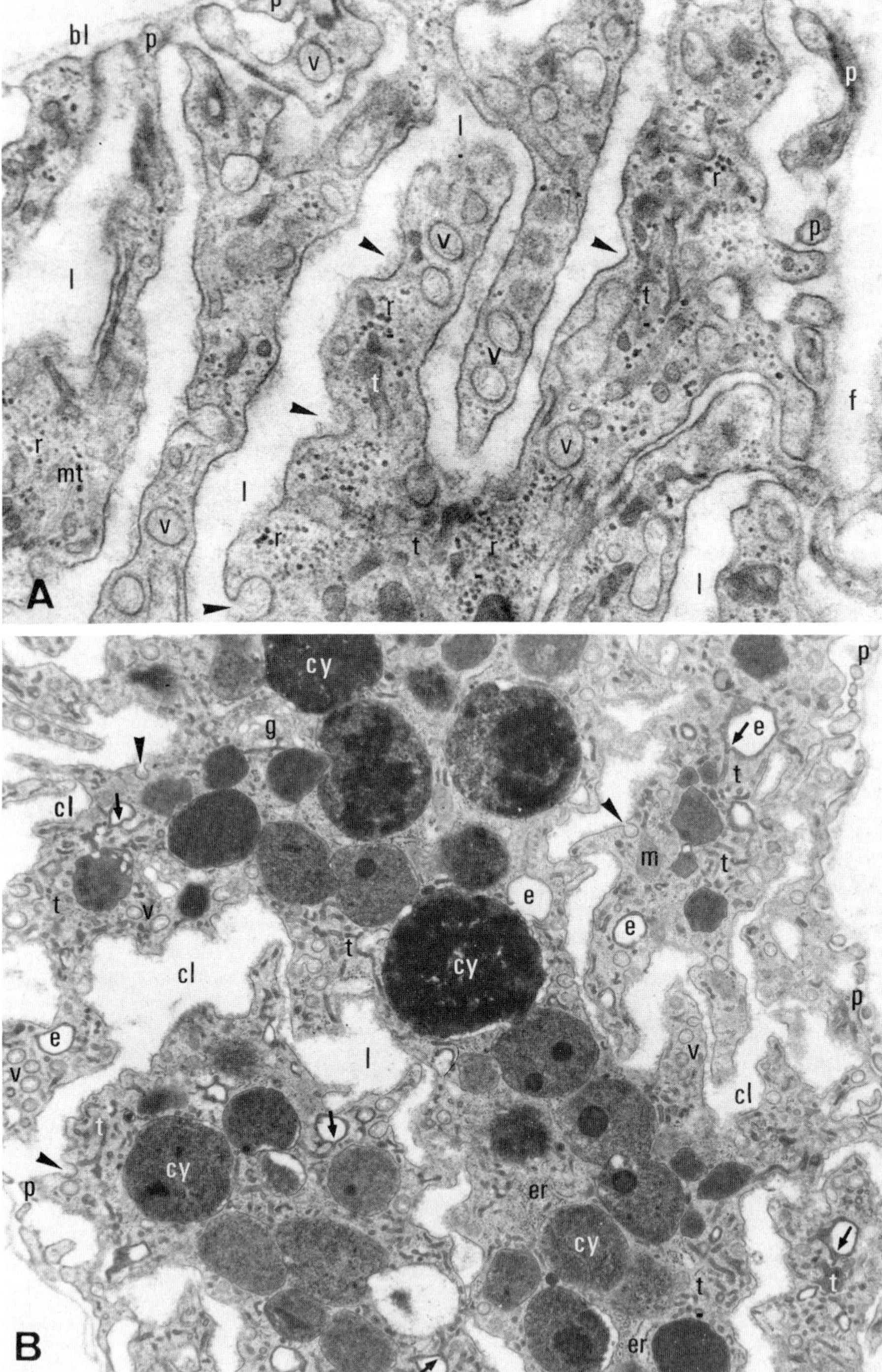

each podocyte (Fig. 6.14) is the nucleus, with a localized nucleolus. Its dispersed chromatin indicates a high transcription rate. The nucleus is surrounded by mitochondria, large tubules of endoplasmic reticulum, and numerous dictyosomes. Cytosomes are frequently found with heterogeneous but predominantly electron-dense content. The variability may depend on the stage of metabolism.

Molting can be initiated if the setae bundles are torn off (Seifert, 1966). At 25–26°C after 24 h, the fine structure of the collar gland changes (El-Hifnawi and Seifert, 1972). The volume of its cells increases, and thus the extracellular system, the intercellular clefts, and the cortical labyrinth are narrowed (Fig. 6.14B). The cisternae of the endoplasmic reticulum are swollen in some areas and filled with flocculent material (Fig. 6.15); in addition, granular and smooth membranes of endoplasmic reticulum are found. The number of free ribosomes, often aggregated into polysomes, however, increases. The dictyosomes are surrounded by a high number of vesicles and show enlarged cisternae, which are subdivided into pearl-like structures, owing to raised vesiculation. Mitochondria seem to be more frequent, and dumbbell- or rosette-shaped figures indicate divisions. Between single mitochondria and tubulus of the endoplasmic reticulum (Fig. 6.15A,B), typically there are narrow but open connections. The cytosomes are more variable (Fig. 6.15C), and most of them are translucent at this stage, whereas some lack an intact vesicular membrane and their content seems to pass into the cytoplasm. Occasionally, there are large bodies reaching the size of a nucleus. These look very similar to the concrement vacuoles in the podocytes of the "lymphatic tissue" of scolopendromorphs and geophilomorphs. El-Hifnawi and Seifert (1972), however, regarded them as cytolysosomes. Their occurrence always corresponds to the start of molting. Receptor-coupled endocytosis seems to remain unchanged. Furthermore, coated vesicles and tubular structures for receptor transport are formed and predominate the podocyte cortex. Ultrastructural investigations of further events of the molting cycle are still lacking.

FIGURE 6.13. Electron micrographs of superficial sections through podocytes of the collar gland of *P. lagurus:* (A) Periphery of the gland with typical structures of receptor-coupled endocytosis. (B) Cortical region flanking an area with cytosomes. *Key:* bl = basal lamina; cl = cortical labyrinth; cy = cytosome; e = endosome; er = endoplasmic reticulum; f = fold; g = Golgi dictyosome; i = intracellular cleft; m = mitochondrium; mt = microtubules; p = pedicells with diaphragms; r = free ribosomes; t = tubular structures; v = coated vesicles; arrows point to developing tubular structures at the endosomes; arrowheads, to coated pits along the cortical labyrinth. A, ×34,800; B, ×14,100.

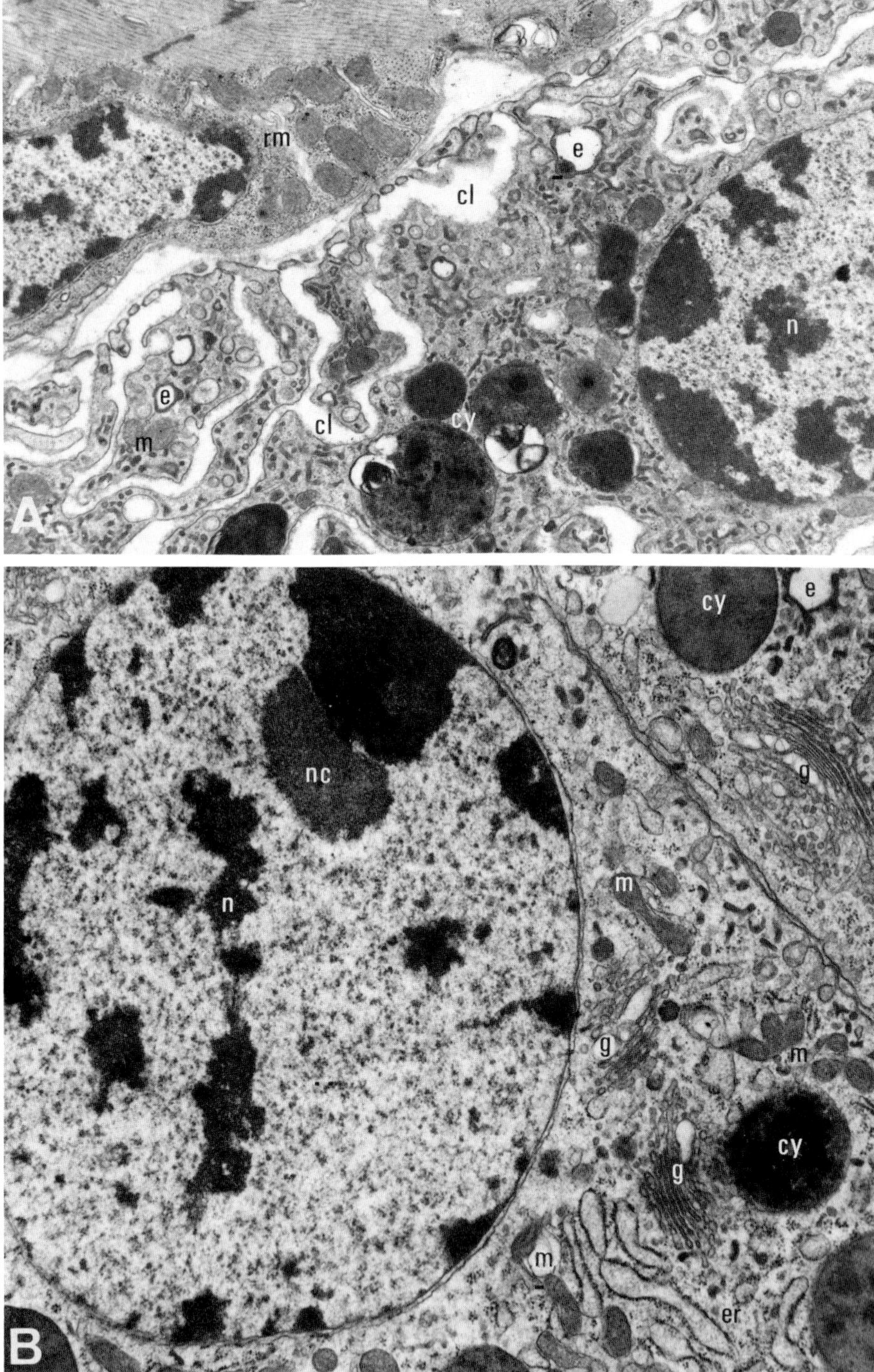

The median anterior part of the collar gland runs mostly lateral to the nervus glandulae perioesophagealis; the nerve and the gland are enclosed in a common basal lamina (Fig. 6.16). It is in this area that glial cells build some layers around the axons. This glial sheath is connected to the neighboring podocytes by desmosomes; wherever the desmosomes are found, the basal lamina is interrupted. Neighboring glial cells also are connected by desmosomes or tight junctions, resulting in a regular narrow intercellular gap of about 20 nm.

The nuclei of glial cells are irregularly shaped and show chromatin plaques peripherally. Their outer enveloping membrane is scattered with ribosomes, and there are short, sometimes bulged, cisternae of granular endoplasmic reticulum within the cytoplasm. Furthermore, there are free ribosomes, mitochondria, and glycogen. The microtubules of the glial cells are orientated along the length of the nerve (Fig. 6.16A, arrows).

Some 20–25 axons are surrounded by the glial sheath. Dorsally, a ~2-μm-thick axon is located in the middle of the nerve fiber; this axon is separated from the hemocoel by a single glial process and the basal lamina. Lateral to this "nerve ridge" are the podocytes of the collar gland. At all stages, the thick axon is densely packed with elementary neurosecretory granules (most of which have a diameter of 130 nm) and mitochondria. Ventrally, the median nerve region is separated from the sinus surrounding the foregut by a glial sheath (Fig. 6.16B). Most axons are enclosed by common glial cells and hence are not isolated from each other. Some contain mitochondria and a few neurosecretory droplets. However, neurotubules, irregularly shaped translucent vesicles, and some free ribosomes can be seen more distinctly. Synaptoid structures, indicating exocytosis of neurosecretory products, have not been observed.

The nerve ravels as it runs through the anterior podocytes. No axons have been described from the voluminous collar region. Hence, the nerve terminates within the median part. No information is available to date about the position of the pericarya and the function of the nervus glandulae perioesophagealis (Richter, 1986).

FIGURE 6.14. Electron micrographs of sections through the collar gland of *P. lagurus:* (A) Cortical and central regions of an animal not involved in molting. (B) Central region of an animal in which molting was induced by removing the end bundles of the setae 2 days before fixation. *Key:* cl = cortical labyrinth; cy = cytosomes; e = endosome; er = endoplasmic reticulum; g = Golgi complexes; m = mitochondria of the podocyte; n = nucleus; nc = nucleolus; rm = circular muscle cell of the pharynx. A, ×14,100; B, ×17,400.

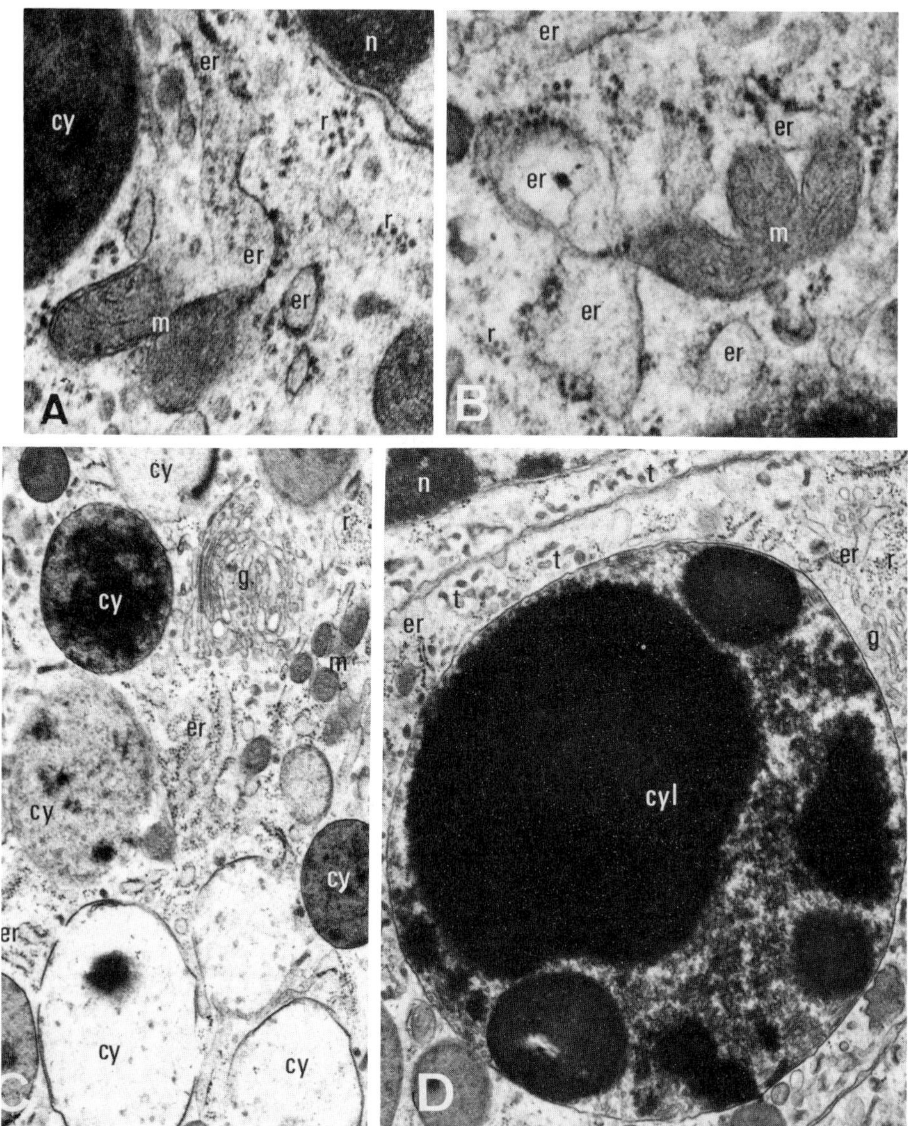

FIGURE 6.15. Electron micrographs from the region of the podocyte nucleus after inducing molting in *P. lagurus:* (A&B) Mitochondria shown attached to granular endoplasmic reticulum. (C) Area with various cytosomes. (D) A region of a large cytolysosome. *Key:* cy = cytosome; cyl = cytolysosome; er = endoplasmic reticulum; g = Golgi dictyosome; m = mitochondria; n = profile of a nucleus; r = free ribosomes; t = tubular structures. A&B, ×52,200; C&D, ×17,400.

6.4. Ecdysial Glands in Pauropoda

Most species of Pauropoda molt five times after hatching from the egg until the start of maturity. In *Pauropus silvaticus*, postembryonic development takes about 100 days (Tiegs, 1947). Adults of some species molt one more time. The life span of most species, however, exceeds one year.

Nothing is known about the control of molting. No putative molting organs, composed of podocytes, have yet been found in the two species investigated, *Pauropus huxleyi* and *Allopauropus vulgaris* (Zanger, 1981, 1986). Many morphological characteristics indicate that pauropods are dwarfed forms; hence, the possibility of an aberrant control of molting cannot be excluded, and probably typical ecdysial glands are lacking. It would be interesting to know whether molting of pauropods is also controlled by ecdysteroids and which organs produce these hormones. However, the molting physiology of pauropods can hardly be expected to be completely different.

6.5. Ecdysial Glands in Symphyla

Information on postembryonic development in Symphyla is available only for species of the genera *Scutigerella* and *Hanseniella*. There are six juvenile stages before the animals reach their maturity ("twelve-legged animals"). Molting, however, continues in adults. Michelbacher (1938) observed at least 40 molts in adults, which have been shown to live for at least 6 years (Verhoeff, 1934). Kaestner (1963) reported 52 molts of adults. The regulation of molting too is unknown, and no organ that might produce ecdysteroids has been described. Perhaps one could find such an organ among podocytes, which have not yet been investigated.

6.6. Summary

Within the four classes of Myriapoda, organs that are thought to synthesize the hormones controlling molting have been investigated only in a few chilopod and diplopod species. These organs are (a) the ecdysial glands of Lithobiomorpha, (b) the head glands of *Scutigera coleoptrata*, (c) with lower probability the "lymphatic strands" of geophilomorph and scolopendromorph chilopods, and (d) the periesophageal (or collar) gland in the diplopod *Polyxenus lagurus*. In Pauropoda and Symphyla, there is no hint of any such organ.

Among Myriapoda, only in *Lithobius forficatus* has the ecdysteroid character of the molting hormone been established recently, as ecdysteroid-binding sites were shown in the cytosomes of its ecdysial glands using

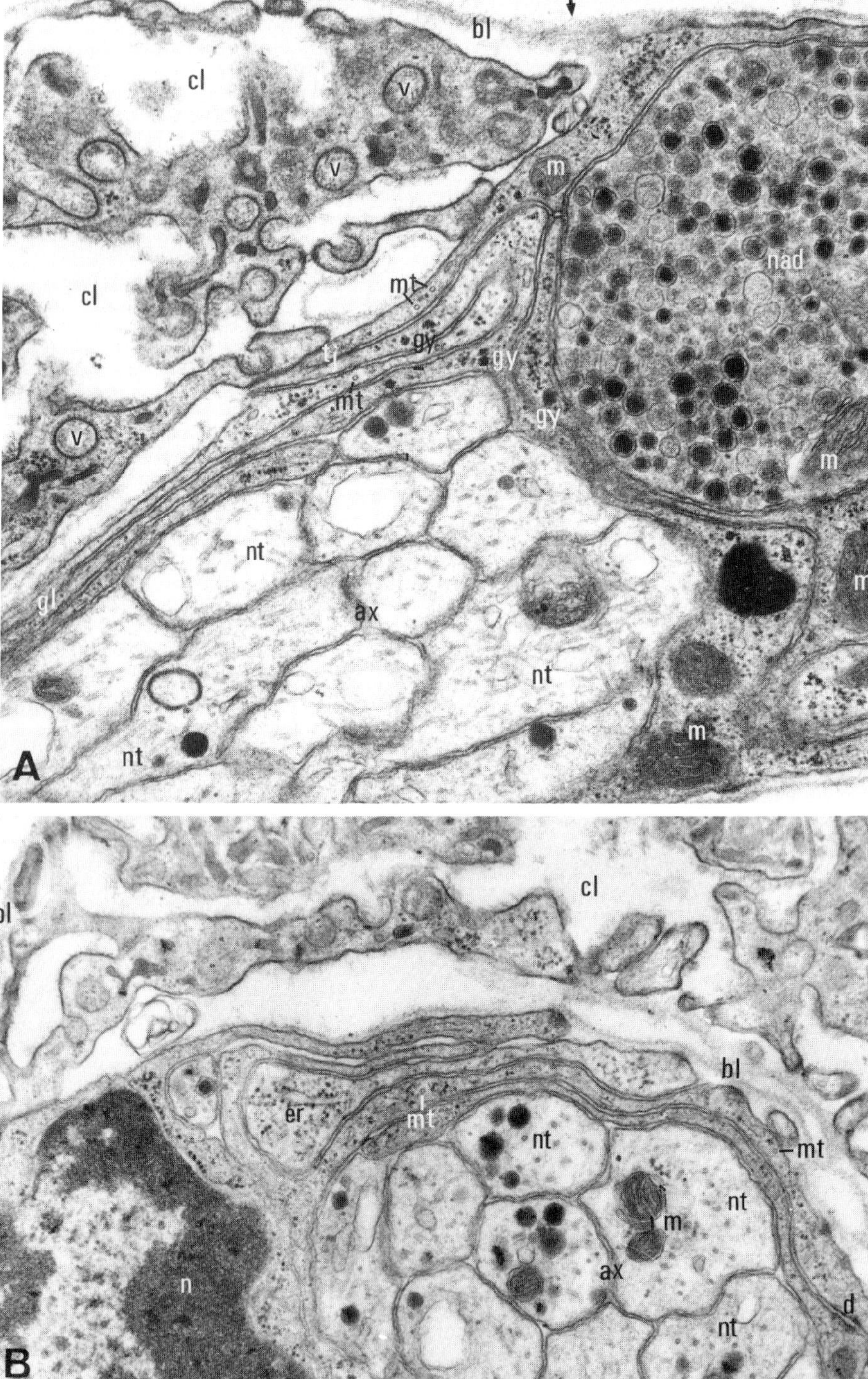

immunohistochemical techniques. Equivalent results for all other organs described as molting glands are still lacking. The ecdysial gland of *Polyxenus lagurus*, however, showed similar changes in its fine structure during the molting cycle, as observed in *Lithobius*; furthermore, preliminary experiments showed ecdysteroid-binding sites in the nuclei of various organs. In the other chilopods mentioned above, the only indication that those organs regulate molting results from their nearly identical fine structure.

All these organs are composed of podocytes, which are surrounded by a common basal lamina. The basal cytoplasmic membrane forms a large cortical labyrinth. The openings of this labyrinth are narrowed by pedicells, which are interconnected by diaphragms. The hemolymph is filtered by the basal lamina and the diaphragms into the labyrinth. Receptor-coupled endocytoses occur alongside the membrane of the labyrinth, as indicated by coated pits and coated vesicles within the nearby cytoplasm. Thus, the metabolites necessary for the synthesis of ecdysone are incorporated into the cell. In the cell cortex, coated vesicles fuse, losing their clathrin sheath, and form endosomes, which split to produce vesicular phagosomes, containing the ligands, and tubular structures, responsible for the recycling of receptors and their exocytotic reincorporation into the cortical plasma membrane. The phagosomes, however, fuse with lysosomes to form cytosomes, which are presumed to be the sites of synthesis and deposition of the ecdysteroid hormone, as indicated by immunohistochemistry. Most cytosomes are located at the transition zone between the cell center and the cortex. All organelles, the nuclei, and the presence of glycogen deposits and lipid droplets give the impression of a high level of metabolic activity. Especially in adults occur vacuoles of the size of nuclei in correlation to the molting cycle; these are considered to be cytolysosomes or concrement vacuoles.

FIGURE 6.16 Electron micrographs from sections dorsal to the foregut through the anterior part of the collar gland connected with the nervus glandulae perioesophagealis of *P. lagurus*: (A) The thick median neurosecretory axon (right above). (B) A lateral part of the nerve and its glial envelope. *Key:* ax = bundles of axons, some showing a few neurosecretory droplets; bl = basal lamina; cl = cortical labyrinth of a podocyte; d = desmosome; er = endoplasmic reticulum; gl = glial sheath; gy = glycogen; m = mitochondria; mt = microtubules within glial processes; n = profile of a glial nucleus; nad = dorsal neurosecretory axon; nt = neurotubules; tj = tight junction; v = coated vesicle; the arrow in A points to the connection between the basal laminae of glial cells and podocytes. A&B, ×34,800.

During a molting cycle, changes mostly are observable in the number and size of organelles, especially the cytosomes. This also indicates that they are the sites of ecdysteroid metabolism.

All molting glands are innervated by nerves or single axons. Some of these often contain neurosecretory material. The origin of the nerves and the location of their pericarya are unknown. Synaptoid structures, indicating the release of neurohormones, have never been observed. Thus, no information is available on the function of these nerves or the neuronal control of the ecdysial glands.

Acknowledgments

I am much obliged to my former collaborators Dr. E. El-Hifnawi and Dr. J. Rosenberg for making available some micrographs, to Mrs. H. Schmidt for making the diagrams, to Dr. W. Xylander for translating the manuscript, and to Mrs. A. Diebel and Mrs. A. Hudel for technical assistance.

References

Attems, C. 1926–1930. Chilopoda. Pp. 239–402 *in* W. Kükenthal and T. Krumbach (eds.), *Handbuch der Zoologie*, Vol. 6. De Gruyter, Berlin and Leipzig.

Bidmon, H. J., G. Seifert, and J. Koolman. 1987. Lokalisation von Ecdysteroiden und ihren Rezeptoren in Evertebraten. Verh. Dtsch. Zool. Ges. 80: 157–158.

Duboscq, O. 1898. Recherches sur les Chilopodes. Arch. Zool. Exp. Gén. 6: 481–650.

El-Hifnawi, E. 1972. Histologische und elektronenmikroskopische Untersuchungen über Drüsen in Kopf und Vorderrumpf von Diplopoden. Dissertation, University of Cologne, W. Germany.

El-Hifnawi, E. and G. Seifert. 1972. Elektronenmikroskopische und experimentelle Untersuchungen über die Kragendrüse von *Polyxenus lagurus* (L.) (Diplopoda: Penicillata). Z. Zellforsch. 131: 255–268.

Joly, R. (1961. Déclenchement expérimental de la mue chez *Lithobius forficatus* L. (Myriapode Chilopode). C. R. Acad. Sci. Paris 252:1673–1675.

Joly, R. 1962. Les glandes cérébrales, organes inhibiteur de la mue chez les Myriapodes Chilopodes. C. R. Acad. Sci. Paris 254: 1679–1681.

Joly, R. 1964. Action de l'ecdysone sur le cycle de mue de *Lithobius forficatus* L. (Myriapode Chilopode). C. R. Soc. Biol. France 158: 548–550.

Joly, R. 1966a. Etude expérimental du cycle de mue et de sa régulation endocrine chez les Myriapodes Chilopodes. Gen. Comp. Endocrinol. 6: 519–533.

Joly, R. 1966b. Contribution a l'étude du cycle de mue et de son déterminisme chez les Myriapodes Chilopodes. Thesis, Faculty of Science, University of Lille, France.

Joly, R. and M. Descamps. 1968. Etude comparative du complexe endocrine cephalique chez les Myriapodes Chilopodes. Gen. Comp. Endocrinol. 10: 364–375.

Joly, R., P. Porcheron, and F. Dray. 1979. Etude des variations du taux d'ecdysteroides au cours du cycle de mue dans l'hemolymphe de *Lithobius forficatus* L. (Myriapode Chilopode), par dosage radio-immunoligique. C. R. Acad. Sci. Paris 288: 243–246.

Kaestner, A. 1963. *Lehrbuch der Speziellen Zoologie*, Pt. I/5: Symphyla. Fischer Verlag, Jena.

Kowalevsky, A. 1895. Etude des glandes lymphatiques de quelques Myriapodes. Arch. Zool. Exp. Gén. 3: 592–616.

Leubert, F. 1982. Über Vorkommen und Bildungsorte von Ecdysteroiden bei *Lithobius forficatus* (L.) (Chilopoda). Diss. (A) Päd. Hochschule Erfurt/Mühlhausen.

Leubert, F. 1986. Untersuchungen zur Ecdysteroid-Biosynthese durch das Lymphstrang-Gewebe von *Lithobius forficatus* (L.) (Chilopoda). Wiss. Z. Päd. Hochschule Erfurt/Mühlhausen MNR 22: 73–77.

Leubert, F., H. Eibisch, and H. Scheffel. 1979. Isolierung und Charakterisierung des Häutungshormons von *Lithobius forficatus* (L.) (Chilopoda). Zool. Jahrb. Physiol. 83: 334–339.

Michelbacher, A. E. 1938. The biology of the garden centiped *Scutigerella immaculata*. Hilgardia 11(3): 55–148.

Nguyen Duy-Jacquemin, M. 1972. Régéneration antennaire chez les larves et les adultes de *Polyxenus lagurus* (Diplopode, Penicillate). C. R. Acad. Sci. Paris 274: 1323–1326.

Palm, N. B. 1954. The elimination of injected vital dyes from the blood in Myriapodes. Ark. Zool. 6: 219–246.

Richter, K. 1986. Zur Frage der Innervierung der Häutungsdrüsen bei Arthropoden. Zool. Jahrb. Physiol. 90: 43–63.

Rosenberg, J. 1974. Topographie und Ultrastruktur der endokrinen Kopfdrüsen (Glandulae capitis) von *Scutigera coleoptrata* L. (Chilopoda, Notostigmophora). Z. Morphol. Tiere 79: 311–321.

Rosenberg, J. 1978. Zur Ultrastruktur der Nephrozyten von Erdläufern (Chilopoda: Pleurostigmophora: Geophilomorpha). Entomol. Ger. 4: 24–32.

Rosenberg, J. 1979. Fine structure of the "lymphatic tissue" of *Cryptops hortensis* (Chilopoda: Scolopendromorpha): general organization and intercellular junctions. Pp. 287–294 *in* M. Camatini (ed.), *Myriapod Biology.* Academic Press, Orlando, Florida.

Rosenberg, J. and G. Seifert. 1975. Ist allein die Glandula ecdysalis die Häutungsdrüse von *Lithobius?* Experientia (Basel) 31: 1100–1101.

Scheffel, H. 1961. Untersuchungen zur Neurosekretion bei *Lithobius forficatus* L. (Chilopoda). Zool. Jahrb. Anat. 79: 529–556.

Scheffel, H. 1963. Zur Häutungsphysiologie der Chilopoden. Zool. Jahrb. Physiol. 70: 284–290.

Scheffel, H. 1965a. Über die Wirkung implantierter Cerebraldrüsen auf die Larvenhäutungen von *Lithobius forficatus* L. (Chilopoda). Zool. Anz. 174: 173–178.

Scheffel, H. 1965b. Der Einfluss von Dekapitation und Schnürung auf die Häutung und die Anamorphose der Larven von *Lithobius forficatus* L. (Chilopoda). Zool. Jahrb. Physiol. 71: 359–370.

Scheffel, H. 1969. Untersuchungen über die hormonale Regulation von Häutung und Anamorphose von *Lithobius forficatus* (L.) (Myriapoda, Chilopoda). Zool. Jahrb. Physiol. 74: 436–505.

Scheffel, H. 1987. Häutungsphysiologie der Chilopoden: Ergebnisse von Untersuchungen an *Lithobius forficatus* (L.). Zool. Jahrb. Physiol. 91: 257–282.

Schömann, K. 1956. Zur Biologie von *Polyxenus lagurus* L. (Diplopoda, Pselaphognatha). Zool. Jahrb. Syst. 84: 195–256.

Seifert, G. 1960. Die Entwicklung von *Polyxenus lagurus* L. (Diplopoda, Pselaphognatha). Zool. Jahrb. Anat. 78: 257–312.

Seifert, G. 1966. Häutung verursachende Reize bei *Polyxenus lagurus* L. (Diplopoda, Pselaphognatha). Zool. Anz. 177: 258–263.

Seifert, G. and E. El-Hifnawi. 1972. Eine bisher unbekannte endokrine Drüse von *Polyxenus lagurus* (L.) (Diplopoda, Penicillata). Experientia (Basel) 28: 74–76.

Seifert, G. and J. Rosenberg. 1974. Elektronenmikroskopische Untersuchungen der Häutungsdrüsen ("Lymphstränge") von *Lithobius forficatus* L. (Chilopoda). Z. Morphol. Tiere 78: 263–279.

Seifert, G. and H. J. Bidmon. 1988. Immunohistochemical evidence for ecdysteroid-like material in the putative molting glands of *Lithobius forficatus* (Chilopoda). Cell Tissue Res. 253: 263–266.

Tiegs, O. W. 1947. The development and affinities of the Pauropoda, based on a study of *Pauropus silvaticus.* J. Microsc. Sci. 88: 263–266.

Verhoeff, K. W. 1905. Über die Entwicklungsstufen der Steinläufer, Lithobiiden, und Beiträge zur Kenntnis der Chilopoden. Zool. Jahrb. Suppl. 7: 195–289.

Verhoeff, K. W. 1934. Symphyla. Pp. 1–120 *in Bronns Klassen und*

Ordnungen des Tierreichs, Vol. V, Pt. II, Sec. 3. Akademische Verlagsges., Leipzig.

Zanger, K. 1981. Licht- und elektronenmikroskopische Untersuchungen des Darms von *Pauropus huxleyi* Lubbock und *Allopauropus vulgaris* Hansen (Pauropoda: Tetramerocerata). Diplomarbeit, University of Giessen, W. Germany.

Zanger, K. 1986. Topographie, Histologie und Ultrastruktur des Darmtrakts von *Pauropus huxleyi* (Pauropoda: Tetramerocerata). Zool. Beitr. [N.F.] 29: 453–467.

Anatomy, Histology, Ultrastructure, and Functions of the Prothoracic (or Ecdysial) Glands in Insects

7

JACQUES BEAULATON

7.1. Introduction

The prothoracic glands (or ecdysial glands) were discovered in 1762 by Lyonet, a French anatomist, in an outstanding study of *Cossus cossus*. In that pioneering work, the author described them as "granulated vessels" located on the prothoracic tracheal trunks. Since then, this description has been confirmed and extended to numerous species. The major physiological role of these ductless glands in the endocrine control of postembryonic development in insects was first experimentally demonstrated by Fukuda (1940a,b, 1941, 1944) in the silkworm, *Bombyx mori*. Later, Williams (1946, 1947, 1948, 1949, 1952) confirmed these results on the saturniid moth *Hyalophora cecropia* and revealed that the prothoracic glands are activated by a brain hormone produced by neurosecretory cells of the protocerebrum. Indeed, it has been demonstrated that the prothoracic secretory cells are target cells for a neurohormone (Hiruma et al., 1978; Bollenbacher et al., 1979, 1984; Okuda et al., 1985; Smith and Gilbert, 1986) commonly called prothoracicotropic hormone (PTTH). The PTTH is a primary effector and elicits secretion of the molting hormone, or ecdysone. It is now well established that the prothoracic gland cells are the major source of ecdysone during the postembryonic development of insects (Borst and Engelman, 1974; Chino et al., 1974; Hoffmann and Koolman, 1974; King and Marks, 1974; King et al., 1974; Okuda et al., 1985; Ciancio et al., 1986). Because ecdysone is not the major active form of molting hormone, several authors regarded it as a prohormone or prehormone (Moriyama et al., 1970; King et al., 1974). In fact, this hormone is converted into a more active form (20-hydroxyecdysone, or ecdysterone) in peripheral tissues such as fat body and Malpighian tubules (King and Siddall, 1969; Koolman, 1982; Hoffmann, 1986). It activates the epidermal cells and subsequently elicits the secretion of a new layer of cuticle (see reviews of Filshie, 1982; Sedlak, 1984). Numerous experimental studies have contributed to establishing the classical mechanism of endocrine control of growth and metamorphosis in which the prothoracic glands play a central role in an orderly sequence of events, via a complex network of neural and humoral interactions.

The purpose of this review is to describe morphological, histological, and ultrastructural aspects of the prothoracic glands during embryonic and postembryonic development, as well as their degeneration during metamorphosis. This comparative study will concentrate on the secretory activity of the prothoracic gland cells in relation to their endocrine environment and to control mechanisms.

7.2. Terminology

These ductless glands were named "ecdysial glands" by Gorbman and Bern (1962) and Herman (1967) to designate type structures (the so-called molting glands, or cephalothoracic glands according to Bounhiol, 1980) controlling ecdysis. However, this generic name has previously been used in insects to designate exocrine glands: the "dermal glands," secreting the cement layer at ecdysis (see Wigglesworth, 1933, 1970). To prevent possible confusion with the latter glands, most authors (Gilbert, 1964; Wigglesworth, 1970; Sedlak, 1984, 1985) attribute the term "prothoracic glands" to the molting glands despite their varying anatomic locations among insect species. It can be argued in favor of this term that, as noted above, a neuropeptide activating these glands is commonly termed prothoracicotropic hormone. Thus, it seems appropriate to utilize the term sanctioned by common usage, i.e., "prothoracic glands," extending it in a broad sense to the "ventral glands" of primitive insects and the "peritracheal glands" of Diptera. In opting for this term, however, we should bear in mind that certain "prothoracic" glands may not be strictly confined to the prothorax, as is frequently found to be the case in Apterygota and hemimetabolous insects.

7.3. Embryonic Prothoracic Glands

7.3.1. *Embryonic Origin*

The prothoracic glands are known to be ectodermal derivatives. As a rule, they arise from paired invaginations of the lateral ectoderm. Toyama (1902) first described their embryonic development in a detailed embryological study of the silkworm, *Bombyx mori*, and called them "hypostigmatic glands." According to this author, they arise from a pair of ectodermal ingrowths of the labial segment. Each lateral pair of invaginations forms elongate cell bulks which are rudiments of the prothoracic glands. During shortening of the somites for head formation, the lateral ingrowths of the labial segment migrate backward into the prothorax. Finally, in the early postembryonic stages, each lateral invagination yields a trilobated band attached to the ventral prothoracic tracheae (Toyama, 1902).

Since that pioneering work, embryological studies in insects have revealed broad variations in the position of gland buds among various species, as observed in the corpora allata (Cassier, 1979). The prothoracic glands originate from either the cephalognathal region or the thorax before katatrepsis. For various insect embryos, the situation can be summarized as follows:

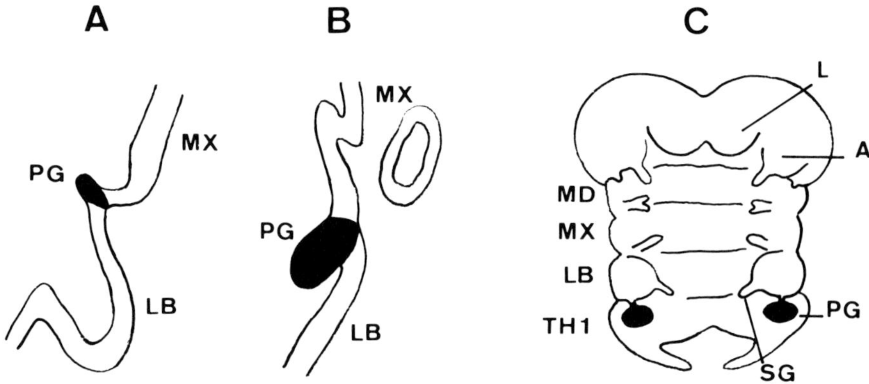

FIGURE 7.1. Stages in the embryonic development of prothoracic glands: (A&B) Diagrams of sagittal sections in *Locusta migratoria*. (A) Stage 4c showing an invagination located between the maxillary and labial segments and giving rise to the prothoracic gland bud. (From Maltête, 1962.) (B) Stage 6g illustrating a prothoracic gland bud in the course of pediculization. (From Maltête, 1962.) (C) The cephalic region of the moth, *Neomicropteryx nipponensis* illustrating the location of developing appendages. The prothoracic gland rudiments arise from the sternal region of the prothorax. (From Kobayashi and Ando, 1983.) *Key:* A = antenna; L = labrum; LB = labial segment; MD = mandibular segment; MX = maxillary segment; PG = prothoracic gland; SG = silk gland; TH1 = prothorax.

(1) In *Pieris rapae* (Eastham, 1930), the *maxillary segment* seems to bear the developing gland rudiments. They appear as tubular invaginations in the vicinity of the posterior arms of the tentorium. However, the maxillary origin of the glands needs confirmation.

(2) The *intermediate region between the maxillary and labial somites* has been well identified as the originating site of the glands (Fig. 7.1A,C) in *Locusta migratoria* (Maltête, 1962) and *Schistocerca gregaria* (Sbrenna-Micciarelli and Sbrenna, 1972). In *Locusta*, each gland rudiment is situated laterally and immediately behind the posterior arm of the tentorium.

(3) In most insects, the prothoracic gland rudiments arise from the *labial segment*. This site is now described among various orders, such as Odonata: *Aeshna cyanea* (Pflugfelder, 1947); Dictyoptera: *Leucophaea maderae* (Darquenne, 1978); Phasmida: *Carausius* (Vignau-Rogueda, 1963, cited in Haget, 1977; Vignau, 1968); Heteroptera: *Dysdercus cingulatus* (Wells, 1954; Jacob and Prabhu, 1985) and *Oncopeltus fasciatus* (Dorn, 1972); Trichoptera: *Stenopsyche griseipennis* (Miyakawa, 1974); Lepidoptera: *Diacrisia virginica* (Johannsen, 1929; Johannsen and Butt, 1941), *Ancylolomia japonica* (Tanaka, 1971, cited in Kobayashi and Ando, 1983),

Parnassius glacialis, Luehdorfia japonica, Byasa alcynous, Papilio protenor (Tanaka, 1985), and *Manduca* (Dorn et al., 1987). In *Diacrisia*, for example, two pairs of invaginations are present in the labial segment. The outer pair forms the prothoracic gland rudiment as in the silkworm. These invaginations originating near the base of the labium are spherical or vesicular before their migration into the prothoracic segment (Johannsen, 1929; Johannsen and Butt, 1941; Tanaka, 1985). The gland lumen is obliterated before the embryonic revolution.

(4) The *intermediate region between the labial segment and the prothorax* has been identified as the glands' originating site in the lepidopteran *Chilo suppressalis*. At stage 13, the gland invaginations are located at the back of the posterior arms of the tentorium near the prothorax (Okada, 1960).

(5) The *sternal region of the embryonic prothorax* (Fig. 7.1C) appears involved in the development of the glands in Odonata, such as *Epiophlebia, Calopteryx, Lestes* (Ando, 1962), as well as in the primitive lepidopteran (*Neomicropteryx nipponensis* (Kobayashi and Ando, 1983). Clearly, this originating site also deserves new investigation.

Most authors assume that the prothoracic glands of insects arise from the labial segment (see reviews by Tombes, 1970; Wigglesworth, 1970; Richards and Davies, 1977; Haget, 1977). However, such an interpretation is difficult to confirm, considering the above-listed variation of originating sites. There are good reasons to think that the glands are not systematically labial derivatives and might well originate from regions anterior or posterior to the labial segment. Nevertheless, as previously indicated, the origins of the glands and especially of the lateral parts of the ring gland remain to clarified despite Poulson's work (1945) on *Drosophila* embryos.

The gland develops simultaneously with the corpora allata (CA), which are considered as intersegmental organs, originating in an anterior area of the maxillary segment (Cassier, 1979; see also his Chapter 2 in this volume). By the fourth day of *Locusta* development (stage 4 according to Maltête, 1962), the gland rudiments become more pronounced than those of the CA and meet with the ventral wall of the labial coelomic sac (Maltête, 1962). On the fifth day (stage 5a), each rudiment displays a vesicular structure as in *Schistocerca* (Sbrenna-Micciarelli and Sbrenna, 1972), contrary to that of the CA. All rudiments are ensheathed by a mesenchymatous tissue perhaps issuing from the labial coelomic sac. At the end of the sixth day (stage 6), each glandular rudiment grows anteriorly and begins to separate from the epidermis. On the seventh day (stage 7), it extends anteriorly and finally cuts off from the ectodermal wall. As development proceeds, only connective strands bind the anterior part of the rudimentary gland to the cephalic ectoderm behind the origin of the adductor mandibular muscles. The gland location in the late embryonic stages appears definitive, since it resembles that in postembryonic instars (Strich, 1954; Strich-Halbwachs, 1959).

7.3.2. Histology and Ultrastructure of Embryonic Prothoracic Glands

The study of the histological and ultrastructural features of the embryonic glands only began during the 1970s. Sbrenna-Micciarelli and Sbrenna (1972) described the changes in nuclear volume at different stages in *Schistocerca*. They concluded that the highest activity of the glands occurred at stage 28 and coincided with a peak in the gland volume. In *Oncopeltus*, Dorn and Romer (1976) were the first to measure the relative DNA content of the gland cells by spectrophotometry on thick sections. They clearly demonstrated the different degrees of ploidy in the gland cells of both embryonic and postembryonic instars. They found that all gland nuclei stained with Feulgen's nuclear reaction remaind *diploid* (see Fig. 7.5, below, in Section 7.4.6.2.2) in all embryonic stages until hatching, in contrast to larval nuclei that underwent different degrees of polyploidization. Mitotic activity was identified only in the early developmental stages of *Oncopeltus* (Dorn and Romer, 1976) and *Schistocerca* (Sbrenna-Micciarelli and Sbrenna, 1972). In *Dysdercus*, Jacob and Prabhu (1985) counted 90–100 individual cells in the glands of a 4-day-old embryo. They reported a rise in cell size with a peak of 96 h (approximately an eightfold increase in volume from 84 h), interpreted as a phase of cyclic activity.

An ultrastructural study in *Oncopeltus* has shown that the embryonic gland cells are loosely associated with each other by means of cell protrusions and a very thin basal lamina or tunica propria (Dorn and Romer, 1976). The developing cells of *Oncopeltus* and *Carausius* (Haget et al., 1982) have some typical features of nearly undifferentiated cells, with a very scanty cytoplasm containing numerous free ribosomes and occasional small mitochondria. Dense bodies, glycogen particles and lipid droplets are often present in the cytoplasm. The nuclei usually have a smooth outline and an irregularly condensed chromatin (Dorn and Romer, 1976).

7.3.3. Secretory Activity and Ecdysteroid Levels During Embryogenesis

Little is known about the gland activity operating during embryonic developement. From ligaturing experiments in post-katatrepsis embryos of locusts, Jones (1956) has suggested that the glands are required for embryonic molting. If such embryos are ligatured immediately behind the head before a critical period, the body fails to molt. At present, ecdysteroid titers have been determined in embryos of several insects (Imboden et al., 1978; Bullière et al., 1979; Cavallin and Fournier, 1981; Imboden and Lanzrein, 1982) and especially in locusts (Gande and Morgan, 1979; Gande et al., 1979; Hoffmann et al., 1980; Lagueux et al., 1979). Correlations between cycles of cuticulogenesis and ecdysone bursts have been established in the embryos of *Blaberus craniifer* (Bullière et al., 1979) and *Locusta migratoria* (Lagueux et al., 1979; Hoffmann et al., 1980). The first ecdysone peaks occurring before differentiation of the glands presumably result from metabolic transformations of conjugated ecdysteroids of maternal origin (Gande and Morgan, 1979; Gande et al., 1979; Lagueux et al., 1979). By contrast, the last peaks of ecdysone (peaks 3 and 4 in *Locusta*) are considered as arising from synthetic activities of the embryonic prothoracic glands.

In Phasmida, Cavallin and Fournier (1981) have reported that early embryos deprived of their glands cannot develop beyond stage VII_{6c}. These results clearly demonstrate a key role of the glands in the late embryonic development of *Clitumnus*. The glands have been also studied in *Carausius* (Vignau-Rogueda, 1963, cited in Haget, 1977; Vignau, 1968), in which the first correlations between ultrastructural data and ecdysteroid levels have been reported (Haget et al., 1982). Changes of ecdysteroid titers during embryogenesis were demonstrated by radioimmunoassay (Louvet et al., 1979). The major result was that three peaks of ecdysteroids occurred coincidentally with three periods of embryonic cuticulogenesis (Haget et al., 1982). It seems that two phases of secretory activity take place at the time of the last two phases of cuticulogenesis. This is especially the case during the third peak of ecdysteroids, coinciding with an enlargement of mitochondrial profiles containing tubular cristae.

7.4. Postembryonic Prothoracic Glands

7.4.1. *Location and Anatomic Types*

As a rule, the glands are paired structures, except in Grylloidea and higher Diptera, located either in the anterior thorax or in the head of all immature instars. In most primitive insects such as Apterygota and lower Pterygota—Odonata, Ephemeroptera, Plecoptera, Phasmida, and Dermaptera—they are situated in the head, lying generally at the base of the labium. They have been termed "ventral glands" of the head (see the review by Wigglesworth, 1964) and are also present in some Orthoptera, such as locusts (Pflugfelder, 1947; Strich, 1954), and Isoptera (see the review by Grassé, 1982) such as *Reticulitermes* (Jucci, 1924) and *Microceroter-*

mes (Pflugfelder, 1947). In holometabolous and in some hemimetabolous insects (Dictyoptera, Heteroptera, and some Isoptera), these glands are found in the anterior thoracic region, chiefly in the prothorax. Sometimes, the glands may be asymmetrical, as is the case in *Papilio demoleus*, where they remain morphologically asymmetrical during larval development (Srivastava et al., 1982). The anatomy of the glands, hitherto described, varies substantially among different orders of insects (see reviews by Herman, 1967; Joly, 1968). However, seven anatomic types of the glands may be recognized:

(a) Cephalic compact type (Fig. 7.2A). The ventral glands are well developed in the posteroventral region of the head in most primitive insects such as Thysanura: *Machilis* (Gabe, 1953a), *Petrobius maritimus* (Gabe, 1953a; Cassier and Fain-Maurel, 1971), *Lepisma saccharina* (Gabe, 1953a), *Thermobia domestica* (Watson, 1963); Odonata: *Aeshna cyanea* (Pflugfelder, 1947; Cazal, 1947; Arvy and Gabe, 1952; Schaller, 1960); Ephemeroptera: *Ecdyonurus* (Pflufgfelder, 1947; Arvy and Gabe, 1950), *Epeorus, Heptagenia, Rhitrogena* (Arvy and Gabe, 1950), and *Oligoneuriella* (Pinet, 1961); Plecoptera: *Dictyogenus* (Pflugfelder, 1947) and *Brachyptera* (Arvy and Gabe, 1953c); as well as Grylloblattodea (Rae and O'Farrel, 1959) and Dermaptera (Pflugfelder, 1947; Lhoste, 1950, 1957). In Psocoptera, Badonnel (1970) has shown that the glands are closely associated with the lateroventral epidermis of the collum. Their form varies appreciably from one order of insects to another: it may be globular, as in Thysanura and Dermaptera, or multilobulated as in Odonata, Ephemeroptera, and Grylloblattodea. The glands consist of a compact structure generally associated with the epidermis either by a large area, as in Thysanura (Gabe, 1953a) and Psocoptera (Badonnel, 1970), or by a stalk-like connection, as in *Aeshna cyanea* (Pflugfelder, 1947; Schaller, 1960), *Grylloblatta campodeiformis* (Rae and O'Farrel, 1959), and *Forficula* (Lhoste, 1950, 1957). In the Diplura *Campodea*, the paired glands, although located in the laterotergal region of the head (Bareth, 1964), are probably homologous with the "ventral glands" described in other primitive insects (Bareth, 1964). Such a gland type is the most primitive known in insects with regard to anatomic locations and structures closely resembling those observed in embryonic development (see Section 7.3.1, above).

(b) Cephalic ribbon-shaped type (Fig. 7.2B). Anatomically, these cephalic glands are composed of paired bands. In *Locusta* and *Stenobothrus*, they are located dorsoventrally in the posterior region of the head, lying behind the mandibular muscles and in the vicinity of the neck (Pflugfelder, 1947; Strich, 1954; Strich-Halbwachs, 1959). The upper part of each gland reaches the head capsule and is attached to the occipital region. The lower

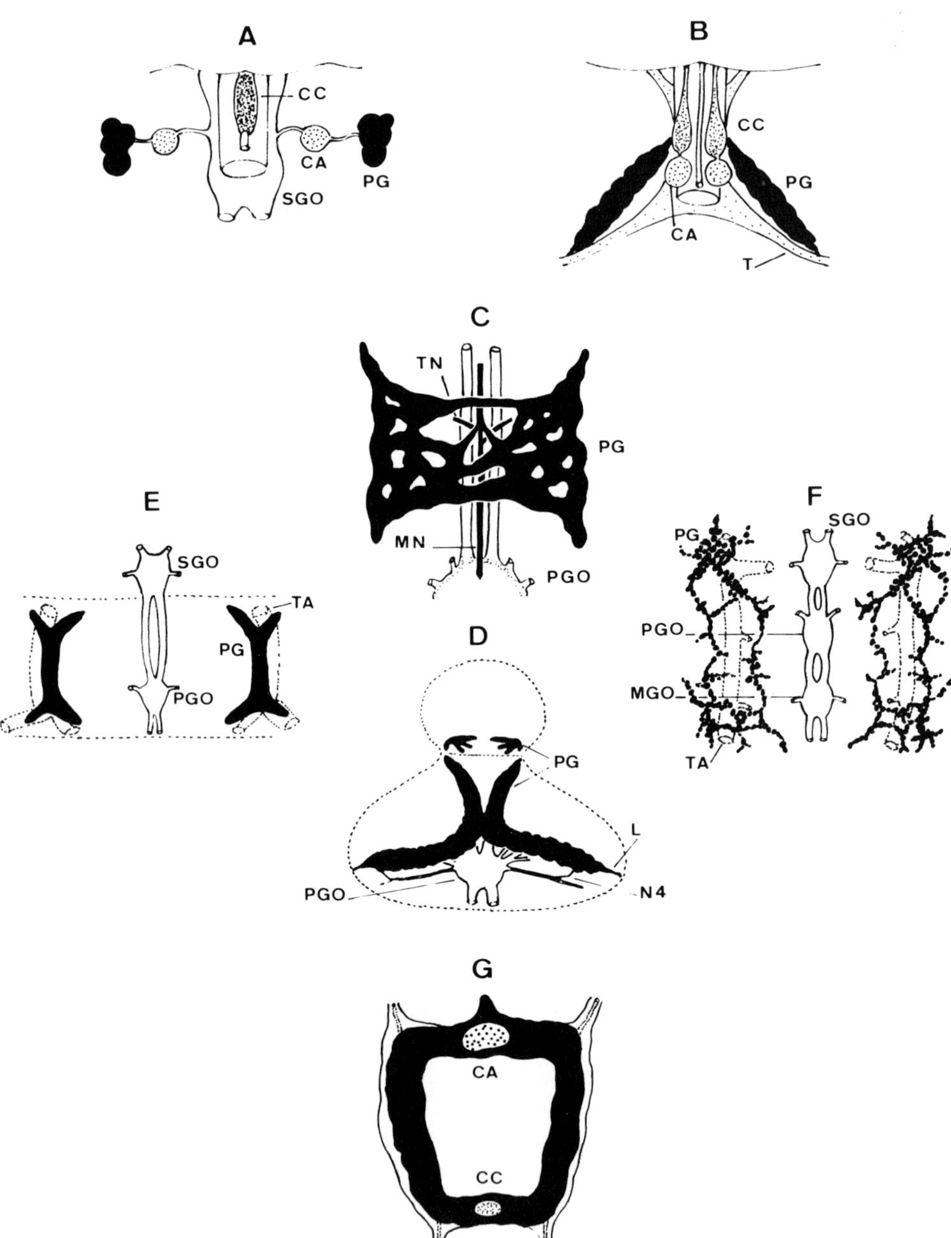

A
CC
CA
SGO
PG

B
CC
PG
CA
T

C
TN
PG
MN
PGO

E
SGO
TA
PG
PGO

F
PG
SGO
PGO
MGO
TA

D
PG
L
PGO
N4

G
CA
CC

part lies externally at the vicinity of the tentorium and is connected to the epidermis by thin strands of connective tissue. Contrary to those of locusts, the homologous paired glands of Phasmida are ventrally located (Pflugfelder, 1938, 1947). For example, in *Bacillus rossii*, each gland occurs longitudinally near the subesophageal ganglion (Boisson, 1950, 1952).

In higher Isoptera (Fig. 7.2B), such as *Reticulitermes*, *Schedorhinotermes* (Rhinotermitidae), *Cubitermes*, and *Macrotermes* (Termitidae), the glands are represented only by the so-called tentorial glands of Jucci (1924), lying ventrolaterally in the head and extending toward the neck (Bernardini-Mosconi, 1959, 1963; Mosconi-Bernardini, 1964; Zuberi and Peeters, 1964; Noirot, 1969; Okot-Kotber, 1980). The gland tissue is attached to muscle fibers considered by Herlant-Meewis and Pasteels (1961) as the dilator muscles of the pharynx.

(c) Unpaired fenestrated type (Fig. 7.2C). In Grylloidea, the glands are found as an unpaired structure located ventrally between the labial region of the head and the prothorax. They normally lie between the esophagus and the ventral nerve cord and form a fenestrated lamina surrounding the esophagus in the neck region (Sellier, 1951). The anterior part of each gland reaches the labial region, whereas the posterior part

FIGURE 7.2. Prothoracic glands and associated organs in nymphal and/or larval instars: (A) Cephalic compact type (Ephemeroptera) showing innervation from the subesophageal ganglion via the corpus allatum (CA). (B) Cephalic ribbon-shaped type (tentorial glands of higher Isoptera). The gland tissue surrounds the pharyngeal dilator muscles attached to the posterior arms of the tentorium (T). (C) Unpaired fenestrated type (*Gryllus bimaculatus*), showing sympathetic innervation from two delicate nerves issuing from the median nerve (MN) near the originating site of the transverse nerves (TN). (From Thomas and Raabe, 1974.) (D) Prothoracic X-shaped type in Dictyoptera: (*Periplaneta americana*), illustrating their innervation from a branch of nerve 4 (N4) arising from the prothoracic ganglion (PGO). Note the cervical or satellite glands located in the head. (From Mosconi-Bernardini, 1964; Richter, 1983.) (E) Prothoracic ribbon-shaped type (Lepidoptera: *Bombyx mori*). Each gland has two anterior arms and two posterior arms spreading over the tracheal trunks. (F) Prothoracic diffuse type (Lepidoptera: *Hyalophora*). (From Herman and Gilbert, 1966.) (G) Annular type (higher Diptera: e.g., *Calliphora*). *Key:* CA = corpus allatum; CC = corpus cardiacum; L = ligament; MN = median nerve; MGO = mesothoracic ganglion; N4 = nerve 4; PG = prothoracic gland; PGO = prothoracic ganglion; SGO = subesophageal ganglion; T = tentorium; TA = trachea; TN = transverse nerve.

ends in the prothorax. They are well demonstrated in *Gryllus, Gryllulus, Pteronemobius* (Sellier, 1951), and *Acheta domestica* (Maleville, 1978a). From an evolutionary standpoint, such glands appear to derive from paired cephalic glands by coalescence and extension of longitudinal bands similar to those of Phasmida.

(d) Prothoracic X-shaped type associated to coxal muscles (Fig. 7.2D).   In Dictyoptera, the paired glands have a ribbon-shaped arrangement forming an X. They are mainly located in the ventral region of the prothoracic segment, lying dorsally on the nerve cord in *Leucophaea maderae* (Scharrer, 1948), *Periplaneta americana* (Bodenstein, 1953; Chadwick, 1956; Mosconi-Bernardini, 1964), *Blabera* (Mosconi-Bernardini, 1964; Brousse-Gaury, 1971a,b) and *Blatella* (Mosconi-Bernardini, 1964; Hazarika and Gupta, 1986). These two bands (Fig. 7.2D) cross anteriorly to the prothoracic ganglion, where they are often connected to each other by a thin strand (Brousse-Gaury, 1971a,b) or a coalescent area (Chadwick, 1956; Bernardini-Mosconi and Cella-Malugani, 1960). Supernumerary branches may occasionally occur (Chadwick, 1956). Additionally, some variation exists in the cephalic extension of the glands among various species of cockroaches. In *Leucophaea* and *Periplaneta,* the anterior part of each gland inserts on the lateral epidermis on both sides of the submentum, whereas in *Blabera* it does not penetrate into the head capsule (Brousse-Gaury, 1971a,b). The posterior part reaches the epidermis in the coxal region (Bodenstein, 1953; Chadwick, 1956).

Additionally, cervical glands have been found in some cockroaches (Mosconi-Bernardini, 1964). They are very likely homologous with the "ventral glands" first described by Pflugfelder (1947) in *Blatta orientalis,* and since then known in *P. americana* and *Blatella germanica,* but not in *Blabera craniifer* (Mosconi-Bernardini, 1964). They must be considered as gland satellite bulks which do not migrate into the prothorax. The presence of such satellite centers (Fig. 7.2D) explains the surprising results of extirpation experiments of the prothoracic glands in *P. americana* (Chadwick, 1956). Because satellite bulks were not removed from the head, the partial extirpations were without effect on molting.

Such a dictyopteran gland type appears to originate from the ribbon-shaped type by a migration of the gland tissue along the coxal muscles, overcrossing in the prothoracic region. Typically, each gland consists of a muscle core containing nerves and tracheae (Scharrer, 1948) and an outer layer of the gland tissue. The same structure is recognized in the satellite centers where the gland tissue surrounds muscle fibers (Mosconi-Bernardini, 1964). Likewise, such an arrangement in two parts is well developed in Mantidea such as *Tenodera sinensis, Stagmomantis carolina*

(Chadwick, 1956), *Orthodera ministralis* (Rae, 1957), and *Mantis religiosa* (Brousse-Gaury, 1969), and also in some Isoptera such as *Kalotermes flavicollis* (Bernardini-Mosconi, 1959; Lüscher, 1960; Herlant-Meewis and Pasteels, 1961; Mosconi-Bernardini, 1964) and *Zootermopsis angusticollis* (Bernardini-Mosconi, 1958, 1959; Gillot and Yin, 1972), for which the two gland parts have been described. By contrast, in a few Isoptera such as *Mastotermes* and *Anacanthotermes*, the prothoracic bands do not cross (Mosconi-Bernardini, 1964).

(e) Prothoracic ribbon-shaped type (Fig. 7.2E). These paired glands are characterized by a flattened band-like structure free of a muscle core, occurring in the prothorax. Anatomically, each gland has a variable, usually branched structure. Such a pleomorphic type is often encountered with a Y-, double-Y-, or H-shaped structure among various Coleoptera (Nunez, 1954; Srivastava, 1958, 1959; de Wilde et al., 1980); Megaloptera (Rahm, 1952a,b), Mecoptera (Schwinck, 1951); in most Lepidoptera such as Pieridac, Papilionidae, and Galleriidae (Lee, 1948; Kaiser, 1949; Lahargue, 1951; Yokoyama, 1956; Bernardini-Mosconi, 1961; Srivastava and Singh, 1968; Karlinsky and Srihari, 1973; Yin and Chippendale, 1975; Sen and Gangrade, 1977; Granger, 1978; Srivastava et al., 1982); some Hymenoptera such as *Myrmica* (Weir, 1959) and *Formica* (Schmidt, 1964); Diptera: Chironomidae (Possompès, 1946) and Culicidae (Rempel and Rueffel, 1964); and Heteroptera such as *Dysdercus* (Wells, 1954). In *Bombyx mori*, for example, each gland (Fig. 7.2E) mainly consists of five parts: a main body from which four branches arise, two anterior arms (anterodorsal and anteroventral arms), and two posterior arms (posterodorsal and posteroventral arms). These branches spread over the tracheal trunks (Bounhiol, 1949; Yokoyama, 1956; Bernardini-Mosconi, 1961), running from the prothoracic spiracles, as in *Tenebrio molitor* where each gland is shaped like the prothoracic tracheal trunks (Romer, 1971). On the other hand, in Curculionidae such as *Pentarthrum* or *Calandra*, the gland is reduced to a main body without branches and is attached to neighboring tracheae (Srivastava, 1959). An unusual arrangement occurs in at least one species of Diptera (Cychorrhapha) in which the ring gland is absent. Indeed, in *Protophormia terrae-novae*, Fraser (1957) has shown that the prothoracic glands are paired structures separated from the aorta and attached to each side of the corpus cardiacum (CC). Finally, it is of interest to note that in mosquito larvae each gland is closely associated with both the CC and the CA (Rempel and Rueffel, 1964).

In Collembola, the presence of paired glands, considered by Lauga-Reyrel (1983) as prothoracic glands, has been recently recognized in *Hypogastrura*. These glands consist of two longitudinal bands enclosed within the fat body and extending from the posterior part of the head to the metathorax. However, in the absence of any experimental evidence this example is still hypothetical.

(f) Prothoracic diffuse type (Fig. 7.2F).   This paired gland type consists of cell clusters that are loosely held together. Such diffuse arrangements of the gland cells occur in Hymenoptera Apidae (Schaller, 1955; Formigoni, 1956); large-sized Lepidoptera such as Sphingidae, Lasiocampidae, and Saturniidae (Lee, 1948; Williams, 1948, 1949; Herman and Gilbert, 1966; Beaulaton, 1968c); Diptera Tipulidae (Pihan, 1966); and some Heteroptera such as *Rhodnius prolixus* (Wigglesworth, 1952a) and *Chrysocoris* (Srivastava and Singh, 1969). They most frequently form bead-like structures (Fig. 7.2F), such as in *Marumba* (Lee, 1948) and *Hyalophora cecropia* (Herman and Gilbert, 1966). Anatomic variations are also known in that type. For example, in *Tipula flavolineata*, the glands have a paired arrangement in larval instar and become unpaired in the pupal instar (Pihan, 1966). As has been previously recognized by Herman (1967), such a diffuse type is encountered only in higher insects. It may improve the exchange gland surface with the hemolymph by comparison with a compact arrangement.

(g) Annular type (Fig. 7.2G).   In higher Diptera (Brachycera), the prothoracic glands typically consist of an unpaired, annular structure surrounding the aorta immediately behind the brain, as in Orthorrhapha: *Tabanus* (Thomsen, 1951; Possompès, 1953), or just above the hemispheres in Cyclorrhapha such as *Calliphora erythrocephala* (Thomsen, 1951; Possompès, 1953), *Drosophila melanogaster* (King et al., 1966) and *Musca domestica* (Ranade, 1964). They include the major part of the so-called ring gland or Weismann's ring. Classically, three regions may be recognized in the ring gland: dorsally the CA, ventrally the CC (except in Orthorrhapha, where there are two CC), and laterally the homologous glands of the "peritracheal glands" (Possompès, 1953). Therefore, it is suggested that the annular prothoracic gland may derive phylogenetically from coalescence of paired glands (as in Culicidae and *Protophormia terrae-novae*) on both sides of the aorta.

7.4.2.   *Sexual Dimorphism*

In most insects, no striking difference exists in the prothoracic glands between males and females during postembryonic development, as for example in *Petrobius maritimus* (Cassier and Fain-Maurel, 1971), *Locusta migratoria* (Fain-Maurel and Cassier, 1968a), and *Diatraea grandiosella* (Yin and Chippendale, 1973). Nevertheless, sexual dimorphism in size has been reported in *Ephemera danica* (Kaiser, 1978). In this species, the gland volumes between females and males are significantly different. The female glands have a higher volume than the male glands, particularly at the last larval instar, where the former are three times more voluminous than the latter.

7.4.3. *Volume and Growth*

During postembryonic development, the prothoracic gland volume displays two types of growth, namely, an allometric growth closely combined with a cyclic growth, as in the CA (Cassier, 1979).

7.4.3.1. ALLOMETRIC GROWTH

The growth of the prothoracic glands has been analyzed during postembryonic development of a few Hemimetabola: *Locusta migratoria* (Strich, 1954; Strich-Halbwachs, 1959), *Macrotermes michaelseni* (Okot-Kotber, 1980, 1982), and *Forficula auricularia* (Lhoste, 1957), as well as Holometabola such as *Pieris brassicae* (Karlinsky and Srihari, 1973) and *Galleria mellonella* (Slama and Mala, 1984). The gland growth is due to a *proliferation of secretory cells, or hyperplasia,* and/or an *increase of cell volume without mitoses, or hypertrophy.* Indeed, in Dermaptera, Lepidoptera, Hymenoptera, and Heteroptera, gland growth occurs only by hypertrophy, unlike that of Paleoptera, Blattopteroidea, and Orthopteroidea, where both hyperplasia and hypertrophy take place. Such gland enlargement is related to global insect growth.

In *L. migratoria*, the gland growth has been compared to the length increase of the third femur (its growth is isometric to that of body length) during all larval instars and at the onset of imaginal life. Strich-Halbwachs (1959) has shown that it is approximately isometric (see Fig. 5 in Strich-Halbwachs, 1959), since the slope is nearly 1 ($b = 0.83$). The maximal size of the glands in *Locusta* is reached prior to imaginal molt (Strich, 1954). Similarly, it has been demonstrated that in *P. brassicae* (Karlinsky and Srihari, 1973) and *G. mellonella* (Slama and Mala, 1984), gland growth and body growth are correlated ($r = 0.66$ in the fourth instar of *Pieris*, and $r = 0.68$ in *Galleria*). In holometabola, the largest gland volume is reached prior to the last larval molt. Changes in the gland volume of *M. michaelseni* have been compared to those of the CA during nymphal development by Okot-Kotber (1980). The author shows that the prothoracic gland volume increases more steadily than that of the CA (see Fig. 7 in Okot-Kotber, 1980) with a 3.5- to 4.5-fold increase in gland volume occurring after the first nymphal molt.

7.4.3.2. CYCLIC GROWTH

It is well known that insect growth proceeds cyclically, achieved in relation to molt cycles. Indeed, the prothoracic glands also show cyclical volume changes during each instar of postembryonic development in Apterygota as well as Pterygota.

In *P. brassicae*, Karlinsky and Srihari (1973) studied the changes in the gland size at 12-h intervals after the second molt to adult ecdysis. At each instar, they demonstrated a gradual increase up to a peak before ecdysis. According to these authors, the increase in the gland size parallels that of hemolymph ecdysteroids. Therefore, they infer that the gland size may be a measure of secretory activity. These results have been confirmed during the penultimate and last larval instars of *Nauphoeta* (Lanzrein, 1975) and *Galleria* (Slama and Mala, 1984), as well as in heavier larvae of *Macrotermes* (Okot-Kotber, 1982). However, in the last larval instar of *Galleria*, gland volume rises less than in *Pieris*.

At the pupal instar of *Pieris*, Karlinsky and Srihari (1973) found a global decrease in size with the lowest level before molting. In an adult instar of *Thermobia domestica*, Watson (1964) reported changes in gland volume. During the first half of intermolt, it increases until a peak at day 5, attaining three times its volume at resting state. This step is followed by a steep decrease until the eighth day. According to Watson (1964), such a cyclic change in gland volume is not generated by a mitotic activity.

7.4.3.3. GROWTH AND POLYMORPHISM

Changes in gland size have been described in relation to the castes of termites and social Hymenoptera as well as the phases of locusts.

7.4.3.3.1. Social Polymorphism

In *Anoplotermes pacificus*, Kaiser (1955, 1956) reported that the gland size varies in larvae according to the caste, as in other higher Isoptera (Jucci, 1956; Zuberi and Peeters, 1964; Noirot, 1969). The glands of future reproductives are larger than those of future workers, attaining twice their volume. In nymphs, the increase of the prothoracic gland is more important than that of the CA (Kaiser, 1955; Noirot, 1969). There is evidence that the prothoracic glands are only maintained in the neuter castes, i.e., workers, soldiers, and pseudergates (Bernardini and Palestra, 1956; Springhetti, 1957; Lüscher, 1960; Noirot, 1969; Grassé, 1982). In the higher termite *Macrotermes michaelseni*, Okot-Kotber (1980) showed that the glands persist in rudimentary form in workers, as well as in soldiers. Striking changes in gland volume have been demonstrated during the development of paired, female, final-instar nymphs adopted by female reproductives (Okot-Kotber, 1982). In heavier nymphs committed to develop into minor presoldiers, the glands—like the CA—become substantially larger than in lighter nymphs that differentiate into workers. Because in heavier nymphs the maximal volume of the CA precedes an increase in the prothoracic gland volume, it is suggested that a prothoracicotropic effect of juvenile hormone (JH) occurs (see Section 7.4.7.4.16, below).

Similarly, glandular dimorphism exists in the castes of Hymenoptera. In *Apis*, Lukoschus (1952) reported that the glands of queen larvae are larger than those of worker larvae at the onset of pupal molt. This fact was planimetrically confirmed by Schaller (1955) on histological sections of the glands in the three castes of *Apis mellifica*. He showed that the gland surface area of developing queens is 2.5 times that of the glands of developing workers and males. Likewise, according to Formigoni (1956), the size of the gland cells is largest in queens reaching 30–35 μm, against 20–25 μm in workers, and 15–18 μm in males.

In histological sections of *Formica polytecna*, Schmidt (1964) calculated the prothoracic gland : CA surface area ratios on days 11–12 of adult development. In males, the mean area of the prothoracic glands is 3.75 times larger than that of the CA. This ratio reaches 3 in females and 5 in workers.

7.4.3.3.2.   Phase Polymorphism

Carlisle and Ellis (1963) in their study on the polymorphism in locusts (*Schistocerca, Locusta*) noted that the fate of the glands is related to phase differences. In the solitary nonmigratory phase, they are larger than in the gregarious migratory phase, like the CA. However, Wilson and Morgan (1978) found similar levels of ecdysteroids between the gregarious and solitary phases of *Schistocerca gregaria*. According to Carlisle and Ellis (1959), the glands persist some 6–8 weeks after metamorphosis in the solitary adults of *Locusta migratoria* and *S. gregaria*, contrary to the gregarious phase. But, Cassier and Fain-Maurel (1968c) reported that the gregarious adults of *L. migratoria* reared (from hatching or after metamorphosis) in short photoperiod (12 h) have the same glands as solitary adults. The glands degenerate only in gregarious locusts maintained in long photoperiod. Additionally, these authors reported that the fate of the glands in solitary locusts is also controlled by humidity levels (Cassier and Fain-Maurel, 1969a). The glands persist only in locusts reared in high relative humidity (green solitary adults). These observations support the interpretation of Wigglesworth (1964, 1970), according to which the solitary adult locusts represent partially neotenic forms. By contrast, in beige solitary locusts reared in low relative humidity, the glands degenerate after adult ecdysis.

7.4.4.   *Tracheal Supply*

In Pterygota, the prothoracic glands are usually supplied with tracheae and tracheoles. This tracheal supply has been reported in several insects, such as *Aeshna cyanea* (Schaller, 1960), *Rhodnius prolixus* (Wigglesworth, 1952a; Gras, 1977), *Dysdercus cingulatus* (Wells, 1954), *Antheraea pernyi* (Beaulaton, 1964a), and *Tipula flavolineata* (Pihan, 1966). The abundance of tracheae and tracheoles supplying the glands appears in relation to their oxygen demand (Wigglesworth, 1954; Gras, 1977; Noirot and Noirot-Timothée, 1982), as was reported in an implanted CA lacking oxygen and attracting the neighboring tracheoles (Wigglesworth, 1954). After injection with cobalt sulfide or cobalt naphthenate, the tracheal supply has been well demonstrated, for example, in *Rhodnius* (Wigglesworth, 1952a, 1954) and *Dysdercus* (Wells, 1954), where a rich tracheal network of small tracheoles surrounds the gland tissue (Gras, 1977). It is similar to that found in the oak silkworm (Beaulaton, 1964a).

7.4.5. *Innervation*

Anatomic study of gland innervation has often been a difficult task, since most nerves are minute and very delicate to dissect, especially in small insects requiring time-consuming histological studies. Lyonet (1762) first mentioned the nerves supplying the prothoracic glands in *Cossus cossus*. Gland innervation has been found in most orders of insects, including representatives of Apterygota. In primitive insects such as Thysanura (Gabe, 1953a), as well as in lower Pterygota (Gabe, 1953b), the glands are connected to the subesophageal ganglion by a pair of nerves. However, in the latter subclass, the anatomic connections vary considerably with the taxonomic position of insects and can be very complex.

Ephemeroptera

In Ephemeroptera, Arvy and Gabe (1953a) showed that the neurosecretory cells lying in the ventral region (toward the caudal third) of the subesophageal ganglion, give rise to the nervi corporis allati. These nerves pass through the CA to supply the glands (Fig. 7.2A) and ramify there.

Odonata

The connections are quite different in Odonata, where the glands are supplied by nerves arising from the subesophageal ganglion. Each nerve seems to contain processes of neurosecretory cells, which are ventrally located in the subesophageal ganglion (Arvy and Gabe, 1952).

Dictyoptera

Some observations in cockroaches have shown that a branch of the third (Scharrer, 1948) or fourth segmental nerve (Chadwick, 1956; Birkenbeil and Agricola, 1980; Hazarika and Gupta, 1986) arises from the prothoracic ganglion and runs to the posterior part of the X-shaped gland. However,

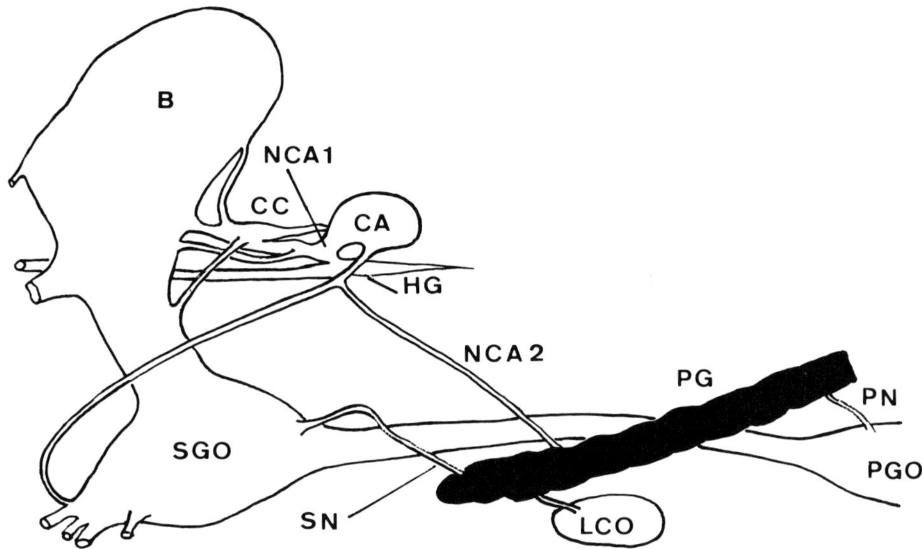

FIGURE 7.3. Innervation of the prothoracic gland of *Blabera fusca* (X-shaped type): lateral view in the last nymphal instar. Relations with the retrocerebral and laterocervical organs. *Key:* B = brain; CA = corpus allatum; CC = corpus cardiacum; HG = hypocerebral ganglion; LCO = laterocervical organ; NCA1 = nervus corporis allati 1; NCA2 = nervus corporis allati 2; PG = prothoracic gland; PGO = prothoracic ganglion; PN = prothoracic nerve of the prothoracic gland; SGO = subesophageal ganglion; SN = subesophageal nerve of the prothoracic gland. (From Brousse-Gaury, 1971a.)

in a careful investigation of the nerve supply of *Blabera fusca* (Fig. 7.3) and *Periplaneta americana*, Brousse-Gaury (1971a, 1972) found that they are connected (a) to the brain and CA by the nervi corporis allati 2 (NCA2), (b) to lateroposterior parts of the subesophageal ganglion by the subesophageal nerves (NSGP), and (c) to the prothoracic ganglion by a pair of prothoracic nerves. However, according to the aforementioned author, there are no direct relationships between the CA and the glands at the NCA2 level. Such a mixed-type nerve, originating from the brain and subesophageal ganglion, contains both cardiac fibers arising from the proto- and deutocerebrum (NCA1) and subesophageal fibers from the subesophageal branch of the NCA2. It is likely that the first two pairs of nerves supplying the prothoracic glands consist of ordinary and neurosecretory fibers. Neurosecretory products have been stained in such nerves with paraldehyde–fuchsin (Brousse-Gaury, 1971a, 1972). By cobalt chloride iontophoresis, Birkenbeil and Agricola (1980) demonstrated that in *P. americana* four neurons laterally located in the prothoracic ganglion give rise to the anterior branch of the fourth segmental nerve supplying each gland (Fig. 7.2D). This nerve enters into the posterior part of the gland. In addition, these glands receive a sensory innervation. In *Periplaneta* and *Blabera*, Brousse-Gaury (1971a) showed that a sensory nerve arises from the laterocervical hair plates and supplies the glands (Fig. 7.3).

In Mantidea, the nervous connections are rather similar to those seen in Blattidea. Brousse-Gaury (1969) described in *Mantis religiosa* three pairs of nerves as follows: (a) a branch of NCA2; (b) a subesophageal nerve (third tractus according to Brousse-Gaury, 1969) issuing posteriorly from the subesophageal ganglion; and (c) a prothoracic nerve (fifth tractus) arising laterally from the prothoracic ganglion. As in *Periplaneta* and *Blabera*, a pair of sensory nerves (fourth tractus, according to Brousse-Gaury) arises from the laterocervical hair plates and runs to the glands.

Plecoptera

The only description known to me of the nerve supply in plecopterous glands is that of Arvy and Gabe (1953c) in *Brachyptera risi*. They clearly showed that paired nerves issuing from the laterocerebral parts of the subesophageal ganglion supply the glands and ramify inside.

Grylloblattodea

No nerve supply of the prothoracic glands was observed in *Grylloblatta campodeiformis* (Rae and O'Farrel, 1959).

Phasmida

In *Bacillus rossii*, Boisson (1952) described a pair of nerves emerging from the subesophageal ganglion and supplying the glands laterally (see Fig. 4 in Boisson, 1952).

Orthoptera

In Orthoptera, the nerve supply of the glands is well established, especially in Gryllidae, consisting of two branches of an unpaired median nerve arising from the prothoracic ganglion (Sellier, 1951). According to Thomas and Raabe (1974), this sympathetic nerve emerges posteriorly from the subesophageal ganglion and connects anteriorly to the prothoracic ganglion. It runs between the interganglionic connectives and gives off a pair of sympathetic transverse nerves. In *Gryllus bimaculatus*, two delicate sympathetic nerves arise from the site of the branching point and join the gland anteriorly (Fig. 7.2C). Additionally, this gland receives a second pair of delicate nerves issuing dorsally from the subesophageal ganglion (Thomas and Raabe, 1974). Note that in *Gryllodes sigillatus* (Nandchahal, 1967), as in *Periplaneta* and *Blabera* (Brousse-Gaury, 1971a, 1972), the gland is supplied by a ramification of the nerve connecting the CA and the subesophageal ganglion.

In the acridian *Stenobothrus rufipes*, Pflugfelder (1947) identified only a pair of nerves emerging laterodorsally from the subesophageal ganglion. Surprisingly, in *Locusta migratoria*, no innervation of the glands was detected by several investigators (e.g., Strich, 1954; Strich-Halbwachs, 1959; Fain-Maurel and Cassier, 1968a), except for Thomas and Raabe (1974), who reported that the prothoracic perisympathetic organ (Chalaye, 1967) supplies the gland. According to Thomas and Raabe (1974), there is anatomic evidence that a delicate branch of the transverse nerve reaches the gland.

Dermaptera

Lhoste (1957) has noted only that the glands of *Forficula* are connected to the subesophageal ganglion by a pair of nerves. However, according to Raabe (1972), a long median nerve arising from the prothoracic ganglion divides into two branches, innervating the glands. She has shown that the paired prothoracic perisympathetic organs are also associated with these gland nerves.

Coleoptera

Arvy and Gabe (1953b) have identified the anterior nerve supply of the glands of *Tenebrio molitor*. It consists of a pair of nerves issuing laterally from the subesophageal ganglion and containing processes of the neurosecretory cells located in two median groups. However, according to Romer (1971), the glands receive three pairs of nerves: (a) a first pair of subesophageal nerves (previously described by Arvy and Gabe, 1953b); (b) a second pair arising by branching from the connectives between the subesophageal and prothoracic ganglia; and (c) a third pair arising lateroanteriorly from the prothoracic ganglion.

Lepidoptera

The gland innervation is well documented in Lepidoptera in at least 20 species representing different families (Lee, 1948; Yokoyama, 1956; Herman and Gilbert, 1966; Beaulaton, 1968a,c; Srivastava and Singh, 1968; Karlinsky and Srihari, 1973; Yin and Chippendale, 1973; Singh, 1975; Srivastava et al., 1977, 1982; Granger, 1978; Singh and Sehnal, 1979). It seems established that the glands are innervated from the subesophageal, prothoracic, and mesothoracic ganglia, except in *Diatraea grandiosella* (Yin and Chippendale, 1973), in which the subesophageal ganglion does not supply these organs, contrary to other pyralids previously described by Lee (1948). The pattern of gland innervation may vary in different families (Lee, 1948), but also in the same species, such as *Pieris brassicae* (Karlinsky and Srihari, 1973) and *Galleria mellonella* (Granger, 1978; Singh and Sehnal, 1979). These authors note a certain degree of variability in the pathways of innervation. Accordingly, it is difficult to establish a general,

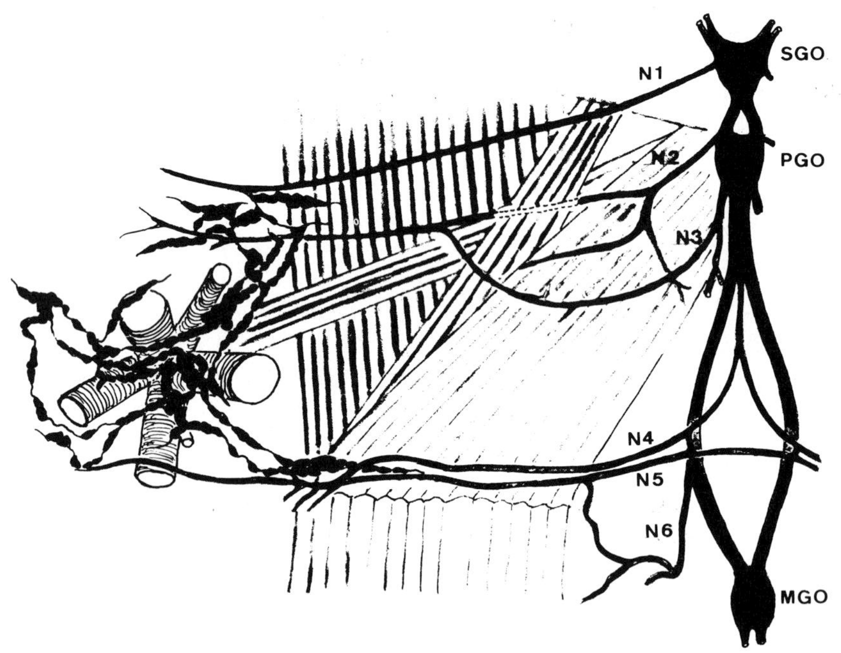

FIGURE 7.4. Innervation of the prothoracic gland of *Antheraea pernyi* (diffuse type) in the last larval instar after methylene blue nerve staining. *Key:* MGO = mesothoracic ganglion; N1–N6 = nerves of the prothoracic gland; PGO = prothoracic ganglion; SGO = subesophageal ganglion. (From Beaulaton, 1968c.)

definite outline of distribution. Nevertheless, we propose a tentative scheme for innervation. Each gland (Fig. 7.4) appears to be supplied by the following: (a) the first nerve (N1), a branch of the cervical nerve arising laterally from the subesophageal ganglion; (b) the N2 issuing from the interganglionic connectives between the subesophageal and prothoracic ganglia (N2 is designated as "nervus prothoracalis anterior" by Singh and Sehnal, 1979); (c) the N3 emerging posteriorly from the prothoracic ganglion and customarily joining the N2, forming a common nerve which supplies the anterior part of the gland as in *Antheraea pernyi* (Beaulaton, 1968c), or the middle part of the gland as in *Papilio demoleus* (Srivastava et al., 1977) or *G. mellonella* (Singh and Sehnal, 1979); (d) the transverse nerve, or N4 (sympathetic transverse nerve), a branch of the median nerve issued from the prothoracic ganglion between the interganglionic connectives; (e) the N5 arising from the interganglionic connective between the pro- and mesothoracic ganglia near the latter ganglion [frequently, this nerve joins the N4 before supplying the posterior part of the gland (Lee, 1948; Yokoyama, 1956; Herman and Gilbert, 1966; Beaulaton, 1968c; Srivastava et al., 1977; Singh and Sehnal, 1979)]; and (f) a branch of the transverse nerve, or N6 (perisympathetic nerve), emerging posteriorly from the mesothoracic ganglion [N6 has been reported only in pupae of *Hyalophora cecropia* (Herman and Gilbert, 1966)]. Some of the six nerves may be absent in the gland, such as N1, N3, and N6 in *Diatraea grandiosella* (Yin and Chippendale, 1973).

Diptera

According to Possompès (1953), in *Chironomus plumosus*, the nervus corporis allati passes through each gland but does not enter the connective sheath. However, it may be noted that each gland is medially closely associated with the CC and perhaps directly receives via this connection a nerve supply with neurosecretory products. In *Ptychoptera* and *Tipula*, no innervation of the gland has been found by Thomsen (1951) and Pihan (1966). In the Brachycera (*Tabanus*, *Thereva*, *Drosophila*, and *Calliphora*), each gland is very closely associated with the CC and CA in the ring gland (Thomsen, 1951; Possompès, 1953; King et al., 1966; Aggarwal and King, 1969). However, no specific innervation of the gland was reported by the authors.

Heteroptera

Ewen (1962) found that the gland of *Adelphocoris lineolatus* receives a pair of nerves arising from the subesophageal–prothoracic ganglion. Each nerve ramifies along the gland. However, no innervation of the glands has been described in 12 species of Heteroptera so far studied (Wigglesworth, 1952a; Wells, 1954; Srivastava and Singh, 1969; Dorn and Romer, 1976).

Psocoptera

In various species of Psocoptera, Badonnel (1970) noted that the glands are apparently connected to the prothoracic ganglion by a pair of nerves arising from the anterior part.

In summary, these anatomic data show that gland innervation has evolved in different ways in the various orders of insects. A primitive type occurs in Apterygota, Paleoptera, Plecoptera, Phasmida, and Dermaptera, with a nerve supply arising only from the subesophageal ganglion. An intermediate type is represented by innervation from the subesophageal and prothoracic ganglia, as in Orthoptera (including sympathetic nerves), Heteroptera, and Coleoptera. A more complex type is found in Dictyoptera (Blattidea and Mantidea) and contains, in addition to the aforementioned connections, those with the brain, the CA, and the laterocervical organs. The most complex type occurs in Lepidoptera, where the glands are innervated from the subesophageal, prothoracic, and mesothoracic ganglia. However, in several orders, such as Mecoptera, Trichoptera, and Hymenoptera, we as yet have no morphological data on the subject.

7.4.6. Histology

7.4.6.1. HISTOLOGICAL STRUCTURE AND HISTOLOGICAL TYPES

In all insect species examined until now, the prothoracic glands consist of only one secretory cell type, which is associated with tracheal end cells (see Section 7.4.7.2, below), interstitial cells, glial cells, neurosecretory fibers (see Section 7.4.7.3, below), and adherent hemocytes such as granulocytes and spherule cells (Beaulaton, 1968b). In some cases, the glands may adhere closely to neighboring tissues such as the salivary gland system and the thoracic fat body in Heteroptera (Wigglesworth, 1952a; Wells, 1954; Srivastava and Singh, 1969), the skeletal muscles and nerves in Blattidea (Scharrer, 1948; Chadwick, 1956; Mosconi-Bernardini, 1964), Mantidea (Rae, 1957; Brousse-Gaury, 1969) and Isoptera (Bernardini-Mosconi, 1959; Mosconi-Bernardini, 1964; Noirot, 1969; Gillot and Yin, 1972; Okot-Kotber, 1980), or the CA and CC in some Diptera (Possompès, 1953; Rempel and Rueffel, 1964; Ranade, 1964; King et al., 1966; Aggarwal and King, 1969).

The following three types of histological structure can be distinguished from the available literature on prothoracic glands:

(a) Compact epithelioid type with small cells. This type consists of small cells densely packed together with an individual size of some 1–25 μm, depending on larval instars and physiological state. The tissue is usually compact and frequently comprises several cell layers. It has been described in Apterygota (Gabe, 1953a; Watson, 1964; Cassier and Fain-Maurel, 1971), Paleoptera (Cazal, 1947; Arvy and Gabe, 1950, 1952); Dictyoptera (Scharrer, 1948, 1964), Isoptera (Gillot and Yin, 1972), Plecoptera (Arvy and Gabe, 1953c), Phasmida (Boisson, 1952), and Orthoptera (Sellier, 1951; Strich, 1954; Fain-Maurel and Cassier, 1968a; Maleville, 1978a,b), and in Dermaptera beyond the first two nymphal instars (Lhoste, 1957), Coleoptera (Nunez, 1954; Srivastava, 1959, 1960; Glitho, 1977; Chu et al., 1980; de Wilde et al., 1980).

The resting gland cells have smoothly rounded or ovoid nuclei (6–20 μm in diameter) with a scanty, highly basophilic cytoplasm. Nevertheless, in the glands of Orthoptera, investigators have recognized two or three size classes of nuclei. In Gryllidae, Sellier (1951) reported the presence of giant nuclei next to small nuclei, as in *Stenobothrus rufipes* (Pflugfelder, 1947). This fact was confirmed by Strich (1954) in *Locusta* (three types of nuclei, 8–10, 16, and 20–29 μm in diameter, respectively) and by Maleville (1978a,b) in *Acheta domestica*. The latter author showed that the giant cells contain a polytene nucleus. In brief, the glands of Orthoptera display intermediate features between the compact and the giant cell epithelioid type (see below).

(b) Vesicular epithelioid type. Lhoste (1957) reported a vesicular structure in the glands of *Forficula auricularia* during the first two nymphal instars. This type is characterized by one layer of gland cells surrounding a central lumen and contains about 100 cells per gland in *Anisolabis maritima* (Ozeki, 1968). Because the embryonic development of the Dermaptera gland includes a vesicular stage (see Section 7.3.1, above), such a transitory structure probably represents a residual embryonic arrangement.

(c) Giant cell epithelioid type. This gland type consists of large, associated secretory cells beneath the tunica propria. They are often arranged in bands, strings, or clusters. This type is well known in higher Pterygota: Lepidoptera (Kaiser, 1949; Ichikawa et al., 1955; Beaulaton, 1961; Herman and Gilbert, 1966; Mala et al., 1974; Wolbert, 1979), Heteroptera (Wells, 1954; Dorn and Romer, 1976; Smith and Nijhout, 1982a), and at least in higher Diptera (Thomsen, 1951; Possompès, 1953; Fraser, 1957; King et al., 1966; Aggarwal and King, 1969). Only a few histological studies on the hymenopterous glands have indicated the presence of large cells (Schaller, 1955; Formigoni, 1956; Weir, 1959; Schmidt, 1964). Generally, the giant cell epithelioid type contains fewer cells than the compact epithelioid type. Each gland contains about 33 cells in *Diatraea grandiosella* (Yin and Chippendale, 1975); 40–70 cells in *Galleria mellonella* (Mala et al., 1974; Blazsek et al., 1975); 51–62 cells in *Spodoptera litura* (Sen and Gangrade, 1977); 60 cells in *Drosophila melanogaster* (Poulson, 1945; King et al., 1966); 150 cells in *Oncopeltus fasciatus* (Smith and Nijhout, 1982a); and 242 cells in the giant silkworm, *Hyalophora cecropia* (Herman and Gilbert, 1966). In contrast, the compact gland type comprises 800–1000 cells in *Tenebrio molitor* (Srivastava, 1959), and presumably even more cells in Dictyoptera and Odonata.

The giant cell epithelioid type may be considered as the most specialized and evolved. It is of interest to note that a similar histological type has been recognized in the CA of Holometabola (Cassier, 1979).

7.4.6.2. HISTOLOGICAL CHANGES: MITOSIS, ENDOMITOSIS, GROWTH, AND DEGENERATION

One of the most characteristic histological features of glands of the small cell type is their mitotic activity. In contrast, the giant cell type is generally recognized to have lost its capacity for mitosis. Surprisingly, however, the case of apterygotes remains unclear because no mitosis has been reported in their compact glands (Gabe, 1953a; Watson, 1963, 1964; Bareth, 1964; Lauga-Reyrel, 1983).

7.4.6.2.1.   Mitoses

Mitoses were found in the compact epithelioid glands of Hemimetabola such as Odonata (Cazal, 1947; Schaller, 1960), Dictyoptera (Scharrer, 1948; Bodenstein, 1953; Rae, 1957), Plecoptera (Arvy and Gabe, 1953c), Phasmida (Boisson, 1952), and Orthoptera (Pflugfelder, 1947; Sellier, 1951; Strich, 1954; Strich-Halbwachs, 1959; Fain-Maurel and Cassier, 1968a; Maleville, 1975, 1978a). In *Locusta migratoria*, the mitotic activity of the glands has been analyzed during the fourth and fifth nymphal instars (Strich, 1954; Strich-Halbwachs, 1959). In both instars, the mitotic rate increased steadily after ecdysis, reaching the highest peak late in the second day (at 43–44 h). At that time, the rates were, respectively, 4.1% and 5%. After that there was a slow decline until the fifth day (level 0). Strich-Halbwachs (1959) also showed that the maximum mitotic rate precedes an increase of that of the epidermal cells. In both instars, the mitotic rate of epidermal cells peaked about 2 days after that of the glands. Maleville (1978a) studied the mitotic rate in small gland nuclei of *Acheta domestica* during the penultimate nymphal instar at 24-h intervals. She found two peaks, the first occurring 24 h after ecdysis and the second 8 days later (about 12 h before the next molt). According to Maleville (1978a), the two mitotic crises were not related to the hemolymph ecdysteroid level, which remained very low during the first day and the ninth day of the penultimate instar. Similar to *Stenobothrus rufipes* (Pflugfelder, 1947) and *Gryllus campestris* (Sellier, 1951), the glands of *Acheta* contain some giant cells exhibiting endomitoses (Maleville, 1978a).

7.4.6.2.2.   Polyploidy and Polyteny

Histological changes in the glands of the giant cell type were found in some species of Lepidoptera, Diptera, and Heteroptera. In the gland cells of *Oncopeltus fasciatus*, Dorn and Romer (1976) have demonstrated cytophotometrically a polyploidization process during the last nymphal instar. Estimations of the degree of polyploidy derived from the diploid nuclei of ganglion cells revealed a gland cell DNA peak of 128 C (Fig. 7.5). On the other hand, wild type larvae of *Drosophila melanogaster* contained 64-C DNA at the 96- and 120-h stages following oviposition (Welch, 1957). Oberlander et al. (1965) showed that the gland cells of *Philosamia cynthia* and *Antheraea polyphemus* synthesized DNA during the third, fourth, and fifth larval instars, whereas such syntheses were absent in developing adults. Certainly, such DNA synthesis in the absence of cell division reflects an increase in the level of ploidy. This activity is associated with a

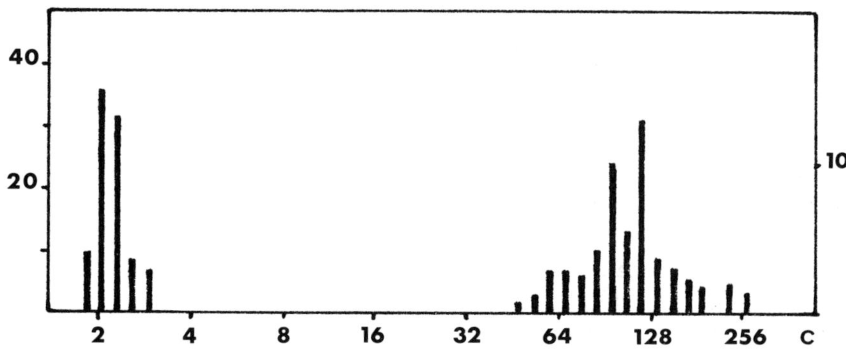

FIGURE 7.5. Distribution of DNA content of the prothoracic gland nuclei in *Oncopeltus fasciatus*. Comparison between embryonic glands (on the left) and nymphal glands of the 5th instar (on the right). Cytophotometric data; in abscissa: estimates of degrees of polyploidy. (From Dorn and Romer, 1976.)

large increase of cell volume: at least 300 times in *Drosophila melanogaster* (Aggarwal and King, 1969) and 700 times in *Pieris* from the first larval instar to pupae (Kaiser, 1949). However, Aggarwal and King (1969) reported mitoses in the prothoracic gland of *Drosophila*, occurring only during the first larval instar. After the first larval molt, these cells grow without division, as in all species with a giant cell type gland. After several replications, polytene chromosomes of *Drosophila* become visible during the late third instar and persist through the prepupal instar (King et al., 1966; Aggarwal and King, 1969). Polyploid cells are described in all other works published to date on that gland type, as for example in *Panorpa communis* (Schwinck, 1951), *Cerura vinula* (Hintze-Podufal, 1971), *Galleria mellonella* (Blazsek et al., 1975; Gersch et al., 1975), and *Spodoptera littoralis* (Zimowska et al., 1985), as well as in Heteroptera.

7.4.6.2.3. Degeneration

The degeneration of the glands of Hemimetabola at metamorphosis has commonly been observed. Numerous examples of gland involution are known in Paleoptera (Pflugfelder, 1947; Arvy and Gabe, 1950, 1952; Schaller, 1960), Dictyoptera (Scharrer, 1948; Bodenstein, 1953; Mosconi-Bernardini, 1966), Plecoptera (Pflugfelder, 1947; Arvy and Gabe, 1953c), Phasmida (Boisson, 1952), Orthoptera (Pflugfelder, 1947; Sellier, 1951; Strich, 1954; Strich-Halbwachs, 1959; Maleville, 1973), Dermaptera (Ozeki, 1968), Heteroptera (Wigglesworth, 1952a, 1955; Wells, 1954; Gras, 1977), and primary and supplementary reproductives of Isoptera (Pflugfelder, 1947; Springhetti, 1957; Lüscher, 1960; Herlant-Meewis and

Pasteels, 1961; Zuberi and Peeters, 1964; Lebrun, 1967; Noirot, 1969; Gillot and Yin, 1972). In *Leucophaea maderae* and *Blatella germanica*, the autolysis begins during the first hours after adult molt (Scharrer, 1948, 1966; Hazarika and Gupta, 1986). By 24 h after ecdysis, clusters of pycnotic nuclei are identified near unaltered cells. As in other cases of programmed cell death, the degeneration proceeds asynchronously, continuing for the next 4 or 5 days. Scharrer (1948, 1966) observed that such an involution was completed within 8 days. In contast, Mosconi-Bernardini (1966) reported that in the same species (*L. maderae*) the process persisted until 16 days post-ecdysis in males and 18–20 days in females. The latter histological observations are consistent with the physiological results of Richter (1984), showing that the ecdysteroid production *in vitro* decreases after ecdysis and remains at a low level up to 10 days post-ecdysis. Thereafter, from 10 to 15 days post-ecdysis, the ecdysteroid level decreases to zero.

Interestingly, the glands of Heteroptera degenerate promptly in the adult. For example, the first signs of collapse appear soon after the final molt in *Rhodnius prolixus* (Wigglesworth, 1952a, 1955; Gras, 1977) and in *Dysdercus cingulatus* (Wells, 1954), and disintegration is complete 48–50 h later. By contrast, in *Oncopeltus fasciatus* (Smith and Nijhout, 1982a) and in *Kalotermes* (Herlant-Meewis and Pasteels, 1961), the glands completely collapsed at adult ecdysis. Similarly, Maleville (1973) observed in *Acheta* that the first signs of cell death, such as the appearance of pycnotic nuclei, occur prior to adult ecdysis. By 24 h after emergence, the glands have completely disappeared.

As a rule, the glands of Holometabola degenerate precociously during the pharate adult instar, as in Coleoptera (Nunez, 1954; Srivastava, 1960; Glitho, 1977), Lepidoptera (Williams, 1948; Ichikawa et al., 1955; Herman and Gilbert, 1966; Mala et al., 1974; Srivastava et al., 1982), Diptera (Possompès, 1953; Burgess, 1967) and Hymenoptera (Formigoni, 1956; Schmidt, 1964). During the adult development of *Hyalophora cecropia* (21 days), Herman and Gilbert (1966) found pycnotic nuclei and nuclear fragments in the glands at day 12. Nuclear residues covered with hemocytes were seen before emergence. Generally, the gland cells are completely lost just after emergence, as in *Papilio demoleus* (Srivastava et al., 1982). A more precocious degeneration is known in the glands of *Tenebrio molitor* (Glitho et al., 1979), where the cells disappear prior to the ecdysteroid burst triggering adult development; Glitho et al. suggest that ecdysone is produced from another source once the gland cells are lost. In many species of Hemimetabola, as well in Holometabola, involution of the different gland cells has been recognized to proceed asynchronously. Fain-Maurel and Cassier (1969) reported a dorsoventral gradient of degeneration in *Locusta*.

One might ask why the prothoracic glands of Pterygota degenerate, except those of soldiers and workers of termites (Lüscher, 1960; Herlant-Meewis and Pasteels, 1961; Lebrun, 1967; Okot-Kotber, 1980) and solitary locusts (see Section 7.4.3.3, above). It has long been known that the glands are retained by the implantation of the active CA into supernumerary nymphs (Wigglesworth, 1952a) and by topical application or injection of JH or JH analogues (Gilbert, 1962; Lanzrein and Lüscher, 1970; Smith and Nijhout, 1982b; Hazarika and Gupta, 1979). In addition, there is evidence that the CA of solitary locusts are hypertrophic (see the review by Cassier, 1979). The role of JH was demonstrated by implantation of one pair of the active CA (from males) into fifth-instar gregarious nymphs reared in conditions amenable for gland breakdown (Cassier and Fain-Maurel, 1970). Thus, the presence of active CA and therefore of JH for the maintenance of the prothoracic glands has clearly been confirmed. In gregarious locusts, as in nonpolymorphic Pterygota, the gland degeneration appears to be programmed by the absence of JH. Lanzrein (1975) demonstrated in *Nauphoeta cinerea* that the gland collapse begins some 6–7 days before adult ecdysis. Her transplantation experiments have suggested the existence of degeneration-inducing factors in nymphs. Lanzrein suggests that PTTH as well as the absence of JH from the sixth day are important degeneration-inducing factors. On the other hand, she agrees with Bodenstein (1953) that protecting factors are released from the CC. Recent results have shown that the glands of *Oncopeltus* cultured *in vitro* on the second day of the fifth instar may be induced to degenerate after 24-h exposure to 20-hydroxyecdysone, with doses similar to those previously used for other tissues (Smith and Nijhout, 1983). The authors show that by day 3 a fall of JH initiates cell death of the glands, whereas a rise of JH prevents their degeneration. Undoubtedly, the gland involution is initiated by hormonal events and especially by low titers of JH and rising concentrations of 20-hydroxyecdysone.

7.4.7.   *Cytology, Ultrastructure, Cytochemistry, and Cell Physiology*

7.4.7.1.   TUNICA PROPRIA AND ATTACHMENT STRANDS

The prothoracic gland cells are always separated from the hemolymph by the acellular tunica propria (Beaulaton, 1964a,b, 1968a,b; Romer, 1971; Gersch et al., 1975; Beaulaton et al., 1984) as in all organs of insects (Ashhurst, 1968, 1982; François, 1976). This extracellular matrix has been referred to by various terms: connective tissue sheath (Scharrer, 1964; Whitten, 1964; Hintze-Podufal, 1971), stroma (Normann, 1965; Scharrer,

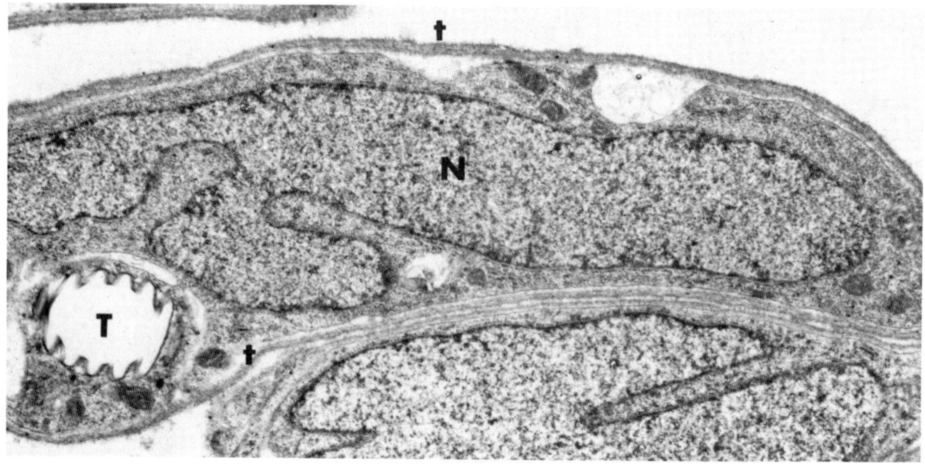

FIGURE 7.6. Prothoracic gland of *Rhodnius prolixus* in the fourth nymphal instar (fasting period). View of two cells in resting state, showing a high nucleocytoplasmic ratio. *Key:* N = nucleus; t = tunica propria; T = tracheole. ×18,000. (Courtesy of Dr R. Gras.)

Key to abbreviations in Figs. 7.6–7.20: ax = axon; cp = coated pit; FB = fat body; g = glial cytoplasm; G = the Golgi apparatus; gl = glycogen; hd = hemidesmosome; is = interstitial space; L = lysosome; m = mitochondrion; mt = microtubule; N = nucleus; n = nucleolus; p = cytoplasmic projection; r = ribosomes; RER = rough endoplasmic reticulum; SER = smooth endoplasmic reticulum; t = tunica propria; T = tracheole.

1966), acellular tunic (Aggarwal and King, 1969), basement membrane (Blazsek et al., 1975; Dorn and Romer, 1976; McDaniel et al., 1976; Glitho et al., 1979; Smith and Nijhout, 1982a), basal membrane (Cassier and Fain-Maurel, 1971; Romer, 1971), basal lamina (Sedlak et al., 1983), covering membrane (Ura, 1983), and peripheral or outer membrane (Beaulaton, 1961; Herman, 1967). It has sometimes been regarded as an outer layer of the cell coat (Langer, 1978). However, in the glands, the tunica propria is always continuous with the basal lamina of the tracheae, nerves, or muscles connected with them (Figs. 7.6 and 7.7A).

In *Antheraea pernyi*, the tunica propria especially thickens, particularly during advanced larval development (Beaulaton, 1968b). Just after hatching, it appears as a thin layer, about 100 nm thick or less, as in *Leucophaea* (Scharrer, 1964). During the next instars, the tunica propria progressively increases in thickness, attaining about 700 nm at the last nymphal instar, as in *Rhodnius* (Gras, 1977). There are also variations in matrix

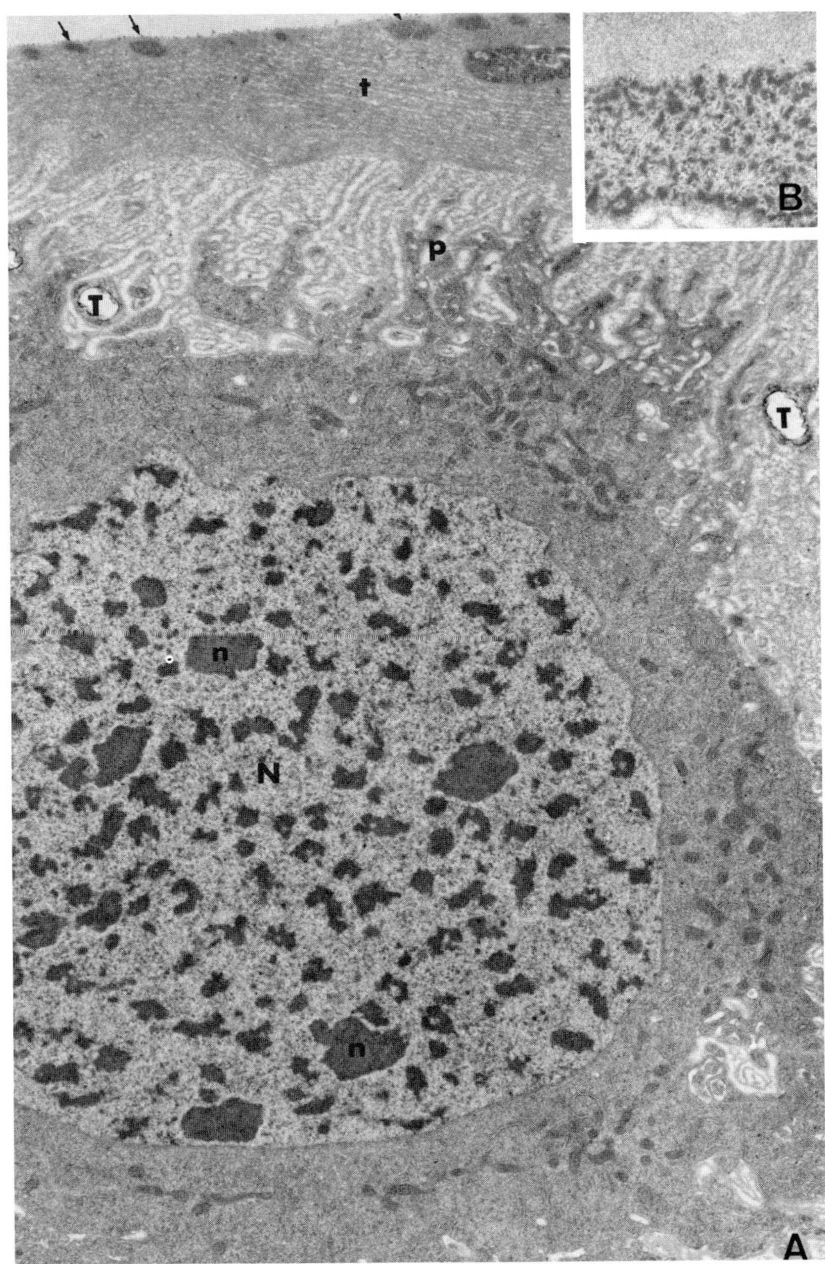

FIGURE 7.7. (A) Low magnification micrograph of a portion of one prothoracic gland cell in *Bombyx mori* in the fifth larval instar. The multilayered tunica propria (t) contains few dense fibrils (arrows) in the peripheral area. Note that the interstitial spaces between the cytoplasmic projections (p) of the plasma membrane are partially occupied with loose layers of the tunica propria (t) and tracheoles (T). *Key:* N = nucleus; n = nucleolus. ×7392. (B) Detail of the tunica propria in relation to an attachment strand. Note the presence of a network of dense fibrils. ×15,840.

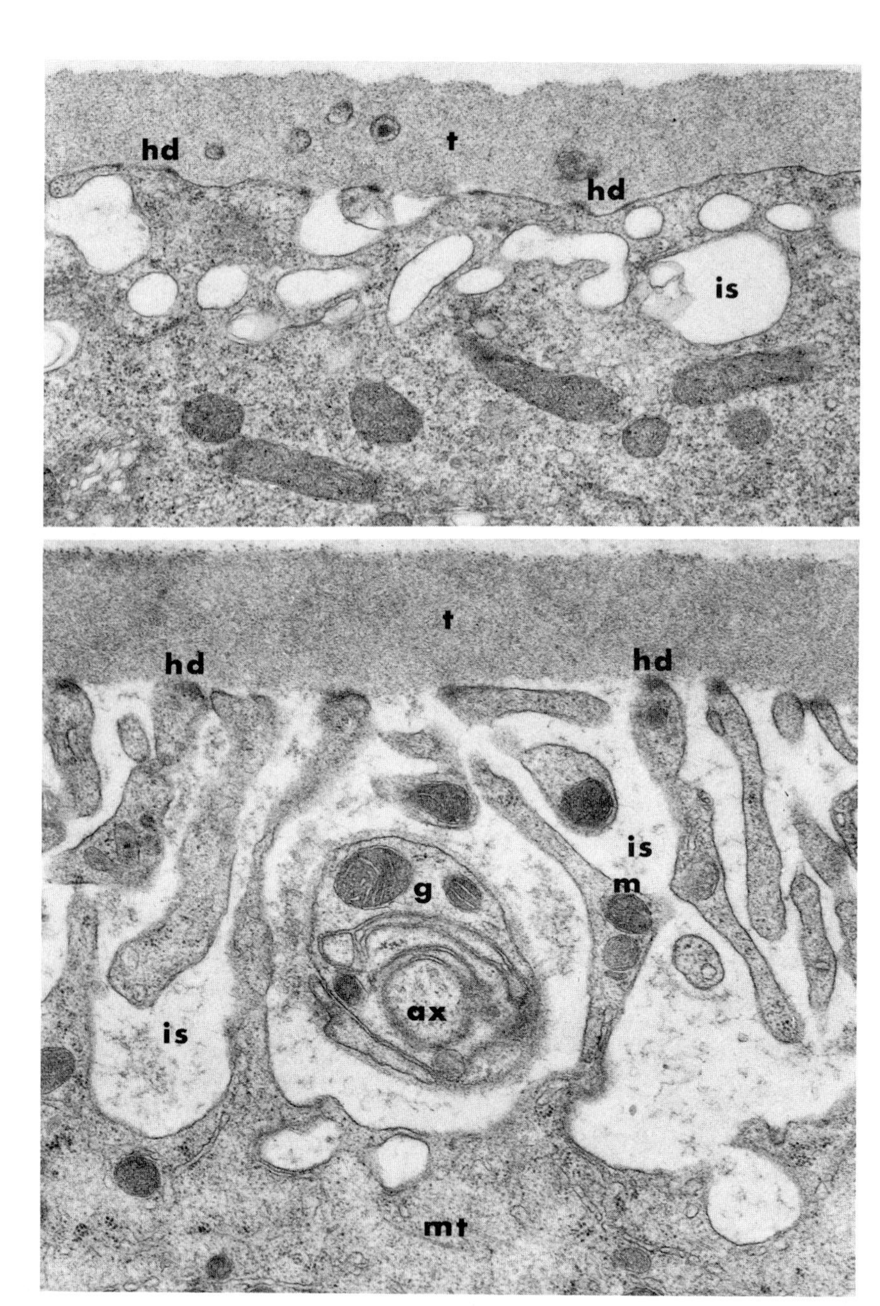
hd
t
hd
is
t
hd
hd
is
m
g
ax
is
mt

thickness in various species: 50–200 nm in *Tenebrio* (Romer, 1971); 100–250 nm in *Oncopeltus* (Herman, 1967; Dorn and Romer, 1976); 200–700 nm in *Rhodnius* (Gras, 1977); and 2–3 μm in *Bombyx mori* (Fig. 7.7A) or *Saturnia pyri* (Beaulaton, 1968b).

Interestingly, the tunica propria thickens especially during periods of inactivity (5–6 μm in diapausing pupae of *Philosamia cynthia*: Beaulaton, 1961) and invaginates along the cytoplasmic processes of gland cells in *Hyalophora* (Herman, 1967; McDaniel et al., 1976) or *Philosamia*. During that physiological state, the interstitial spaces between invaginations of the plasma membrane are completely obliterated by an ultrastructurally rearranged extracellular matrix. Consequently, its permeability may change as in pathological basement membranes. In addition, McDaniel et al. (1976) showed that the tunica propria of *H. cecropia* is disrupted during early phase of adult development.

The tunica propria often consists of two layers (as in *Antheraea*: Beaulaton, 1968b): a thin inner coat, the lamina rara, or lamina lucida, loosely linked to the outer surface of the plasma membrane; and an outer layer, the external lamina, or lamina densa. The latter may be subdivided into two or three layers in *Petrobius* (Cassier and Fain-Maurel, 1971), one to four layers in *Rhodnius* (Gras, 1977), five to six layers in *Locusta* (Fain-Maurel and Cassier, 1968a), or more in *Bombyx* (Fig. 7.7A). Frequently, one can recognize within large interstitial spaces, the lamina rara (30–60 nm thick in *Antheraea*), which is irregularly attached to the plasma membrane and separated from the lamina densa. However, the distinction between these two layers was not found in all the cells. Alkaline phosphatase activity has been revealed in *Antheraea pernyi* at these 200- to 250-nm-long

FIGURES 7.8. & 7.9. Peripheral portions of prothoracic gland cells in *Antheraea pernyi*, illustrating the enlargement of the interstitial spaces in relation with secretory activity: *Fig. 7.8(top)*—First day after ecdysis (fourth larval instar). The greater part of the plasma membrane is applied to the tunica propria (t) and closely associated at the level of hemidesmosomes (hd). There are few interstitial spaces (is) consisting of relatively narrow channels. ×20,470. *Fig. 7.9(bottom)*—First day of spinning period (fifth instar). The plasma membrane is deeply invaginated and appears only in contact with the tunica propria (t) at the level of hemidesmosomes (hd). The interstitial spaces (is) expand and consist of interconnecting channels. Note the presence of mitochondria (m) within the sinuous polymorphic projections of the plasma membrane. An interstitial space contains a small axon (ax) devoid of neurosecretory granules and ensheathed by several layers of glial cytoplasm (g). ×20,470.

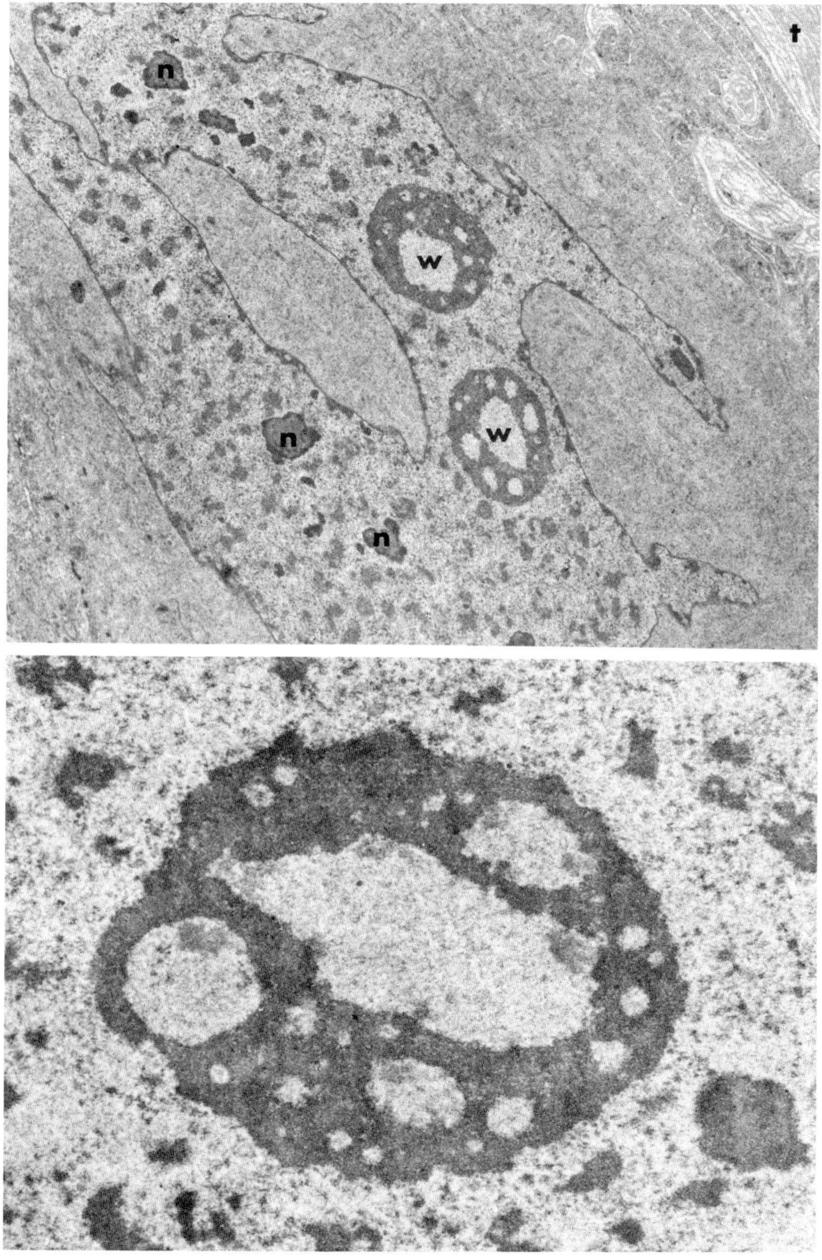

FIGURE 7.10. Prothoracic gland cell in *Bombyx mori* (female fourth larval instar). Portion of a nucleus containing three nucleoli (n) and two large chromocenters showing an alveolar structure. Such chromocenters are sex chromatin (W polytene chromosome) and cannot be confused with nucleoli. ×7,920.

FIGURE 7.11. Detail of sex chromatin showing an alveolar structure (*B. mori*, fifth instar). ×15,840.

sites of contact (Beaulaton, unpublished data). The dense reaction product was exclusively located within insterstitial spaces at attachment sites on the outer side of membranes, suggesting that phosphatase molecules may assemble into specialized domains of the plasma membrane. Such a distribution of enzyme activity in relation to attachment sites of the tunica propria to the plasma membrane might either transmit a signal toward the matrix or play a role in its formation. However, no alkaline phosphatase activity was observed at the sites of hemidesmosomes anchoring the plasma membrane (most frequently at apexes of the cytoplasmic processes) in the tunica propria, as in *Antheraea* (Beaulaton, 1968d), *Galleria* (Blazsek and Mala, 1978), *Manduca* (Sedlak et al., 1983), *Heliothis* (Mamdouh, 1985), *Oncopeltus* (Dorn and Romer, 1976), or *Rhodnius* (Gras, 1977).

Closer examination of electron microscopic sections at higher magnification has revealed that the lamina densa of *Antheraea* (Figs. 7.8 and 7.9) is always composed of an intricate meshwork of fine fibrils (about 2 nm in diameter) and small granules, which appear more frequently at the hemolymph interface. It has also been shown that the glands are related to attachment strands (Beaulaton, 1968b; Gras, 1977), forming connections between the glands and other organs or the body wall. In *Antheraea*, the core of each attachement strand contains a bundle of thick fibrils about 30–35 nm in diameter. They penetrate and spread into the lamina densa of the glands, forming a fibrous sheath (François, 1976). It is worth noting that no banding pattern of collagen fibrils has so far been seen. This fibrous component (Fig. 7.7B) appears ultrastructurally similar to that of elastic fibers described in *Calpodes* connective tissue (Locke and Huie, 1975). Such an arrangement is very probably involved in the elasticity of the tunica propria, contributing to the shape and cohesiveness of the glands in relation to attachment strands that join them to other organs.

Early cytochemical studies in the glands of *Bombyx* (Kobayashi, 1956) and *Antheraea* (Beaulaton, 1968b) have indicated that the extracellular matrix consists of neutral mucopolysaccharides or glycoproteins visualized by the periodic acid–Schiff (PAS) and lead tetraacetate–Schiff methods, as in other basal laminae of insects (Ashhurst, 1968; Holter, 1970; François, 1976). The occurrence of proteins rich in arginine residues has been demonstrated in the tunica propria of *Antheraea* (Beaulaton, 1968b), cockroaches (Harper et al., 1967) and other basement membranes (Kefalides, 1975). These protein polysaccharides presumably include glycoproteins (such as laminin and fibronectin) and proteoglycans. However, the neutral nature of carbohydrate content deserves to be reexamined at different pH, considering available information in other animals. Recently, a cytochemical study of the glands of *Heliothis armigera* has shown that calcium accumulates primarily in the tunica propria and in the interstitial

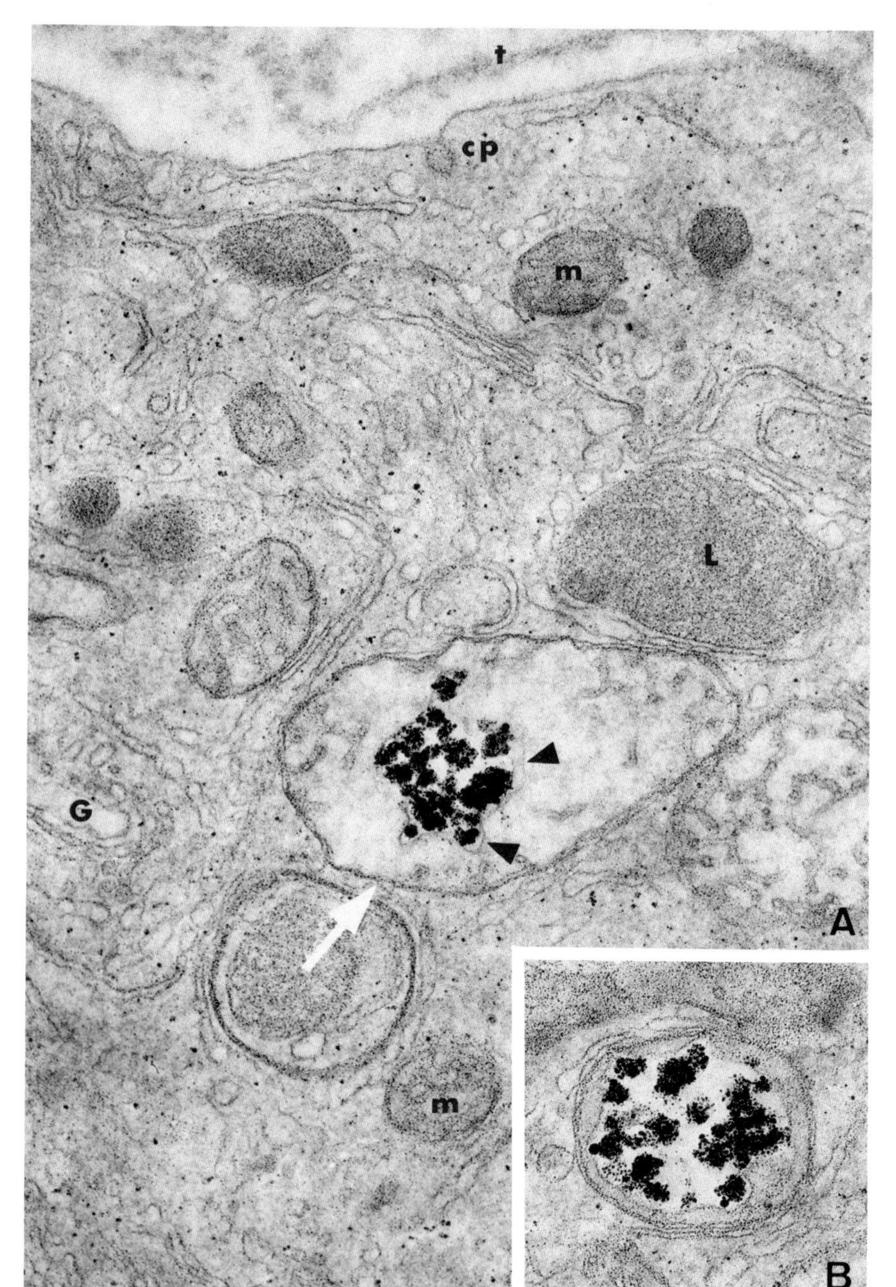

t
cp
m
m
L
G
m
A
B

gland spaces (Mamdouh, 1985). This result awaits confirmation by X-ray microanalysis of the dense reaction product. Nevertheless, it may be assumed that the tunica propria contains Ca-binding sites.

In summary, all results suggest that the extracellular matrix of the prothoracic glands performs structural as well as filtering roles. The tunica propria, by connecting the glands to other organs, acts as a supporting or cementing structure for cell anchorage; moreover, it has elastic properties enhancing gland distensibility and deformability. It presumably also functions as a molecular sieve through an intricate meshwork of fibrils. Doubtless, the tunica propria occupies a strategic location controlling ionic and molecular exchanges between the gland cells and the hemolymph.

7.4.7.2. TRACHEAL SUPPLY AND GLAND CELL REGULATION

Tracheo-glandular junctions have been found in *Antheraea* beneath the basal lamina of the gland where tracheal end cells or stellate cells (also called Holmgren's cells, or tracheoblasts) are distributed (Beaulaton, 1964a). At this level, the tracheae (about 4–5 μm in diameter) branch out into tracheoles, tapering gradually to a diameter of about 0.3–0.5 μm, and penetrating into the peripheral intercellular spaces of the large gland cells (Fig. 7.7A). Such a close association does not exclude the presence of a basal lamina (often very thin) between tracheoles and invaginations of the plasma membrane of the gland cells. Nevertheless, in *Antheraea*, certain tracheoles penetrate into the gland cells up to the perinuclear region where they lack the basal lamina, as in *Rhodnius* (Gras, 1977). Indeed, the adjacent plasma membrane of the tracheoblast and that of the gland cell may be separated by a narrow intercellular gap (Beaulaton, 1968a). This very close association between tracheoles and gland cells is most likely induced by a high oxygen demand owing to high metabolic activity of the latter cells. However, note that in various Coleoptera the prothoracic

FIGURE 7.12. Cytochemical characterization of intramitochondrial glycogen in a prothoracic gland cell of *Antheraea pernyi* (fifth instar, first day of spinning period). Periodic acid–thiocarbohydrazide–silver proteinate method. (A) In a giant mitochondrion (white arrow) note the swelling of the intracristal space (arrowhead) containing glycogen. In contrast to small mitochondria (m), the giant type contains a clear matrix. At the onset of glycogen accumulation, the tubular network of the mitochondrial cristae can be seen in spite of mitochondrial changes. A coated pit (cp) arises from the plasma membrane. The Golgi apparatus (G) and a lysosome (L) are also visible. (B) Detail of a giant mitochondrion at a later stage. Note an increase of the intracristal space loaded with glycogen. A&B, ×44,640.

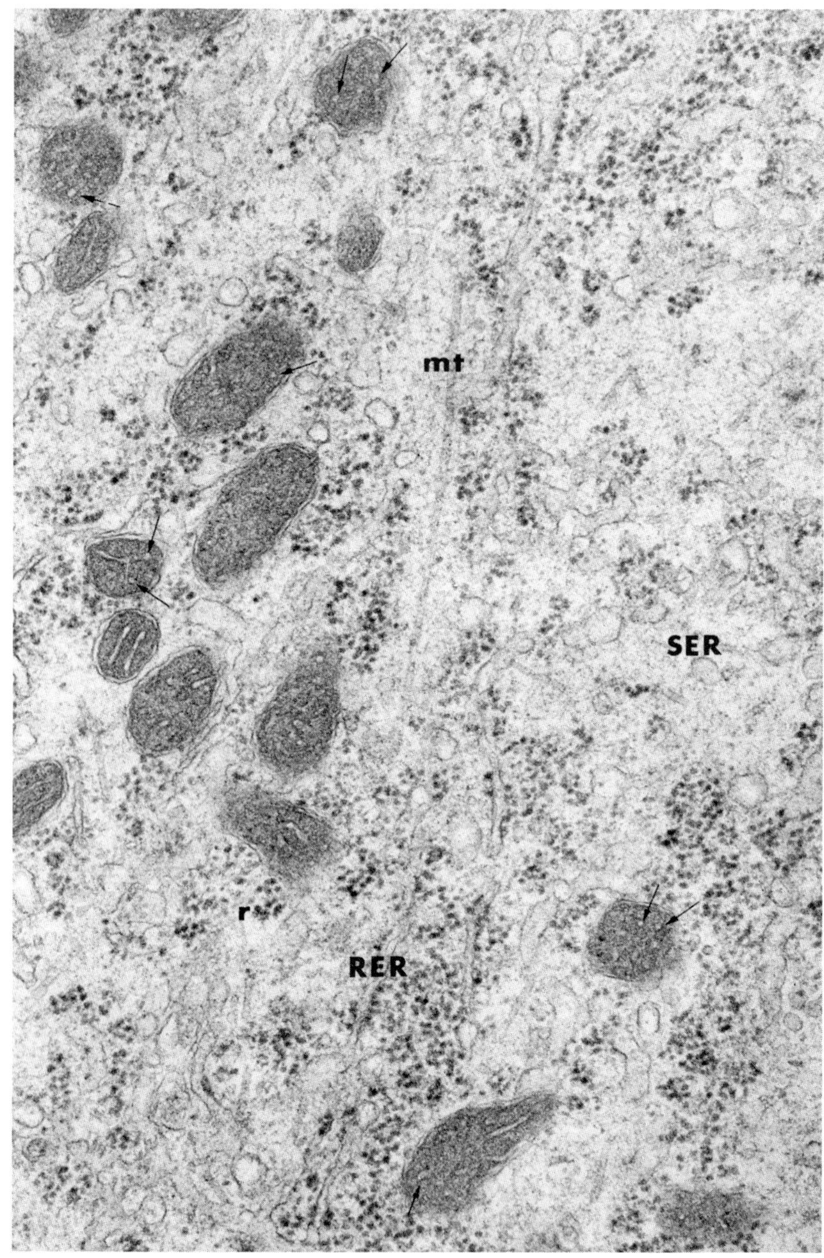

FIGURE 7.13. Cytoplasmic portion of a prothoracic gland cell of *Antheraea pernyi* (on the first day of the spinning period), illustrating the differentiation of smooth-surfaced tubules of the endoplasmic reticulum (SER) from the RER. Note the presence of several profiles of small mitochondria with tubular cristae (small arrows). ×42,240.

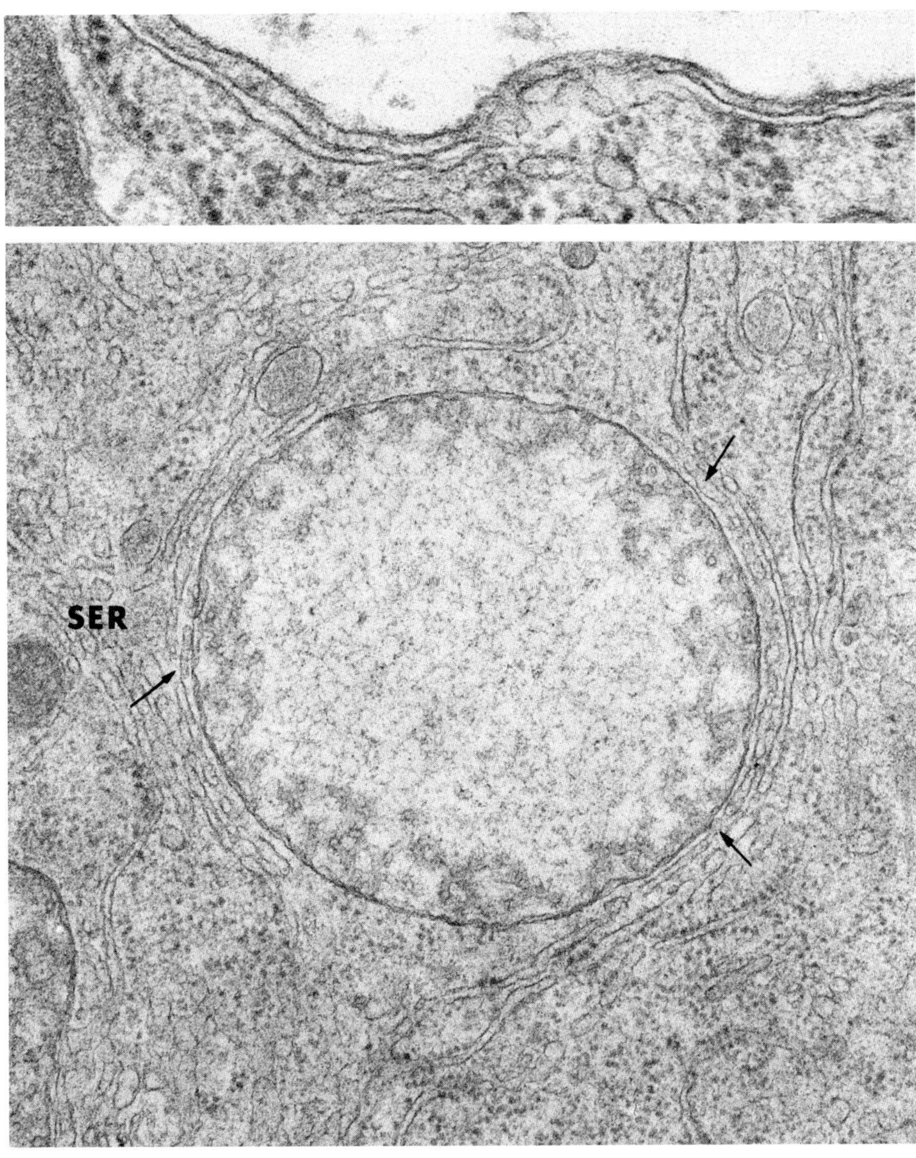

FIGURE 7.14. & 7.15. Prothoracic gland of *Antheraea pernyi* in the last instar (first day of spinning): *Fig. 7.14(top)*—Detail of smooth-surfaced cisternae of the SER parallel to the plasma membrane. ×78,000. *Fig. 7.15(bottom)*— Detail of the cytoplasm, showing a close relationship (arrows) between the SER and a giant mitochondrion. The latter is almost completely surrounded by the smooth-surfaced tubules or tubulocisternae, suggesting exchange activities in relation to ecdysone biosynthesis. Note that the tubular network is only seen at the peripheral region of the giant mitochondrion containing a swollen, clear matrix. ×60,000.

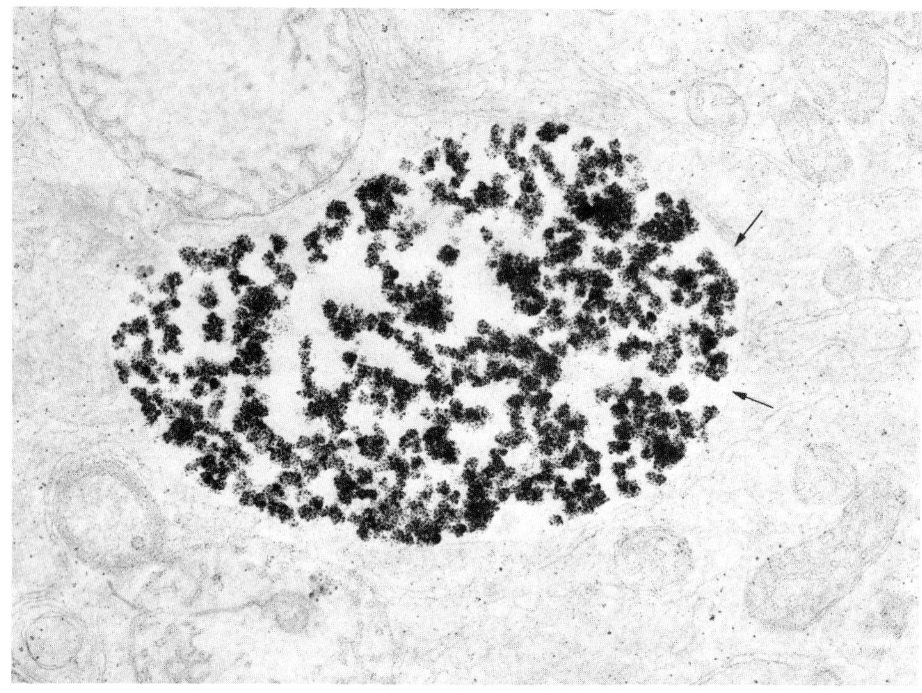

FIGURE 7.16. Cytochemical characterization of a glycogen vacuole in a prothoracic gland cell of *Antheraea pernyi* (fifth instar). The glycogen content is separated from the cytosol by a limiting membrane (arrows). Such vacuoles arise from giant mitochondria by collapse of the inner membrane. ×30,000.

gland does not appear to be supplied with tracheoles (Nunez, 1954; Srivastava, 1959; Glitho, 1977), although they are tightly associated with large tracheal trunks, as in *Tenebrio* (Romer, 1971; Glitho, 1977). This atypical arrangement is presumably related to a lower thickness of the gland tissue, in contrast to the giant gland cells of *Antheraea* where the tracheolization seems especially profuse.

This tracheolization is important for oxygen supply to the gland tissue. Exposure of *Rhodnius* nymphs (1 day after feeding) to a gas mixture poor in oxygen results in a delay in nymphal–nymphal molting, or prevents ecdysis (Wigglesworth, 1952b). Conversely, Takaoka (1957) showed that an interruption of the tracheal system in mature *Drosophila* larvae made at two distinct levels will prevent pupation in the central part of the body, when such experiments were done 2–7 h before the presumptive pupation. The same results were obtained either by ligation or sectioning of the tracheal trunks. Takaoka suggested that the oxygen supply is a

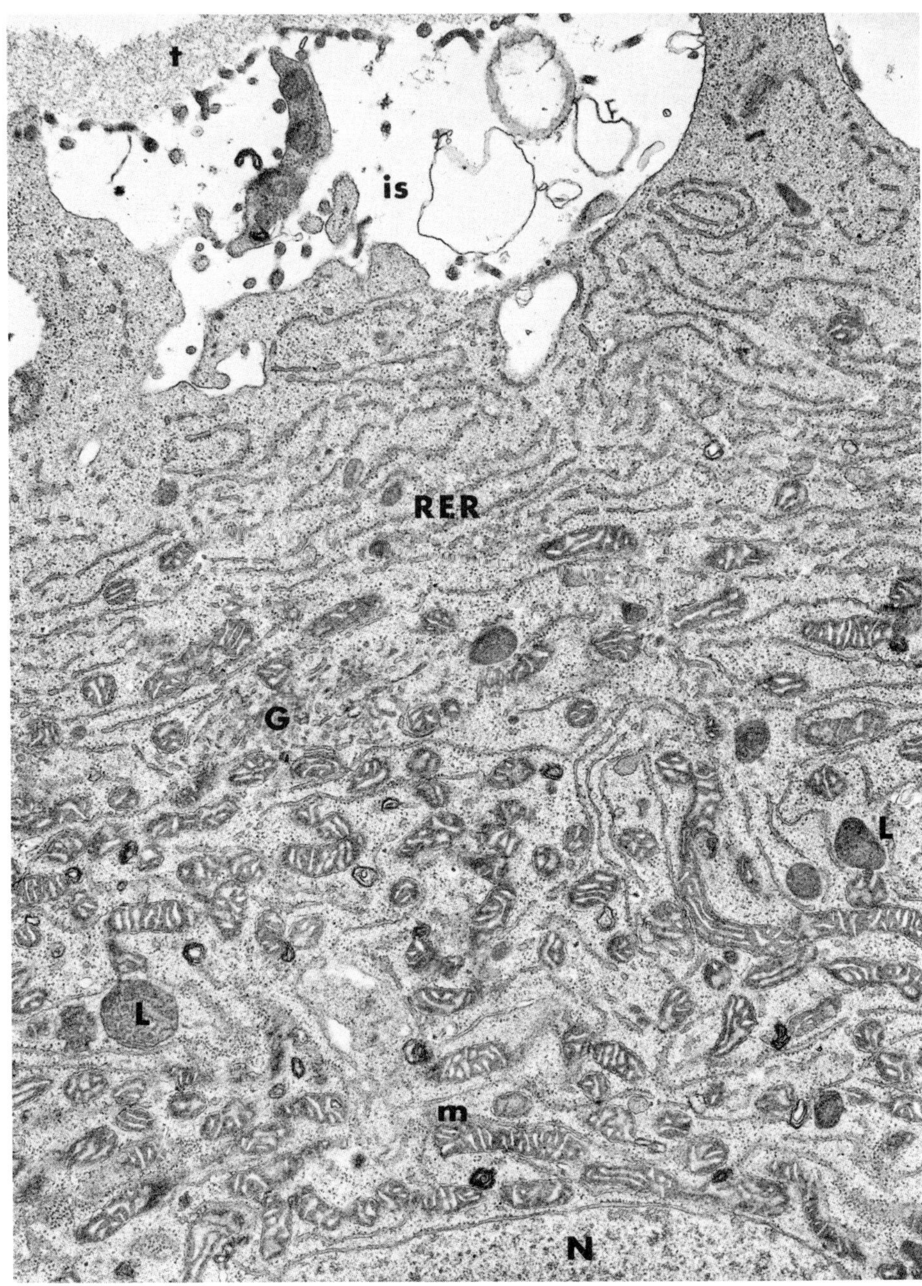

FIGURE 7.17. Peripheral area of a prothoracic gland cell in *Rhodnius prolixus* at the onset of steroidogenic activity (day 6 of the fifth instar): 85 pg hemolymph ecdysteroid (HE) eq/µl. Low magnification, showing the development of the RER, free ribosomes, and polysomes among numerous mitochondria (m) chiefly located in the perinuclear cytoplasm. A lysosome (L), the nucleus (N), and the interstitial space (is) are also visible. ×18,000. (From Beaulaton et al., 1984: unpublished micrograph.)

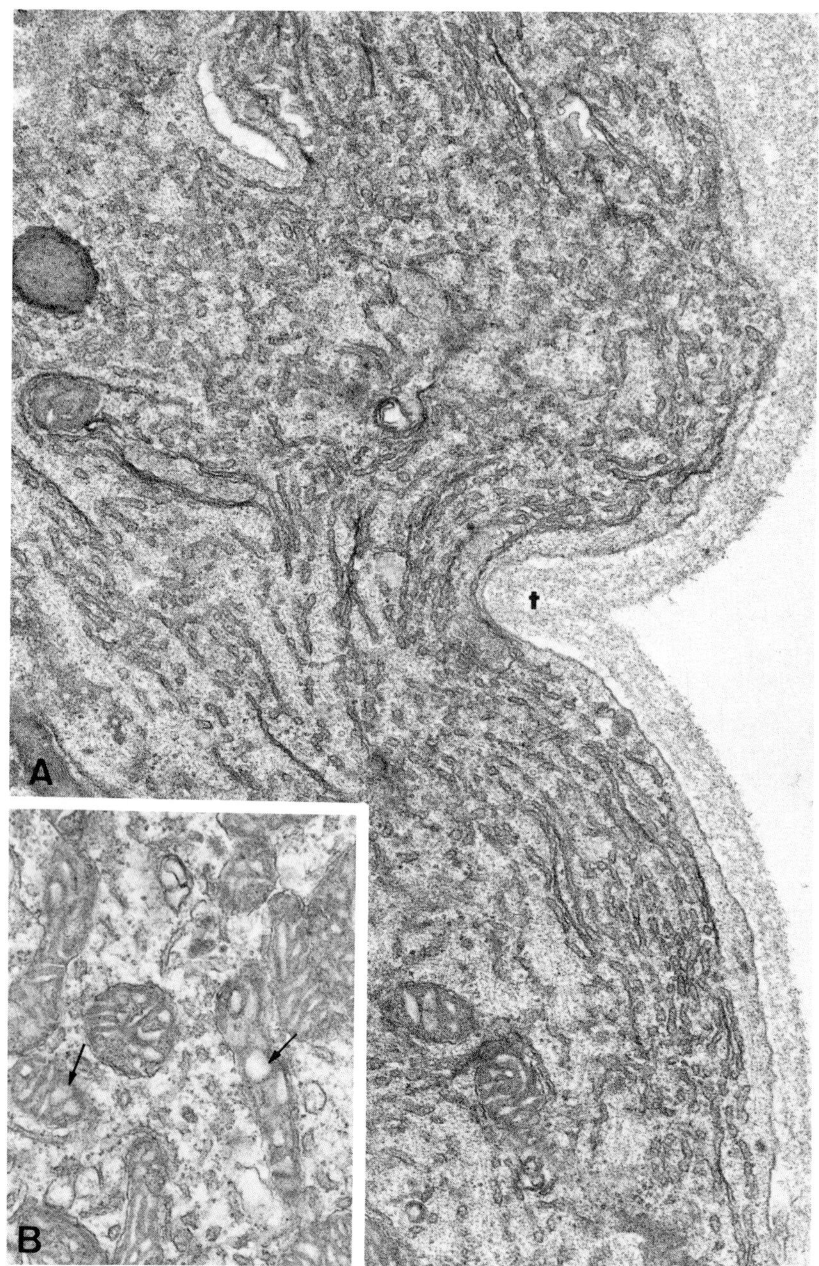

FIGURE 7.18. Prothoracic gland cell of *Rhodnius prolixus* during steroidogenic activity (day 11 of the fifth instar): 240 pg HE eq/µl. (A) Subplasmalemmal region illustrating the maximal development of the SER. It consists in a tubular or tubulocisternal network located in a peripheral area where other organelles are mostly excluded. The tunica propria (t) can be seen. (B) Aspect of the mitochondria with tubulovesicular cristae (arrows). A&B, ×31,680. (From Beaulaton et al., 1984: unpublished micrograph.)

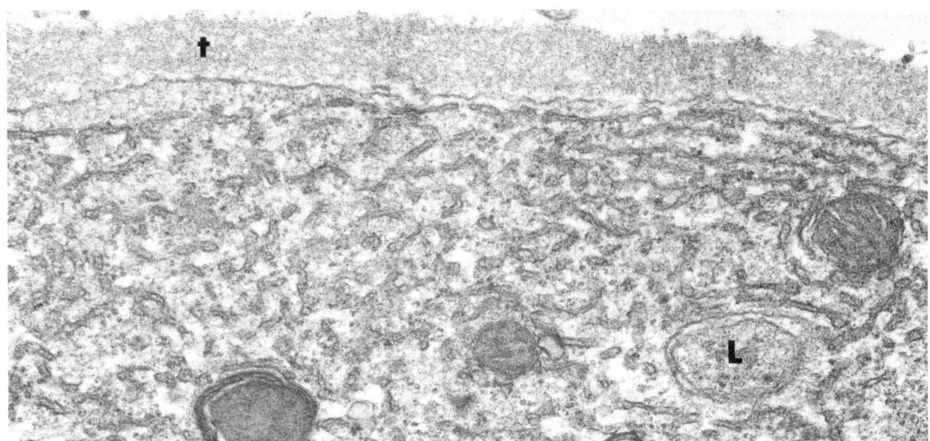

FIGURE 7.19. Prothoracic gland cell of *Rhodnius prolixus* during the peak of steroidogenic activity (day 12 at the fifth instar). 1000 pg HE eq/μl. Network of the tubular SER located in the peripheral area. The tunica propria (t) and a lysosome are visible. ×36,800. (From Beaulaton et al., 1984.)

limiting factor for molting. The latter author reached a similar conclusion in pupation experiments with mature larvae of *Drosophila* exposed to low partial pressure of oxygen. Pupation required a minimum oxygen tension of 1.5% at 25°C (Takaoka, 1959). Nevertheless, in all these experiments, the oxygen demand in the gland was not analyzed. Note that the ecdysone production by the prothoracic glands of *Bombyx* cultured *in vitro* increases four times with an increase of the partial pressure of oxygen from 0.2 to 0.5 (Chino et al., 1974). This fact clearly indicates that the secretory activity of the glands requires large amounts of oxygen. Such experimental results indirectly provide evidence that the tracheal supply of the glands plays a key role in *in vivo* ecdysone biosynthesis.

7.4.7.3. NEUROGLANDULAR JUNCTIONS

In Gryllidae, Thomas and Raabe (1974) have demonstrated the neurohemal feature of some parts of the sympathetic nerves, and especially that of the transverse and median nerves, which are designated as perisympathetic organs of a primitive mediotransverse type. Examination of transverse sections of *Acheta domestica* at the level of the prothoracic perisympathetic organ shows that the glands become adjacent to this organ, in which two types of neurosecretory products (A and C type) have been revealed (Thomas and Raabe, 1974). Again, in the same species, Maleville (1978a) found that the perisympathetic organs may be internalized in the glands. Thomas and Raabe (1974) suggested that these close relationships

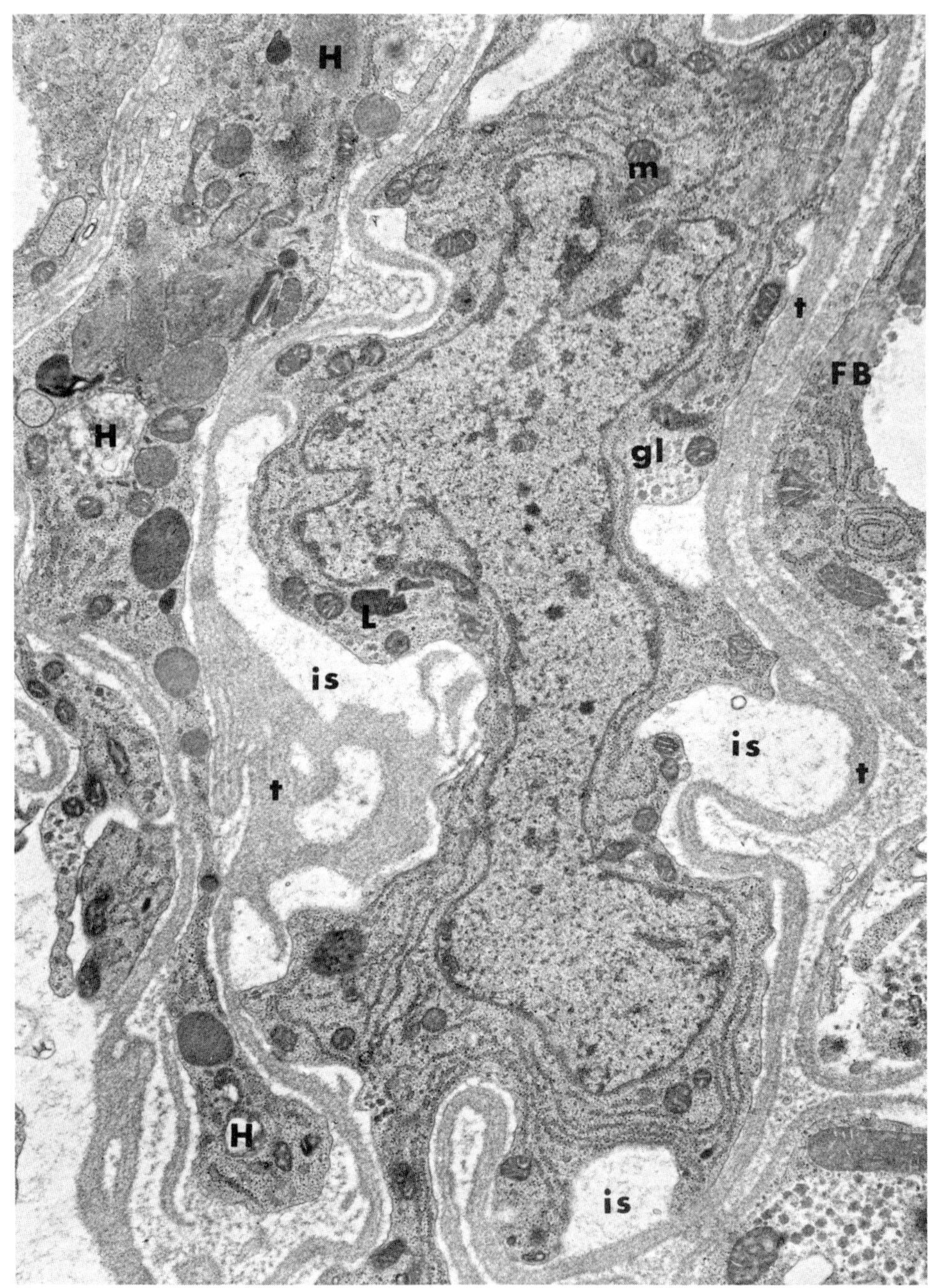

FIGURE 7.20. Decapitated nymph of *Rhodnius prolixus:* day 11 of the fifth instar: 1.6 pg HE eq/μl. Survey picture of a prothoracic gland cell, showing a resting appearance (compare with Fig. 7.6). The cytoplasm is much reduced and contains some cisternae of the RER, few mitochondria (m), lysosomes (L), and a glycogen area (gl). Note that a hemocyte (H) (granulocyte) adheres to the tunica propria (t) of the gland. *Key:* FB = fat body; is = interstitial space. ×12,500. (From Beaulaton et al., 1984: unpublished micrograph.)

between perisympathetic organs and prothoracic glands meet functional requirements. Thus, in Orthoptera, neurosecretory products may migrate to the level of target organs where they are released.

In higher Diptera, we have previously seen (Section 7.4.5) that no specific innervation has been found in the gland. Using a cobalt-filling technique, Giebultowicz and Denlinger (1985) have shown in *Sarcophaga* that three to five axons traverse each lateral part of the ring gland to the CA. According to Whitten (1964), the tunica propria of the ring gland forms a common sheath with the basal lamina of nerves, providing a complex network of channels in which neurosecretory products are directly transported from neuron terminals to target cells. Such a concept postulates that the neurosecretory products would not diffuse immediately through the tunica propria into the hemolymph.

By cobalt filling, Granger (1978), Singh and Sehnal (1979), and Singh et al. (1986) obtained evidence in *Galleria* that the cervical nerve contains axons issuing from three ventromedial neurons (group 1) in which at least one kind of neurosecretory cell has been identified. The perisympathetic transverse nerves of the thoracic ganglia include axons from ventromedial (group 2) and ventrolateral (group 3) neurons. According to Singh and Sehnal (1979), these two groups contain neurosecretory cells giving rise to peptidergic neurosecretory processes which have been reported in the perisympathetic transverse-type organs found in Lepidoptera (Raabe et al., 1974). There is evidence that some neurons revealed by cobalt filling participate in the nerve supply of the glands. Thus, peptidergic neurosecretory products, especially those of the group-3 cells, reach interstitial spaces of the glands by traveling along the axons of the transverse nerve.

In vivo microscopic observations in *Antheraea pernyi* by methylene blue nerve staining have revealed that interstitial cells, often closely associated with tracheoles, are located in the interstitial spaces beneath the tunica propria of the glands (Beaulaton, 1968c). Such cells may be present in either bipolar or multipolar form; so far no neurosecretory products have been found therein. In some instances, the interstitial cells are linked to form a cellular ring easily revealed by methylene blue. Such a nerve ring has previously been described by Yokoyama (1956) in the gland trunk. This ring consists perhaps of nonsecretory neurons (Fig. 7.9); thus, the interstitial cells of silkworms (Beaulaton, 1968c) might be peripheral neurons. Indeed, a single multipolar neuron has recently been identified in each gland of *Agrotis segetum* by Griffiths and Finlayson (1982). This peripheral neuron, fluorescing orange after staining with acridine orange, is regarded by these authors as a neurosecretory cell with possible aminergic synapses. Electron microscopy has revealed that the perikaryon, ensheathed in glial laminae, is closely associated with the gland through the tunica propria (Griffiths and Finlayson, 1982, 1983). In *A. segetum,* axon terminals penetrating into the tunica propria of the glands contain numerous membrane-bounded, electron-dense neurosecretory granules 150–200 nm in diameter. Such neurosecretory terminals, usually devoid of a glial sheath, are also known in the glands of *Leucophaea* (Scharrer, 1964), *Tenebrio* (Romer, 1971), *Rhodnius* (Gras, 1977), *Cerura* (Hintze-Podufal, 1970), *Diatraea* (Yin and Chippendale, 1975), and *Antheraea* (Beaulaton, 1968c), the last of these having larger, peptidergic neurosecretory granules (120–300 nm diameter). Benedeczky et al. (1980) have found that axon terminals in *Galleria* can be located close to the gland cells in which the extracellular space between the adjacent plasma membranes (axolemma and plasma membrane of the gland cell) is only 20–30 nm wide without basal lamina. One type of neurosecretory electron-dense granules (120–300 nm diameter) usually occurs in terminals supplying the gland cells, except in *Calliphora* where two types have been found (Normann, 1965). Such axon terminals may thus have both central and peripheral origins (Griffiths and Finlayson, 1982).

7.4.7.4.   GLAND CELLS

7.4.7.4.1.   Plasma Membrane, Cell Association, and Interstitial Spaces

The first ultrastructural observations of the prothoracic glands (Table 7.1) (Beaulaton, 1961, 1964a,b,c; Scharrer, 1964) have not confirmed the syncytial interpretation reported by some authors (Williams, 1948; Boisson, 1950, 1952). It is now well established that in various orders of insects the glands have a cellular structure with plasma membranes readily recognizable in ultrathin sections. In most glands, the secretory cells display a definite structural polarity. Indeed, the adjacent gland cells are bound together by interdigitations (Scharrer, 1964; Fain-Maurel and Cassier 1968a) and/or cell junctions. Two types are known in these glands. Tight junctions, or zonulae occludentes (terminal bars according to Fain-Maurel and Cassier, 1968a), were found in association with a few small desmosomes in *Petrobius* (Cassier and Fain-Maurel, 1971), *Locusta* (Fain-Maurel and Cassier, 1968a), *Antheraea* (Beaulaton, 1968d), and *Rhodnius* (Gras, 1977); this type was also reported in *Leucophaea* (Scharrer, 1964) and *Pyrgomorpha* (Guelin, 1971). In *Bombyx mori,* long chains of several desmosomes (presumably maculae adherentes rather than zonulae) were described between adjacent cells. Each desmosome may extend to nearly 2 μm.

We have previously seen (Section 7.4.7.1) that the plasma membrane immediately adjacent to the tunica propria frequently has polymorphic and sinuous projections (Figs. 7.7A and 7.9). Between them exists a network of anostomosing or interstitial channels known as pericellular or

TABLE 7.1. Ultrastructural Studies of the Prothoracic Glands

Taxa	Instars[a]			References
Thysanura				
Petrobius maritimus			A	Cassier and Fain-Maurel (1971)
Ephemeroptera				
Ephemera danica	N	SI	A	Kaiser (1978)
Dictyoptera				
Leucophaea maderae	N		A	Scharrer (1964, 1966)
Phasmida				
Carausius morosus	E			Haget et al. (1982)
Orthoptera				
Locusta migratoria	N		A	Cassier and Fain-Maurel (1968a,b,c, 1969a,b,c) Fain-Maurel and Cassier (1968a,b, 1969)
Pyrgomorpha conica	N			Guelin (1971)
Acheta domestica	N			Maleville (1976, 1978a,b)
Coleoptera				
Tenebrio molitor	L	P		Romer (1971); Glitho (1977) Glitho et al. (1979)
Xyleborus ferrugineus		P		Chu et al. (1980)
Heteroptera				
Oncopeltus fasciatus	E	N		Dorn and Romer (1976); Smith and Nijhout (1982)
Rhodnius prolixus		N		Gras (1977); Gras and Beaulaton (1979); Beaulaton and Gras (1980); Beaulaton et al. (1984)
Lepidoptera				
Antheraea pernyi	L			Beaulaton (1964a,b,c, 1967a,b, 1968a,b,c,d,e,f,g,h,i)
Bombyx mori	L			Beaulaton (1968a,b,c,d,e,f)
Cerura vinula	L			Hintze (1968); Hintze-Podufal (1971)

(*continued*)

TABLE 7.1. *(Continued)*

Taxa	Instars[a]	References
Galleria mellonella	L P	Blazsek et al. (1975, 1978); Gersch et al. (1975)
Manduca sexta	L	Sedlak et al. (1983); Sedlak (1984, 1985)
Hyalophora cecrophi	P	McDaniel et al. (1976)
Heliothis armigera	L	Mamdouh (1985)
Spodoptera littoralis	L	Karaçali (1978)
Diptera		
Drosophila melanogaster	L	King et al. (1966); Aggarwal and King (1969)

[a]E = embryos; L = larvae; N = nymphs; P = pupae; SI = subimagoes; A = adults.

interstitial spaces. They are presumably filled with a distinct interstitial fluid. Although this fluid can be assumed to be closely related to the hemolymph, it must be an intermediate medium for the passage of hemolymph nutrients, hormonal precursors, and chemical signals (such as PTTH and JH) into the gland cells. Conversely, secretory products such as ecdysone and metabolic wastes released from the gland cells can move with the interstitial fluid toward the hemolymph (Beaulaton, 1968d).

As has been reported by Sedlak (1984, 1985), the interstitial spaces (or peripheral intercellular spaces) show large variations in depth and width throughout larval development. In resting glands and at the beginning of the secretory cycle, the interstitial spaces are often narrow (Fig. 7.8), or even lacking in some cases, such as *Galleria* (Gersch et al., 1975). They increase in size as the secretory cycle progresses, and very large channels (Fig. 7.9) can be seen during the release of hemolymph ecdysteroids in *Manduca* (Sedlak et al., 1983) and *Rhodnius* (Gras, 1977). It is thus clear that the gland cells have their maximal surface area when their ecdysteroid production attains the highest level, thereby suggesting dramatic changes in exchange activities of metabolites (precursors, ecdysone) across the plasma membrane. Changes of the membrane potential have been demonstrated in *Galleria* during intermolt cycles (Gersch and Birkenbeil, 1973; Gersch et al., 1976). Gündel et al. (1982) have found that such membrane potentials are caused by changes of the intracellular potassium concentration between inactive and active phases of the cells.

Coated vesicles frequently occur on the plasma membrane lining the interstitial spaces (Scharrer, 1964; Beaulaton, 1968d; Fain-Maurel and Cassier, 1968a; Blazsek and Mala, 1978; Mamdouh, 1985). They presumably provide a pathway for internalization of specific molecules. It is possible that hemolymph lipoproteins containing steroid precursors, similar to low-density lipoprotein (LDL) complexes of vertebrates, are internalized via coated pits and vesicles, as suggested by Blazsek and Mala (1978). In *Manduca*, these spaces may contain multivesicular sacs with smooth-surfaced vesicles (Sedlak et al., 1983; Sedlak, 1984, 1985). They originate from blebs of the plasma membrane sequestering smooth vesicles, as in *Locusta migratoria*, where they were considered to be secretory vesicles involved in the release of ecdysone (Cassier and Fain-Maurel, 1968b; Fain-Maurel and Cassier, 1968a). However, similar surface protrusions containing smooth-walled vesicles have recently been found in *Calpodes* epidermal cells during metamorphic changes (Locke, 1985). Similarly, such multivesicular sacs budding from the plasma membrane into intercellular spaces were observed by the author in various insect tissues, such as hemocytopoietic organs, hemocytes, or degenerating intersegmental muscles (*Antheraea* and *Manduca*), previously fixed with glutaraldehyde. These observations on various non-steroid-secreting cells suggest that such structures most likely are aldehyde-induced artifacts (see the review by Hayat, 1981). An alternative approach to this problem would be to use ultrarapid freezing to avoid chemical fixation.

Blazsek and Mala (1978), studying the effect of digitonin on the prothoracic glands of *Galleria*, have shown that tubular crystals of the digitonin complex are located in the interstitial spaces close to clear microvesicles. However, it is known that digitonin may form tubular crystals from reaction with lecithin. There is no evidence to support specific reaction *in situ* with sterols, such as cholesterol and ecdysteroids, having a free hydroxy in position 3 (see the discussion by Hatier, 1973; also by Hayat, 1981).

7.4.7.4.2. Nucleocytoplasmic Ratio and Cytoplasmic Growth

A stereological analysis of the prothoracic gland cells has not yet been concluded. However, these cells might be an interesting model with respect to changes of the nucleus–cytoplasm ratio within intermolt cycles. In *Rhodnius* an ultrastructural study during the last two larval instars has provided some rough estimates (Gras, 1977; Beaulaton et al., 1984). A high nucleus–cytoplasm ratio is always observed during the fasting period after each molt (Fig. 7.6). At increasing intervals after feeding, this ratio gradually decreases up to mid-intermolt, as in *Oncopeltus* (Smith and Nijhout, 1982a) and *Galleria* (Blazsek et al., 1975). Conversely, after the ecdysteroid production peak, the ratio increases until molting (Beaulaton et al., 1984).

In *Drosophila*, Aggarwal and King (1969) estimated that the cytoplasmic volume of the gland cells increases by at least 100 times during larval development. From these data, it is apparent that such cell growth is cyclical and parallel to that of the glands (see Section 7.4.3.2, above), at least in the giant cell epithelioid type.

7.4.7.4.3. Nucleus and Nuclear Envelope

Intermolt-related changes in the shape and size of cell nuclei have frequently been observed in the glands, as in the CA (Cassier, 1979). For example, in *Antheraea pernyi*, the nuclei show ovoid outlines just after hatching; they become lobulated (Fig. 7.6) in the next larval instars while their size gradually increases (Beaulaton, 1968d). In the last instar, the nuclei have branched outlines (Fig. 7.10) similar to those seen in the silk glands (Akai, 1983). Sex differences have not been observed (Scharrer, 1964), except in Lepidoptera for the female sex heterochromatin (see the next subsection). Cyclic changes in the nuclear volume have been found in the last two nymphal instars of *Dysdercus* (Wells, 1954). Similarly, in *Rhodnius*, Gras (1977) has reported an increase in the size of the nuclei, which become deeply lobulated during the first two-thirds of intermolt. Thereafter, until molt, this process appears reversed. Circadian changes in the nuclear volume of prothoracic gland cells were reported by Rensing (1966) in *Drosophila* larvae. Note that bimodal changes occur in the cells and closely coincide with those of the brain neurosecretory cells and CA cells (Rensing et al., 1965). Such results demonstrate synchronic changes among different endocrine cells and suggest an overall circadian rhythmicity (Brady, 1974; Saunders, 1981; Scharrer, 1987).

In *Antheraea*, the nuclear envelope exhibits a typical organization with a perinuclear space (about 11–20 nm) and numerous nuclear pores of about 100 nm in diameter. The chemical nature of the pore complexes has been studied by ultracytochemistry (Beaulaton, 1968i). On tangential section, the pore complex frequency was estimated as about 25 pores/μm^2 of membrane area (Beaulaton, 1968d). Such a high pore frequency may be correlated to active nucleocytoplasmic transport through each pore. Additionally, these activities may be increased when the nuclear surface becomes ramified.

7.4.7.4.4. Nucleoli, Nuclear Bodies and Sex Chromatin

Each nucleus contains a variable number of polymorphic nucleoli (Figs. 7.7A and 7.10). There are one or two nucleoli in *Drosophila* (Aggarwal and King, 1969) and *Locusta* (Fain-Maurel and Cassier, 1968a); two in *Tenebrio* (Glitho, 1977); or more in *Hypogastrura* (Lauga-Reyrel, 1983), *Leucophaea*

(Scharrer, 1964), *Oncopeltus*, (Dorn and Romer, 1976), *Philosamia, Antheraea,* and *Bombyx* (Beaulaton, 1961, 1968d). Their size and appearance change during each intermolt cycle. Good examples of these changes can be seen in *Leucophaea* (Scharrer, 1964) and especially in *Rhodnius* (Gras, 1977). In resting cells, each nucleolus (about 1.6 μm in diameter in *Leucophaea* and 2 μm in *Rhodnius*) has a compact structure with a dense fibrillar core surrounded by a granular component. At increasing intervals after molt in *Leucophaea,* or after feeding in *Rhodnius,* the nucleoli enlarge and their peripheral material becomes dispersed. During this phase of progressive expansion and fragmentation, they show an irregular shape, while reaching 2.8 μm in diameter in *Leucophaea* (Scharrer, 1964), and 3–4 μm in *Rhodnius.* Such nucleolar changes are closely related to those of the cytoplasm and reflect high biosynthetic activities in connection with ribosome production.

Some nuclear inclusions have been observed in the glands of Dictyoptera (Scharrer, 1964), Heteroptera (Gras, 1977), and silkworms (Beaulaton, 1968d). In *Leucophaea,* Scharrer (1964) has reported the presence of electron-dense spherical inclusions (0.5 μm in diameter), as in the CA. Their significance is unknown. In *Rhodnius,* the nuclei contain spherical bodies of the same electron density as that of the fibrillar nucleolar component (Gras, 1977).

In *Bombyx mori,* Kobayaski (1953a,b) described nuclear vesicles (2.6–4.5 μm in diameter) distinct from nucleoli. Such Feulgen-positive inclusions were redescribed as prominent nucleoli (about 10 μm in diameter) or annular nucleoli in *Hyalophora* (Herman and Gilbert, 1966; McDaniel et al., 1976), *Galleria* (Blazsek et al., 1975), and *Heliothis* (Mamdouh, 1985) without cytochemical characterization. In the silkworms *Antheraea pernyi* and *Philosamia cynthia,* the present author has confirmed their high Feulgen positivity and has shown their failure to be digested with ribonuclease (Beaulaton, 1968d). These chromocenters are only present in the female gland cells (Fig. 7.10 and 7.11), as in other lepidopterous polyploid cells such as nurse cells (Guelin, 1968, 1970, 1978), silk gland cells (Gillot, 1968), and mandibular gland cells (Traut and Scholz, 1978). In female Lepidoptera, there commonly are one or two heterochromatic bodies per gland cell (Fig. 7.10). Such nuclear inclusions display an electron-dense alveolar or annular structure (about 1–10 μm in diameter), enclosing clear material with sometimes granular bulks (Beaulaton, 1968d). Each chromocenter, ultrastructurally composed of heterochromatin (Fig. 7.11), represents sex chromatin, presumably the W polytene chromosome, as in other lepidopterous polyploid nuclei (Guelin, 1978; Traut and Scholz, 1978). Owing to their ultrastructural and cytochemical features (Beaulaton, 1968d), they cannot be confused with nucleoli.

7.4.7.4.5. Mitochondria and Mitochondrial Changes

The mitochondria are generally pleomorphic, and their structure as well as their numbers are particularly dependent on the physiological state (Figs. 7.12A, 7.13, and 7.15). In resting cells, they are not very numerous (Fig. 7.6) and show rounded, oblong, or elongated profiles (0.3–0.6 μm in diameter and 1.0–1.5 μm in length). Their matrix is often electron-dense, and the cristae have a variable shape. Lamellar or tubulolamellar cristae are reported in *Locusta* (Fain-Maurel and Cassier, 1968a) and *Rhodnius* (Gras, 1977). In other insects, tubular, tubulosaccular, or tubulovesicular cristae are described, for example, as in *Petrobius* (Cassier and Fain-Maurel, 1971), *Antheraea* (Figs. 7.12–7.15), *Bombyx* (Beaulaton, 1964b, 1968e), and *Spodoptera* (Karaçali, 1978).

Ultrastructural changes of the mitochondria have been demonstrated in active cells, but only in a few species of insects. It is well known that changes in mitochondrial structure can be induced by changes in osmotic pressure but also by fixation procedures (Munn, 1974; Hayat, 1981). Since such prothoracic gland cells, and especially the mitochondria, frequently are poorly preserved (e.g., Joly et al., 1969; Maleville, 1976, 1978a), the effects of major technical factors on the cell ultrastructure have been investigated in *Rhodnius* (Gras and Beaulaton, 1978; Beaulaton and Gras, 1980). Cytophysiological studies on this species showed that one of the main features in active cells is a substantial increase of the mitochondrial population (cf. Figs. 7.6 and 7.17). It is accompanied by ultrastructural changes in the cristae, becoming tubular or sac-like (see Fig. 7.18B, below, in Section 7.4.7.4.4), especially during the production of ecdysteroids (Gras, 1977; Beaulaton et al., 1981, 1984). At that time, in *Bombyx* and *Petrobius*, some mitochondria elongate (up to 3–5 μm in length) and undergo structural changes, resulting in cup-shaped profiles with tubular or vesicular cristae and concentric mitochondrial complexes as large as 1–4 μm (Beaulaton, 1968e; Cassier and Fain-Maurel, 1971). In *Petrobius*, such complexes can be regularly associated with intercalary layers of the endoplasmic reticulum. They turn into autophagic bodies (see Section 7.4.7.4.10, below) during the last part of intermolt.

Another striking change occurs in *Antheraea* with the differentiation of giant mitochondria (macromitochondria, 2–3 μm in diameter) containing a complex network of tubular cristae 20–25 nm in diameter in *Antheraea* (Beaulaton, 1964b, 1968e). Such mitochondrial changes (Figs. 7.12A and 7.15) peak during the releasing phases of ecdysteroids and presumably derive from physiological changes related to secretory activity. Similar observations have been reported in *Spodoptera littoralis* (Karaçali, 1978). In the giant mitochondria, the tubular network is often associated with peculiar lamellae, or narrow tubules of high electron density (*Antheraea*) or crystalloid structures (*Spodoptera*). These mitochondria are very labile organelles, irreversibly undergoing vacuolization after clearing of the matrix and eventually fusing with other adjacent giant mitochondria (Beaulaton, 1964b, 1968e).

A second major mitochondrial change is related to glycogen accumulation and has been documented in *A. pernyi* during secretory cycles. Particles cytochemically identified as glycogen (20–25 nm in diameter) were first demonstrated within the giant mitochondria (Figs. 7.12A,B) of these cells (Beaulaton, 1964c). They often occur as aggregates in enlarged intracristal spaces. A great expansion of the intracristal spaces together with the peripheral space results in the formation of a large glycogen-loaded pseudomatrix (Beaulaton, 1968f,h), as in some other cells (see the references in Munn, 1974). The glycogen accumulation leads to a complete vacuolization of the mitochondria (Fig. 7.16) once the cristae and matrix are lost (Beaulaton, 1968f).

Cytochrome oxidase activity has been localized by ultrastructural cytochemistry in the mitochondria of *A. pernyi*. In the last larval instar, cytochrome oxidase activity in active cells is essentially confined to small mitochondria with a dense matrix. The reactive sites are primarily found at the level of cristae. No activity has been demonstrated in the giant mitochondria, except occasionally on the inner membrane (J. Beaulaton, unpublished data). Evidently, irreversible structural changes in the mitochondria are correlated to physiological changes demonstrated by a loss of cytochrome oxidase activity. One may assume that despite various structural changes the mitochondria, or at least one type among them, are involved in cyclic steroidogenic activities and also in glycogen storage.

7.4.7.4.6.   Rough Endoplasmic Reticulum

In active cells of all known prothoracic glands, the endoplasmic reticulum (ER) is well developed, at least at the onset of activity. The rough ER (RER) usually consists of flattened cisternae (Figs. 7.13, 7.15, and 7.17), which form one large interconnected network together with the nuclear envelope. Some authors have reported a vesicular structure of the RER with or without enclosed flocculent material (Cassier and Fain-Maurel, 1969b; Maleville, 1978a,b; Kaiser, 1978; Lauga-Reyrel, 1983).

In *resting cells*, the cytoplasm is often very restricted (Fig. 7.6), as in *Rhodnius* (Gras, 1977) and *Oncopeltus* (Smith and Nijhout, 1982a), and contains free ribosomes, polysomes, and few cisternae of the RER (Fain-Maurel and Cassier, 1968a; Beaulaton, 1968d; Cassier and Main-Maurel, 1971; McDaniel et al., 1976; Chu et al., 1980; Lauga-Reyrel, 1983; Beaulaton et al., 1984).

By contrast, in *active cells* at the early step of the intermolt cycle, the

RER, free ribosomes, and polysomes augment abruptly, as in *Locusta* (Fain-Maurel and Cassier, 1968a; Joly et al., 1969) and *Drosophila* (Aggarwal and King, 1969), *Oncopeltus* (Smith and Nijhout, 1982a), and in *Rhodnius* (Fig. 7.17) up to the head-critical period (Beaulaton et al., 1984). In pharate adults of *Hyalophora cecropia*, the cisternae of RER vacuolize (McDaniel et al., 1976); McDaniel and associates interpreted the flocculent material within vacuoles and cisternae as being an indicator of a cellular response to PTTH. At the second step of the secretory cycle, the cisternae of the RER are more extensive during an increase of ecdysteroid levels (Fain-Maurel and Cassier, 1968a; Smith and Nijhout, 1982a; Beaulaton et al., 1984). Thiamine pyrophosphatase activity was cytochemically localized within the cisternae of the RER and the nuclear envelope of *Leucophaea* (Osinchak, 1966), but glucose 6-phosphatase activity was absent in *Antheraea* (Beaulaton, 1968f). Finally, with declining ecdysteroid levels, the RER regresses, while the lysosomes and autophagic vacuoles become more numerous (Beaulaton, 1967a,b; Smith and Nijhout, 1982a; Beaulaton et al., 1984). Note that in certain insects (Collembola, Orthoptera, Lepidoptera) most investigators have found two types of secretory cells: dark electron-dense cells with few RER and abundant ribosomes or polysomes, and light electron-transparent cells with a large amount of RER (Fain-Maurel and Cassier, 1968a; Blazsek et al., 1975; Maleville, 1976, 1978a,b; Lauga-Reyrel, 1983). It is still unclear whether dark and light cells respectively are fixation artifacts, resulting from a poor penetration of fixative mixture, as in vertebrate steroid glands (see the discussion by Hayat, 1981), or indicate variable states of cellular hydration or differences of intracellular osmotic pressure. A comparison with other fixation techniques such as freezing methods would be especially useful to clarify this problem. However, according to some authors, the light and dark cells reveal asynchronic secretory activities between the different cells in a given gland (Cassier and Fain-Maurel, 1968a; Maleville, 1978a,b).

7.4.7.4.7. Smooth Endoplasmic Reticulum

Ultrastructural studies of various prothoracic glands have clearly established that many cells contain smooth ER (SER) during active secretory phases. This organelle originates from the cisternal RER by budding of smooth surfaced tubules or tubulocisternae (Fig. 7.13). However, the identification of the SER, easily destabilized by fixation, is related to the problem of satisfactory ultrastructural preservation. According to Hayat (1981), a well-fixed SER appears tubular rather than vesicular. Nevertheless, according to various authors, many prothoracic gland cells would contain the latter type. In the prothoracic glands of *Locusta* (Fain-Maurel and Cassier, 1968a), *Cerura* (Hintze-Podufal, 1971), *Galleria* (Blazsek et al.,

1975), *Bombyx* (J. Beaulaton, unpublished data), and *Drosophila* (King et al., 1966; Aggarwal and King, 1969), only the vesicular form of SER is known. However, this aspect may be a fixation artifact. It is apparent that the presence of vesicular forms remains to be clarified.

The tubular or tubulocisternal form of SER is found in *Ephemera danica* (Kaiser, 1978), *Musca domestica* (Herman, 1967), *Calpodes* (Locke, 1970), *Antheraea pernyi* (Beaulaton, 1968e), and *Rhodnius* (Gras and Beaulaton, 1976, 1979; Gras, 1977; Beaulaton et al., 1984). Both forms of the SER are encountered in *Petrobius* (Cassier and Fain-Maurel, 1971), *Leucophaea* (Scharrer, 1964), and *Tenebrio* (Romer, 1971). We now know that the SER is extensively distributed in at least three models of gland cells: *Drosophila melanogaster* (King et al., 1966; Aggarwal and King, 1969), *Calpodes* (Locke, 1970), and *Rhodnius prolixus* (Gras and Beaulaton, 1976, 1979; Gras, 1977; Beaulaton et al., 1984).

Aggarwal and King (1969) have shown that the volume of SER in *Drosophila* larvae increases 25-fold between hours 50 and 94 of the third instar, followed by another 10-fold increase between hours 94 and 98 (pre-pupa). Simultaneously, the SER:RER volume ratio increases 30-fold at prepupal stage. In the *1(2) gl.* mutant of *Drosophila*, the prothoracic gland cells are smaller than wild-type cells and the cytoplasm is especially deficient in SER. In prepupal cells, the SER volume of the mutant is only 1% of the wild-type volume (Aggarwal and King, 1969). The authors infer that the mutant cells cannot convert the RER into SER, as opposed to wild-type cells. One may tentatively conclude that the mutant cells are deficient in SER-synthesizing enzymes.

In *Rhodnius*, after the development of RER cisternae (the first step of activation), the SER begins to differentiate after the head-critical period (the second phase of PTTH release). Tubular profiles proliferate, forming a tridimensional network (Figs. 7.18 and 7.19) distributed over large areas of the peripheral cytoplasm (Gras and Beaulaton, 1979), similar to those observed in oenocytes (Locke, 1969; Rinterknecht and Matz, 1983; Beaulaton, unpublished data) and in activated Leydig cells (Nussdorfer et al., 1980). The SER proliferation (cf. Fig. 7.17 with Figs. 7.18 and 7.19) coincides with the burst of hemolymph ecdysteroids (Beaulaton et al., 1984). During the last step of intermolt cycle, the SER regresses.

In all other gland cells known so far, the SER seems to be less extensive (see Table 7.2). Interestingly, the SER of certain cells undergoes diurnal changes in its distribution and extension (see the review by Dillon, 1981). To my knowledge, no ultrastructural investigation has been reported indicating a possible circadian rhythm in the secretory activity of these glands. However, evidence exists for a circadian clock controlling the timing of ecdysone release by the glands of *Philosamia cynthia* (Mizoguchi and Ishizaki, 1982, 1984). This aspect deserves more detailed studies.

TABLE 7.2. Cytoplasmic Organelles in the Prothoracic Gland Cells

Taxa	Instar[a]	Mitochondrion Cristae[a]	RER[a]	SER	Fixation	Refs.[b]
Thysanura						
Petrobius maritimus	A	tubules	+++	+ tubules or vesicles	G–O	·(1)
Ephemeroptera						
Ephemera danica	N	?	+++	+++ tubules	GF–O	(2)
Dictyoptera						
Leucophaea maderae	N A	?	++	+ tubules or vesicles	O	(3,4)
Phasmida						
Carausius morosus	E	tubules	—	—		(5)
Orthoptera						
Locusta migratoria	N	tubulolamellae	+++	++ vesicles	G–O	(6,8)
L. migratoria	A	lamellae	+++	++ vesicles	G–O	(7,9)
Acheta domestica	N	tubules & lamella	+++	+	GF–O	(11,12)
Pyrgomorpha conica	N	lamellae	++	?	G–O	(10)
Coleoptera						
Tenebrio molitor	L	?	++	+ tubules or vesicles	O G–O	(13–15)

Species	Stage				Fixation	Refs.
Xyleborus ferrugineus	P	?	+	+	?	(16)
Heteroptera						
Oncopeltus fasciatus	N	?	+++	+++	G–O	(17,18)
Rhodnius prolixus	N	sac-like	+++	+++	GF–O	(19–22)
Lepidoptera						
Antheraea pernyi	L	tubules	+++	+/++ tu-bules	G–O	(23–25)
Hyalophora cecropia	P	?	+++	?	G–O	(26)
Bombyx mori	L	tubules	+++	?	G–O	(27)
Cerura vinula	L	lamellae	++	+++	O	(28,29)
Galleria mellonella	L	tubulovesicles	+++	+++ vesicles	G–O	(30,31)
Spodoptera littoralis	L	tubules	?	?	GF–O	(35)
Heliothis armigera	L	?	++	+/++	G–O	(34)
Manduca sexta	L	tubules	++	?	G–O	(32,33)
Diptera						
Drosophila melanogaster	L	?	+	+++ vesicles	O	(36,37)

Key: A = adult; L = larvae; N = nymphs; P = pupae; Mito. = mitochondria; RER = rough endoplasmic reticulum; SER = smooth endoplasmic reticulum; O = osmium tetroxide; G–O = glutaraldehyde and postfixation with osmium tetroxide; GF–O = glutaraldehyde + formaldehyde and postfixation with osmium tetroxide; + = few; ++ = moderately numerous; +++ = very numerous; ? = unknown.

Refs.: (1) Cassier and Fain-Maurel, 1971; (2) Kaiser, 1978; (3) Scharrer, 1964; (4) Scharrer, 1966; (5) Haget et al., 1982; (6) Cassier and Fain-Maurel, 1968a,b,c; (7) Cassier and Fain-Maurel, 1969a,b; (8) Fain-Maurel and Cassier, 1968a,b; (9) Fain-Maurel and Cassier, 1969; (10) Guelin, 1971; (11) Maleville, 1976; (12) Maleville, 1978a,b; (13) Romer, 1971; (14) Glitho, 1977; (15) Glitho et al., 1979; (16) Chu et al., 1980; (17) Dorn and Romer, 1976; (18) Smith and Nijhout, 1982a; (19) Gras, 1977; (20) Gras and Beaulaton, 1979; (21) Beaulaton and Gras, 1980; (22) Beaulaton et al., 1984; (23) Beaulaton, 1964a,b,c; (24) Beaulaton, 1967a,b; (25) Beaulaton, 1968a,b,c,d,e,f,g,h,i; (26) McDaniel et al., 1976; (27) Beaulaton, 1968a,b,c,d,e,f; (28) Hintze, 1968; (29) Hintze-Podufal, 1971; (30) Blazsek et al., 1975, 1978; (31) Gersch et al., 1975; (32) Sedlak et al., 1983; (33) Sedlak, 1984, 1985; (34) Mamdouh, 1985; (35) Karaçali, 1978; (36) King et al., 1966; (37) Aggarwal and King, 1969.

7.4.7.4.8. Annulate Lamellae

In the gland cells, annulate lamellae have only been found in some lepidopterans: *Antheraea pernyi, Bombyx mori* (Beaulaton, 1968g), *Hyalophora cecropia* (McDaniel et al., 1976), and *Heliothis armigera* (Mamdouh, 1985). They occur as stacks of some 2–40 cisternae, which usually are continuous with the RER. In *Antheraea*, annulate lamellae may be partially located in dense cytoplasmic material similar to that observed during the differentiation of annulate lamellae in *Drosophila* spermatocytes (Kessel, 1981). At the present time, their functional significance remains unclear. The annulate lamellae are usually observed in cells with high metabolic needs.

A resorption process of annulate lamellae by cisternal flattening, collapse of pore complexes, and coalescence of lamellae has been described in *Antheraea* (Beaulaton, 1968g). These morphological changes appear to result in thin lamellar bodies, sequestering a cytoplasmic island (see Section 7.4.7.4.10, below).

7.4.7.4.9. The Golgi Apparatus

Typically, active cells contain numerous dictyosomes in contrast to resting cells, where they are very sparse. Each Golgi apparatus (about 1 μm in diameter in silkworms) consists of a stack of flattened saccules (e.g., two to three in *Rhodnius*, two to six in *Antheraea*) budding smooth-surfaced vesicles (about 20–100 nm in diameter). The stack of fenestrated saccules may be arched and classically is surrounded on the *cis*-face (or forming side) by a smooth-surfaced part of the transitional ER.

In *Rhodnius*, aldehyde fixation with phosphate buffers revealed that the lumen of saccules and peripheral vesicles may be occupied by moderately electron-dense material (Beaulaton and Gras, 1980). More frequently, however, the lumen has a very low density. Acid phosphatase activities have been localized within the Golgi saccules or vesicles in *Leucophaea maderae* (Osinchak, 1966) and *Antheraea pernyi* (Beaulaton, 1967a). Such acid phosphatase–carrying Golgi vesicles can be interpreted as primary lysosomes. Additionally, thiamine pyrophosphatase activity (Osinchak, 1966) and nucleoside di- and monophosphatase activities (Beaulaton, 1967a) were revealed in the Golgi elements at the last larval instar or in early adults.

7.4.7.4.10. Lysosomes, Autophagy, and Cell Death

There are very few cytochemical studies demonstrating lysosomes in the prothoracic gland cells. The first attempts to localize acid phosphatase have been carried out by Osinchak (1966) in *Leucophaea madereae* and Beaulaton (1966) in *Antheraea pernyi*. Acid phosphatase and thiamine pyrophosphatase activities were found in dense bodies.

During the prepupal stage of *Antheraea*, the lysosomes assume various forms as in other cells: dense bodies, vacuolar bodies (200–700 nm in diameter), and multivesicular bodies. This pleomorphic population of lysosomes increases before molting. Activities of thiolacetate esterase at pH 7.0, nucleoside tri- and diphosphatases at pH 7.2, and 5'-nucleotidase at pH 5.0 have also been demonstrated (Beaulaton, 1967a). Furthermore, the autophagic activity increases during the last phase of the secretory cycle (Beaulaton, 1967b; Cassier and Fain-Maurel, 1971; Blazsek et al., 1975; Dorn and Romer, 1976; Glitho, 1977; Gras, 1977; Beaulaton et al., 1984). Beaulaton (1967b) has morphologically and cytochemically analyzed the autophagic activity in *Antheraea* and proposed a tentative scheme of the cycle of events. The sequestration envelope may derive from (a) Golgi elements as in *Leucophaea*, *Blaberus* (Scharrer, 1966), or *Locusta* (Cassier and Fain-Maurel, 1968a, Fain-Maurel and Cassier, 1968a,b, 1969); (b) cisternae of the RER as in *Antheraea* (Beaulaton, 1967b), *Rhodnius* (Gras, 1977), or *Tenebrio* (Glitho, 1977); (c) annulate lamellae after flatenning and coalescence (Beaulaton, 1968g); or (d) mitochondrial complexes with or without association with cisternae of the RER (Beaulaton, 1968e; Cassier and Fain-Maurel, 1971). In a given species, the autophagic activity may simultaneously act through at least two distinct pathways, as demonstrated in *Antheraea*. Acid phosphatase (Osinchak, 1966; Beaulaton, 1967b) and nucleoside di- and triphosphatase activities have been located in autophagic vacuoles at the first two phases of the autophagic process (Beaulaton, 1967b).

During normal degeneration (often occurring before adult emergence), the cells contain many more lysosomes and especially large autophagic vacuoles in which various organelles are sequestered, as in *Leucophaea*, *Blaberus* (Osinchak, 1966; Scharrer, 1966), *Locusta* (Fain-Maurel and Cassier, 1969), *Tenebrio* (Glitho, 1977), *Rhodnius* (Gras, 1977), or *Oncopeltus* (Smith and Nijhout, 1982a). Most authors have noted an asynchrony in the lytic processes between the different cells of a given gland.

Acid phosphatase activity has also been demonstrated in autophagic vacuoles or secondary lysosomes in *Rhodnius* (Gras, 1977) and in *Oncopeltus* (Smith and Nijhout, 1982a) a few hours after emergence. Lysosomal hydrolases seem to be acquired by the autophagic vacuoles either by incorporation of lysosomes or directly from the sequestration envelope. The question then arises as to how the sequence of autophagic activity becomes irreversible in the cell. Possible mechanisms have not yet been found (for a discussion see the review by Beaulaton and Lockshin, 1982). At advanced phases of degeneration, the nuclei become pycnotic before or during the cytoplasmic fragmentation in *Ephemera* (Kaiser, 1978), *Locusta* (Fain-Maurel and Cassier, 1969), and *Rhodnius* (Gras, 1977).

In *Rhodnius*, the gland cells are completely fragmented by 36 h after adult ecdysis, and their residual structures remain ensheathed by the tunica propria.

7.4.7.4.11.   Glycogen and Glycogenic Vacuoles

Glycogen has been cytochemically demonstrated in the gland cells of a few insects: *Bombyx mori* (Kobayashi, 1956), *Philosamia cynthia* (Beaulaton, 1962), *Antheraea pernyi* (Beaulaton, 1964b, 1968f), and *Diatraea grandiosella* (Yin and Chippendale, 1975) at larval and pupal instars. Since their first ultrastructural identification (beta particles of 15–30 nm and alpha particles of 100–150 nm) in the cytosol of prothoracic gland cells in diapausing pupae (Beaulaton, 1962), glycogen granules have been found in the same site and in the same forms in various insects such as *Blaberus* (Scharrer, 1964), *Locusta,* (Fain-Maurel and Cassier, 1968a), *Acheta* (Maleville, 1978a), *Tenebrio* (Romer, 1971; Glitho, 1977; Glitho et al., 1979), *Xyleborus* (Chu et al., 1980), *Oncopeltus* (Herman, 1967; Dorn and Romer, 1976; Smith and Nijhout, 1982a), *Rhodnius* (Beaulaton et al., 1984), *Musca* (Herman, 1967), *Drosophila* (Aggarwal and King, 1969), *Hyalophora* (Herman, 1967), *Diatraea* (Yin and Chippendale, 1975), and *Heliothis* (Mamdouh, 1985).

Another major site of massive accumulation of glycogen has been identified ultracytochemically to be large vacuoles in *Antheraea* larvae (Beaulaton, 1968f). Such glycogen-storing vacuoles are separated from the cytosol by a single membrane (Fig. 7.16), which frequently shows close associations with the tubular network of the SER, as in giant mitochondria (Beaulaton, 1968e). It may provide a pathway for transfer of certain molecules. The glycogenic vacuoles form during mitochondrial transformation (Fig. 7.17A,B) from collapse of the inner membrane (see Section 7.4.7.4.5, above) and accumulation of glycogen particles. Most likely, only the outer mitochondrial membrane is preserved to form the vacuolar membrane. Lastly, glycogen may also be stored in the cisternae of RER, as in *Locusta* nymphs (Fain-Maurel and Cassier, 1968a) or in vacuoles deriving from the RER, as in *Tenebrio* (Romer, 1971).

Some authors have shown that the glycogen storage in the prothoracic glands is a cycle-dependent process (Kobayashi, 1956; Beaulaton, 1968f; Glitho, 1977). In *Bombyx* and *Antheraea,* it begins after feeding and decreases before molting after cessation of feeding. The decline in glycogen stores during the last phase of each cycle is apparently correlated with the peak oxygen-consumption phase (Mosconi-Bernardini and Laudani, 1966).

7.4.7.4.12.   Lipid Droplets

Lipid droplets are seen in the gland cells of various insects such as Thysanura (Cassier and Fain-Maurel, 1971), Orthoptera (Guelin, 1971),

Coleoptera (Romer, 1971; Glitho, 1977), Heteroptera (Gras, 1977), and Lepidoptera (Beaulaton, 1968d; Blazsek et al., 1975). By contrast to the CA (Cassier, 1979), the lipid inclusions in the gland cells, sometimes described as "secretory droplets," appear rather less numerous (see Fig. 14 in Aggarwal and King, 1969; also King et al., 1966). Such lipid droplets consist of globular, coalescent inclusions (0.2–2.0 μm) of moderate electron density, often surrounded by a limiting membrane. Their content stains with Sudan black B (Beaulaton, 1961, 1964b, 1968d). Note that certain lipid droplets of *Antheraea* are closely associated with the tubular SER, as are glycogenic vacuoles (see Section 7.4.7.4.11, above) and giant mitochondria (see Section 7.4.7.4.5, above). At this level, each droplet is cup shaped, as in other cases, suggesting a transport of lipids to the SER (see Fig. 9 in Beaulaton, 1968d). According to this hypothesis, such cup-shaped inclusions seem to appear during the course of extraction and absorption by the SER.

Frequently, lipid droplets can be associated with mitochondria, as was seen in *Rhodnius* (Gras, 1977), and also with glycogen areas, as was seen in *Tenebrio* (Romer, 1971; Glitho, 1977). In the first case, such associations may indicate a mobilization of lipids such as fatty acids, classically known to be oxidized by the mitochondria, or cholesterol, which may be transported from lipid droplets to mitochondria.

Further, it has been determined at least in *Tenebrio* (Glitho, 1977) and *Rhodnius* (Gras, 1977) that the population of lipid droplets increases during each maximal phase of secretory activity, coinciding with the burst of hemolymph ecdysteroids.

7.4.7.4.13.   Centrioles and Microtubules

The centrioles are only known in *Antheraea* and *Rhodnius*. They occur in a juxta-nuclear area with a typical diplosome structure (Beaulaton, 1968d). Microtubules are present in the pericentriolar area and in the peripheral cytoplasm, especially at the level of desmosomes and cytoplasmic processes.

7.4.7.4.14.   Cyclical Steroidogenic and Proteosynthetic Activities

It is now well demonstrated through *in vitro* cultures that the prothoracic glands are the predominant site of ecdysone biosynthesis during postembryonic development (see reviews by Smith et al., 1980; Hoffmann, 1986). In all cases, the glands can synthesize ecdysone from cholesterol, considered as a common precursor of ecdysone (Romer et al., 1974; Hoffmann et al., 1975; Marks, 1976; Hoffmann, 1986). The currently known main natural intermediates between cholesterol and ecdysone are presumably 5β-ketodiol (Smith et al., 1980) and 2-deoxyecdysone. Note that we now have some data on the subcellular distribution of the enzyme systems involved in ecdysone biosynthesis. The side chain hydroxylation systems (C-25 and C-22) are presumably located in the microsomal compartment, whereas those of the C-2 hydroxylation are probably present in the mitochondrial compartment (see the review by Hoffmann, 1986). As in vertebrate steroidogenic cells, it appears that both microsomal and mitochondrial compartments are involved in the biosynthetic pathways of ecdysone. These facts are consistent with the ultrastructural observations showing mitochondrial changes (tubular or vesicular cristae, giant mitochondria) and a differentiation of SER forming networks of tubular or tubulocisternal profiles (Fig. 7.13, 7.15, 7.18, and 7.19) in the active cells. Such ultrastructural changes, characterizing steroid secreting cells, very probably reflect a steroidogenic activity of the prothoracic gland cells in response to effectors. Recent immunocytochemical results support the participation of SER in ecdysone synthesis (Birkenbeil, 1983). Additionally, the close relationships observed between SER and mitochondria suggest their role in the transport of sterol intermediates. Earlier, we recalled the importance of oxygen in ecdysone biosynthesis of *in vitro* gland cultures. The oxygen supply presumably is a rate-limiting factor in the successive hydroxylations of precursors to ecdysone.

The question then arises: how is ecdysone secreted by the prothoracic glands? As in vertebrate steroidogenic cells, the mode of release of steroids remains unclear. Some authors like Blazsek and Mala (1978) or Sedlak (1985) suggest that supposedly ecdysone-containing microvesicles are exocytosed. The first attempts of the immunocytochemical detection of ecdysone did not provided really convincing evidence that microvesicles are involved in ecdysone transfer (Porcheron, 1979; Birkenbeil et al., 1979; Gersch, 1980). Birkenbeil (1983) found an immunocytochemically reactive subpopulation of electron-dense pleomorphic bodies without acid phosphatase activity in the glands of *Galleria*. However, this result needs confirmation. It is generally recognized that under *in vitro* conditions ecdysone is not stored in the gland cells. Carrier proteins possibly transport ecdysone and precursors such as cholesterol from lipid droplets to the SER or mitochondria within the cell, as has been shown for other steroids in vertebrates (Hall, 1984; Sleight, 1987).

Ultrastructural studies suggest that another cyclic activity of the glands deserves consideration. Most authors have found a sharp increase of ribosomes, polysomes, and RER, beginning at activation (see Section 7.4.7.4.6, above). These facts suggest that the gland cells are cyclically involved in a high biosynthetic activity of proteins (Hoffmann and Weins, 1974; Beaulaton et al., 1984), related to the production of carrier proteins and enzymes, especially those used for ecdysone biosynthesis (Gras, 1977; Beaulaton, 1968d).

7.4.7.4.15. Neural Regulation

What is the physiological significance of the neuroglandular junctions mentioned above? Do they affect secretory activity? In *Periplaneta*, Richter (1983) has studied by electrophysiological methods the efferent neural activity of the gland nerves (anterior branch of the fourth segmental nerve emerging from the prothoracic ganglion). In the last two nymphal instars, this author has recorded continuous sequences of efferent potentials, issuing from the prothoracic ganglion. Similarly, Richter and Gersch (1983) have shown that the level of neural activity is precisely correlated with hemolymph ecdysteroid concentrations, suggesting a neurosecretory control of prothoracic gland activity. When the gland nerves are sectioned before day 15 of the last nymphal instar, the rise in ecdysteroid production, normally beginning at the day 15, is prevented. Recently, Richter (1985) demonstrated that sectioning of the connectives between the subesophageal and prothoracic ganglia causes an increase in the neural activity in the prothoracic gland nerve. This *in vitro* experiment corroborates the previous results obtained *in vivo* (Richter, 1983) and indicates that inhibitory signals (prothoracostatin is perhaps involved) are delivered to the prothoracic gland by neurons of the subesophageal ganglion. By contrast, in *Galleria*, the isolated glands are stimulated by the subesophageal ganglion but are inhibited by the prothoracic and mesothoracic ganglia (Mala et al., 1977). Excitation signals are received by the prothoracic glands of *Periplaneta* via the gland nerves. Indeed, Richter (1985) has recently demonstrated that after denervation at a certain phase of intermolt cycle (day 13 of the last instar), only the first peak of ecdysteroids is prevented. Evidently, the regulation of gland activity in *Periplaneta* is bimodal, with a neural control on the first burst of ecdysteroids and presumably a hormonal control on the second burst (maximal production). However, we are not aware of products released at the level of such neuroglandular junctions located beneath the tunica propria. They presumably consist of neuropeptides, owing to the ultrastructural appearance of the large dense-cored granules.

In Blattidea, extirpation experiments of the cervical organs revealed direct relationships between the peripheral nervous system and the glands (Brousse-Gaury, 1971b). Likewise, experiments by unilateral cauterization of ocelli morphometrically showed an asymmetrical development of the paired glands in cockroaches. Brousse-Gaury (1971b) demonstrated that external stimuli promote an activation first of brain neurosecretory cells and secondarily of the CC–CA system, and that they finally modulate synthetic activities of the gland cells. Thus, sensory stimulation leads to a control of the prothoracic gland function through protocephalic neurosecretory pathways in a cascade of physiological events.

7.4.7.4.16.   Endocrine and Humoral Regulation

It is well known that the primary effector of the glands is a brain neuropeptide, the so-called prothoracicotropic hormone (PTTH) produced by the protocerebrum. Several forms of PTTH are now identified and characterized. Agui et al. (1979) have demonstrated that the PTTH of *Manduca* is synthesized in the type III protocerebral neurosecretory cells (homologous of the A-5 cells of *Galleria*, according to Muszynska-Pytel, 1986) and then transported via neurosecretory axons to neurohemal areas. However, in *Locusta*, electrical stimulation has revealed a dual control by median and lateral neurosecretory cells of the protocerebrum (Girardie and de Reggi, 1978). Interestingly, the major neurohemal center for release of brain neurosecretory products in Lepidoptera and Coleoptera resides in an area surrounding each CA (Carrow et al., 1981; Raabe, 1983). PTTH is released from axon terminals ensheathed by the tunica propria (Agui et al., 1979, 1980; Gilbert et al., 1981) during the so-called head-critical period (HCP), or gate (a specific period of photophase, according to Truman, 1972). As has been well established *in vitro*, the PTTH stimulates the gland cells and elicits ecdysone secretion (Agui et al., 1979, 1983; Gilbert et al., 1980, 1981; Roberts et al., 1984). In the last larval instar of *Manduca*, there are two HCP—the first inducing the small peak of commitment presumably responsible for a cellular reprogramming from larval to pupal syntheses, and the second triggering a high burst responsible for molting.

At the present time, the molecular and cellular mechanisms of action of PTTH are being actively studied, especially in *Manduca*. According to Smith and Gilbert (1986), the PTTH acts on Ca^{2+}-dependent adenylate cyclase, as does ACTH (adrenocorticotropic hormone) on the adrenal cortex in vertebrates. These authors assume that the membrane-bound receptors for PTTH are functionally coupled to Ca^{2+}-gated channels owing to the primary role played by Ca^{2+} in stimulating adenylate cyclase. Previous experiments have revealed high adenylate cyclase activities in the gland cells of *Manduca* (Vedeckis and Gilbert, 1973; Vedeckis et al., 1974). Measurements of adenosine $3':5'$-cyclic monophosphate (cAMP) levels have shown that an increase of cAMP precedes the secretory activity (Vedeckis et al., 1976). The level of cAMP in activating cAMP-dependent protein kinase elicits enzyme activations, ultimately stimulating the cells to synthesize ecdysone (Smith and Gilbert, 1986).

Electrophysiological measurements have revealed a higher membrane potential in the active gland cells of *Galleria* than in inactive cells (Gersch and Birkenbeil, 1973). In *Periplaneta*, two prothoracicotropic factors have been separated from extracts of the brain and CC. Only the activator factor II elicits a rise of membrane potential in the gland cells, presumably related to changes of membrane permeability. The activator factor I causes a significant increase of RNA synthesis (all kinds of RNA), as under normal physiological conditions. This event stimulates in turn ecdysone production in the gland cells (Bräuer, 1973; Gersch and Bräuer, 1974; Bräuer et al., 1976, 1977; Gersch, 1980). There is good evidence that steroidogenic activity requires a high rate of RNA synthesis, accompanied in turn by a rise in protein synthesis (Oberlander et al., 1965; Gersch, 1980). These data are consistent with those of an ultrastructural study in *Rhodnius* showing that the activation phase consists of differentiation of the RER followed by that of the SER (Figs. 7.18 and 7.19), with proliferation of ribosomes, polysomes, and other cytoplasmic organelles (Beaulaton et al., 1984). Conversely, it has been demonstrated by these authors that the proliferation of the SER is prevented in the cells of larvae debrained (Fig. 7.20) before the HCP. In *Rhodnius,* the second period of PTTH release must be an induction of steroidogenic activity correlated with the differentiation of the SER.

Several secondary effectors have also been recognized. The regulatory effects of JH on the glands have frequently been reported. It appears that JH can inhibit or stimulate the glands (Nijhout and Williams, 1974; Chippendale and Yin, 1976; Hiruma et al., 1978; Hiruma 1982; Cymborowski and Stolarz, 1979; Porcheron, 1979; Sehnal et al., 1981; Sakurai, 1984; Gruetzmacher et al., 1984b), depending on developmental states. These effects may be indirect, as in *Manduca,* where the release of the fat body stimulatory factor is elicited by JH (Gruetzmacher et al., 1984a). Another stimulatory protein has recently been identified in the hemolymph. It increases ecdysone synthesis approximately fivefold (Watson et al., 1986; Ciancio et al., 1986). This hemolymph stimulatory factor acts *in vitro* with PTTH and causes a 10-fold increase in ecdysone synthesis. The aforementioned authors have also revealed that the response depends on gland competency. However, in diapausing pupae the glands do not respond to the stimulatory factor (Ciancio et al., 1986). In brief, these secondary effectors presumably modulate the action of PTTH on the competent cells.

7.5.   Concluding Remarks

This survey of the prothoracic glands reveals the diversity of anatomic arrangements among various groups of insects and the variability of their location and innervation. Large gaps in our knowledge exist about certain orders of insects such as Trichoptera, Mecoptera, Planipenna, and Homoptera. The assumption that the cephalic glands of Apterygota are homologous to the prothoracic glands of Pterygota is based only on morphological criteria. Indeed, there is some uncertainty about the functional significance of the cephalic glands in Collembola and Diplura. Although their study seems technically difficult, it may be expected that in the years to come the improvement of *in vitro* culture systems will permit investigators to obtain enough products from such glands for the microdetermination of ecdysteroids by radioimmunoassays.

On the other hand, the cytophysiological correlations between gland ultrastructure and hemolymph ecdysteroid levels have revealed cyclic, steroidogenic activities of the cells, especially in identical larvae of *Rhodnius*. Some studies have unequivocally shown that the active gland cells display the typical major characteristics of steroidogenic cells known in vertebrates (e.g., Nussdorfer et al., 1978), as well as in invertebrates other than insects (e.g., Bonaric, 1980; Buchholz and Adelung, 1980). We are now beginning to understand how gland cell activity is regulated. Recent studies have explored this field and especially that of regulatory substances. Their action appears more complex than previously assumed. In spite of the relative complexity, much has been learned about the molecular mechanism of the action of prothoracicotropic neuropeptides (PTTH), providing evidence that Ca^{2+} and cAMP play a role in intracellular signal transfer, as in vertebrate cells.

Significant advances in our knowledge may be expected in the future on the identification of steroidogenic pathways and the purification of enzymes involved in ecdysone synthesis. More cell biology studies on ecdysteroidogenesis remain to be done, especially on the subcellular localization of enzyme systems.

7.6.   Summary

Among insects, the prothoracic glands are also known as ecdysial glands, molting glands, ventral glands in primitive taxa, tentorial glands in Isoptera, and peritracheal glands in Diptera. As is the case in the CA, they appear as paired invaginations of the lateral ectoderm. The originating sites of the glands are surprisingly variable, although in numerous insects they arise from the labial segment. In all embryonic stages of *Oncopeltus*, the gland cells remain diploid, in contrast to postembryonic instars where they are polyploid. At the present time, there is little information about their secretory activity during embryonic development.

The prothoracic glands are paired structures, except in Grylloidea and higher Diptera. In most primitive insects, the glands are situated in the head, whereas in evolved taxa they are found in the anterior thoracic region. Seven anatomic types of these glands have been recognized in various insects: the cephalic compact type (in most primitive taxa); the cephalic ribbon-shaped type (in locusts, Phasmida, and higher Isoptera);

the unpaired fenestrated type (in Grylloidea); the prothoracic X-shaped type associated with coxal muscles (in Dictyoptera and some Isoptera); the prothoracic ribbon-shaped type (especially in various Holometabola); the prothoracic diffuse type (in higher Hymenoptera and Lepidoptera); and the annular type (in higher Diptera). The allometric growth of the prothoracic glands is due to a proliferation of secretory cells (hyperplasia) and/or an increase of cell volume without mitoses (hypertrophy). They also present cyclical volume changes during each postembryonic instar. The abundant tracheal supply of the glands is easily demonstrated and appears to be related to their high oxygen demand. Gland innervation is known in most insects, and the most complex type occurs in Lepidoptera. Three histological types of the glands have been reported: the compact epithelioid type with small cells; the vesicular epithelioid type; and the giant cell epithelioid type. Mitoses are frequently observed in the compact epithelioid type. In contrast a polyploidization process occurs in the giant cell type. Gland degeneration usually occurs during metamorphosis and is apparently programmed by low titers of JI I.

Ultrastructural studies of the prothoracic glands have suggested that the extracellular matrix (or tunica propria) performs supporting as well as filtering functions. This matrix provides a complex network of channels between the polymorphic projections of the gland cells in which neurosecretory terminals, devoid of glial sheaths, have been revealed. In *Galleria*, excitation signals are received by the gland cells via the gland nerves arising from the subesophageal ganglion, and inhibitory signals are released via the nerves issuing from the prothoracic and mesothoracic ganglia. Ultrastructural data acquired over the last 20 years on the changes of the gland cells during each molting cycle are compatible with both cyclical proteosynthetic and steroidogenic activities.

Acknowledgments

I thank Professor C. Noirot for communication of his manuscript and for kindly providing some papers on termites, Janine Roudeix for excellent technical assistance, and Jean-Luc Vincenot for help in preparing photographic prints.

References

Aggarwal, S. K. and R. C. King. 1969. A comparative study of the ring glands from wild type and *1(2)gl* mutant *Drosophila melanogaster*. J. Morphol. 129(2): 171–199.

Agui, N., W. E. Bollenbacher, and L. I. Gilbert. 1983. *In vitro* analysis of prothoracicotropic hormone specificity and prothoracic gland sensitivity in Lepidoptera. Experientia (Basel) 39: 984–988.

Agui, N., W. E. Bollenbacher, N. A. Granger, and L. I. Gilbert. 1980. Corpus allatum is release site for insect prothoracicotropic hormone. Nature (Lond.) 285: 669–670.

Agui, N., N. A. Granger, L. I. Gilbert, and W. E. Bollenbacher. 1979. Cellular localization of the insect prothoracicotropic hormone: *in vitro* assay of a single neurosecretory cell. Proc. Natl. Acad. Sci. U.S.A. 76(11): 5694–5698.

Akai, H. 1983. The structure and ultrastructure of the silk gland. Experientia (Basel) 39: 443–449.

Ando, H. 1962. *The Comparative Embryology of Odonata with Special Reference to a Relic Dragonfly* Epiophlebia superstes *Selys.* Japan Society for the Promotion of Science, Tokyo.

Arvy, L. and M. Gabe. 1950. Données histophysiologiques sur les formations endocrines rétrocérébrales chez les Ecdyonuridae (Ephéméroptères). Bull. Soc. Zool. Fr. 75: 267–285.

Arvy, L. and M. Gabe. 1952. Données histophysiologiques sur les formations endocrines rétro-cérébrales de quelques Odonates. Ann. Sci. Nat. Zool. [11] 14: 345–374.

Arvy, L. and M. Gabe. 1953a. Données histophysiologiques sur la neurosécrétion chez les Paléoptères (Ephéméroptères et Odonates). Z. Zellforsch. Mikrosk. Anat. 38: 591–610.

Arvy, L. and M. Gabe. 1953b. Particularités histophysiologiques des glandes endocrines céphaliques chez *Tenebrio molitor* L. C. R. Acad. Sci. Paris 237D: 844–846.

Arvy, L. and M. Gabe. 1953c. Données histophysiologiques sur les glandes endocrines céphaliques de *Brachyptera risi* Morton (Plécoptère). Arch. Zool. Exp. Gén. 90: 105–122.

Ashhurst, D. E. 1968. The connective tissues of insects. Annu. Rev. Entomol. 13: 45–74.

Ashhurst, D. E. 1982. The structure and development of insect connective tissues. Pp. 313–350 *in* R. C. King and H. Akai (eds.), *Insect Ultrastructure*, Vol. 1. Plenum Press, New York.

Badonnel, A. 1970. Sur les glandes ecdysiales des Psocoptères. Bull. Soc. Zool. Fr. 95(4): 861–868.

Bareth, C. 1964. Une formation endocrine nouvelle de la tête des Diploures Campodéidés. C. R. Acad. Sci. Paris 258D: 5982–5984.

Beaulaton, J. 1961. Observations sur la cytologie de la glande prothoracique des chrysalides en diapause de *Philosamia cynthia* Drury (Lépidoptère, Attacide). C. R. Acad. Sci. Paris 253D: 2126–2128.

Beaulaton, J. 1962. La morphologie du glycogène dans la glande prothoracique de *Philosamia cynthia* Drury, au cours de la diapause nymphale. J. Microsc. (Paris) 1: 469–472.

Beaulaton, J. 1964a. Les ultrastructures des trachées et de leurs ramifications dans la glande prothoracique du ver à soie tussor (*Antheraea pernyi* Guér. Lépidoptère, Attacide). J. Microsc. (Paris) 3: 91–104.

Beaulaton, J. 1964b. Evolution du chondriome dans la glande prothoracique du ver à soie tussor (*Antheraea pernyi* Guér.) au cours du cycle sécrétoire pendant les quatrième et cinquième stades larvaires. J. Microsc. (Paris) 3: 167–186.

Beaulaton, J. 1964c. Sur l'accumulation intramitochondriale de glycogène dans la glande prothoracique du ver à soie du chêne *Antheraea pernyi* Guér. pendant les quatrième et cinquième stades larvaires. C. R. Acad. Sci. Paris 258D: 4139–4141.

Beaulaton, J. 1966. Sur la localisation ultrastructurale d'activités de phosphatases dans la glande prothoracique du ver à soie du chêne: *Antheraea pernyi*. Pp. 89–90 *in* R. Uyeda (ed.), *Sixth International Congress of Electron Microscopy*, Vol. 2. Maruzen, Tokyo.

Beaulaton, J. 1967a. Localisation d'activités lytiques dans la glande prothoracique du ver à soie du chêne (*Antheraea pernyi* Guér.) au stade prénymphal. I. Structures lysosomiques, appareil de Golgi et ergastoplasme. J. Microsc. (Paris) 6: 179–200.

Beaulaton, J. 1967b. Localisation d'activités lytiques dans la glande prothoracique du ver à soie du chêne (*Antheraea pernyi* Guér.) au stade prénymphal. II. Les vacuoles autolytiques (cytolysomes). J. Microsc. (Paris) 6: 349–370.

Beaulaton, J. 1968a. Etude des glandes prothoraciques de vers à soie: ultrastructure et cytochimie. Doctoral thesis (Nat. Sci.), University of Clermont-Ferrand, France, Vol. 1, 233 pp.; Vol. 2, 54 plates.

Beaulaton, J. 1968b. Etude ultrastructurale et cytochimique des glandes prothoraciques de vers à soie aux quatrième et cinquième âges larvaires. I. La *tunica propria* et ses relations avec les fibres conjonctives et les hémocytes. J. Ultrastruct. Res. 23: 474–498.

Beaulaton, J. 1968c. Etude ultrastructurale et cytochimique des glandes prothoraciques de vers à soie aux quatrième et cinquième âges larvaires. II. Les cellules interstitielles et les fibres nerveuses. J. Ultrastruct. Res. 23: 499–515.

Beaulaton, J. 1968d. Etude ultrastructurale et cytochimique des glandes prothoraciques de vers à soie aux quatrième et cinquième âges larvaires. III. Les cellules sécrétrices. J. Ultrastruct. Res. 23: 516–536.

Beaulaton, J. 1968e. Modifications ultrastructurales des cellules sécrétrices de la glande prothoracique de vers à soie au cours des deux derniers âges larvaires. I. Le chondriome et ses relations avec le réticulum agranulaire. J. Cell Biol. 39: 501–525.

Beaulaton, J. 1968f. Modifications ultrastructurales des cellules sécrétrices de la glande prothoracique de vers à soie au cours des deux derniers âges larvaires. II. Le glycogène, ses relations avec le chondriome et le réticulum endoplasmique. J. Microsc. (Paris) 7(5): 673–692.

Beaulaton, J. 1968g. Modifications ultrastructurales des cellules sécrétrices de la glande prothoracique de vers à soie, au cours des deux derniers âges larvaires. III. Les lamelles annelées et leur dégradation. J. Microsc. (Paris) 7(6): 895–906.

Beaulaton, J. 1968h. Sur l'identification cytochimique du glycogène intramitochondrial dans la glande prothoracique du ver à soie du chêne *Antheraea pernyi.* Pp. 217–218 *in* D. V. Bocciarelli (ed.), *Fourth European Regional Conference on Electron Microscopy, Rome,* Vol. 2. Tipografia Poliglotta Vaticana, Rome.

Beaulaton, J. 1968i. Sur l'action d'enzymes au niveau des pores nucléaires et d'autres structures de cellules sécrétrices prothoraciques incluses en Epon. Z. Zellforsch. Mikrosk. Anat. 89: 453–461.

Beaulaton, J. and R. Gras. 1980. Influence of aldehyde fixatives on the ultrastructure of the prothoracic gland cells in *Rhodnius prolixus* (Insecta, Heteroptera) with special reference to their osmotic effects. Microsc. Acta 82(4): 351–366.

Beaulaton, J., R. Gras, and P. Porcheron. 1981. Variations du taux hémolymphatique d'ecdystéroïdes et évolution ultrastructurale des glandes prothoraciques de *Rhodnius prolixus:* nouvelles données sur les corrélations cytophysiologiques, 2 pp. 6e Colloq. Physiol. Insecte, Les Eyzies, University of Paris VI.

Beaulaton, J. and R. Lockshin. 1982. The relation of programmed cell death to development and reproduction: comparative studies and an attempt at classification. Int. Rev. Cytol. 79: 215–235.

Beaulaton, J., P. Porcheron, R. Gras, and P. Cassier. 1984. Cytophysiological correlations between prothoracic gland activity and hemolymph ecdysteroid concentrations in *Rhodnius prolixus* during the fifth larval instar: further studies in normal and decapitated larvae. Gen. Comp. Endocrinol. 53(1): 1–16.

Benedeczky, I., J. Mala, and F. Sehnal. 1980. Ultrastructural study on the innervation of prothoracic glands in *Galleria mellonella.* Gen. Comp. Endocrinol. 41: 400–407.

Bernardini-Mosconi, P. 1958. Le ghiandole endocrine e le cellule neurosecretici protocerebrali di ninfa di *Zootermopsis angusticollis.* Symp. Genet. 6: 129–139.

Bernardini-Mosconi, P. 1959. La ghiandola protoracica in alcune specie di isotteri. Boll. Zool. 26(2): 307–310.

Bernardini-Mosconi, P. 1961. Alcune ricerche sulla endocrinologia dei bachi da seta: tentativo di trapianto di ghiandole protoraciche. Genet. Agrar. 14(1/2): 129–135.

Bernardini-Mosconi, P. 1963. Ricerche istologiche e istochimiche sul sistema endocrino del *Reticulitermes lucifugus* (Rhinotermitidae). Symp. Genet. Biol. Ital. 10: 22–28.

Bernardini-Mosconi, P. and L. Cella-Malugani. 1960. La ghiandola protoracica in alcune specie di Blatte. Symp. Genet. 7: 44–58.

Bernardini, P. and A. M. Palestra. 1956. Le giandole tentoriali in varie caste e stadi di *Calotermes flavicollis* (Calotermitidae—Isoptera). Boll. Zool. 23(2): 727–734.

Birkenbeil, H. 1983. Ultrastructural and immunocytochemical investigation of ecdysteroid secretion by the prothoracic gland of the waxmoth *Galleria mellonella*. Cell Tissue Res. 229(2): 433–441.

Birkenbeil, H. and H. Agricola. 1980. Die Innervierung der Prothorakaldrüse von *Periplaneta americana* L. Zool. Anz. 204: 331–336.

Birkenbeil, H., M. Eckert, and M. Gersch. 1979. Electron microscopical–immunocytochemical evidence of ecdysteroids in the prothoracic gland of *Galleria mellonella*. Cell Tissue Res. 200: 285–290.

Blazsek, I. and J. Mala. 1978. Steroid transport through the surface of the prothoracic gland cells in *Galleria mellonella* L. Cell Tissue Res. 187: 507–513.

Blazsek, I., A. Balazs, V. J. A. Novak, and J. Mala. 1975. Ultrastructural study of the prothoracic glands of *Galleria mellonella* L. in the penultimate, last larval and pupal stages. Cell Tissue Res. 158(2): 269–280.

Bodenstein, D. 1953. Studies on the humoral mechanisms in growth and metamorphosis of the cockroach, *Periplaneta americana*. II. The function of the prothoracic gland and the *corpus cardiacum*. J. Exp. Zool. 123: 413–433.

Boisson, C. 1950. Mode de sécrétion des *corpora incerta* de *Bacillus rossii* (Fabr.). C. R. Soc. Biol. 114: 784–786.

Boisson, C. 1952. Les glandes endocrines céphaliques ventrales ou *corpora incerta* de *Bacillus rossii* Fab. Ann. Sci. Nat. Zool. [11] 14: 453–472.

Bollenbacher, W. E., N. Agui, N. A. Granger, and L. I. Gilbert. 1979. *In vitro* activation of insect prothoracic glands by the prothoracicotropic hormone. Proc. Natl. Acad. Sci. U.S.A. 76(10): 5148–5152.

Bollenbacher, W. E., E. J. Katahira, M. O'Brien, L. I. Gilbert, M. K. Thomas, N. Agui, and A. H. Baumhorer. 1984. Insect prothoracicotropic hormone: evidence for two molecular forms. Science (Wash., D.C.) 224(4654): 1243–1245.

Bonaric, J. C. 1980. Contribution à l'étude de la biologie du développement chez l'araignée *Pisaura mirabilis* (Clerck): approche physiologique des phénomènes de mue et de diapause hivernale. Doctoral thesis, University of Montpellier II, France.

Borst, D. W. and F. Engelman. 1974. *In vitro* secretion of α-ecdysone by prothoracic glands of hemimetabolous insect, *Leucophaea maderae* (Blattaria). J. Exp. Zool. 189: 413–419.

Bounhiol, J. J. 1949. La glande prothoracique du ver à soie. 68e Congr. Assoc. Fr. Av. Sci., Clermont-Ferrand, pp. 141–143.

Bounhiol, J. J. 1980. *Larves et Métamorphoses.* Presses Universitaires de France, Paris.

Brady, J. 1974. The physiology of insect circadian rhythms. Adv. Insect Physiol. 10: 1–115.

Bräuer, R. 1973. Untersuchungen zur neurohormonalen Steuerung der RNS-Synthese der Prothoracaldrüse von *Periplaneta americana* L. Zool. Jahrb. Physiol. 77(2): 107–144.

Bräuer, R., M. Gersch, and E. Baumann: 1976. Specific action of activation factor I on RNA synthesis in prothoracic glands in insect. J. Insect Physiol. 22: 147–151.

Bräuer, R., M. Gersch, G. A. Böhm, and E. Baumann. 1977. *In vitro–*Stimulierung der Synthese von Ecdyson in Prothoracaldrüsen von *Periplaneta americana* durch den Aktivationsfaktor I. Zool. Jahrb. Physiol. 81: 1–12.

Brousse-Gaury, P. 1969. Corpora allata et glandes de mue de *Mantis religiosa* L. reçoivent une innervation partant des sensilles cervicales. Ann. Sci. Nat. Zool. [12] 11: 559–567.

Brousse-Gaury, P. 1971a. Influence de stimuli externes sur le comportement neuro-endocrinien de blattes. I. Les organes sensoriels céphaliques, point de départ de réflexes neuro-endocriniens. Ann. Sci. Nat. Zool. [12] 13(2): 181–332.

Brousse-Gaury, P. 1971b. Influence de stimuli externes sur le comportement neuro-endocrinien de blattes. II. Histophysiologie des voies reflexes neuro-endocriniennes. Ann. Sci. Nat. Zool. [12] 13(3): 333–450.

Brousse-Gaury, P. 1972. Innervation, cérébrale et sous-oesophagienne, des formations endocrines rétrocérébrales, chez *Blabera fusca* Br. et *Periplaneta americana* L. Bull. Biol. Fr. Belg. 106(4): 315–336.

Buchholz, C. and D. Adelung. 1980. The ultrastructural basis of steroid production in the Y-organ and the mandibular organ of the crabs *Hemigrapsus nudus* (Dana) and *Carcinus maenas* L. Cell Tissue Res. 206: 83–94.

Bullière, D., F. Bullière, and M. de Reggi. 1979. Ecdysteroid titres during ovarian and embryonic development in *Blaberus craniifer*. Wilhelm Roux's Arch. Entwicklungsmech. Org. 186: 103–114.

Burgess, L. 1967. Pycnosis of the nuclei of ecdysial gland cells in *Aedes aegypti* L. (Diptera, Culicidae). Can. J. Zool. 45(6): 1294–1295.

Carlisle, D. B. and P. E. Ellis. 1959. La persistance des glandes ventrales céphaliques chez les criquets solitaires. C. R. Acad. Sci. Paris 249: 1059–1060.

Carlisle, D. B. and P. E. Ellis. 1963. Prothoracic gland and gregarious behaviour in locusts. Nature (Lond.) 200(4906): 603–604.

Carrow, G. M., R. L. Calabrese, and C. M. Williams. 1981. Spontaneous and evoked release of prothoracicotropin from multiple neurohemal organs of the tobacco hornworm. Proc. Natl. Acad. Sci. U.S.A. 78(9): 5866–5870.

Cassier, P. 1979. The corpora allata of insects. Int. Rev. Cytol. 57: 1–74.

Cassier, P. and M. A. Fain-Maurel. 1968a. Origine golgienne et évolution des corps vacuolaires impliqués dans la dégénérescence des glandes de mue de *Locusta migratoria migratorioides* (R. et F.). C. R. Acad. Sci. Paris 266D: 1290–1292.

Cassier, P. and M. A. Fain-Maurel. 1968b. Etude infrastructurale de la genèse et de la sécrétion de l'ecdysone, hormone stéroïde dans les glandes de mue de *Locusta migratoria migratorioides* R. et F. (Insecte, Orthoptère). C. R. Acad. Sci. Paris 266D: 2477–2479.

Cassier, P. and M. A. Fain-Maurel. 1968c. Influence de la photopériode sur la persistance ou la dégénérescence des glandes de mue chez les imagos grégaires du criquet migrateur: étude expérimentale et infrastructurale. C. R. Acad. Sci. Paris 267D: 646–648.

Cassier, P. and M. A. Fain-Maurel. 1969a. Etude infrastructurale des glandes de mue de *Locusta migratoria migratorioides* (R. et F.). III. Sur la persistance ou la dégénérescence des glandes ventrales chez les imagos solitaires. Arch. Zool. Exp. Gén. 110(2): 203–224.

Cassier, P. and M. A. Fain-Maurel. 1969b. Etude infrastructurale des glandes de mue de *Locusta migratoria migratorioides* (R. et F.). IV. Evolution des glandes de mue au cours de la maturation sexuelle et des cycles ovariens chez les solitaires verts. Arch. Zool. Exp. Gén. 110(2): 267–278.

Cassier, P. and M. A. Fain-Maurel. 1969c. Nouvelle observation sur la genèse et la localisation de microtubules dans l'ergastoplasme chez le criquet migrateur, *Locusta migratoria migratorioides* (R. et F.). C. R. Acad. Sci. Paris. 268D: 537–539.

Cassier, P. and M. A. Fain-Maurel. 1970. Contrôle plurifactoriel de l'évolution post-imaginale des glandes ventrales chez *Locusta migratoria* L.: données expérimentales et infrastructurales. J. Insect Physiol. 16: 301–318.

Cassier, P. and M. A. Fain-Maurel. 1971. Modalités de l'évolution et du renouvellement du chondriome au cours des cycles d'activités des glandes de mue de *Petrobius maritimus* Leach (insecte aptérygote). Arch. Zool. Exp. Gén. 112: 457–470.

Cavallin, M. and B. Fournier. 1981. Characteristics of development and variations in ecdysteroid levels in *Clitumnus* embryos deprived of their cephalic endocrine glands. J. Insect Physiol. 27(8): 527–534.

Cazal, P. 1947. Recherches sur les glandes endocrines rétrocérébrales des Insęctes. II. Odonates. Arch. Zool. Exp. Gén. 85: 55–82.

Chadwick, L. E. 1956. Removal of prothoracic glands from the nymphal cockroach. J. Exp. Zool. 131: 291–305.

Chalaye, D. 1967. Neurosécrétion au niveau de la chaîne nerveuse ventrale de *Locusta migratoria migratorioides* R. et F. (Orthoptère, Acridien). Bull. Soc. Zool. Fr. 92: 87–108.

Chino, H., S. Sakurai, T. Ohtaki, N. Ikekawa, H. Miyazaki, M. Ishibashi, and H. Abuki. 1974. Biosynthesis of α-ecdysone by prothoracic glands *in vitro*. Science (Wash., D.C.) 183: 529–530.

Chippendale, G. M. and C.-M. Yin. 1976. Endocrine interactions controlling the larval diapause of the southwestern corn borer, *Diatraea grandiosella*. J. Insect Physiol. 22: 989–995.

Chu, H. M., D. M. Norris, and K. D. P. Rao. 1980. Ultrastructure of the prothoracic gland of variously aged female pupae of *Xyleborus ferrugineus* and associated ecdysteroid titers. Cell Tissue Res. 213: 1–8.

Ciancio, M. J., R. D. Watson, and W. E. Bollenbacher. 1986. Competency of *Manduca sexta* prothoracic glands to synthesize ecdysone during development. Mol. Cell. Endocrinol. 44(2): 171–178.

Cymborowski, B. and G. Stolarz. 1979. The role of juvenile hormone during larval–pupal transformation of *Spodoptera littoralis:* switchover in the sensitivity of the prothoracic gland to juvenile hormone. J. Insect Physiol. 25: 939–942.

Darquenne, J. 1978. Développement embryonnaire des glandes prothoraciques chez *Leucophaea maderae* (Fabr.), (Dictyoptera: Blaberidae). Int. J. Insect Morphol. Embryol. 7(3): 215–220.

de Wilde, J., T. H. Hsiao, and C. Hsiao. 1980. The regulation of the metamorphic moults in the Colorado potato beetle *Leptinotarsa decemlineata* Say. Pp. 83–98 *in* J. A. Hoffmann (ed.), *Progress in Hormone Research.* Elsevier/North-Holland Publ., Amsterdam and New York.

Dillon, L. S. 1981. *Ultrastructure, Macromolecules and Evolution.* Plenum Press, New York.

Dorn, A. 1972. Die endokrinen Drüsen im Embryo von *Oncopeltus fasciatus* Dallas (Insecta, Heteroptera). Z. Morphol. Oekol. Tiere 71(1): 52–104.

Dorn, A. and F. Romer. 1976. Structure and function of prothoracic glands and oenocytes in embryos and last larval instars of *Oncopeltus fasciatus* Dallas (Insecta, Heteroptera). Cell Tissue Res. 171(3): 331–350.

Dorn, A., S. T. Bishoff, and L. I. Gilbert. 1987. An incremental analysis of the embryonic development of the tobacco hornworm, *Manduca sexta*. Int. J. Invert. Reprod. Dev. 11: 137–158.

Eastham, L. E. S. 1930. The embryology of *Pieris rapae:* organogeny. Philos. Trans. R. Soc. Lond. 219B: 1–50.

Ewen, A. B. 1962. Histophysiology of the neurosecretory system and retrocerebral endocrine glands of the alfalfa plant bug, *Adelphocoris lineolatus* (Goeze) (Hemiptera, Miridae). J. Morphol. 111(3): 255–273.

Fain-Maurel, M. A. and P. Cassier. 1968a. Etude infrastructurale des glandes de mue de *Locusta migratoria migratorioides*. (R. et F.). Arch. Zool. Exp. Gén. 109(3): 445–476.

Fain-Maurel, M. A. and P. Cassier, 1968b. Rôle de l'appareil de Golgi dans l'involution des glandes de mue du criquet migrateur (*Locusta migratoria migratorioides* R. et F.). Pp. 233–234 *in* D. V. Bocciarelli (ed.), *Fourth European Regional Conference on Electron Microscopy*, Vol. 2. Tipografia Poliglotta Vaticana, Rome.

Fain-Maurel, M. A. and P. Cassier. 1969. Etude infrastructurale des glandes de mue de *Locusta migratoria migratorioides* (R. et F.). II. Analyse morphologique des étapes de la dégénérescence chez les imagos grégaires. Arch. Zool. Exp. Gén. 110(1): 91–124.

Filshie, B. K. 1982. Fine structure of the cuticule of insects and other arthropods. Pp. 281–312 *in* R. C. King and H. Akai (eds.), *Insect Ultrastructure*, Vol. 1. Plenum Press, New York.

Formigoni, A. 1956. Neurosécrétion et organes endocrines chez *Apis mellifica* L. Ann. Sci. Nat. Zool. [11] 18: 283–291.

François, J. 1976. Le tissu conjonctif des Insectes. Pp. 491–516 *in* P. P. Grassé (ed.), *Traité de Zoologie*, Vol. 8, Fasc. 4. Masson, Paris.

Fraser, A. 1957. The retrocerebral endocrine organs of the larva of *Protophormia terrae-novae* Bobineau-Desvoidy (Diptera, Cyclorrhapha). Proc. R. Entomol. Soc. Lond. 32A: 40–46.

Fukuda, S. 1940a. Induction of pupation in silkworm by transplanting the prothoracic gland. Proc. Imp. Acad. (Tokyo) 16: 414–416.

Fukuda, S. 1940b. Hormonal control of molting and pupation in the silkworm. Proc. Imp. Acad. (Tokyo) 16: 417–420.

Fukuda, S. 1941. Role of the prothoracic gland in differentiation of the imaginal characters in the silkworm pupa. Annot. Zool. Jpn. 20: 9–13.

Fukuda, S. 1944. The hormonal mechanism of larval molting and metamorphosis in the silkworm. J. Fac. Sci. Imp. Univ. Tokyo 6: 477–532.

Gabe, M. 1953a. Données histologiques sur les glandes endocrines céphaliques de quelques Thysanoures. Bull. Soc. Zool. Fr. 78: 177–193.

Gabe, M. 1953b. Quelques acquisitions récentes sur les glandes endocrines des Arthropodes. Experientia (Basel) 9: 352–356.

Gande, A. R. and E. D. Morgan. 1979. Ecdysteroids in the developing eggs of the desert locust, *Schistocerca gregaria*. J. Insect Physiol. 25(3): 289–293.

Gande, A. R., E. D. Morgan, and I. D. Wilson. 1979. Ecdysteroid levels throughout the life cycle of the desert locust, *Schistocerca gregaria*. J. Insect Physiol. 25(8): 669–675.

Gersch, M. 1980. Control of activity of the prothoracic glands in insects at the cellular level. Pp. 111–123 *in* J. Hoffmann (ed.), *Progress in Ecdysone Research*. Elsevier/North-Holland Publ., Amsterdam and New York.

Gersch, M. and H. Birkenbeil. 1973. Das Membranpotential der Prothorakaldrüsenzellen von *Galleria mellonella* in Beziehung zum Entwicklungsstadium und nach Einwirkung von Neurohormonen. Zool. Jahrb. Physiol. 77(1): 1–24.

Gersch, M., H. Birkenbeil, and G. Klare. 1976. Das Membranpotential des Prothorakaldrüsenzellen von *Galleria mellonella*. Zool. Jahrb. Physiol. 80(1): 16–23.

Gersch, M., H. Birkenbeil, and J. Ude. 1975. Ultrastructure of the prothoracic gland cells of the last instar of *Galleria mellonella* in relation to the state of development. Cell Tissue Res. 160: 389–397.

Gersch, M. and R. Bräuer. 1974. *In vitro*–Stimulation der Prothoracaldrüsen von Insekten als Testsystem (Prothoracaldrüsentest). J. Insect Physiol. 20: 735–741.

Giebultowicz, J. M. and D. L. Denlinger. 1985. Identification of neurons innervating the ring gland of the flesh fly larva, *Sarcophaga crassipalpis* (Diptera: Sarcophagidae). Int. J. Insect Morphol. Embryol. 14(3): 155–161.

Gilbert, L. I. 1962. Maintenance of the prothoracic glands by the juvenile hormone in insects. Nature (Lond.) 193: 1205–1207.

Gilbert, L. I. 1964. Physiology of growth and development: endocrine aspects. Pp. 149–225 *in* M. Rockstein (ed.), *Physiology of Insecta*, Vol. 1. Academic Press, Orlando, Florida.

Gilbert, L. I., W. E. Bollenbacher, and N. A. Granger. 1980. Insect endocrinology: regulation of endocrine glands, hormone titer, and hormone metabolism. Annu. Rev. Physiol. 42: 493–510.

Gilbert, L. I., W. E. Bollenbacher, N. Agui, N. A. Granger, B. J. Sedlak, D. Gibbs, and C. M. Buys. 1981. The prothoracicotropes: source of the prothorcicotropic hormone. Am. Zool. 21: 641–653.

Gillot, C. and C.-M. Yin. 1972. Morphology and histology of endocrine glands of *Zootermopsis angusticollis* Hagen (Isoptera). Can. J. Zool. 50: 1537–1545.

Gillot, S. 1968. Hétérogénéités fonctionnelles dans l'ADN de noyaux géants: étude autoradiographique sur la glande séricigène de *Bombyx mori* L. Exp. Cell Res. 50: 388–402.

Girardie, A. and M. de Reggi. 1978. Moulting and ecdysone release in response to electrical stimulation of protocrebral neurosecretory cells in *Locusta migratoria*. J. Insect Physiol. 24(12): 797–802.

Glitho, I. 1977. Evolution de la glande de mue pendant la métamorphose de *Tenebrio molitor* L. (Ins. Col.): rapports entre les taux d'ecdystérone et les étapes du développement. Doctoral thesis, third cycle, University of Dijon, France.

Glitho, I., J. P. Delbecque, and J. Delahambre. 1979. Prothoracic gland involution related to moulting hormone levels during the metamorphosis of *Tenebrio molitor* L. J. Insect Physiol. 25: 187–191.

Gorbman, A. and H. A. Bern. 1962. *A Textbook of Comparative Endocrinology.* Wiley, New York.

Granger, N. A. 1978. Innervation of the prothoracic glands in *Galleria mellonella* larvae (Lepidoptera: Pyralidae). Int. J. Insect Morphol. Embryol. 7(4): 315–324.

Gras, R. 1977. Etude ultrastructurale et cytophysiologique des glandes thoraciques (glandes ecdysiales) de *Rhodnius prolixus* Stål. Doctoral thesis, third cycle, University of Clermont-Ferrand II, France.

Gras, R. and J. Beaulaton. 1976. Observations préliminaires sur les modifications ultrastructurales des glandes thoraciques (glandes ecdysiales) de *Rhodnius prolixus* (Ins. Heteroptera) au cours du cycle de mue. J. Microsc. Biol. Cell. 26(2/3): 15a.

Gras, R. and J. Beaulaton. 1978. Données méthodologiques sur la préservation ultrastructurale des cellules endocrines ecdysiales (glandes thoraciques) de *Rhodnius* (Ins. Hétéroptère). Biol. Cell. 33: 22a.

Gras, R. and J. Beaulaton. 1979. Développement cyclique du réticulum endoplasmique agranulaire dans les glandes thoraciques (glandes ecdysiales) de *Rhodnius prolixus* Stål (Insecte, Hétéroptère) aux deux derniers stades larvaires. Experientia (Basel) 35: 386–388.

Grassé, P. P. 1982. Les glandes endocrines. Pp. 429–442 *in* P. P. Grassé (ed.), *Termitologia*, Vol. 1. Masson, Paris.

Griffiths, A. C. and L. H. Finlayson. 1982. Extra-ganglionic neurosecretory and non-neurosecretory neurons in the larva of the moth, *Agrotis segetum* Schiff. (Lepidoptera, Noctuidae). Int. J. Insect Morphol. Embryol. 11(5/6): 341–349.

Griffiths, A. C. and L. H. Finlayson. 1983. Ultrastructural changes in, and evolution of, the ecdysial gland neuron at metamorphosis in *Agrotis, Spodoptera,* and *Manduca* (Lepidoptera). Int. J. Insect Morphol. Embryol. 12: 225–233.

Gruetzmacher, M. C., L. I. Gilbert, and W. E. Bollenbacher. 1984a. Indirect stimulation of the prothoracic glands of *Manduca sexta* by juvenile hormone: evidence for a fat body stimulatory factor. J. Insect Physiol. 30(10): 771–778.

Gruetzmacher, M. C., L. I. Gilbert, N. A. Granger, W. Goodman, and W. E. Bollenbacher. 1984b. The effect of juvenile hormone on prothoracic gland function during the larval–pupal development of *Manduca sexta:* an *in situ* and *in vitro* analysis. J. Insect Physiol. 30(4): 331–340.

Guelin, M. 1968. Données préliminaires sur l'existence de vésicules chromatiniennes différenciées dans les noyaux des trophocytes d' *Ephestia kühniella* Z. (Lépidoptère). C. R. Acad. Sci. Paris 266D: 1740–1742.

Guelin, M. 1970. Etude ultrastructurale de la chromatine liée au sexe dans les noyaux des trophocytes d'*Ephestia kühniella* Z. (Lépidoptère): quelques données cytochimiques. C. R. Acad. Sci. Paris 270D: 984–987.

Guelin, M. 1971. Comparaison de l'ultrastructure des glandes ventrales à l'état "actif" et au cours de la diapause larvaire chez *Pyrgomorpha conica* Oliv. (Orthoptère *Acrididae*). C. R. Acad. Sci. Paris 273D: 764–767.

Guelin, M. 1978. Propriétés de l'hétérochromatine d'un Insecte Lépidoptère: étude ultrastructurale de la chromatine sexuelle dans les noyaux polyploïdes des cellules nourricières d'*Ephestia kühniella* Z., au cours de l'ovogenèse. Doctoral thesis (Nat. Sci.), P. and M. Curie University, Paris VI, Vol. 1, 320 pp.; Vol. 2, 72 plates.

Gündel, von J., A. Müller, E. Müller, and M. Gersch. 1982. Änderungen der Kaliuminnenkonzentrationen als Ursache für die entwicklungsabhängigen Änderungen der Ruhepotentiale von Prothorakaldrüsenzellen von *Galleria mellonella*. Zool. Jahrb. Physiol. 86: 183–192.

Haget, A. 1977. L'embryologie des Insectes. Pp. 1–262 *in* P. P. Grassé (ed.), *Traité de Zoologie*, Vol. 8, Fasc. 5B. Masson, Paris.

Haget, A., A. Ressouches, and J. Rogueda. 1982. Ultrastructural detection of endocrine activities in the ventral glands of the head of *Carausius* (Phasmida: *Lonchodidae*) during embryonic development. Int. J. Insect Morphol. Embryol. 11(2): 99–108.

Hall, P. F. 1984. Cellular organization for steroidogenesis. Int. Rev. Cytol. 86: 53–95.

Harper, E., S. Seifter, and B. Scharrer. 1967. Electron microscopic and biochemical characterization of collagen in blattarian insects. J. Cell Biol. 33(2): 385–393.

Hatier, R. 1973. Etude histochimique ultrastructurale des inclusions lipidiques des cellules corticosurrénaliennes de l'embryon de poulet au cours des stades précoces du développement. C. R. Soc. Biol. 167(6/7): 930–933.

Hayat, M. A. 1981. *Fixation for Electron Microscopy*. Academic Press, Orlando, Florida.

Hazarika, L. K. and A. P. Gupta. 1986. Structure, innervation, persistence, and effects of juvenile hormone on the prothoracic glands in adult *Blattella germanica* (L.) (Dictyoptera, Blattellidae). Zool. Sci. 3: 145–150.

Herlant-Meewis, H. and J. M. Pasteels. 1961. Les glandes de mue de *Calotermes flavicollis* F. (Insecte, Isoptère). C. R. Acad. Sci. Paris 253D: 3078–3080.

Herman, W. S. 1967. The ecdysial glands of arthropods. Int. Rev. Cytol. 22: 269–347.

Herman, W. S. and L. I. Gilbert. 1966. The neuroendocrine system of *Hyalophora cecropia* L. (Lepidoptera: Saturniidae). I. The anatomy and histology of the ecdysial glands. Gen. Comp. Endocrinol. 7: 275–291.

Hintze, C. 1968. Histologische Untersuchungen über die Aktivität der inkretorischen Organe von *Cerura vinula* L. (Lepdoptera) während der Verpuppung. Wilhelm Roux' Arch. Entwicklungsmech. Org. 160: 313–343.

Hintze-Podufal, Ch. 1970. The innervation of the prothoracic glands of *Cerura vinula* L. (Lepidoptera). Experientia (Basel) 26: 1269–1271.

Hintze-Podufal, Ch. 1971. Studies on the fine structure of prothoracic gland cells of *Cerura vinula* (Lepidoptera). J. Zool. 164: 425–428.

Hiruma, K. 1982. Factors affecting change in sensitivity of prothoracic glands to juvenile hormone in *Mamestra brassicae*. J. Insect Physiol. 28(2): 193–199.

Hiruma, K. and N. Agui. 1982. Larval–pupal transformation of the prothoracic glands of *Mamestra brassicae*. J. Insect Physiol. 28(2): 89–95.

Hiruma, K., H. Shimada, and S. Yagi. 1978. Activation of the prothoracic gland by juvenile hormone and prothoracicotropic hormone in *Mamestra brassicae* (Lep., Noctuidae). J. Insect Physiol. 24(3): 215–220.

Hoffmann, J. A. 1986. Ten years of ecdysone workshops: retrospect and perspectives. Insect Biochem. 16(1): 1–9.

Hoffmann, J. A. and J. Koolman. 1974. Prothoracic glands in the regulation of ecdysone titres and metabolic fate of injected labelled ecdysone in *Locusta migratoria* (Orth., Acridiidae). J. Insect Physiol. 20(8): 1593–1601.

Hoffmann, J. A., J. Koolman, and C. Beyler. 1975. Rôle des glandes prothoraciques dans la production d'ecdysone au cours du dernier stade larvaire de *Locusta migratoria* L. C. R. Acad. Sci. Paris 280D: 733–736.

Hoffmann, J. A., M. Lagueux, C. Hetru, M. Charlet, and F. Goltzene. 1980. Ecdysone in reproductively competent female adults and in embryos of insects: Pp. 431–465 *in* J. A. Hoffmann (ed.), *Progress in Ecdysone Research* Elsevier/North-Holland Publ., Amsterdam and New York.

Hoffmann, J. A. and M.-J. Weins. 1974. Activité protéosynthétique des glandes prothoraciques et titre d'ecdysone chez des larves permanentes de *Locusta migratoria* obtenues par irradiation sélective du tissu hématopoïétique. Experientia (Basel) 30(7): 821–822.

Holter, P. 1970. Regular grid-like substructures in the midgut epithelial basement membrane of some Coleoptera. Z. Zellforsch. Mikrosk. Anat. 110: 373–385.

Ichikawa, M., J. Nishiitsutsuji, and K. Yashika. 1955. Studies on the insect metamorphosis. IV. Prothoracic glands of *Ephestia cautella*. Mem. Coll. Sci. Univ. Kyoto, Ser. B 22: 1–9.

Imboden, H. and B. Lanzrein. 1982. Investigations on ecdysteroids and juvenile hormones and on morphological aspects during early embryogenesis in the ovoviviparous cockroach *Nauphoeta cinerea*. J. Insect Physiol. 28: 37–46.

Imboden, H., B. Lanzrein, J. P. Delbecque, and M. Lüscher. 1978. Ecdysteroids and juvenile hormone during embryogenesis in the ovoviviparous cockroach *Nauphoeta cinerea*. Gen. Comp. Endocrinol. 36: 628–635.

Jacob, M. and V. K. K. Prabhu. 1985. Development, endocrine organs and moulting in the embryos of *Dysdercus cingulatus* (Heteroptera: Pyrrhocoridae). Proc. Indian Acad. Sci. (Anim. Sci.) 94(5): 489–501.

Johannsen, O. A. 1929. Some phases in the embryonic development of *Diacrisia virginica* Fabr. (Lepidoptera). J. Morphol. Physiol. 48: 493–541.

Johannsen, O. A. and F. H. Butt. 1941. *Embryology of Insects and Myriapods.* McGraw-Hill, New York.

Joly, L., P. Joly, and A. Porte. 1969. Remarques sur l'ultrastructure de la glande ventrale de *Locusta migratoria* L. (Orthoptère) en population dense. C. R. Acad. Sci. Paris 269D: 917–918.

Joly, P. 1968. *Endocrinologie des Insectes.* Masson, Paris.

Jones, B. M. 1956. Endocrine activity during insect embryogenesis: function of the ventral head glands in locust embryos (*Locustana pardalina* and *Locusta migratoria*, Orthoptera). J. Exp. Biol. 33: 174–185.

Jucci, C. 1924. Su la differenziazione de le caste ne la società dei Termitidi: I Neotenici (Reali veri e neotenici, l'escrezione nei reali neotenici—la fisiologia e la biologia). Atti Accad. Naz. Lincei, Cl. Sci. Fis. Mat. Nat. [5] 14: 269–500.

Jucci, C. 1956. Ricerche sulle ghiandole endocrine nelle Termiti un nuevo campo di studio: la endocrinologia comparata degli Isotteri. Insectes Soc. 3: 283–284.

Kaiser, H. 1978. Light and electron microscopic studies on ventral glands of *Ephemera danica* Mull. (Ephemeroptera, Ephemeridae) during metamorphosis. Int. J. Insect Morphol. Embryol. 7(4): 377–386.

Kaiser, P. 1949. Histologische Untersuchungen über die *Corpora allata* und Prothoraxdrüsen der Lepidopteren in bezug auf ihre Funktion. W. Roux' Arch. Entwicklungsmech. Org. 144: 99–131.

Kaiser, P. 1955. Die Inkretorgane der Termiten im Kasten-differenzierungsgeschehen. Naturwissenschaften 42: 303–304.

Kaiser, P. 1956. Uber die Hormonalorgane der Insekten. Mikrokosmos 45: 97–99.

Karaçali, S. 1978. Crystalloid structures within giant mitochondria of the prothoracic glands of *Spodoptera littoralis* Bois. (Lepidoptera). Cell Tissue Res. 191(2): 357–362.

Karlinsky, A. and T. Srihari. 1973. La glande prothoracique au cours du développement post-embryonnaire chez *Pieris brassicae* L. (Lépidoptère). Bull. Soc. Zool. Fr. 98(2): 243–262.

Kefalides, N. A. 1975. Basement membranes: current concepts of structure and synthesis. Dermatologica 150: 4–15.

Kessel, R. G. 1981. Origin, differentiation and possible functional role of annulate lamellae during spermatogenesis in *Drosophila melanogaster*. J. Ultrastruct. Res. 75: 72–96.

King, D. S., W. E. Bollenbacher, D. W. Borst, W. V. Vedeckis, J. D. O'Connor, P. I. Ittycheriah, and L. I. Gilbert. 1974. The secretion of α-ecdysone by the prothoracic glands of *Manduca sexta in vitro*. Proc. Natl. Acad. Sci. U.S.A. 71: 793–796.

King, D. S. and E. P. Marks. 1974. The secretion and metabolism of α-ecdysone by cockroach (*Leucophaea madera*) tissues *in vitro*. Life Sci. 15: 147–154.

King, D. S. and J. B. Siddall. 1969. Conversion of α-ecdysone to β-ecdysone by crustaceans and insects. Nature (Lond.) 221: 955–956.

King, R. C., S. K. Aggarwal, and D. Bodenstein. 1966. The comparative submicroscopic morphology of the ring gland of *Drosophila melanogaster* during the second and third larval instars. Z. Zellforsch. Mikrosk. Anat. 73: 272–285.

Kobayashi, M. 1953a. Cytohistological studies on the prothoracic gland of the domestic silkworm (*Bombyx mori* L.). I. On the changes of the nucleic acid and the intrnuclear vesicle during moulting. J. Sericult. Sci. Jpn. 22: 155–162.

Kobayashi, M. 1953b. Cytohistological studies on the prothoracic gland of the domestic silkworm (*Bombyx mori* L.). II. On the changes of the nucleic acid and the intranuclear vesicle during pupation. J. Sericult. Sci. Jpn. 22: 163–170.

Kobayashi, M. 1956. Cyto-histological studies on the prothoracic gland of the silkworm, *Bombyx mori*. III. On the histochemical changes of carbohydrates and the physiological significance of the secretions during larval and pupal moulting. Bull. Sericult. Exp. Stn. 14(9): 451–469.

Kobayashi, Y. and H. Ando. 1983. Embryonic development of the alimentary canal and ectodermal derivatives in the primitive moth, *Neomicropteryx nipponensis* (Lepidoptera, Micropterygidae). J. Morphol. 176(3): 289–314.

Koolman, J. 1982. Ecdysone metabolism. Insect Biochem. 12: 225–250.

Lagueux, M., C. Hétru, F. Goltzène, C. Kappler, and J. A. Hoffmann. 1979. Ecdysone titre and metabolism in relation to cuticulogenesis in embryos of *Locusta migratoria*. J. Insect Physiol. 25(9): 709–723.

Lahargue, J. 1951. La glande prothoracique de Toyama chez quelques Lépidoptères: sa persistance chez les nymphes de la Piéride du chou en hivernage prolongé. Proc. Verb. Soc. Linn. Bordeaux 95: 22–27.

Langer, G. A. 1978. The structure and function of the myocardial cell surface (review). Am. J. Physiol. 235(5): H461–H468.

Lanzrein, B. 1975. Programming, induction, or prevention of the breakdown of the prothoracic gland in the cockroach, *Nauphoeta cinerea*. J. Insect Physiol. 21: 367–389.

Lanzrein, B. and M. Lüscher. 1970. Experimentelle Untersuchungen über die Degeneration der Prothorakaldrüsen nach der Adulthäutung bei der Schabe *Nauphoeta cinerea*. Rev. Suisse Zool. 77(3): 616–620.

Lauga-Reyrel, F. 1983. L'écomorphose chez *Hypogastrura tullbergi* (Collomboles): manifestations cytophysiologiques et déterminisme endocrinien. Doctoral thesis (Nat. Sci.), P. Sabatier University, Toulouse, France, Vol. 1, 224 pp; Vol. 2, 62 plates.

Lebrun, D. 1967. La détermination des castes du termite à cou jaune (*Calotermes flavicollis* Fabr.). Bull. Biol. 101(3): 139–217.

Lee, H. T.-Y. 1948. A comparative morphological study of the prothoracic glandular bands of some lepidopterous larvae with special reference to their innervation. Ann. Entomol. Soc. Am. 41: 200–205.

Lhoste, J. 1950. Les modifications de structure des glandes ventrales chez *Forficula auricularia* L. au cours du développement. Bull. Soc. Zool. Fr. 75: 285–292.

Lhoste, J. 1957. Données anatomiques et histophysiologiques sur *Forficula auricularia* L. (Dermaptère). Arch. Zool. Exp. Gén. 95: 75–252.

Locke, M. 1969. The ultrastructure of the oenocytes in the molt/intermolt cycle of an insect. Tissue Cell 1(1): 103–154.

Locke, M. 1970. The control of activities in insect epidermal cells. Mem. Soc. Endocrinol. 18: 285–299.

Locke, M. 1985. The structure of epidermal feet during their development. Tissue Cell 17(6): 901–921.

Locke, M. and P. Huie. 1975. Staining of the elastic fibers in insect connective tissue after tannic acid/glutaraldehyde fixation. Tissue Cell 7(1): 211–216.

Louvet, J. P. 1974. Observations en microscopie électronique des cuticules édifiées par l'embryon et discussion du concept de "mue embryonnaire" dans le cas du phasme *Carausius morosus* Br. (Insecta, Phasmida). Z. Morphol. Tiere 78: 159–179.

Louvet, J. P., M. Cavallin, and B. Fournier. 1979. Evolution des ecdysteroïdes au cours du développement embryonnaire des phasmes *Carausius* et *Clitumnus*. 5e Colloq. Physiol. Insecte, Aix-en-Provence, France.

Lukoschus, F. 1952. Über die Prothoraxdrüse der Honigbiene (*Apis mellifica* L.). Naturwissenschaften 39: 116.

Lüscher, M. 1960. Hormonal control of caste differentiation in termites. Ann. N.Y. Acad. Sci. 89: 549–563.

Lyonet, P. 1762. *Traité anatomique de la chenille qui ronge le Bois de Saule*. The Hague.

Mala, J., N. A. Granger, and F. Sehnal. 1977. Control of prothoracic gland activity in larvae of *Galleria mellonella* (Lep., Galleriidae). J. Insect Physiol. 23(3): 309–316.

Mala, J., V. J. A. Novak, I. Blazsek, and A. Balazs. 1974. The effects of juvenile hormone on the prothoracic glands in *Galleria mellonella* L. I. Morphology of the glands in the course of the postembryonic development (Lep., Galleriidae). Acta Biol. Acad. Sci. Hung. 25(1/2): 85–95.

Maleville, A. 1973. Dégénérescence de la glande prothoracique de l'Orthoptère Gryllide *Acheta domestica* L. C. R. Acad. Sci. Paris 277D: 1803–1804.

Maleville, A. 1975. Description de la glande prothoracique d'*Acheta domestica* L. au cours de l'avant dernier stade larvaire. C. R. Acad. Sci. Paris 280D(10): 1281–1283.

Maleville, A. 1976. Etude ultrastructurale de la glande prothoracique d'*Acheta domestica* L. C. R. Acad. Sci. Paris 282D: 373–375.

Maleville, A. 1978a. Recherches sur la glande prothoracique d'*Acheta domestica* L., en relation avec la mue dans des conditions normales et expérimentales. Doctoral thesis (Nat. Sci.), University of Science and Technology of Languedoc, Montpellier, France.

Maleville, A. 1978b. Cytophysiologie de la glande prothoracique au cours

de l'avant dernier stade larvaire du grillon *Acheta domestica* L. et durant le cycle de régénération. Arch. Anat. Microsc. Morphol. Exp.
67(4): 201–211.

Maltête, F. 1962. Contribution à l'étude chronologique de l'embryogenèse chez *Locusta migratoria migratorioides* R. et F.: développement
des corps allates et des glandes ventrales de la tête. Doctoral thesis,
third cycle (Anim. Biol.), University of Bordeaux, France.

Mamdouh, E. 1985. Relations cytophysiologiques entre la glande prothoracique et le tissu sanguin durant le dernier stade larvaire d'*Heliothis armigera* (Ins. Lep. Noctuidae). Doctoral thesis (Nat. Sci.),
University of Science and Technology of Languedoc, Montpellier,
France.

Marks, E. P. 1976. The uses of cell and organ cultures in insect endocrinology. Pp. 117–132 *in Invertebrate Tissue Culture: Research Applications.* Academic Press, Orlando, Florida.

McDaniel, C. N., E. Johnson, T. Saum, and S. J. Berry. 1976. Ultrastructure of active and inhibited prothoracic glands. J. Insect Physiol.
22(3): 473–481.

Miyakawa, K. 1974. The embryology of the caddisfly *Stenopsyche griseipennis* MacLachlan (Trichoptera: Stenopsychidae). III. Organogenesis:
ectodermal derivatives. Kontyu 42(3): 305–324.

Mizoguchi, A. and H. Ishizaki. 1982. Prothoracic gland of the saturniid
moth *Samia cynthia ricini* possess a circadian clock controlling gut
purge timing. Proc. Natl. Acad. Sci. U.S.A. 79: 2726–2730.

Mizoguchi, A. and H. Ishizaki. 1984. Further evidence for the presence of
a circadian clock in the prothoracic glands of the saturniid moth
Samia cynthia ricini: decapitated larvae can respond to light–dark
changes. Dev. Growth Differ. 26(6): 607–611.

Moriyama, H., K. Nakanishi, D. S. King, T. Okauchi, J. B. Siddall, and W.
Hafferl. 1970. On the origin and metabolic fate of α-ecdysone in insects. Gen. Comp. Endocrinol. 15(1): 80–87.

Mosconi-Bernardini, P. 1964. Morphologie de la glande ventrale et de la
glande prothoracique chez les blattes et les termites. Ann. Endocrinol. 25(5, Suppl.): 72–78.

Mosconi-Bernardini, P. 1966. Différents aspects de la dégénérescence de
la glande prothoracique chez *Leucophea maderae* et *Pycnoscelus sur-
inamensis.* Ann. Endocrinol. 27(3bis, Suppl.): 367–370.

Mosconi-Bernardini, P. and U. Laudani. 1966. La consommation d'oxy-
gène de la glande prothoracique de *Leucophea maderae* et *Periplaneta
americana* d'après l'aspect histologique des organes endocrines. J.
Insect Physiol. 12(10): 1289–1294.

Munn, E. A. 1974. *The Structure of Mitochondria.* Academic Press, Orlando, Florida.

Muszynska-Pytel, M. 1986. Changes in the brain neurosecretory cells of the wax moth, *Galleria mellonella* L. (Lepidoptera), during larval–pupal development. Folia Biol. (Cracow) 34(4): 435–448.

Nandchahal, N. 1967. Stomodeal nervous system of *Gryllodes sigillatus* Walker. Bull. Entomol. 8: 43–47.

Nijhout, H. F. and C. M. Williams. 1974. Control of moulting and metamorphosis in the tobacco hornworm, *Manduca sexta* (L.): cessation of juvenile hormone secretion as a trigger for pupation. J. Exp. Biol. 61: 493–501.

Noirot, C. 1969. Glands and secretions. Pp. 89–123 *in* K. Krishna and F. M. Weesner (eds.), *Biology of Termites*, Vol. 1. Academic Press, Orlando, Florida.

Noirot, C. and C. Noirot-Timothée. 1982. The structure and development of the tracheal system. Pp. 351–381 *in* R. C. King and H. Akai (eds.), *Insect Ultrastructure*, Vol. 1. Plenum Press, New York.

Normann, C. 1965. The neurosecretory system of the adult *Calliphora erythrocephala*. I. The fine structure of the corpus cardiacum with some observations on adjacent organs. Z. Zellforsch. Mikrosk. Anat. 67: 461–501.

Nunez, J. A. 1954. Über das Vorkommen von Prothoraxdrüsen bei *Anisotarsus cupripennis* (Coleoptera, Carabidae). Biol. Zentralbl. 73: 602–610.

Nussdorfer, G. G., G. Mazzocchi, and V. Meneghelli. 1978. Cytophysiology of the adrenal zona fasciculata. Int. Rev. Cytol. 55: 291–365.

Nussdorfer, G. G., C. Robba, G. Mazzocchi, and P. Rebuffat. 1980. Effects of human chorionic gonadotropins on the interstitial cells of the rat testis: a morphometric and radioimmunological study. Int. J. Androl. 3(3): 319–332.

Oberlander, H., S. J. Berry, A. Krishnakumaran, and H. A. Schneiderman. 1965. RNA and DNA synthesis during activation and secretion of the prothoracic glands of saturniid moths. J. Exp. Zool. 159(1): 15–32.

Okada, M. 1960. Embryonic development of the rice stem-borer, *Chilo suppressalis*. Sci. Rep. Tokyo Kyoiku Daigaku, Sect. B 9: 243–296.

Okot-Kotber, B. M. 1980. Histological and size changes in corpora allata and prothoracic glands during development of *Macrotermes michaelseni* (Isoptera). Insectes Soc. 27(4): 361–376.

Okot-Kotber, B. M. 1982. Correlation between larval weights, endocrine gland activities and competence period during differentiation of workers and soldiers in *Macrotermes michaelseni* (Isoptera, Termitidae). J. Insect Physiol. 28(11): 905–910.

Okuda, M., S. Sakurai, and T. Ohtaki. 1985. Activity of the prothoracic gland and its sensitivity to prothoracicotropic hormone in the penultimate and last-larval instar of *Bombyx mori. J. Insect Physiol.* 31(6): 455–461.

Osinchak, J. 1966. Ultrastructural localization of some phosphatases in the prothoracic gland of the insect *Leucophaea maderae. Z. Zellforsch. Mikrosk. Anat.* 72: 236–248.

Ozeki, K. 1968. Experimental studies on the regression of the ventral glands of the earwig, *Anisolabis maritima,* during metamorphosis. *Sci. Pap. Coll. Gen. Educ. Univ. Tokyo (Biol.)* 18: 199–219.

Pflugfelder, O. 1938. Weitere experimentelle Untersuchungen über die Funktion der Corpora allata von *Dixippus morosus. Z. Wiss. Zool.* 151: 149–191.

Pflugfelder, O. 1947. Über die Ventraldrüsen und einige andere inkretorische Organe des Insektenkopfes. *Biol. Zentralbl.* 66: 211–235.

Pihan, J. C. 1966. La glande péritrachéenne de *Tipula flavolineata* Meigen. *Arch. Zool. Exp. Gén.* 107: 261–274.

Pinet, J. M. 1961. Etude anatomique et biologique de l'Ephémère *Oligoneuriella rhenana* Imkoff. Diplôme Et. Sup. Zool., Faculty of Science, University of Paris.

Porcheron, P. 1979. L'hormone de mue des Arthropodes: dosage radioimmunologique, production, divers aspects de son rôle physiologique. Doctoral thesis (Nat. Sci.), Pierre et Marie Curie University, Paris.

Possompès, B. 1946. Les glandes endocrines post-cérébrales des Diptères. I. Etude chez la larve de *Chironomus plumosus* L. *Bull. Soc. Zool. Fr.* 71: 99–109.

Possompès, B. 1953. Recherches expérimentales sur le déterminisme de la métamorphose de *Calliphora erythrocephala. Arch. Zool. Exp. Gén.* 89: 203–364.

Poulson, D. F. 1945. On the origin and nature of the ring gland (Weismann's ring) of the higher Diptera. *Trans. Conn. Acad. Arts Sci.* 36: 449–469.

Raabe, M. 1972. Les organes périsympathiques des Dermaptères. *C. R. Acad. Sci. Paris* 275D: 2925–2928.

Raabe, M. 1983. The neurosecretory–neurohaemal system of insects: anatomical, structural and physiological data. *Adv. Insect Physiol.* 17: 205–303.

Raabe, M., N. Baudry, J. P. Grillot, and A. Provansal. 1974. The perisympathetic organs of insects. Pp. 59–71 *in Neurosecretion: The Final Neuroendocrine Pathway* (Proc. 6th Int. Symp. Neurosecretion). Springer-Verlag, Berlin and New York.

Rae, C. A. 1957. Prothoracic glands in a mantid (*Orthodera ministralis* Fabr.). *Aust. J. Sci.* 19: 229.

Rae, C. A. and A. F. O'Farrel. 1959. The retrocerebral complex and ventral glands of the primitive Orthopteroid *Grylloblatta campodeiformis*. Proc. R. Entomol. Soc. Lond. 34A: 76–82.

Rahm, U. H. 1952a. Über Bau und Funktion der Prothoraxdrüse von *Sialis lutaria* L. (Megaloptera). Experientia (Basel) 8: 62.

Rahm, U. K. 1952b. Die innersekretorische Steuerung der postembryonalen Entwicklung von *Sialis lutaria* L. (Megaloptera). Rev. Suisse Zool. 59: 173–237.

Ranade, D. R. 1964. Observations on the ring gland of the larva of *Musca domestica nebulo* Fabr. (Diptera, Cyclorrhapha, Muscidae). Sci. Cult. 30(4): 205–206.

Rempel, J. G. and P. G. Rueffel. 1964. The retrocerebral glands of mosquito larvae. Can. J. Zool. 42: 39–51.

Rensing, L. 1966. Zur circadianen Rhythmik des Hormonsystems von *Drosophila*. Z. Zellforsch. Mikrosk. Anat. 74: 539–558.

Rensing, L., B. Thach, and V. Bruce. 1965. Daily rhythms in the endocrine glands of *Drosophila* larvae. Experientia (Basel) 21: 103–104.

Richards, O. W. and R. G. Davies. 1977. *Imms' General Textbook of Entomology* (10th ed.), Vol. 1: *Structure, Physiology and Development*. Chapman & Hall, London.

Richter, K. 1983. Experimentelle Untersuchungen der nervösen Steuerung der Prothorakaldrüse bei *Periplaneta americana*. Zool. Jahrb. Physiol. 87: 65–78.

Richter, K. 1984. Die Prothorakaldrüse im Imaginalstadium von *Periplaneta americana*. Zool. Jahrb. Physiol. 88: 433–440.

Richter, K. 1985. Physiological investigation on the role of the innervation in the regulation of the prothoracic gland in *Periplaneta americana*. Arch. Insect Biochem. Physiol. 2(3): 319–329.

Richter, K. and M. Gersch. 1983. Electrophysiological evidence of nervous involvement in the control of the prothoracic gland in *Periplaneta americana*. Experientia (Basel) 39(8): 917–918.

Rinterknecht, E. and G. Matz. 1983. Oenocyte differentiation correlated with the formation of ectodermal coating in the embryo of a cockroach. Tissue Cell 15(3): 375–390.

Roberts, B., L. I. Gilbert, and W. E. Bollenbacher. 1984. *In vitro* activity of dipteran ring glands and activation by the prothoracicotropic hormone. Gen. Comp. Endocrinol. 54: 469–477.

Romer, F. 1971. Die Prothorakaldrüsen der Larve von *Tenebrio molitor* L. (Tenebrionidae, Coleoptera) und ihre Veränderungen während eines Häutungszyklus. Z. Zellforsch. Mikrosk. Anat. 122: 425–455.

Romer, F., H. Emmerich, and J. Nowock. 1974. Biosynthesis of ecdysones in isolated prothoracic glands and oenocytes of *Tenebrio molitor in vitro*. J. Insect Physiol. 20: 1975–1987.

Sakurai, S. 1984. Temporal organization of endocrine events underlying larval–pupal metamorphosis in the silkworm, *Bombyx mori*. J. Insect Physiol. 30(8): 657–664.

Saunders, D. S. 1981. Insect photoperiodism—the clock and the counter: a review. Physiol. Entomol. 6: 99–116.

Sbrenna-Micciarelli, A. and G. Sbrenna. 1972. The embryonic apolyses of *Schistocerca gregaria* (Orthoptera). J. Insect Physiol. 18: 1027–1037.

Schaller, F. 1955. Etude comparative de la glande prothoracique dans les trois castes de l'Abeille (*Apis mellifica* L.). C. R. Soc. Biol. 149: 1487–1490.

Schaller, F. 1960. Etude du développement post-embryonnaire d'*Aeschna cyanea* Müll. Ann. Sci. Nat. Zool. [12] 2: 751–868.

Scharrer, B. 1948. The prothoracic glands of *Leucophaea maderae* (Orthoptera). Biol. Bull. (Woods Hole) 95: 186–198.

Scharrer, B. 1964. The fine structure of blattarian prothoracic glands. Z. Zellforsch. Mikrosk. Anat. 64: 301–326.

Scharrer, B. 1966. Ultrastructural study of the regressing prothoracic glands of blattarian insects. Z. Zellforsch. Mikrosk. Anat. 69: 1–21.

Scharrer, B. 1987. Insects as models in neuroendocrine research. Annu. Rev. Entomol. 32: 1–16.

Schmidt, G. H. 1964. Aktivitätsphasen bekannter Hormondrüsen während der Metamorphose von *Formica polyctena* Foerst. (Hym. Ins.). Insectes Soc. 2(1): 41–57.

Schwinck, I. 1951. Veränderungen der Epidermis, der Perikardialzellen und der Corpora allata in der Larven-Entwicklung von *Panorpa communis* L. unter normalen und experimentellen Bedingungen. Wilhelm Roux' Arch. Entwicklungsmech. Org. 145: 62–108.

Sedlak, B. J. 1984. The ultrastructure of interacting endocrine and target cells. Pp. 269–302 *in* R. C. King and H. Akai (eds.), *Insect Ultrastructure*, Vol. 2. Plenum Press, New York.

Sedlak, B. J. 1985. Structure of endocrine glands. Pp. 25–60 *in* G. A. Kerkut and L. I. Gilbert (eds.), *Comprehensive Insect Physiology, Biochemistry and Pharmacology*, Vol. 7: *Endocrinology*, Pt. I. Pergamon Press, Oxford and Elmsford, New York.

Sedlak, B. J., L. Marchione, B. Devorkin, and R. Davino. 1983. Correlations between endocrine gland ultrastructure and hormone titers in the fifth larval instar of *Manduca sexta*. Gen. Comp. Endocrinol. 52(2): 291–310.

Sehnal, F., P. Maroy, and J. Mala. 1981. Regulation and significance of ecdysteroid titre fluctuations in lepidopterous larvae and pupae. J. Insect Physiol. 27(8): 535–544.

Sellier, R. 1951. La glande prothoracique des Gryllides. Arch. Zool. Exp. Gén. 88: 61–72.

Sen, S. K. and G. A. Gangrade. 1977. Endocrine organs of different larval instars of *Spodoptera (Prodenia) litura* Fabricius (Lepidoptera: Noctuidae). Experientia (Basel) 33(9): 1247–1249.

Singh, H. H. and F. Sehnal. 1979. Lack of specific neurons in the ventral nerve cord for the control of prothoracic glands. Experientia (Basel) 35: 1117–1119.

Singh, H. H., K. P. Srivastava, and R. L. Katiyar. 1986. Neuronal distribution in the ganglia of the cervico-thoracic segment of the ventral nerve cord of the larva of waxmoth, *Galleria mellonella*, as revealed by cobalt-filling technique. Curr. Sci. 55(1): 49–51.

Singh, Y. N. 1975. The anatomy and innervation of the ecdysial glands of the mature larva of castor silk moth *Philosamia ricini* Hutt. Experientia (Basel) 31(1): 40–42.

Slama, K. and J. Mala. 1984. Growth and respiratory enzyme activity in the prothoracic glands of *Galleria*. J. Insect Physiol. 30(11): 877–882.

Sleight, R. G. 1987. Intracellular lipid transport in eukaryotes. Annu. Rev. Physiol. 49: 193–208.

Smith, S. L., W. E. Bollenbacher, and L. I. Gilbert. 1980. Studies on the biosynthesis of ecdysone and 20-hydroxyecdysone in the tobacco hornworm, *Manduca sexta*. Pp. 139–162 *in* J. A. Hoffmann (ed.), *Progress in Hormone Research*. Elsevier/North-Holland Publ., Amsterdam and New York.

Smith, W. A. and L. I. Gilbert. 1986. Cellular regulation of ecdysone synthesis by the prothoracic glands of *Manduca sexta*. Insect Biochem. 16(1): 143–147.

Smith, W. A. and H. F. Nijhout. 1982a. Ultrastructural changes accompanying secretion and cell death in the molting glands of an insect (*Oncopeltus*). Tissue Cell 14(2): 243–252.

Smith, W. A. and H. F. Nijhout. 1982b. Synchrony of juvenile hormone–sensitive periods for internal and external development in last-instar larvae of *Oncopeltus fasciatus*. J. Insect Physiol. 28(9): 797–803.

Smith, W. A. and H. F. Nijhout. 1983. *In vitro* stimulation of cell death in the moulting glands of *Oncopeltus fasciatus* by 20-hydroxyecdysone. J. Insect Physiol. 29(2): 169–176.

Springhetti, A. 1957. Ghiandole tentoriali (ventrali, protoraciche) e corpora allata in *Kalotermes flavicollis* Fabr. Symp. Genet. Biol. Ital. 5: 333–349.

Srivastava, K. P., R. L. Katiyar, R. K. Tiwari, H. H. Singh, and V. K. Awasthi. 1982. Anatomical and histological changes in the prothoracic glands of the lemon-butterfly, *Papilio demoleus* L. (Lepidoptera) during the larval–pupal–adult development. Entomon 7(3): 349–359.

Srivastava, K. P. and H. H. Singh. 1968. On the innervation of the prothoracic glands in *Papilio demoleus* L. (Lepidoptera). Experientia (Basel) 24: 838–839.

Srivastava, K. P., R. K. Tiwari, and P. Kumar. 1977. Effect of sectioning the prothoracic gland nerves in the larva of the lemon butterfly, *Papilio demoleus* L. Experientia (Basel) 33(1): 98–99.

Srivastava, U. S. 1958. Prothoracic glands in *Tenebrio molitor* L. (Coleoptera: Tenebrionidae). Nature (Lond.) 181: 1668.

Srivastava, U. S. 1959. The prothoracic glands of some coleopteran larvae. Q. J. Microsc. Sci. 100: 51–64.

Srivastava, U. S. 1960. Secretory cycle and disappearance of the prothoracic glands in *Tenebrio molitor* L. (Coleoptera: Tenebrionidae). Experientia (Basel) 16(10): 445.

Srivastava, U. S. and M. P. Singh. 1969. Observations on the morphology and secretory activity of the thoracic glands in *Chrysocoris stollii* Wolff (Heteroptera: Pentatomidae). Beitr. Entomol. 19(3–6): 537–542.

Strich, M. C. 1954. Etude de la glande ventrale chez *Locusta migratoria migratoroides* L. (Orth. Acridoidea). Ann. Sci. Nat. Zool. [11] 16: 399–411.

Strich-Halbwachs, M. C. 1959. Contrôle de la mue chez *Locusta migratoria*. Ann. Sci. Nat. Zool. [12] 1: 483–570.

Takaoka, M. 1957. Studies of the metamorphosis in insects. II. Relation between pupation and the tracheal system in *Drosophila melanogaster*. Embryologia 3: 343–352.

Takaoka, M. 1959. Studies of the metamorphosis in insects. III. Relation between pupation and oxygen tension in *Drosophila melanogaster*. Embryologia 4: 237–246.

Tanaka, M. 1985. Embryonic development of prothoracic glands of *Parnassius glacialis* Butler, *Luehdorfia japonica* Leech, *Byasa (Atrophaneura) alcinous alcinous* (Klug) and *Papilio protenor demetrius* Cramer (Lepidoptera, Papilionidae). Kontyu 53(4): 671–676.

Thomas, A. and M. Raabe. 1974. Les organes périsympathiques des Orthoptères. Bull. Soc. Zool. Fr. 99(1): 187–206.

Thomsen, M. 1951. Weismann's ring and related organs in larvae of Diptera. K. Dan. Vidensk. Selsk. Biol. Skr. 6: 1–32.

Tombes, A. S. 1970. *An Introduction to Invertebrate Endocrinology.* Academic Press, Orlando, Florida.

Toyama, K. 1902. Contributions to the study of silkworms. I. On the embryology of the silkworm. Bull. Coll. Agric. Tokyo Imp. Univ. 5: 73–118.

Traut, W. and D. Scholz. 1978. Structure, replication and transcriptional activity of the sex-specific heterochromatin in a moth. Exp. Cell Res. 113: 85–94.

Truman, J. W. 1972. Physiology of insect rhythms. I. Circadian organization of the endocrine events underlying the moulting cycle of larval tobacco hornworms. J. Exp. Biol. 57: 805–820.

Ura, T. 1983. Changes in the prothoracic gland cells of the silkworm, *Bombyx mori*, during metamorphosis from larvae to pupae. Zool. Mag. 92(3): 267–275.

Vedeckis, W. V., W. E. Bollenbacher, and L. I. Gilbert. 1974. Cyclic AMP as a possible mediator for prothoracic gland activation. Zool. Jahrb. Physiol. 78(3): 440–448.

Vedeckis, W. V., W. E. Bollenbacher, and L. I. Gilbert. 1976. Insect prothoracic glands: a role for cyclic AMP in the stimulation of ecdysone secretion. Mol. Cell. Endocrinol. 5: 81–88.

Vedeckis, W. V. and L. I. Gilbert. 1973. Production of cyclic AMP (adenosine 3',5'-cyclic monophosphate) and adenosine by the brain and prothoracic glands of *Manduca sexta* (Lep., Sphingidae). J. Insect Physiol. 19(12): 2445–2457.

Vignau, J. 1968. Origine embryonnaire des glandes ventrales de la tête chez *Carausius morosus* Br. (Précisions expérimentales). Bord. Méd. 9: 1693–1694.

Watson, J. A. L. 1963. The cephalic endocrine system in the Thysanura. J. Morphol. 113: 359–376.

Watson, J. A. L. 1964. Moulting and reproduction in the adult firebrat, *Thermobia domestica* (Packard) (Thysanura, Lepismatidae). I. The moulting cycle and its control. J. Insect Physiol. 10: 305–317.

Watson, R. D., L. R. Whisenton, W. E. Bollenbacher, and N. A. Granger. 1986. Interendocrine regulation of the corpora allata and prothoracic glands of *Manduca sexta*. Insect Biochem. 16(1): 149–155.

Weir, J. S. 1959. Changes in the retro-cerebral endocrine system of larvae of *Myrmica*, and their relation to larval growth and development. Insectes Soc. 6: 375–386.

Welch, R. M. 1957. A developmental analysis of the lethal mutant *1(2)gl* of *Drosophila melanogaster* based on cytophotometric determination of nuclear desoxyribonucleic acid (DNA) content. Genetics 42: 544–559.

Wells, M. J. 1954. The thoracic glands of Hemiptera Heteroptera. Q. J. Microsc. Sci. 95: 231–244.

Whitten, J. M. 1964. Connective tissue membranes and their apparent role in transporting neurosecretory and other secretory products in insects. Gen. Comp. Endocrinol. 4: 176–192.

Wigglesworth, V. B. 1933. The physiology of the cuticle and of ecdysis in *Rhodnius prolixus* (Triatomidae, Hemiptera); with special reference to the function of the oenocytes and of the dermal glands. Q. J. Microsc. Sci. 76: 269–318.

Wigglesworth, V. B. 1952a. The thoracic gland in *Rhodnius prolixus* (Hemiptera) and its role in moulting. J. Exp. Biol. 29: 561–570.

Wigglesworth, V. B. 1952b. Hormone balance and the control of metamorphosis in *Rhodnius prolixus* (Hemiptera). J. Exp. Biol. 29: 620–631.

Wigglesworth, V. B. 1954. Growth and regeneration in the tracheal system of an insect, *Rhodnius prolixus* (Hemiptera). Q. J. Microsc. Sci. 95(1): 115–137.

Wigglesworth, V. B. 1955. The breakdown of the thoracic gland in the adult insect, *Rhodnius prolixus*. J. Exp. Biol. 32: 485–491.

Wigglesworth, V. B. 1964. The hormonal regulation of growth and reproduction in insects. Adv. Insect Physiol. 2: 247–336.

Wigglesworth, V. B. 1970. *Insect Hormones*. Oliver & Boyd, Edinburgh.

Williams, C. M. 1946. Physiology of insect diapause: the role of the brain in the production and termination of pupal dormancy in the giant silkworm, *Platysamia cecropia*. Biol. Bull. (Woods Hole) 90: 234–243.

Williams, C. M. 1947. Physiology of insect diapause. II. Interaction between the pupal brain and prothoracic glands in the metamorphosis of the giant silkworm: *Platysamia cecropia*. Biol. Bull. (Woods Hole) 93: 89–98.

Williams, C. M. 1948. Physiology of insect diapause. III. The prothoracic glands in the cecropia silkworm, with special reference to their significance in embryonic and postembryonic development. Biol. Bull. (Woods Hole) 94: 60–65.

Williams, C. M. 1949. The prothoracic glands of insects in retrospect and in prospect. Biol. Bull. (Woods Hole) 97: 111–114.

Williams, C. M. 1952. Physiology of insect diapause. IV. The brain and prothoracic glands as an endocrine system in the cecropia silkworm. Biol. Bull. (Woods Hole) 103: 120–138.

Wilson, I. D. and E. D. Morgan. 1978. Variations in ecdysteroid levels in 5th instar larvae of *Schistocerca gregaria* in gregarious and solitary phases. J. Insect Physiol. 24: 751–756.

Wolbert, P. 1979. Cytologische Veränderungen der Prothoraxdrüse und der Corpora allata in der Puppe der grossen Wachsmotte, *Galleria mellonella* L. Zool. Jahrb. Physiol. 83(1): 43–56.

Yin, C. M. and G. M. Chippendale. 1973. Endocrine system of mature diapause and non-diapause larvae of the southwestern corn borer, *Diatraea grandiosella*. Ann. Entomol. Soc. Am. 66: 943–947.

Yin, C. M. and G. M. Chippendale. 1975. Insect prothoracic glands: function and ultrastructure in diapause and non-diapause larvae of *Diatraea grandiosella*. Can. J. Zool. 53(2): 124–131.

Yokoyama, T. 1956. The morphology and innervation of the prothoracic gland in the silkworm, *Bombyx mori*. J. Sericult. Sci. Jpn. 25: 87–94.

Zimowska, G., A. M. Handler, and B. Cymborowski. 1985. Cellular events in the prothoracic glands and ecdysteroid titres during the last-larval instar of *Spodoptera littoralis*. J. Insect Physiol. 31(4): 331–340.

Zuberi, H. and P. Peeters. 1964. A study of the neurosecretory cells and the endocrine glands of *Cubitermes exiguus* Mathot. Pp. 87–105 *in* A. Bouillon (ed.), *Etudes sur les termites africains*. Masson, Paris.

Morphology, Histology, Ultrastructure, and Organogenesis of the Androgenic Glands in Crustacea

8

GENEVIÈVE G. PAYEN

8.1. Introduction

Male malacostracans are the only crustaceans to possess an androgenic gland (AG), source of the masculinizing hormone. Among the malacostracans, these glands have been studied in amphipods, isopods, and decapods. The latter are of economic interest, and experimental work on them is beginning to be conducted.

Since its identification in the amphipod *Orchestia gammarella* by Charniaux-Cotton (1954), the AG has been observed in most orders of Malacostraca (Charniaux-Cotton, 1956a; Charniaux-Cotton et al., 1966; Zerbib, 1967). Search of the literature revealed that Cronin (1947) depicted the AG in the crab *Callinectes sapidus* as a "ductless accessory gland" accompanying "the posterior vas deferens from near the bar almost to the outside shell" but could not determine its function from direct observation. The AG has not as yet been described in the following four orders: Bathynellacea (Syncarida), Thermosbaenacea (Pancarida), Spelaeogriphacea (Peracarida), and Amphionidacea (Eucarida). In spite of a variety of studies, the AG has not been detected in classes of Crustacea other than Malacostraca.

The position of the AG is the same in the subclasses Hoplocarida, Phyllocarida, and Eumalacostraca. Located outside of the gonad, the AG is joined with the subterminal ejaculatory region of the vas deferens. As we shall see in Section 8.2.2., in certain oniscoid isopods, the androgenic tissue appears in the form of a cellular mass enclosed within the suspensory filament of the testicular utricles. The AG induces not only secondary characteristics but also testicular differentiation (see Part 3, Chapter 12 by Payen).

8.2. Differentiation and Morphology in Malacostracans

Study of AG differentiation is closely linked to sexual differentiation of the genital apparatus in the young malacostracan. It has been shown that the AG is part of the male genital tract (see reviews in Charniaux-Cotton, 1960b, 1972; Legrand and Juchault, 1972; Charniaux-Cotton and Payen, 1985). In this respect, it must be recalled that sexual differentiation of the genital apparatus may begin before or after that of the external characteristics, depending on the species. When the young malacostracan acquires its specific morphology, i.e., after embryonic or larval life, it is sexually undifferentiated. This phase lasts for a few molt cycles. Then, male and female characteristics appear in a well-defined order and develop at the time of the successive molts. This development of sexual dimorphism can last throughout the entire lifetime.

8.2.1. Amphipods

AG differentiation has been followed in detail in a study of the postembyonic development of the genital apparatus in the beach flea *Orchestia gammarella*, an amphipod (Talitridae) in which this gland was discovered (Charniaux-Cotton, 1954, 1959; Zerbib, 1964; Hort-Legrand et al., 1974). At hatching, the genital apparatus in this animal is paired and formed by two thin strands of mesodermal cells extending from the second to the seventh thoracic segment, under the pericardial septum and on both sides of the digestive tract.

The anterior region of each cellular strand contains gonia, whereas the posterior region is the anlage of the sperm duct. In the fifth thoracic segment, a separate thin strand represents the beginning of the oviduct. In the last thoracic segment, each strand is directed toward the sternite and ends in the mesenchymatic tissue of the coxopodite. Some mesenchymatic cells near the genital strand are the presumptive cells of the AG (Fig. 8.1). The sexual differentiation of the genital apparatus begins with the second postembronic intermolt. In males, it is characterized by the multiplication of cells that represent rudiments of the AG. Spermatogenesis starts during the third intermolt, and the rudiment of a sperm duct forms as a hollow tube during the fifth intermolt, following which the posterior end grows toward the last thoracic sternite. Genital papillae appear at the fifth molt, and the AGs become differentiated. In the eighth intermolt, the genital apparatus exhibits a functional structure.

Study of the genital apparatus shows that the AG originates from cells situated at the extremities of the rudiments of the sperm ducts. When the gland becomes distinct, it appears as a flattened pyramid whose base is joined by connective tissue to the ejaculatory region of the vas deferens (Fig. 8.2). Each gland measures about 420 μm at its base and is 250 μm high and 50 μm thick in a 1.5-cm-long male (Charniaux-Cotton, 1957); it consists of cellular strands of one cell type that are folded upon themselves. Cells situated near the vas deferens multiply and later increase in size. In the apical region, some cells degenerate. A few small and isolated cellular masses are often joined anteriorely to the sperm duct and accompany the main mass of the gland.

A modification in the shape of the AG has been observed in male *O. gammarella*, bearing the mutation "male sterile 1" (Ginsburger-Vogel, 1983). These mutants, which have a blind vas deferens and no genital papillae, are sterile. The morphogenesis of their AGs is abnormal. These organs do not resemble a pyramid, but appear as a long flattened cord which is irregularly folded against the vas deferens and runs toward the anterior part of this latter.

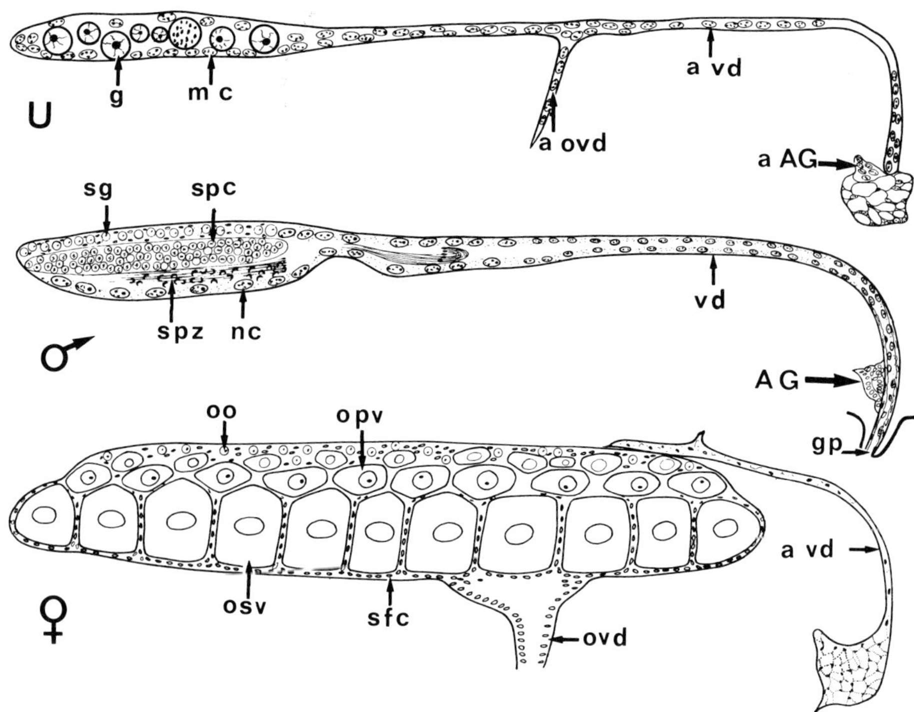

FIGURE 8.1. Organogenesis of the genital apparatus in the amphipod, *Orchestia gammarella*. *Key:* U = undifferentiated stage; ♂ = juvenile male; ♀ = female in (secondary) vitellogenesis. The scale for U is larger than that for ♂ and ♀; AG = androgenic gland; a AG = anlage of AG; a ovd = anlage of oviduct; a vd = anlage of vas deferens; g = protogonia; gp = genital papilla; mc = nucleus of mesodermal cell; nc = nucleus of nurse cell; oo = oogonia (embedded in a stroma of mesodermal cells); opv = oocyte in previtellogenesis (or primary vitellogenesis); osv = oocyte in (secondary) vitellogenesis; ovd = oviduct; sfc = nucleus of (secondary) follicular cell; sg = spermatogonia (embedded in a stroma of mesodermal cells); spc = spermatocyte; spz = spermatozoa; vd = vas deferens. Schematic drawings not to scale. (Redrawn from Charniaux-Cotton, 1965.)

In *O. cavimana* and *O. mediterranea,* the AGs are relatively smaller than in *O. gammarella;* there are also few supplementary cellular masses along the vas deferens, (Veillet and Graf, 1958; Graf, 1958; Charniaux-Cotton, 1959).

In addition to the genus *Orchestia,* the AG has been described or mentioned in four other genera: *Gammarus, Niphargus, Talitrus,* and *Caprella*

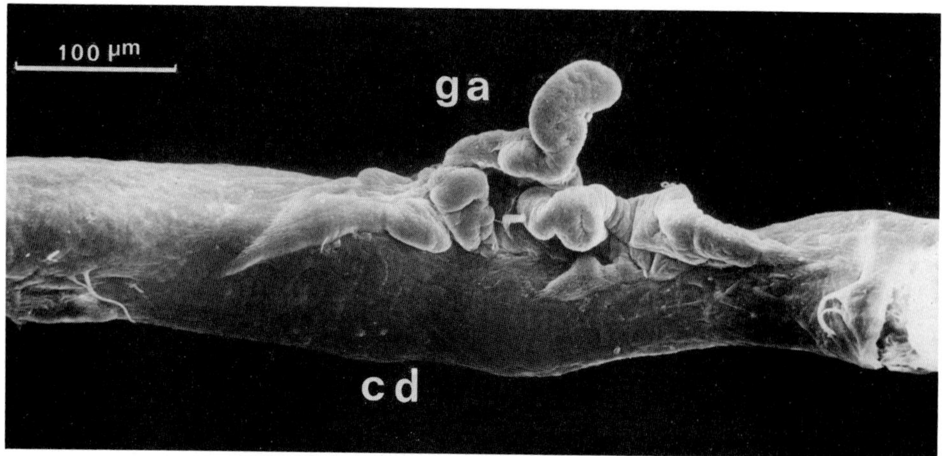

FIGURE 8.2. Scanning electron micrograph of the androgenic gland (GA) of male *Orchestia gammarella*. *Key:* cd = vas deferens. (From Ginsburger-Vogel, 1983.)

(Husson and Graf, 1961; Charniaux-Cotton, 1963; Peyrot and Trilles, 1964). In all these amphipods, the AGs display the same morphology as those in *Orchestia* and are located on the external side of the subterminal portion of the sperm duct, in the seventh thoracic segment.

8.2.2. Isopods

For some time following the discovery of the AGs in amphipods, the terrestrial isopods (oniscoids) seemed to lack AG. Indeed, depending on the species, the androgenic tissue can be located in six different thoracic segments from the second to the seventh, in front of/or along the testicular utricles, against the seminal vesicle, or/and against the vas deferens. From a phylogenetic point of view, the AG seems to disperse along the genital apparatus, from the posterior (as in the Asellidae or Cirolanidae) to the anterior region (as in the Oniscidea, i.e., the Synocheta and Crinocheta sections), which approximates the progression from the primitive to the more evolved forms. In reality, some species belonging to various groups, including oniscoids (e.g., *Helleria brevicornis*), have the AG attached to the extremity of the vas deferens, in the seventh thoracic segment, as in the amphipods. In *Asellus aquaticus*, androgenic cells form short strands along the entire length of the genital tract. The most remarkable situation exists in the Crinocheta (Oniscoidea: e.g., *Porcellio*, *Armadillidium*) where androgenic cells form a gland at the constricted end of each testicular utricle (Fig. 8.3). Finally, various species belonging to the

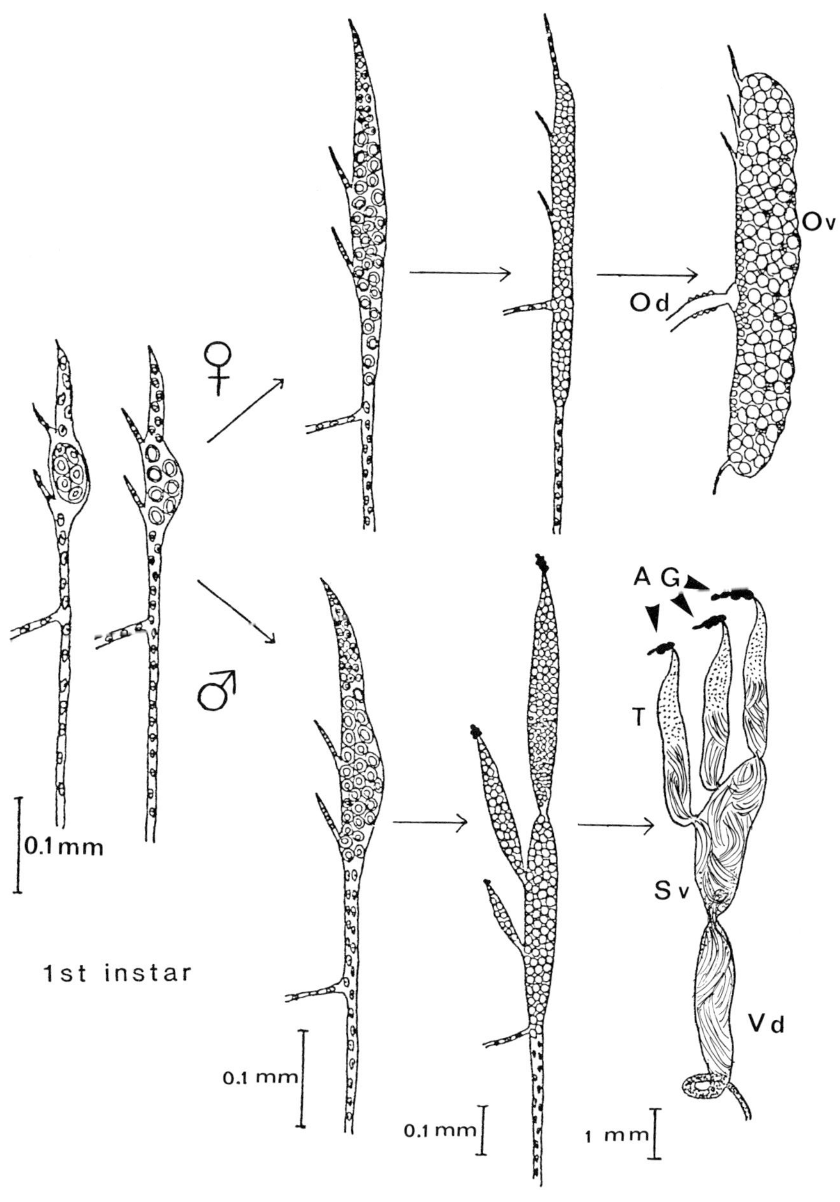

FIGURE 8.3. Postembryonic development of left internal reproductive organs of the terrestrial isopod *Armadillidium vulgare* (Oniscidea). *Key:* AG = androgenic gland; Od = oviduct; Ov = ovary; Sv = seminal vesicle; T = testis; Vd = vas deferens. (Redrawn from Hasegawa and Katakura, 1981.)

suborder Oniscidea show localization of the androgenic tissue both at the end of the testicular utricles and along the vas deferens.

The first description of an AG in an isopod was given by Balesdent-Marquet (1958) in the freshwater species *Asellus aquaticus* (Asellidae). The gland is made of several clusters of cells along the testicular utricles. The same year (1958), Legrand described the AG of the oniscoids *Armadillidium vulgare* and *Porcellio dilatatus* at a location different from the one observed until that time in other malacostracans; the AGs in these animals appear in the form of cellular masses enclosed within the suspensory filament of each testicular utricles. By grafting experiments, the latter investigator demonstrated that they are endocrine glands responsible for differentiation of male characteristics.

Five major types of location of the AG have been distinguished (see the review by Legrand and Juchault, 1972: (1) At the level of the thoracic segments II, III, and IV, the AG has been identified as cellular masses of variable size included in the suspensory filaments of each of the three testicular utricles, essentially in the Oniscidea: Crinocheta. It has also been recognized in a few species of Synocheta, belonging to the families Trichoniscidae and Buddelundiellidae. (2) The AG has been located essentially along the vas deferens in the seventh thoracic segment in species representing five suborders: Oniscidea (Mesoniscidae and Trichoniscidea families), Asellota (Asellidae), Flabellifera (Cirolanidae), Gnathiidae, and Valvifera (Idoteidae). The organ generally displays one single pyramidal or lenticular mass placed next to the surface of the vas deferens, often on its external side. (3) In the thoracic segments II, III, IV, and VII, the AG has been found in the suspensory filaments and along the external side of the vas deferens, essentially in the Oniscidea belonging to the *Synocheta* section and in the family *Trichoniscidae*. (4) The AG has been located exclusively in the thoracic segment V, as in some species representing four suborders: Oniscidea (Tylidae), Flabellifera (Sphaeromatidae), Anthuridea (Anthuridae), and Epicaridea (Bopyridae). In all species, the AG shows a thread-like shape joined to the external side of the seminal vesicle, and its course corresponds to that of the oviduct in a female. (5) The AG has been found in the thoracic segments V and VII, i.e., joined to the seminal vesicle and extending in a thread-like manner in the fifth thoracic segment, as well as located next to the external side of the vas deferens, as a cellular mass in a few species representing three suborders: Oniscidea (Tylidae), Flabellifera (Sphaeromatidae), and Anthuridea (Paranthuridae). A more or less similar location of the androgenic tissue has been found in the hermaphroditic parasite *Meinertia oestroides* (Flabellifera: Cymothoïdae) (Berreur-Bonnenfant, 1962). The AG almost entirely borders the vas deferens, including the seminal vesicle, and a small cluster of androgenic cells is visible on the subterminal portion of the sperm duct (Fig.

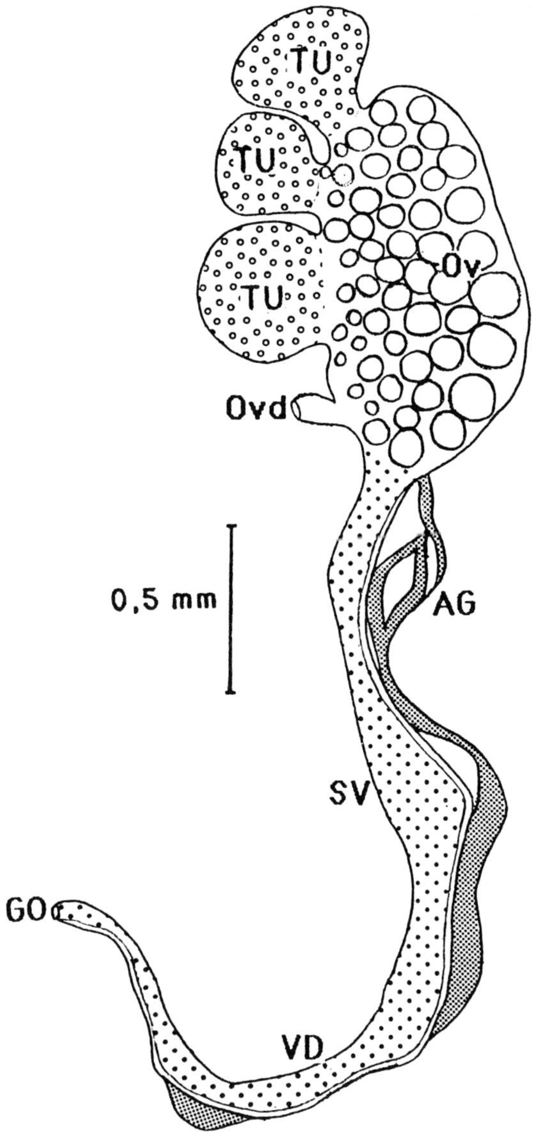

FIGURE 8.4. Genital apparatus and androgenic gland (AG) of the protandrous hermaphroditic isopod, *Meinertia oestroïdes* (Flabellifera) during the male phase. *Key:* GO = genital opening; Ov = ovarian region; Ovd = oviduct; SV = seminal vesicle; TU = testicular utricle; VD = vas deferens. (Redrawn from Berreur-Bonnenfant, 1962.)

8.4). Thus, in this species, the gland is visible in the thoracic segments V, VI, and VII.

In most cases, the male genital tract of isopods is formed by three testicular tubes (utricles), which empty into a seminal vesicle. Differentiation of the genital apparatus has been studied in detail in three oniscoids, *Porcellio dilatatus, Helleria brevicornis,* and *Armadillidium vulgare* (Legrand and Vandel, 1948; Legrand and Juchault, 1960a; Juchault, 1967; Hasegawa and Katakura, 1981; Katakura, 1982, 1984). In these species, the AGs differ in location: at the extremity of the testes in *P. dilatatus* and *A. vulgare,* and along the subterminal portion of the vas deferens in *H. brevicornis.*

An undifferentiated sexual stage in a crustacean was first described in *Porcellio dilatatus* (Legrand and Vandel, 1948). The genital rudiment differs from that of amphipods in having three pairs of thin cellular filaments that emerge from the external side of the anterior region. According to the authors, there is one pair of filaments in each of the second, third, and fourth thoracic segments. After the third molt, sexual differentiation of the genital apparatus starts. In general, as in amphipods, external sexual differentiation begins after differentiation of the genital apparatus.

In *P. dilatatus* and *A. vulgare,* rudiments of the AG are first visible in the form of small cellular masses lying at the end of the three anterior filaments. In *Helleria brevicornis,* they have been recognized at hatching as both a very thin cord of about 30 cells, lying at the level of the seminal vesicle, and a cluster of some 10 cells, located on the median part of the vas deferens. Juchault (1963) indicated that this cluster corresponds to a bulge of the wall of the sperm duct that later pulls away from it. This observation, together with similar ones performed on certain Cumacea, enabled the author to attribute a mesodermal origin—identical to that of the gonad—to the androgenic tissue.

The difference in location of rudiments of the AG in *P. dilatatus* and *H. brevicornis* does not seem to influence the differentiation of the genital apparatus, which occurs at the same time as does development of the androgenic rudiments. At the fourth intermolt, cells of the three anterior strands proliferate; gonia migrate into these strands, which become the testicular tubes. The genital tract, empty of germ cells, differentiates into a seminal vesicle and vas deferens. At the fifth molt, the two deferent canals reach the exterior by only one genital papilla.

According to Lane (1977), the AG of the oniscoids *A. vulgare* and *Porcellionides pruinosus* originate independently of the gonads. In fact, all observations show that in isopods, as in amphipods and decapods, the AGs become distinguishable through multiplication of certain mesodermal cells of the genital tract. This proliferation takes place at the posterior end of the genital rudiment in amphipods and decapods, and at various levels in isopods.

8.2.3. Decapods

In decapods, the AG was first described in the hermit crab *Clibanarius erythropus* and the shore crab *Carcinus maenas* (Charniaux-Cotton, 1956a). Later, it was found in other sub- and infraorders (Charniaux-Cotton et al., 1966).

In Brachyura, belonging to the *Brachyrhyncha* section, the AG is a lengthened organ more or less coiled around the ejaculatory portion of the sperm duct, located in the endoskeletal space among the muscles of the coxopodite of the last pair of walking legs. It never penetrates the penis. It is made up of sinuous and entwined strands of cells. In *Carcinus maenas*, with about a 5- to 6-cm-broad cephalothorax, the AG can be as much as 1 cm long. In superfamilies such as Grapsidoidea or Ocypo-doidea, the AG is located between the branched diverticula of the vas deferens, called accessory glands, and the penis.

It is interesting to note that the androgenic strand measures about 40 μm in diameter in the Astacidea, e.g., *Nephrops norvegicus*, and from 40 to 80 μm in *Homarus vulgaris*, in which it is very sinuous and folded upon itself. Moreover, the AG is relatively smaller in *Homarus* than in *Carcinus*.

In shrimps and prawns, the AG is of variable shape; it lies generally along the bulging extremity of the sperm duct (Figs. 8.5 and 8.6) and consists of both anastomosed strands of cells and compact cellular masses. Among the Caridea, two protandrous hermaphroditic species, *Lysmata seticaudata* (Hyppolytidea) and *Pandalus borealis* (Pandalidae), have been studied (Charniaux-Cotton, 1958; Carlisle, 1959; Veillet, 1958). In the first phase of their life, these shrimps function as males and possess a well-developed AG. When they reach a certain size, the gland degenerates and disappears. This disappearance leads to sex reversal in these animals (Charniaux-Cotton, 1960a,b, 1965).

As in amphipods and isopods, the genital apparatus of the decapods exhibits a sexually undifferentiated phase. However, young crayfishes and young crabs differ from peracarids in the following characteristics: (1) there is only one rudiment for the gonoducts; (2) no rudiment of an AG is visible at the site of these organs; and (3) male differentiation begins before a discrete AG is recognized. Young hermit crabs resemble peracarids in that their genital apparatus has the rudiments of genital ducts of both sexes.

The sexually undifferentiated phase of the genital apparatus persists until the two postembryonic intermolts of crayfishes (Payen, 1973) and during the entire larval life of hermit crabs (Le Roux, 1976); in the crabs studied (*Callinectes sapidus*, *Rhithropanopeus harrisii*, and *Menippe mercenaria*), this phase extends to the second crab stage (Payen, 1974).

Postembryonic development of the genital rudiment of crabs is difficult

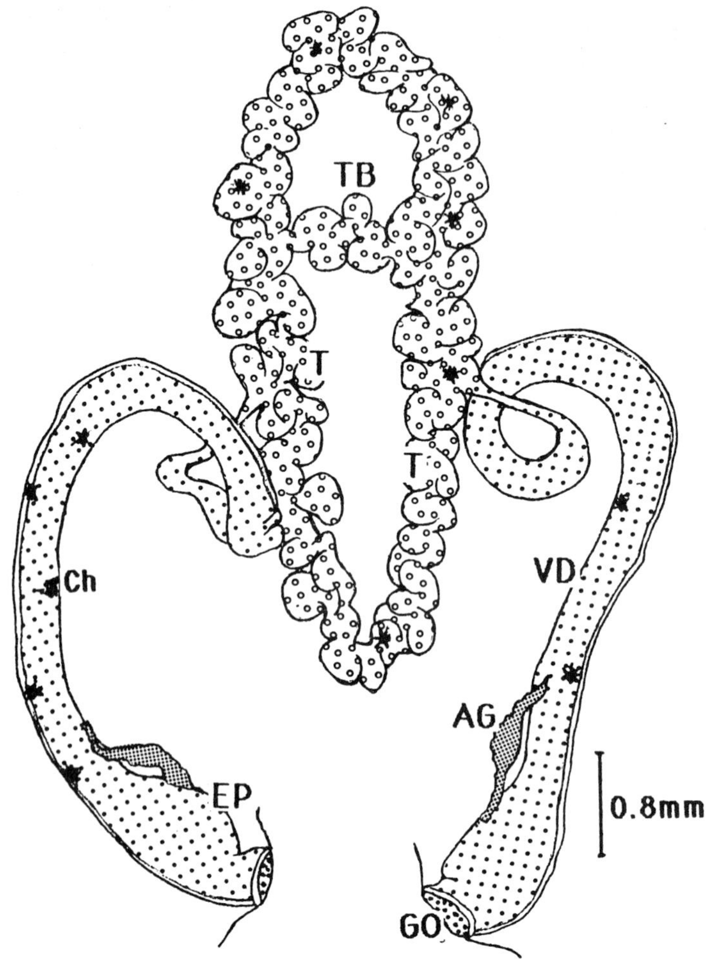

FIGURE 8.5. Genital apparatus and androgenic gland (AG) of a male shrimp *Palaemonetes varians* (Caridea, Palaemonidae). *Key:* Ch = chromatophores; EP = ejaculatory part of the vas deferens (VD); GO = genital opening; T = testis; TB = testicular bridge.

to follow. Initially localized between the third and fifth thoracic segments, on both sides of the digestive tube and suspended from the pericardial septum, it grows in the form of an H during the first two crab stages in all individuals.

Differentiation of the gonads in reptantian Decapoda occurs later in males than in females. The male sex of the genital apparatus is recognizable in young crayfishes (in the third postembryonic stage) and in young

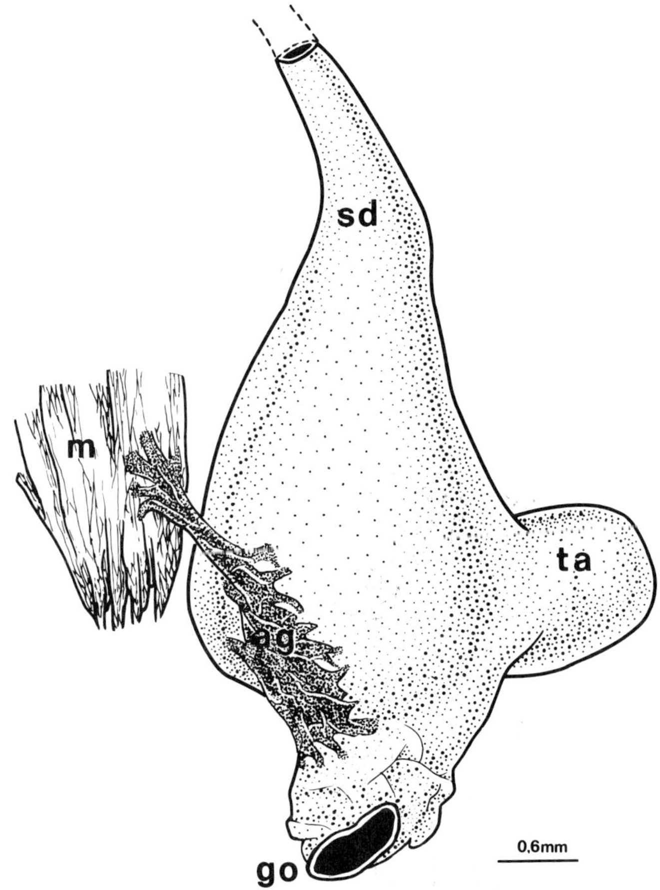

FIGURE 8.6. Distal part of the genital apparatus of a penaeid prawn, *Penaeus japonicus*, showing the location of the androgenic gland (AG). *Key:* go = genital opening; m = muscles of the coxopodite of the fifth pereopod; Sd = sperm duct; ta = terminal ampoule. (From G. Payen, L. Chim, and A. Laubier-Bonichon, unpublished work.)

crabs (in the third crab stage) only by the progression of a single pair of deferent ducts toward the eighth thoracic sternite (bearing the fifth pair of pereopods).

Beginning with the second postembryonic stage, the gonads of the European crayfish *Pontastacus leptodactylus leptodactylus* exhibit a dimorphism that has not been observed in crabs: the mesodermal tissue is of more consequence in males than in females. The AG becomes visible from the seventh stage of postembryonic development in the form of cellular masses situated at the extremity of the deferent ducts; it is distinguishable at the time when spermatogenesis begins (Payen, 1973). The androgenic rudiment of crabs can generally be detected as soon as the sperm ducts are completely formed. At that time, it consists of a single row of several cells (diameter: ~4.5 μm), extending along the distal part of the male ducts. This rudiment was not distinguishable during either the first three crab stages in *R. harrisii* or the first five in *Callinectes sapidus* (Payen, 1974). In young *Carcinus maenas* (9 mm cephalothoracic length), the gland consists of one or two straight, thread-like strands, 8–9 μm thick (Démeusy, 1960). The androgenic rudiment then develops into several distinct rows of cells that coil back up around the subterminal region of the vas deferens. Study of organogenesis of the male genital apparatus in crabs has revealed that, with the differentiation of the AGs (which apparently are derived from the gonoducts, as in peracarids), clustered regions of the sperm ducts are formed, the genital papillae differentiate, and spermatogenesis begins (Démeusy, 1960; Payen, 1974). The synchronism of these changes leads to genital maturity. A similar synchronism between organogenesis of the AG and onset of spermatogenesis occurs in the penaeid prawn *Penaeus japonicus* (Laubier et al., 1985).

An analysis of the developmental stages of the AG in the protandric shrimp *Pandalus platyceros* indicates that the AG cells originate from the differentiating epithelium of the ejaculatory bulb of the sperm duct (Hoffman, 1969); this has not been confirmed in other species, however.

Vascularization of the AG has been studied only in the crayfish *Orconectes limosus* (Graf, 1962). The gland receives ramifications from both the arteria genitalis and the arteria pedalis. Arterioles run along the AG and send to it hemolymph capillaries.

8.2.4. Other Malacostracans

Among the Phyllocarida, the AG of *Nebalia geoffroyi* was studied by Duveau (1957). As in most malacostracans, it is contained in the coxopodite of the last pereopod and placed against the distal part of the vas deferens. The AG of the stomatopod *Squilla mantis* (Hoplocarida) has the same location as in *Nebalia* but, in functional males, appears almost totally separated from the vas deferens by a muscle. Charniaux-Cotton (1960c) indicated that in young individuals the gland may form a single mass closely joined with the genital apparatus. In addition to Amphipoda and Isopoda, the AG has been described in three other orders of Peracarida: Mysidacea, Cumacea (Meusy, 1963; Juchault, 1963), and Tanaidacea (Juchault, 1963). In the first two orders the gland appears either in the form of a pyramid-like structure (Mysidacea) or as a continuous cord (Cumacea) along the subterminal portion of the vas deferens, whereas it presents a variable form and location along the genital apparatus in Tanaidacea.

The AG of a Syncarida species belonging to the order Anaspidacea has been detected (Zerbib, 1967). Its characteristics resemble those of the amphipod *Orchestia gammarella*.

Among the Eucarida, the Euphausiacea have also been investigated (Zerbib, 1967). Their AG consists of a cellular cord rather isolated from the sperm duct and adhering to the sternal epithelium.

8.3. Histology and Ultrastructure

Histological study of the AG reveals a rather wide-ranging variety of structures, apparently having no relationship with the classification of the considered species. Thus, the AG of *Paramysis nouveli* (Peracarida: Mysidacea) is comparable, both in terms of its structure and form, to that of *Palaemon serratus* (Eucarida, Decapoda). In contrast, the AG of this shrimp does not resemble that of the shore crab *Carcinus maenas* (Eucarida, Decapoda) (Charniaux-Cotton et al., 1966). The arrangement of cells in the AG is quite varied: (a) Sinuous cords of cells are found in certain isopods, e.g., *Paragnathia formica* (Gnathiidea), *Helleria brevicornis* and *Armadillidium vulgare* (Oniscidea), *Sphaeroma rugicauda* and *S. serratum* (Flabellifera), and *Idotea chelipes* (Valvifera) (Bonnenfant, 1961; Katakura, 1961; Legrand and Juchault, 1960a,b, 1961). A slight modification of the arrangement of the androgenic cords into simple strands of aligned discoidal cells is found in most brachyuran decapods, e.g., *Carcinus maenas*, *Pachygrapsus crassipes*, *Macropipus depurator*, *Macropipus puber*, *Cancer pagurus*, and *Callinectes sapidus* (Charniaux-Cotton, 1956a; King, 1964; Charniaux-Cotton et al., 1966; Payen et al., 1971), as well as in the caridean shrimp *Palaemon dayanus* (Thampy and John, 1972). (b) A mixed constitution of both anastomosed and entwined strands of cells can be observed in Talitridae amphipods (Meusy, 1965), the crayfish *Orconectes nais*, (Carpenter and De Roos, 1970) and several natantians such as the shrimps *Palaemon serratus*, *Crangon crangon*, *Lysmata seticaudata*, *Pandalus borealis*, *P. platyceros*, and *Atyaephyra desmaresti* (Huguet and Huguet, 1971; Charniaux-Cotton, 1958; Carlisle, 1959; Hoffman, 1969; Huguet, 1968). Moreover, the AG of Ocypodidae and Xanthidae crabs (Thampy and John, 1972; Payen, 1972; Payen et al., 1971), of Penaeoidae prawns (Huguet and Huguet, 1971; Payen et al., 1982), as well as of *Paramysis nouveli* (Mysidacea) (Meusy, 1963) and of some isopods such as *Anilocra* and *Meinertia* (Remy and Veillet, 1961; Bonnenfant, 1961), are close to this type. (c) Lobules have been described in *Maja* (Oxyrhyncha decapods) and *Eriocheir sinensis* (Brachyrhyncha) (unpublished data of S. Paoli and G. Péron quoted by Charniaux-Cotton et al., 1966). (d) Spherules appear to be a feature of the AG of the stomatopod *Squilla mantis* (Charniaux-Cotton, 1960a). A more or less well-developed connective tissue surrounds these different cellular groupings.

8.3.1. Amphipods

The AG of *Orchestia gammarella* (Talitridae) has been observed by light microscopy (Charniaux-Cotton, 1960b; Meusy, 1965), showing that the cells are arranged transversely to the cords and present a variable aspect: the cytoplasm, which is slightly positive to periodic acid–Schiff (PAS), appears sometimes dense and very basophilic or sometimes highly vacuolated. The two aspects often exist together in the same gland. The nuclei, generally elongated, have a size of about 3.8×6.5 μm; areas of the gland where some of them are pycnotic can be detected.

At the ultrastructural level (Meusy, 1965), the endoplasmic reticulum consists of abundant small vesicles ranging from 50 to 130 nm in diameter, externally bordered with ribosomes. Some cisternae of granular endoplasmic reticulum are also visible. The mitochondria possess few transverse cristae. The Golgi vesicles vary between 35 to 120 nm in diameter, the smallest being electron-dense. Various inclusion bodies and organelles have also been mentioned: lysosomes, lipid droplets, microtubules, multivesicular bodies, and sometimes glycogen. The nuclei display large areas of electron-dense chromatin arranged at their periphery. The nucleolus is about 1 μm in diameter. These ultrastructural features point to a peptidic nature of the androgenic secretion (Meusy, 1965), but the mechanism of its discharge has not been elucidated.

8.3.2. Isopods

Histological observations of the AG of isopods have been rather modest in scope. The major peculiarities appear related to the nuclear polymorphism. A variation in the diameter of the nuclei within the same species, as well as between species, has been noticed (Juchault, 1967). Thus, in marine Sphaeromatidea, Cirolanidae, Anthuridae, Valvifera, and certain Oniscoidea, such as those representing the genus *Armadillidium*, the variations are small. On the other hand, in certain species such as *Oniscus asellus* and *Armadillo officinalis*, the nuclear diameter can vary from 6 to 35 μm. Such polymorphism has been correlated to the activity of the gland. Degenerating and dividing cells are seldom observed, and no evidence of a holocrine-type functioning of the cells, as initially described by Charniaux-Cotton (1957) in *O. gammarella*, was found.

Histochemical studies have been conducted on the AG of *Idotea balthica basteri* (Valvifera) (Reidenbach, 1971). The cytoplasm, not very abundant, is rich in RNA, as shown by its affinity to Pyronine and gallocyanine shellac, and by its high basophilia, detected with the usual techniques. The elongated nuclei possess one or two prominent nucleoli and an abundant supply of peripheral chromatin. Their diameter size varies from 3 to 6 μm. The smallest nuclei are located near the base of the gland; their chromatin is homogeneous and the nucleolus is not visible. According to Reidenbach (1971), the histological criteria in favor of a holocrine secretion are not evident, because no degenerating figure has been seen.

Examination of the AG of *Porcellio scaber* (Oniscoidea) (Radu and Cracium, 1976) indicates that sometimes one may observe nuclei that are in a more or less advanced state of pycnosis. No distinct categories of cells were observed. In this species, the mean values of the diameters of the reniform nuclei are between 5 and 10 μm. Some large nuclei (about 12 × 22 μm in diameter) are irregular in shape, often with deep incisions. Amitotic division has been suggested, based on observation of two or several nuclei in the cells. Nucleoli are numerous, often more than five. The large nuclei are the richest in chromatin granules and in nucleoli. As already pointed out in the isopods cited above, the diverse forms of the nuclei reflect different functional stages. Occasional intercellular spaces appear occupied by blood sinuses. Some chromatophores with dispersed pigment granules can be seen on the surface of the gland.

Ultrastructural studies were conducted on a few terrestrial species (oniscoids): *Oniscus asellus, Porcellio dilatatus, Armadillidium vulgare,* and *Porcellio scaber* (Malo, 1970; Malo and Juchault, 1970; Chaigneau and Juchault, 1970; Radu and Cracium, 1976). Based on various observations, it seems that the AG shows an abundant rough endoplasmic reticulum (RER), often arranged in parallel lamellar fascicles, and a well-developed Golgi apparatus, whose vesicles in the vicinity of the dictyosomes contain an electron-dense material. No exocytotic phenomenon has been ever observed. The major differences between the considered species seem to concern (a) the presence or lack of juxtaposed cells with dark or clear shading of the cytoplasm, and (b) the development of degenerating processes. A detailed analysis of the fine structure of the androgenic cells of *P. scaber* allowed Radu and Cracium (1976) to suggest that the content of the RER is fluffy or finely granular. This substance presents varying degrees of electron opacity. At the final phase of the secretory activity of the cell, the ergastoplasmic lamellae show constrictions so that they appear like "pearl necklaces." A functional evolution of the chondriome has also been noticed. The strong vacuolization of the cytoplasm, observed in the advanced stages of secretory activity, seems to be the result of the development of clear vesicles from the Golgi apparatus, along with modifications of both the endoplasmic reticulum and the mitochondria, as well as with the appearance of nuclear protrusions. According to the authors, the organelles undergo hypertrophy during secretion and finally disappear. The secretory cells that cannot regenerate are holocrine. However, no proof is given as to whether the cellular changes correspond to a normal process of secretion or are only aspects of cell degeneration.

8.3.3. Decapods

Several histological studies have been carried out on the AG of decapods. They concern the crabs *Carcinus maenas*, *Pachygrapsus crassipes*, *Ocypoda platytarsis*, *Ocypode quadrata*, *Rhithropanopeus harrisii*, and *Callinectes sapidus* (Charniaux-Cotton, 1956a; Démeusy, 1960; Meusy, 1965; King, 1964; Thampy and John, 1970; Payen, 1972; Payen et al., 1971), the crayfishes: *Astacus fluviatilis*, *Orconectes nais*, *Procambarus clarki* (Brodzicki, 1964; Carpenter and De Roos, 1970; Taketomi, 1986), the caridean shrimps: *Lysmata seticaudata*, *Pandalus platyceros*, *Palaemon dayanus*, *P. serratus*, and *Crangon crangon* (Veillet and Graf, 1965; Hoffman, 1969; Thampy and John, 1972; Huguet and Huguet, 1971; Veith and Malecha, 1983), and the penaeid prawns: *Aristeus antennatus* and *Penaeus japonicus* (Huguet and Huguet, 1971; Payen et al., 1982).

Basically, the histological aspects of the AGs of decapods resemble those of peracarids. Depending on the species, the cells measure from 10 to 20 μm in diameter, and the nuclei are generally prominent and ovoid in shape, about 5–15 μm along their long axis.

In pubertal *Carcinus maenas* (≥24 mm cephalothoracic length), the AG is made up of various types of strands (Démeusy, 1960): (1) large strands (25 μm average diameter) in which the cells contain round or slightly ovoid nuclei (diameter: 4.9 × 5.4 μm); (2) lacunary and large strands with smaller and elongated nuclei (diameter: 3.2 × 4.6 μm); (3) extensive degenerated areas close to the lacunary ones; and (4) surrounding areas filled with small, ovoid nuclei (diameter: 2.7 × 4.4 μm) that resemble those of a young gland.

Histochemical tests revealed that the gland is encased in a thin PAS-positive sheath. In the cytoplasm, the presence of proteins, lipid inclusions and an intense acid phosphatase activity in certain granules have been pointed out in the AG of *Pachygrapsus crassipes* (King, 1964). In *Macrobrachium rosenbergii*, 3 β-hydroxysteroid dehydrogenase (HSD) activity appears to be confined to partially vacuolated cell types (Veith and Malecha, 1983). Note that previous studies conducted on *Homarus americanus* and *Callinectes sapidus* (Gilgan and Idler, 1967; Tcholakian and Eik-Nes, 1969, 1971) attributed an important role to the AG in the metabolism of androstenedione and progesterone *in vitro*. However, this role has not been confirmed in *Carcinus maenas*, where the AG appears less active than testis or vas deferens in reducing androstenedione to testosterone *in vitro* (Blanchet et al., 1972). In fact, steroid metabolism is chiefly active in the genital apparatus, which is the main target tissue of the androgenic hormone.

The holocrine mode of secretion of the AG of decapods has been supported by some authors on the bases of histological observations of degenerating zones examined at times when the glands are considered to be functional (Tcholakian and Reichard, 1964; Carpenter and De Roos, 1970; Thampy and John, 1972; Veith and Malecha, 1983). However, such a hypothesis appears difficult to establish, or even uncertain, for some other authors (Veillet and Graf, 1965; Hoffman, 1969; Frechette et al., 1970). Indeed, according to these last authors, cytolysis occurs only after a period of activity.

The ultrastructural investigations are still rather scarce. They are limited to a few crabs, *Pachygrapsus crassipes*, *Carcinus maenas*, *Rhithropanopeus harrisii*, *Callinectes sapidus*, and *Ocypode quadrata* (King, 1964; Meusy, 1965; Payen et al., 1971; Payen, 1972); one crayfish, *Procambarus clarki* (Miyawaki and Taketomi, 1978; Taketomi, 1986); and one penaeid, *Penaeus japonicus* (Payen et al., 1982). As in peracarids, the common characters are as follows: well-developed lamellar RER; many free ribosomes; mitochondria with flat, transverse cristae; presence of more or less discrete Golgi apparatus; numerous lysosomes, microtubules; absence of detectable secretory material; and cellular connections of the macula adherens type (desmosomes) (Fig. 8.8A). Variation in the electron density of the cytoplasm has been reported in the crab *Pachygrapsus crassipes* and the crayfish *Procambarus clarki* (King, 1964; Taketomi, 1986). In *Penaeus japonicus* (Fig. 8.7A,B), some additional interesting features have been observed: numerous dispersed β-particles of glycogen (revealed by periodic acid oxidation–thiocarbohydrazide and silver proteinate treatment) and presence of lysosome-like dense bodies (slightly digested after pronase treatment). In this decapod, attachment plates, gap junctions, and cellular extensions, acting as clamps, establish contact between cells. Involuted processes, carrying on total cellular autodestruction, have been particularly described in the ghost crab, *Ocypode quadrata* (Payen, 1972). Coexistence of degenerating cells with normal cells indicates a cellular asynchronism. The most striking modifications leading to cellular breakdown can be summarized as follows: substitution of RER by marked development of SER (smooth endoplasmic reticulum), vacuolization of mitochondria, differentiation of septate junctions, elaboration of vacuolar bodies and autolytic vacuoles and then of residual bodies preceding processes of cellular fragmentation and nuclear pycnosis (Fig. 8.8B). In *Penaeus japonicus*, hemocytes phagocytose degenerative androgenic cells

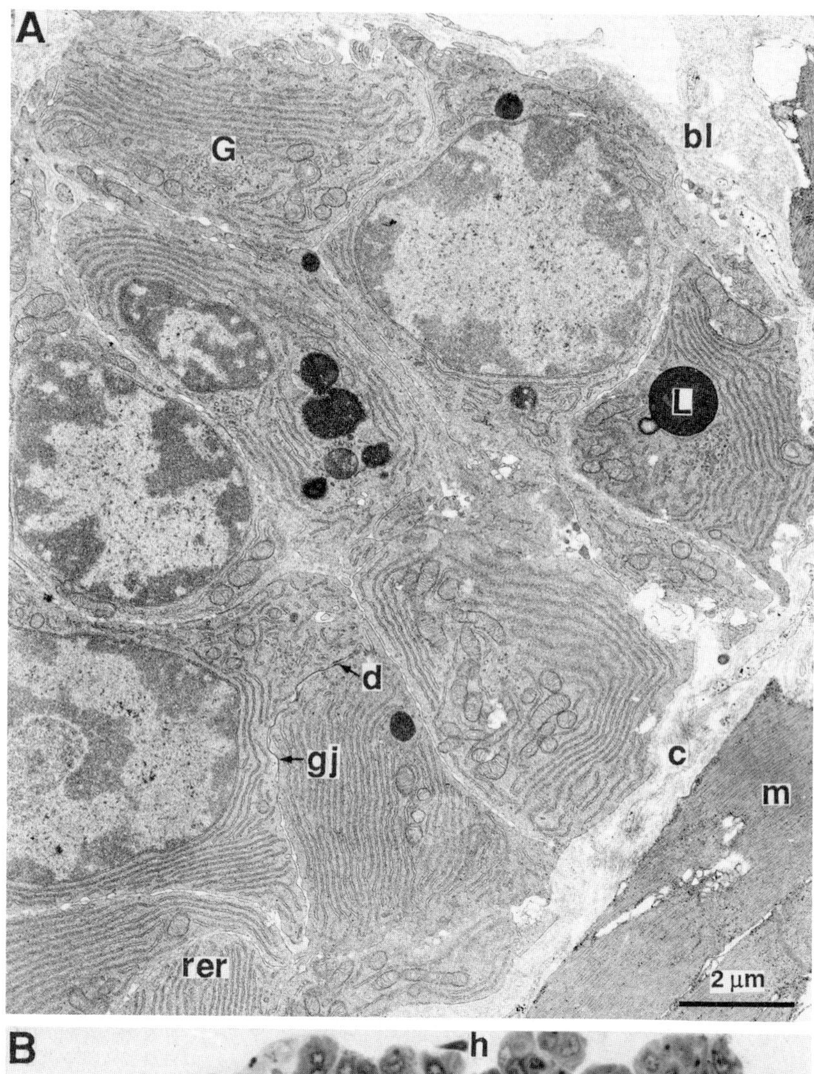

A
G
bl
L
d
gj
c
m
rer
2 µm
B
h
dac
sfac
25 µm

(Payen et al., 1982). An accumulation of glycogen and the development of bundles of collagenous fibrils between cells characterize this cytolysis. In both decapods, cellular degeneration does not seem to result from an accumulation of secretory material. However, in *Procambarus clarki*, a degradative process associated with holocrine secretion would seem to occur in low-electron-density androgenic cells that present large perinuclear space (Miyawaki and Taketomi, 1978; Taketomi, 1986).

8.3.4. *Seasonal Variations*

Seasonal variations of the ultrastructure of the androgenic cells have been observed in quite few species. Thus, in *Orchestia gammarella*, the endoplasmic reticulum is poorly represented in November and December, whereas it is abundant in February, March, and April. During the latter three months, the Golgi apparatus is also voluminous (Meusy, 1965). In the crayfish *Procambarus clarki*, Taketomi (1986) indicates that, among the two cell types constituting the AG, one with a cell matrix of high electron density, containing a well-developed RER, is abundant during summer, whereas the other, with a wide perinuclear space, is dominant during winter. According to the latter author, because the water temperature remains almost constant throughout the year, it is not the main environmental factor that influences the structure and functioning of the AG.

Histological examination of the AG in another crayfish, *Orconectes nais*, has led Carpenter and De Roos (1970) to establish that the AGs increase greatly in size during summer. They attain their maximum size a week or two prior to the population molt from the sexually inactive form to the sexually active one that occurs in late August and early September. At that time, the glands change from a sheet-like arrangement to one of entwined cords. Moreover, the cells appear hypertrophied and extensive areas of cellular degeneration develop. The AGs regress during winter.

FIGURE 8.7. Androgenic gland of a functional male prawn, *Penaeus japonicus:* (A) Ultrathin section of a portion of the gland near the muscles (m) of the fifth pereopod. Note the abundant rough endoplasmic reticulum (rer), the lysosomes (L), and the numerous mitochondria; Golgi apparatus (G) is generally discrete with small vesicles; contact between cells is ensured by gap junctions (gj) and desmosones (d). *Key:* bl = basal lamina; c = collagenous fibrils. (B) Semithin section showing the arrangement of the functional cells in a mixed constitution: strands of androgenic cells (sfac) folded upon themselves and more or less anastomosed. Hemocytes (h) are sometimes present between the cellular masses. Degenerating androgenic cells (dac) are frequently encountered. (From G. Payen, unpublished work.)

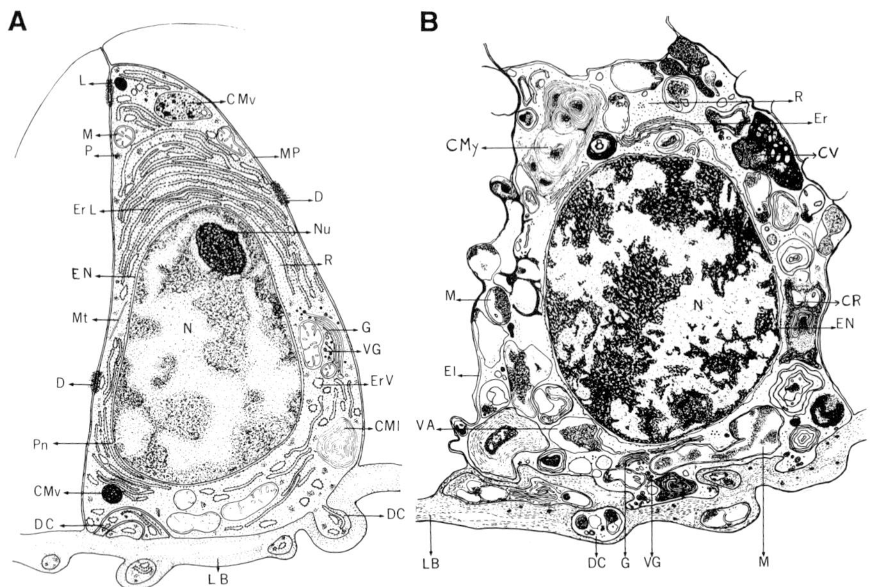

FIGURE 8.8. Schematic representation of a functional (A) and degenerating (B) androgenic cell in decapods. *Key:* CMl = multilamellar body; CMv = multivesicular body; CMy = myelin-like body; CR = residual body; CV = vacuolar body; D = desmosome; DC = cellular digitation; EI = intercellular space; EN = nuclear envelope; ErL = lamellar esgastoplasm; Er V = vesicular ergastoplasm; G = Golgi apparatus; L = lipid inclusion; LB = basal lamina; M = mitochondria; MP = plasmic membrane; Mt = microtubule; N = nucleus; Nu = nucleolus; P = polysome; Pn = nuclear pore; R = free ribosomes; VA = autophagic vacuole; VG = Golgi vesicles. Schematic drawings not to scale. [(A) From Payen et al., 1971. (B) From Payen, 1972.]

The spring molt from the active form back to the inactive one occurs when the AG are small and contain only small cells.

The transformations of the AG cells that occur in the protandric shrimp *Pandalus platyceros* (Hoffman, 1969) are also related to changes of seasons. Thus, six developmental stages can be recognized: first, a stage of proliferation where mitotic figures are abundant; this is followed by two stages found during the active male phase (i.e., winter, spring, and early summer months) and characterized by cytoplasmic hypertrophy preceding vacuolization; then, a stage of cellular breakdown closely associated with the previous stage; next, a stage of transitional atrophy in transitional males during late summer and early autumn months; finally, the ghost stage, characterized by empty sheaths of androgenic cords, occurring at the end of autumn and early winter in transforming males whose gonads lack testicular elements and contain only ovarian cells.

The correlations between modifications of the AG and reproductive activity are presented in Chapter 12 of Part 3.

8.3.5. Modifications Due to Experimental Conditions

The influence of two kinds of factors, environmental and endocrine, on the structure and functioning of the AG have been investigated. They concern respectively: (1) the variations of the rearing temperature on the terrestrial isopod *Oniscus asellus* (Malo, 1970; Legrand and Juchault, 1972); (2) the effects of eyestalk excision in some decapods, e.g., *Carcinus maenas*, *Pandalus platyceros*, *Rhithropanopeus harrisii*, and *Callinectes sapidus*, and *Lysmata setivcaudata* (Démeusy, 1960; Meusy, 1965; Meusy, 1968; Hoffman, 1968; Foulks and Hoffman, 1974; Payen, 1970; Payen et al., 1971; Touir, 1973), and of the ablation of the optic lobes or of the protocerebrum, or both, in various isopods, e.g., *Idotea balthica*, *Porcellio dilatatus*, *Oniscus asellus*, and *Armadillidium vulgare* (Reidenbach, 1966; Legrand et al., 1968; Malo and Juchault, 1970; Chaigneau and Juchault, 1979). In addition, the consequences of the implantation of a brain into the protandrous hermaphroditic isopod *Nerocila orbignyi* were also described (Trilles, 1963).

The action of elevated temperature (20°C for 3–9 weeks) on the ultrastructure of the androgenic cells of *O. asellus* (Malo, 1970; Legrand and Juchault, 1972) in genital repose leads to three major modifications: a swelling of the RER with accumulation of a fibrous material; the development of contacts between the RER and the Golgi apparatus, which releases small electron-dense vesicles; and a distension of the intercellular spaces. In contrast, when the animals are maintained at 10°C for the same length of time, the cell organization shows a reduction of the RER, which displays no electron-opaque material that is easily detectable, while the number of free ribosomes increases; the extrusion process from the Golgi apparatus is reduced; and the intercellular space is narrowed. According to the authors, the variations of both the Golgi and the ergastoplasmic systems lend support for the likelihood of protein synthesis by the androgenic tissue.

The effect of ablation of protocerebral areas, including eyestalks in decapods, is generally interpretated as a hyperactivity of the AG, especially in juvenile individuals (Fig. 8.9). Indeed, hypertrophy of the AG involving hyperplasia and an increase in size of androgenic cells occurs simultaneously with a precocious acquiring of male physiology (Démeusy, 1960; Reidenbach, 1966, 1971; Legrand et al., 1968; Malo and

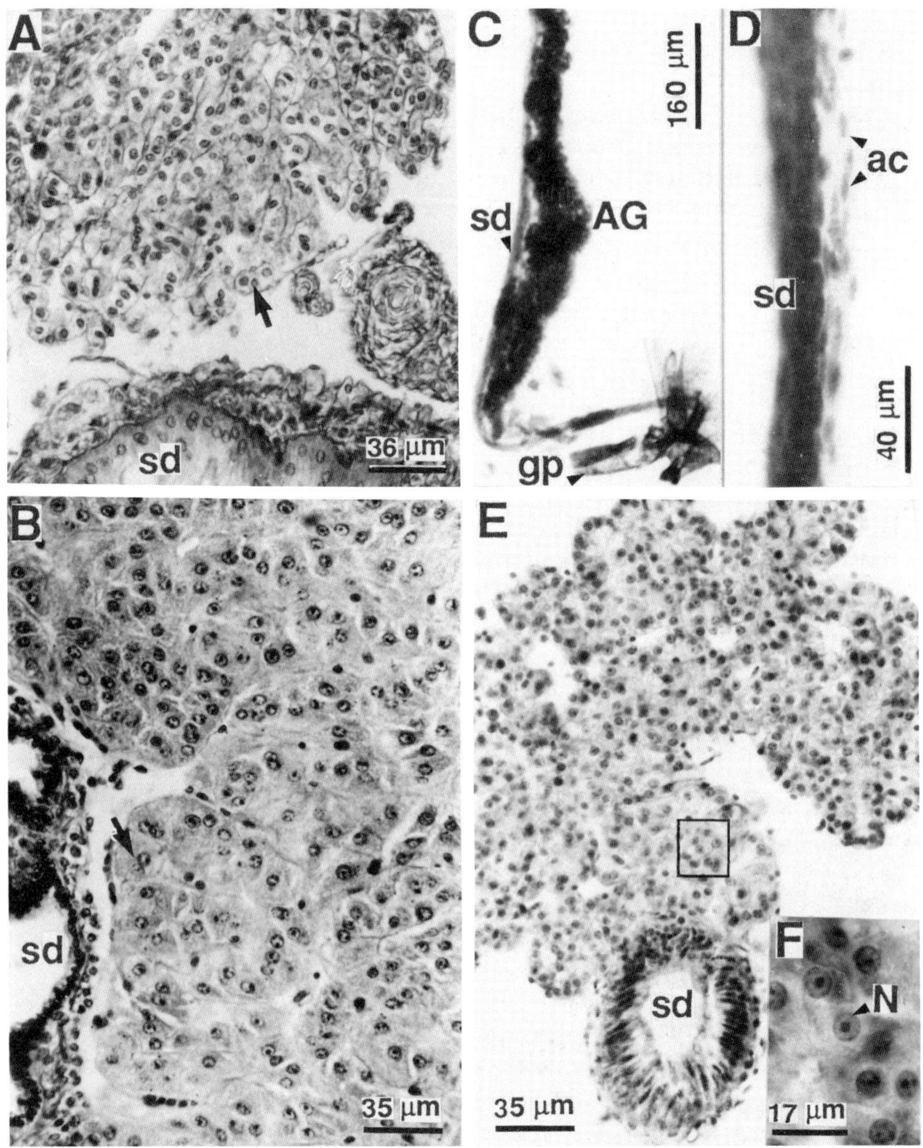

FIGURE 8.9. Androgenic gland of the crabs, *Callinectes sapidus* (A,B,C) and *Rhithropanopeus harrisii* (D,E,F): (A) Cross section in a pubertal male, *C. sapidus*. (B) Cross section in a juvenile male at the seventh postlarval intermolt, following bilateral eyestalk ablation at the megalops stage. (C) Whole-mount of the subterminal part of the sperm duct in a destalked individual in the same condition as B. Note the hypertrophy of the AG and of the androgenic cords (arrows) in the young destalked crab. (D) Control male, *R. harrisii* at the fourth postlarval intermolt. (E) Juvenile crab at the same stage as D, following bilateral eyestalk ablation at the zoeal stage. As in B, the hypertrophy of the AG is obvious. (F) Detail of a protion selected on E. *Key:* ac = androgenic cord; gp = genital papilla; N = nucleus; sd = sperm duct. (From Payen et al., 1971.)

Juchault, 1970; Legrand and Juchault, 1972; Payen, 1970; Payen et al., 1971). For instance, in *Rhithropanopeus harrisii*, after destalking in the zoeal stage, the AGs that are about 4.5 μm thick at the fourth crab stage become even thicker than in normal pubertal males. Thus, they reach 250 μm in young destalked crabs, whereas the average thickness is 43 μm in mature males (Payen et al., 1971). Ultrastructural observations carried out on these young destalked crabs can be summarized as follows: extensive development of the RER; mitochondria grouped in a juxtanuclear area; rounding of the nuclei; and accumulation of electron-dense material in intercellular spaces. In young and pubertal isopods deprived of the median part of their protocerebrum, ultrastructural variations of the hypertrophied AG concern essentially the RER and the nuclear envelope, which distends; furthermore, the dictyosomes become electron dense (Malo and Juchault, 1970; Legrand and Juchault, 1972). In *Armadillidium vulgare* 3 weeks after ablation, electron-opaque androgenic cells show a more developed endoplasmic reticulum than do the electron-lucent cells, as well as numerous lysosomes (Chaigneau and Juchault, 1979).

Note that mitotic figures are more frequently encountered in the destalked animals than in controls. In pubertal crabs such as *Carcinus maenas*, although the hypertrophy resulting from eyestalk ablation leads to a marked increase in the width of the androgenic strands (of 50–150%), the modification seems to be correlated to hyposecretion, since the endoplasmic reticulum proliferates and degranulates (Meusy, 1965). On the other hand, in *Rhithropanopeus harrisii*, the operation leads to no distinctive variation (Payen et al., 1971).

Hypertrophy of the AG has also been observed in the protandrous hermaphroditic shrimps *Pandalus borealis*, *P. platyceros*, and *Lysmata seticaudata*, when eyestalk removal was performed at the time when the glands were atrophying and the gonad transforming into an ovary (Charniaux-Cotton, 1967; Hoffman, 1968; Touir, 1973). According to Touir, the increase of spermatogenic activity in males following destalking is a consequence of the hyperactivity of the AG. These glands appear well developed also in the hermaphrodite isopod *Nerocila orbignyi* (Cymothoïdae), in which a brain was implanted at the time of sex reversal (Trilles, 1963). Thus, in this species, the brain would have a positive action on the maintenance of the AG. An increase in RNA synthesis has been noted by Foulks and Hoffman (1974) in the AG of *P. platyceros* after bilateral destalking and administration of [5-³H]uridine. This increase is similar to the one that occurs after administration of exogenous 20-OH-ecdysone and [5-³H]uridine. With both treatments, the greater incorporation of [5-³H]uridine was found during the active male phase, i.e., into actively secreting androgenic cells.

8.3.6. Modifications Due to Symbionts and Parasites

Thelygenic lines of the isopod Oniscidea species *Armadillidium vulgare*, due to a feminizing symbiotic bacteria belonging to the chlamydial group, comprise neo-females and sterile intersex males (iM) (Legrand and Juchault, 1972; Martin et al., 1973; Juchault and Legrand, 1985). These iMs possess testes surmounted by voluminous AGs which are composed of giant cells. Observations of semithin sections of the hypertrophied AGs show that the nuclei display irregular contours. Ultrastructural analysis reveals an extensive development of the RER with stacks of tight and parallel saccules, as well as numerous dictyosomes. The presence of abundant lysosomes is interpreted as a crinophagic resorption of the androgenic hormone (Chaigneau and Juchault, 1979). Indeed, the symbiotic bacteria of the iM individuals act in a comparable way to brain destruction concerning hypertrophy (see Section 8.3.5, above), but the release mechanism of the hormone, which as yet remains unknown, appears different in both cases. Grafting experiments and injections of hemolymph in normal females have shown that the AGs of iMs are the source of an important synthesis of male hormone. However, the lack of androgenic hormone activity in the thelygenic lines is supposed to result from the action of bacteria on the androgenic hormone receptors (Juchault and Legrand, 1985).

Histological studies of the AG of crabs parasitized with rhizocephalans have been conducted by a number of authors who observed that these organs become temporarily hypertrophied before they degenerate (Charniaux-Cotton, 1956b; Veillet and Graf, 1959; Rubiliani-Durozoi et al., 1980). I shall summarize the more recent investigations that describe the modifications induced by *Sacculina carcini* in two host crabs: *Carcinus maenas* and *C. aestuarii* (Fig. 8.10). Roots of the parasite, which are gathered in the surrounding muscles, can reach the AG and sometimes penetrate between the strands, as soon as the parasite is in its "internal stage." In most cases, the roots do not impair the androgenic strands and

FIGURE 8.10. Androgenic gland (AG) of the crab, *Carcinus* parasitized with the rhizocephalan, *Sacculini carcini:* (A) External *Sacculina* stage. In spite of the contact with a root (R), the androgenic cords (ac) have a normal aspect. (B) Internal *Sacculina* stage. The AG is hypertrophied. Compare the diameter of the androgenic cords with those in A. (C) Degenerating AG in a crab with an external *Sacculina*. *Key:* dac = degenerating androgenic cords; R = root of the parasite; sd = sperm duct; asterisk (white star) = vacuolization of the cytoplasm. (From Rubiliani-Durozoi et al., 1980.)

A
R
ac
sd
R
40 µm
B
sd
ac
40 µm
C
dac
R
sd
40 µm

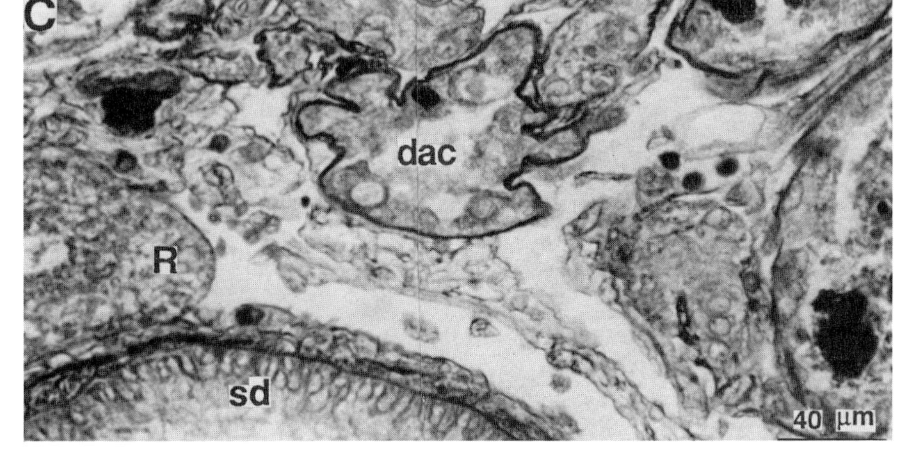

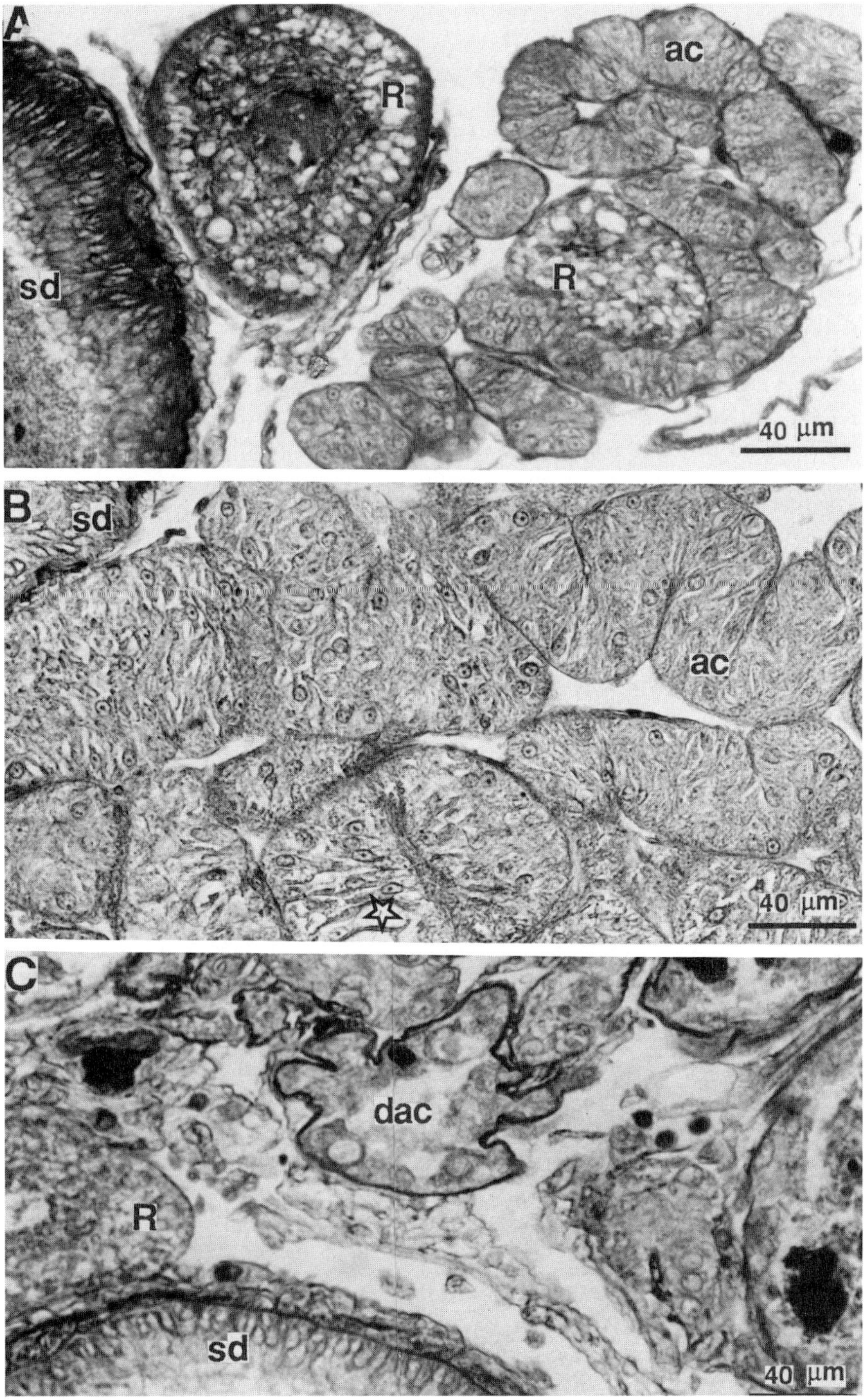

there is only a partially tight contact between the two structures. The progressive degeneration of the AG begins by an hypertrophy of the androgenic strands, whose diameter is about 35 μm, instead of 18.5 μm in the nonparasitized crabs. There is vacuolization of the cytoplasm, and the nuclei are no longer ovoid but spherical, with an average diameter of 9 μm, instead of 6 μm (in their maximum axis) in healthy males.

When *Sacculina* becomes external, the hypertrophy is visible only in some portions of the androgenic strands, whereas in others the diameter is reduced. The cytoplasm becomes dense, but the nuclei remain spherical. At a later stage, degeneration extends to the whole organ. It is characterized by fragmentation of the androgenic strands, which are wrapped with a thick collagenous layer. These fragments contain lipid droplets. It must be emphasized that the successive histological modifications leading to the atrophy of the AG is the same whether the roots are in close contact with the glands or not (Rubiliani-Durozoi et al., 1980).

8.4. Summary

The androgenic gland (AG), which is the source of the masculinizing hormone in Crustacea, was discovered in 1954 in the amphipod *Orchestia gammarella* and has been described in nearly all orders of Malacostraca. In amphipods, some isopods, and decapods, it is located in the last thoracic segment, against the subterminal portion (ejaculatory part) of the sperm duct. In certain oniscoid isopods and in tanaidaceans, the gland is placed in front of/or along the testicular utricules, against the seminal vesicle, or/and against the vas deferens. In addition, it is often closely joined with the genital apparatus. The AG is of genital origin. Male sexual differentiation of the genital apparatus is characterized by the multiplication of cells that represent rudiments of the AG. When the young male acquires its specific morphology, the AG becomes distinct and displays the cytological and cytochemical features of the functioning gland. It is encased in a thin PAS-positive sheath, and its cytoplasm is rich in RNA. An abnormal morphogenesis of the gland can occur in some mutants.

The arrangement of cells is quite varied: sinuous cords of cells in certain isopods; simple strands of aligned discoidal cells in most brachyuran decapods; anastomosed strands of cells in amphipods and prawns; lobules in Majoidae (Decapoda); and spherules in Stomatopoda (Hoplocarida). By contrast, the cytological aspects reveal several common characteristics: a large, central nucleus; an abundant granular endoplasmic reticulum; many free ribosomes; mitochondria with transverse cristae; a partially developed Golgi apparatus; cellular connections of the macula adherens type; and numerous lysosomes. No secretory product is detectable, and no exocytotic phenomenon has been observed. In some oniscoids, the AG shows cellular and nuclear polymorphism. When the RER is well developed, it contains a finely granular material. The presence of areas of degenerating cells as well as of involutive processes, coexisting in the vicinity of normal cells, has been pointed out in amphipods, most isopods, and decapods. These findings have led certain authors to conclude that the destruction of cells terminates the secretory activity, which might be holocrine. In protandrous hermaphroditic species, the AG degenerates and then disappears at the time of sex reversal.

Seasonal transformations of the AG are particularly striking in crayfishes, where the glands regress in winter but reach their maximum size in summer. During the latter period, the RER and the Golgi apparatus are much developed. The effect of elevated temperatures also seem to increase the development of the Golgi system and to induce an accumulation of material in the RER cisternae.

Hypertrophy of the AG can be observed in various circumstances: (1) after ablation of the protocerebrum and/or the optic lobes in isopods; (2) after bilateral eyestalk excision in decapods; (3) under the effect of symbiotic bacteria in intersex males of thelygenic lines in isopods; and (4) in crabs parasitized with rhizocephalans. This hypertrophy results from hyperplasia of the androgenic cells and an increase in their size. In isopods experimentally deprived of the protocerebrum or/and optic lobes, in some thelygenic lines, as well as in eyestalk-ablated gonochoristic and hermaphroditic decapods, this hypertrophy is generally interpreted as an hyperactivity. The Golgi and RER systems develop, followed by an accumulation of material in the extracellular spaces and an increase in RNA synthesis. Moreover, in some cases, the augmentation in the number of lysosomes seems to indicate a crinophagic resorption of the androgenic hormone. In crabs, hypertrophy of the AG following destalking is particularly significant in juvenile males. In functional males, the effect of the operation is variable: either no distinctive modification or hypertrophy leading to hyposecretion. In sacculinized crabs, the hypertrophy induced by the parasite is transitory and precedes degeneration of the gland.

References

Balesdent-Marquet, M. L. 1958. Présence d'une glande androgène chez le Crustacé Isopode *Asellus aquaticus* L. C. R. Acad. Sci. Paris 247: 534–536.

Berreur-Bonnenfant, J. 1962. Glande androgène et différenciation sexuelle mâle et femelle chez le Crustacé Isopode *Meinertia oestroides*. Bull. Soc. Zool. Fr. 87: 253–259.

Bonnenfant, J. 1961. La glande androgène de deux Isopodes: *Paragnathia formica* et *Meinertia oestroides*. C. R. Acad. Sci. Paris 252: 1518–1520.

Blanchet, M. F., R. Ozon, and J. J. Meusy. 1972. Metabolism of steroids, in vitro, in the male crab *Carcinus maenas* Linné. Comp. Biochem. Physiol. 41B: 251–261.

Brodzicki, S. 1964. La glande androgène de *l'Astacus fluviatilis* Fabr. Folia Biol. (Cracow) 12(2): 161–165.

Carlisle, D. B. 1959. On the sexual biology of *Pandalus borealis* (Crustacea Decapoda). 1. Histology of incretory elements. J. Mar. Biol. Assoc. U.K. 38: 381–394.

Carpenter, M. B. and R. De Roos. 1970. Seasonal morphology and histology of the androgenic gland of the crayfish, *Orconectes nais*. Gen. Comp. Endocrinol. 15: 143–157.

Chaigneau, J. and P. Juchault. 1979. Quelques aspects ultrastructuraux de glandes androgène hyperthophiées d'*Armadillidium vulgare* Latreille (Crustacé Isopode Oniscoide), dans différents états de fonctionnement naturels ou induits. Bull. Soc. Sci. Nat. Tunisie 14: 29–34.

Charniaux-Cotton, H. 1954. Découverte chez un Crustacé Amphipode (*Orchestia gammarella*) d'une glande endocrine responsable de la différenciation des caractères sexuels primaire et secondaires mâles. C. R. Acad. Sci. Paris 239: 780–782.

Charniaux-Cotton, H. 1956a. Existence d'un organe analogue à la glande androgène chez un pagure et un crabe. C. R. Acad. Sci. Paris 243: 1168–1169.

Charniaux-Cotton, H. 1956b. Déterminisme hormonal de la différenciation sexuelle chez les Crustacés. Ann. Biol. 32: 371–399.

Charniaux-Cotton, H. 1957. Croissance, régénération et déterminisme endocrinien des caractères sexuels d'*Orchestia gammarella* (Pallas) (Crustacé Amphipode). Ann. Sci. Nat. Zool. [11] 19: 411–559.

Charniaux-Cotton, H. 1958. La glande androgène de quelques Crustacés Décapodes et particulièrement de *Lysmata seticaudata*, espèce à hermaphrodisme protérandrique fonctionnel. C. R. Acad. Sci. Paris 246: 2814–2817.

Charniaux-Cotton, H. 1959. Etude comparée du développement postembryonnaire de l'appareil génital et de la glande androgène chez *Orchestia gammarella* et *Orchestia mediterranea* (Crustacés Amphipodes): autodifférenciation ovarienne. Bull. Soc. Zool. Fr. 84: 105–115.

Charniaux-Cotton, H. 1960a. Physiologie de l'inversion sexuelle chez une crevette hermaphrodisme fonctionnel, *Lysmata seticaudata*. C. R. Acad. Sci. Paris 25: 4046–4048.

Charniaux-Cotton, H. 1960b. Sex determination. Pp. 411–447 *in* T. H. Waterman (ed.), *The Physiology of Crustacea*, Vol. 1. Academic Press, Orlando, Florida.

Charniaux-Cotton, H. 1960c. La glande androgène du Crustacé Stomatopode *Squilla mantis*. Bull. Soc. Zool. France 85: 110–114.

Charniaux-Cotton, H. 1963. Démonstration expérimentale de la sécrétion d'hormone femelle par la testicule inversé en ovaire de *Talitrus saltator* (Crustacé Amphipode): condidération sur la génétique et l'endocrinologie sexuelle des Crustacés supérieurs. C. R. Acad. Sci. Paris 256: 4088–4091.

Charniaux-Cotton, H. 1965. Hormonal control of sex differentiation in invertebrates. Pp. 701–740 *in* R. De Haan and H. Ursprung (eds.), *Organogenesis*. Holt, New York.

Charniaux-Cotton, H. 1967. Arrêt de la spermatogenèse chez la Crevette *Pandalus borealis* (Kroyer) après ablation des pédoncules oculaires. C. R. Soc. Biol. 161: 2100–2104.

Charniaux-Cotton, H. 1972. Recherches récentes sur la différenciation sexuelle et l'activité génitale chez divers crustacés supérieurs. Pp. 128–178 *in* E. Wolff (ed.), *Hormones et Différenciation Sexuelle chez les Invertébrés*. Gordon & Breach, New York.

Charniaux-Cotton, H. and G. Payen. 1985. Sexual differentiation. Pp. 217–299 *in* D. E. Bliss and L. H. Mantel (eds.), *The Biology of Crustacea*. Vol. 9. Academic Press, Orlando, Florida.

Charniaux-Cotton, H., C. Zerbib, and J. J. Meusy. 1966. Monographie de la glande androgène chez les crustacés supérieurs. Crustaceana (Leiden) 10: 113–136.

Cronin, L. E. 1947. Anatomy and histology of the male reproductive system of *Callinectes sapidus* Rathbun. J. Morphol. 81: 209–239.

Démeusy, N. 1960. Différenciation des voies génitales mâles du crabe *Carcinus maenas* Linné: rôle des pédoncules oculaires. Cah. Biol. Mar. 1: 259–278.

Duveau, J. 1957. Données histophysiologiques sur la glande androgène de *Nebalia geoffroyi*. Arch. Anat. Microsc. Morphol. Exp. 46: 199–209.

Foulks, N. B. and D. L. Hoffman. 1974. The effects of eyestalk ablation and β-ecdysone on RNA synthesis in the androgenic glands of the protandric shrimp, *Pandalus platyceros* Brandt. Gen. Comp. Endocrinol. 22: 439–447.

Frechette, J., G. W. Corrivault, and R. Couture. 1970. Hermaphrodisme protérandrique chez une Crevette de la famille des crangonidés, *Argis dentata* Rathbun. Nat. Can. (Que.) 97: 805–822.

Gilgan, M. W. and D. R. Idler. 1967. The conversion of androstenedione to testosterone by some lobster (*Homarus americanus* Milne Edwards) tissues. Gen. Comp. Endocrinol. 9: 319–324.

Ginsburger-Vogel, T. 1983. Etude de la mutation "mâle stérile l" affectant la morphologie de la glande androgène chez le crustacé amphipode *Orchestia gammarellus* Pallas. Int. J. Invertebr. Reprod. 6: 161–170.

Graf, F. 1958. Développement post-embryonnaire des gonades et des glandes androgènes d'*Orchestia cavimana* (Heller) Crustacé Amphipode. Bull. Soc. Sci. Nancy 17: 223–261.

Graf, F. 1962. Intersexualité et glande androgène chez *Orconectes limosus* (Rafinesque) Crustaceana (Leiden) 4: 151–157.

Hasegawa, Y. and Y. Katakura. 1981. Androgenic gland hormone and development of oviducts in the iospod crustacean, *Armadillidium vulgare*. Dev. Growth. Diff. 23(1): 59–62.

Hoffman, D. L. 1968. Seasonal eyestalk inhibition on the androgenic glands of a protandric shrimp. Nature (Lond.) 218: 170–172.

Hoffman, D. L. 1969. The development of the androgenic glands of a protandric shrimp. Biol. Bull. (Woods Hole) 137: 286–296.

Hort-Legrand, C., J. Berreur-Bonnenfant, and T. Ginsburger-Vogel. 1974. Etude anatomique et histologique comparée de la différenciation des gonades mâles et femelles chez *Orchestia gammarella* Pallas (Crustacé Amphipode). C. R. Acad. Sci. Paris 276D: 1891–1894.

Huguet, D. 1968. Description de la glande androgène et des caractères sexuels secondaires chez la crevette d'eau douce, *Atyaephyra desmaresti* Millet (Crustacea Decapoda Natantia). Bull. Mus. Natl. Hist. Nat. Zool. [2] 40: 351–357.

Huguet, D. and P. Huguet. 1971. La glande androgène de *Palaemon serratus* (Pennant) *Crangon crangon* (Linné), *Aristeus antennatus* (Risso), (Crustacés Décapodes Natantia) Description et étude expérimentale. Bull. Mus. Natl. Hist. Nat. Zool. [3] 10: 597–610.

Husson, R. and F. Graf. 1961. Comparaison des glandes androgènes d'Amphipode appartenant à des genres hypogé (*Niphargus*) et épigé (*Gammarus*). C. R. Acad. Sci. Paris 252: 169–170.

Juchault, P. 1963. Sur la glande androgène d'un certain nombre de péracarides (Cumacés, Mysidacés, Tanaïdacés). C. R. Soc. Biol. Paris 157: 613–615.

Juchault, P. 1967. Contribution à l'étude de la différenciation sexuelle mâle chez les Crustacés Isopodes. Ann. Biol. 6: 191–212.

Juchault, P. and J. J. Legrand. 1985. Contribution à l'étude du mécanisme de l'état réfractaire à l'hormone androgène chez les *Armadillidium vulgare* Latr. (crustacé, isopode, oniscoïde) hébergeant une bactérie féminisante. Gen. Comp. Endocrinol. 60: 463–467.

Katakura, Y. 1961. Hormonal control of development of sexual characters in the isopod crustacean, *Armadillidium vulgare*. Annot. Zool. Jpn. 34: 60–71.

Katakura, Y. 1982. Sex differentiation in the pill bug, *Armadillidium vulgare* (in Japanese). Hiyoshi Sci. Rev. 17: 25–56.

Katakura, Y. 1984. Sex differentiation and androgenic gland hormone in the terrestrial isopod *Armadillidium vulgare*. Symp. Zool. Soc. Lond. 53: 127–142.

King, D. S. 1964. Fine structure of the androgenic gland of the crab, *Pachygrapsus crassipes*. Gen. Comp. Endocrinol. 4: 533–544.

Lane, R. L. 1977. A developmental investigation of the reproductive systems of *Armadillidium vulgare* (Latreille) and *Porcellionides pruinosus* (Brandt) (Isopoda). Crustaceana (Leiden) 33: 237–248.

Laubier, A., L. Chim, and G. G. Payen. 1985. Morphogenèse sexuelle et régulation hormonale de l'activité génitale chez la crevette *Penaeus japonicus* en élevage. IFREMER (Inst. Fr. Rech. Exploit. Mer) Actes Colloq. 1: 195–206.

Legrand, J. J. 1958. Mise en évidence histologique et expérimentale d'un tissu androgène chez les Oniscoïdes. C. R. Acad. Sci. Paris 247: 1238–1241.

Legrand, J. J. and P. Juchault. 1960a. Structure et origine de la glande androgène chez *Helleria brevicornis* Ebner (Oniscoïdea, Tylidae). C. R. Soc. Biol. 154: 676–678.

Legrand, J. J. and P. Juchault. 1960b. Mise en évidence anatomique et expérimentale des glandes androgènes de *Sphaeroma serratum* Fabricius (Isopode, Flabellifère). C. R. Acad. Sci. Paris 250: 3401–3402.

Legrand, J. J. and P. Juchault. 1961. Sur la glande androgène d'un certain nombre de Péracarides et en particulier d'Isopodes marins. C. R. Soc. Biol. 155: 1360–1362.

Legrand, J. J. and P. Juchault. 1972. Le contrôle humoral de la sexualité chez les Crustacés Isopodes gonochoriques. Pp. 179–218 *in* E. Wolff (ed.), *Hormones et différenciation sexuelle chez les Invertébrés*. Gordon & Breach, New York.

Legrand, J. J. and A. Vandel. 1948. Le développement post-embryonnaire de la gonade chez les Isopodes terrestres normaux et intersexués: évolution morphologique de la gonade. Bull. Biol. Fr. Belg. 82: 79–95.

Legrand, J. J., P. Juchault, J. P. Mocquard, and G. Noulin. 1968. Contribution à l'étude du contrôle neurohumoral de la physiologie sexuelle mâle chez les Crustacés Isopode terrestres. Ann. Embryol. Morphog. 1: 97–105.

Le Roux, A. 1976. Aspects de la différenciation sexuelle chez *Pisidia longicornis* (Linné) (Crustacea, Decapoda). C. R. Acad. Sci. Paris 283D: 959–962.

Malo, N. 1970. Premières observations ultrastructurales de la glande androgène d'*Oniscus asellus*, Crustacé Isopode, et ses modifications en fonction de la température d'élevage. C. R. Acad. Sci. Paris 270: 2843–2845.

Malo, N. and P. Juchault. 1970. Contribution à l'étude des variations ultrastructurales de la glande androgène des Oniscoïdes supérieurs

(Crustacés Isopodes) à la suite de la décérébration. C. R. Acad. Sci. Paris 271: 230–232.

Martin, G., P. Juchault, and J. J. Legrand. 1973. Mise en évidence d'un microorganisme intracytoplasmique symbiote de l'Oniscoïde *Armadillidium vulgare* Latr. dont la présence accompagne l'intersexualité ou la féminisation totale des mâles génétiques de la lignée thélygène. C. R. Acad. Sci. 276D: 2313–2316.

Meusy, J. J. 1963. Description de la glande androgène chez deux Crustacés Péracarides *Paramysis nouveli* Labat (Mysidacé) et *Eocuma dollfusi* Calman (Cumacé). C. R. Acad. Sci. Paris 256: 5425–5428.

Meusy, J. J. 1965. Contribution de la microscopie électronique à l'étude de la physiologie des glandes androgènes *d'Orchestia gammarella* P. (Crustacé Amphipode) et de *Carcinus maenas* L. (Crustacé Décapode). Zool. Jahrb. Physiol. 71: 608–623.

Meusy, J. J. 1968. Effets de l'ablation des pédoncules oculaires et des organes Y sur les glandes androgènes et sur l'appareil génital chez le crabe mâle *Carcinus maenas* L. (Crustacé, Décapode) pubère. C. R. Acad. Sci. Paris 267: 1861–1863.

Miyawaki, M. and Y. Taketomi. 1978. The occurrence of an extended perinuclear space in androgenic gland cells of the crayfish, *Procambarus clarki.* Cytologia (Tokyo) 43: 351–355.

Payen, G. 1970. Etude ultrastructurale des glandes androgènes hypertrophiées la suite de l'ablation des pédoncules oculaires au premier stade larvaire chez le crabe *Rhithropanopeus harrisii* (Gould). C. R. Acad. Sci. 270D: 1499–1502.

Payen, G. 1972. Etude ultrastructurale de la dégénérescence cellulaire dans la glande androgène du crabe *Ocypoda quadrata* (Fabricius). Z. Zellforsch. Mikrosk. Anat. 129: 370–385.

Payen, G. G. 1973. Etude descriptive des principales étapes de la morphogenèse sexuelle chez un Crustacé Décapode à développement condensé, l'Ecrevisse *Pontastacus leptodactylus leptodactylus* (Eschscholtz, 1823). Ann. Embryol. Morphog. 6: 179–206.

Payen, G. 1974. Morphogenèse sexuelle de quelques Brachyoures (Cyclométopes) au cours du développement embryonnaire, larvaire et postlarvaire. Bull. Mus. Natl. Hist. Nat. Zool. 209(139): 201–262.

Payen, G. G., J. D. Costlow, and H. Charniaux-Cotton. 1971. Etude comparative de l'ultrastructure des glandes androgènes de Crabes normaux et pédonculectomisés pendant la vie larvaire ou après la puberté chez les espèces: *Rhithropanopeus harrisii* (Gould) et *Callinectes sapidus* Rathbun. Gen. Comp. Endocrinol. 17: 526–542.

Payen, G., L. Chim, A. Laubier-Bonichon, and H. Charniaux-Cotton. 1982. The androgenic gland of the shrimp *Penaeus japonicus* Bate: description, role and control by the eyestalks. Gen. Comp. Endocrinol. 4: 384 (abstr.).

Peyrot, S. and J. P. Trilles. 1964. Recherches sur la sexualité et la glande androgène de *Caprella aequilibra* Say (Amphipode, Caprellidae). Bull. Inst. Océanogr. (Monaco) 63(1315): 1–28.

Radu, V. G. and C. Craciun. 1976. The ultrastructure of the androgenic gland in *Porcellio scaber* Latr. (terrestrial isopods). Cell Tissue Res. 175: 245–263.

Reidenbach, J. M. 1966. Mise en évidence d'une intervention du complexe neurosécréteur céphalique dans la physiologie sexuelle mâle chez le Cructacé Isopode marin *Idotea balthica basteri* Audouin. C. R. Acad. Sci. 262D: 682–684.

Reidenbach, J. M. 1971. Les mécanismes endocriniens dans le contrôle de la différenciation du sexe, la physiologie sexuelle et la mue chez le Crustacé isopode marin: *Idotea balthica* (Pallas). Doctoral thesis, CNRS No. 4874, University of France, 335 pp.

Remy, C. and A. Veillet. 1961. Evolution de la glande androgène chez l'Isopode *Anilocra physodes* L. Bull. Soc. Sci. Nancy 21: 53–80.

Rubiliani-Durozoi, M., C. Rubiliani and G. G. Payen. 1980. Déroulement des gamétogenèses chez les crabes *Carcinus maenas* (L.) et *C. mediterraneus* Czerniavsky parasités par la Sacculine. Int. J. Invertebr. Reprod. 2: 107–120.

Taketomi, Y. 1986. Ultrastructure of the androgenic gland of the crayfish, *Procambarus clarki*. Cell Biol. Int. Report 10(2): 131–136.

Tcholakian, R. K. and K. B. Eik-Nes. 1969. Conversion of progesterone to 11-deoxycorticosterone by the androgenic gland of the blue crab (*Callinectes sapidus* Rathbun). Gen. Comp. Endocrinol. 12: 171–173.

Tcholakian, R. K. and K. B. Eik-Nes. 1971. Steroidogenesis in the blue crab *Callinectes sapidus* Rathbun. Gen. Comp. Endocrinol. 17: 115–124.

Tcholakian, R. K. and S. M. Reichard. 1964. A possible androgenic gland in *Callinectes sapidus* Rathbun. Amer. Zool. 4: 383 (abstr.).

Thampy, D. M. and P. A. John. 1970. On the androgenic gland of the ghost crab *Ocypoda platytarsis* M. Edwards (Crustacea: Brachyura). Acta Zool. 51: 203–210.

Thampy, D. M. and P. A. John. 1972. The androgenic gland of the shrimp *Palaemon dayanus*. Mar. Biol. (Berl.). 12: 285–288.

Touir, A. 1973. Influence de l'ablation des pédoncules oculaires sur les glandes androgènes les gonades, les caractères sexuels externes mâles et l'inversion sexuelle chez la Crevette hermaphrodite *Lysmata seticaudata* Risso. C. R. Acad. Sci. Paris 277: 2541–2544.

Trilles, J. P. 1963. Mise en évidence d'une action du complexe céphalique neurosécrétoire sur la glande androgène et les gonades de *Nérocila orbignyi* (Schioedte et Meinert) (Isopode, Cymothoïdae). C. R. Acad. Sci. Paris 257: 1811–1812.

Veillet, A. 1958. Inversion sexuelle et glande androgène chez quelques crustacés. Bull. Soc. Sci. Nancy 17: 200–203.

Veillet, A., and F. Graf. 1958. Développement post-embryonnaire des gonades et de la glande androgène chez le Crustacé Amphipode *Orchestia cavimana* Heller. C. R. Acad. Sci. Paris 246: 3188–3191.

Veillet, A. and F. Graf. 1959. Dégénérescence de la glande androgène des Crustacés Décapodes parasités par les Rhizocéphales. Bull. Soc. Sci. Nancy 18: 123–128.

Veillet, A. and F. Graf. 1965. Inversion sexuelle et glande androgène chez quelques Crustacés. Bull. Acad. & Soc. Lorraines Sci. 5: 295–308.

Veith, W. J. and S. R. Malecha. 1983. Histochemical study of the distribution of lipids, 3α and 3β-Hydroxysteroid Dehydrogenase in the androgenic gland of the cultured prawn *Macrobrachium rosenbergii* (de Man) (Crustacea: Decapoda). S. Afr. J. Sci. 79: 84–85.

Zerbib, C. 1964. Evolution post-embryonnaire de la voie déférente chez la mâle et chez la femelle normale et masculinisée *d'Orchestia gammarella* Pallas (Crustacé Amphipode). Bull. Biol. Fr. Belg. 98: 391–408.

Zerbib, C. 1967. Première observation de la glande androgène chez un Crustacé Syncaride: *Anaspides tasmaniae* Thomson et chez un Crustacé Eucaride: *Meganyctiphanes norvegica* Sars. C. R. Acad. Sci. Paris 265D: 415–418.

Morphology, Histology, and Ultrastructure of the Maxillary Glands in Crustaceans: Their Probable Function in Morphogenesis

9

GERTRUDE W. HINSCH

9.1. Introduction

The antennal or maxillary glands (MGs) are paired segmental excretory organs that are found in most crustaceans. The general morphology of these glands has been reviewed by Goodrich (1945). The morphology of the cells of the different regions of the glands is highly suggestive of cells capable of an active role in solute reabsorption and ultrafiltration of urine. However, their role in pheromone production, and thus indirectly in morphogenesis, is suspected.

These excretory organs are mesodermal in origin. The mesoderm of the antennary or maxillary region becomes thickened on each side and then differentiates into a coiled tube. One end of the tube extends downward through the ectoderm lateral to the developing central nerve cord to form the duct of the gland. The other end of the tube enlarges to form a terminal sac and establishes a connection with the coelom of the first maxillary segment (Kume and Dan, 1968).

Adult entomostracans and some Malacostraca retain the organs of the maxillary segment, the MGs. In the Amphipoda, Mysidacea, Euphausiacea, and Decapoda it is the derivative of the antennary segment, namely, the antennal or green gland, which is retained. Both glands are retained in the Ostracoda. In all forms the organs basically consist of a proximal, closed end-sac, an excretory canal leading from the end-sac, and a distal duct which opens either on the third, antennary segment or on the sixth, maxillary segment.

9.2. Structure

In the barnacle *Balanus balanoides*, paired organs lie in the posterior part of the body below and to either side of the foregut (Fig. 9.1A). The excretory organ consists of the end-sac, efferent duct, and terminal ducts (White and Walker, 1981).

In crabs, the paired glands are located in the antennary segment near the base of the eyestalks and consist of a coelomosac, followed by the labyrinth (a convoluted tubule), which leads into the bladder (Fig. 9.1B). The coelomosac and labyrinth are coiled around each other so that the coelomosac and labyrinth appear interdigitated (Figs. 9.2A,B and 9.3A). The wall of the coelomosac of *Callinectes sapidus* and other brachyurans has many tubular evaginations dispersed in between the branching invaginations of the labyrinth (Johnson, 1980). The excretory pores open at the bases of the antennae. The tissues of the coelomosac and labyrinth are surrounded by blood sinuses that are fed by the antennary artery.

In the crayfish, a freshwater crustacean, the glands consist of an internal end-sac (coelomosac), an excretory canal, and an excretory duct. A

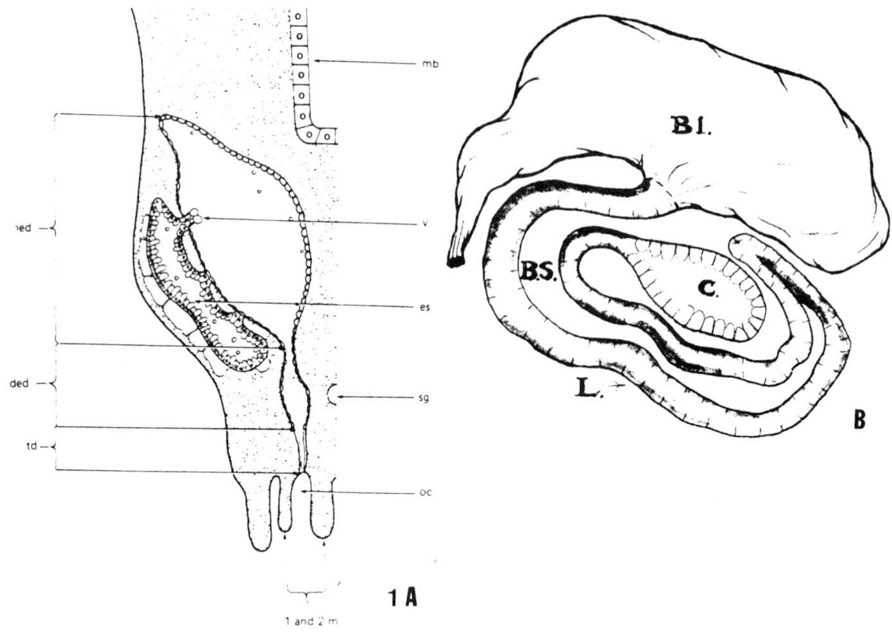

FIGURE 9.1. (A) Diagrammatic section through the body of *Balanus belanoides* showing the regions of an excretory organ. *Key:* ded, distal efferent duct; es, end-sac; 1 and 2, first and second maxilla; mb, midgut; oc, oral cone; ped, proximal efferent duct; sg, subesophageal ganglion; td, terminal duct; v, valve. (B) schematic drawing of the antennal gland of *Uca mordax*. Coelomosac (C) is surrounded by the labyrinth (L), which consists of a single tubule opening into the bladder (BL). Blood sinuses (B.S.) are found at the base of labyrinth and coelomosac cells. [(A) From White and Walker, 1981. (B) From Schmidt-Nielsen et al., 1968.]

FIGURE 9.2. (A) Scanning micrograph of convoluted tubules of the antennal gland of the land hermit crab *Coenobita clypeatus*. Branches of blood vessels are seen adhering to the surface of the tubules. Hemocytes (arrows) are found in the hemal space between the tubules. (B) Scanning micrograph showing the blood vessels in association with tubule of the antennal gland which has been fractured to expose epithelium face from the hermit crab. (From G. W. Hinsch and B. F. Komm, unpublished work.)

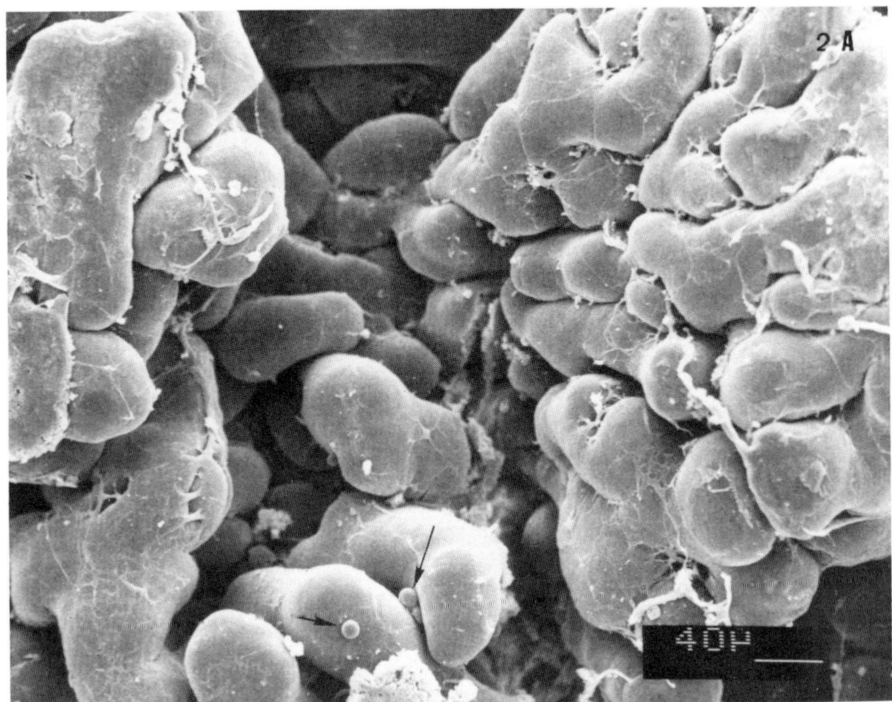
2 A
40P

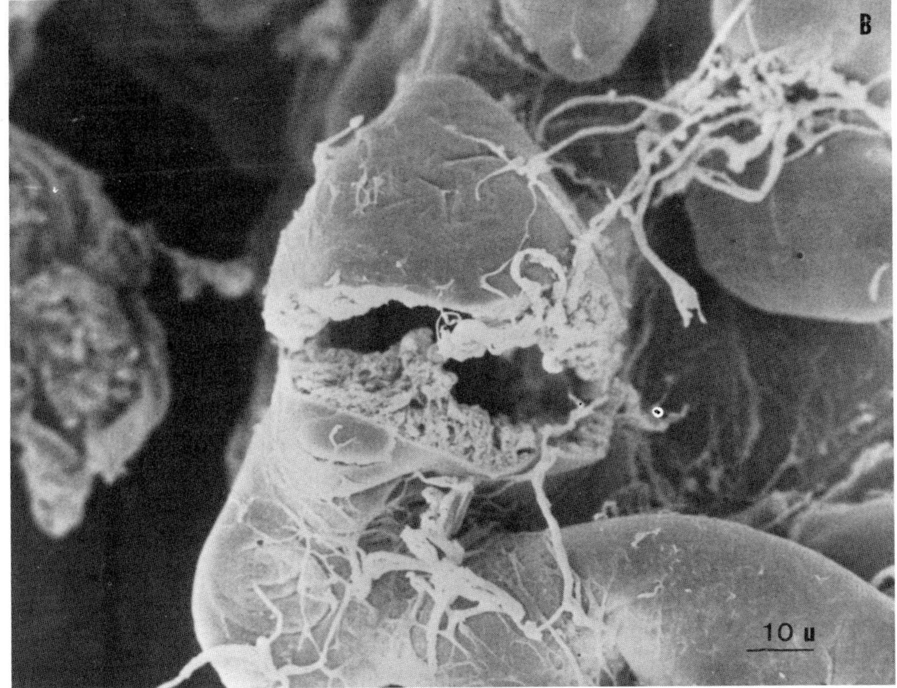
B
10 u

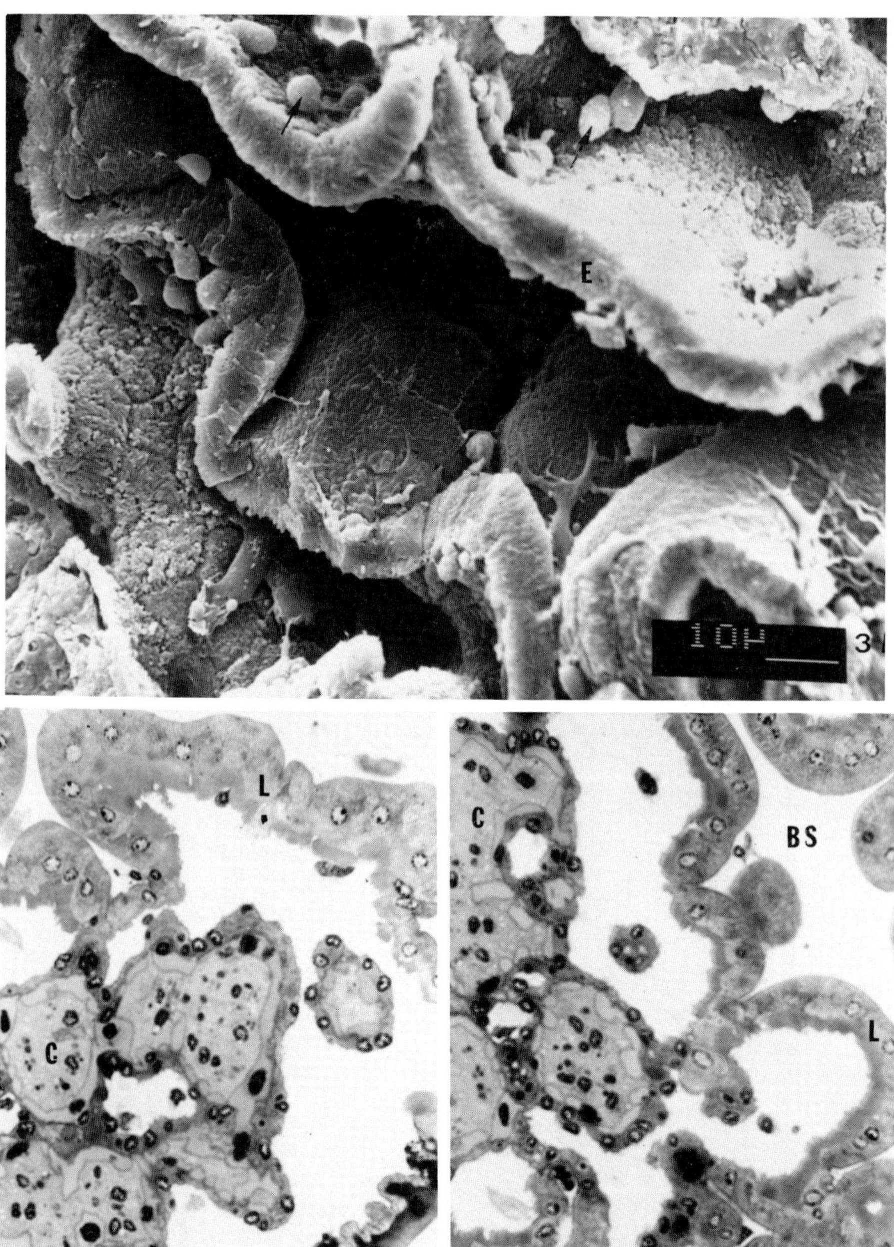
E
10μ
3
L
C
B
C
BS
L
C

collecting bladder formed from the lower part of the excretory canal and the excretory duct may also exist.

In the brine shrimp *Artemia salina,* each gland consists of an end-sac and efferent excretory duct. It possesses a slender terminal duct which opens to the outside via an aperture on the second maxilla (Tyson, 1968). In the amphipod *Corophium volutator,* the coelomosac is linked with the efferent tubule by a valve-like structure (Icely and Nott, 1979).

9.3.   The Coelomosac (or End-sac)

9.3.1.   *Light Microscopy*

The cells of the coelomosac, or end-sac, of the different crustaceans which have been studied are quite similar in structure. They are often highly vaculoated. The nuclei are small and usually located basally in the cells. Granular deposits appear in the cells, and the lumen is often difficult to recognize (Fig. 9.3B,C). In the barnacle (*Balanus* spp.), histochemical studies reveal that only small amounts of RNA are present in coelomosac or end-sac cells (White and Walker, 1981). Lipid is never observed. Large inclusions stain for protein with significant amounts of sulfydryl groups and the amino acids tyrosine and tryptophan (White and Walker, 1981).

9.3.2.   *Ultrastructure*

The epithelial cells of the coelomosac are large and irregular (Fig. 9.4A). The lateral cell membranes are not closely apposed between adjacent cells. No terminal bars appear between cells, and the intercellular spaces are usually open except for frequent desmosome-like attachments (Figs. 9.4B,C). Coated pits may form along the cell surfaces that line the intercellular spaces. These may represent pinocytosis of proteins filtered from the ultrafiltrate.

The basal lamina adjacent to the blood sinuses is composed of granular and fibrous layers. Kummel (1967) described distinctive foot processes of

FIGURE 9.3. (A) Scanning micrograph of hermit crab antennal gland showing fractured lobular surface (E) of the convoluted tubule. Hemocytes (arrows) are visible in several places. (B&C) Light micrographs of hermit crab antennal gland showing the association between the convoluted tubules of the labyrinth (L) and the coelomsac (C). *Key:* BS, blood sinus. (From G. W. Hinsch and B. F. Komm, unpublished work.)

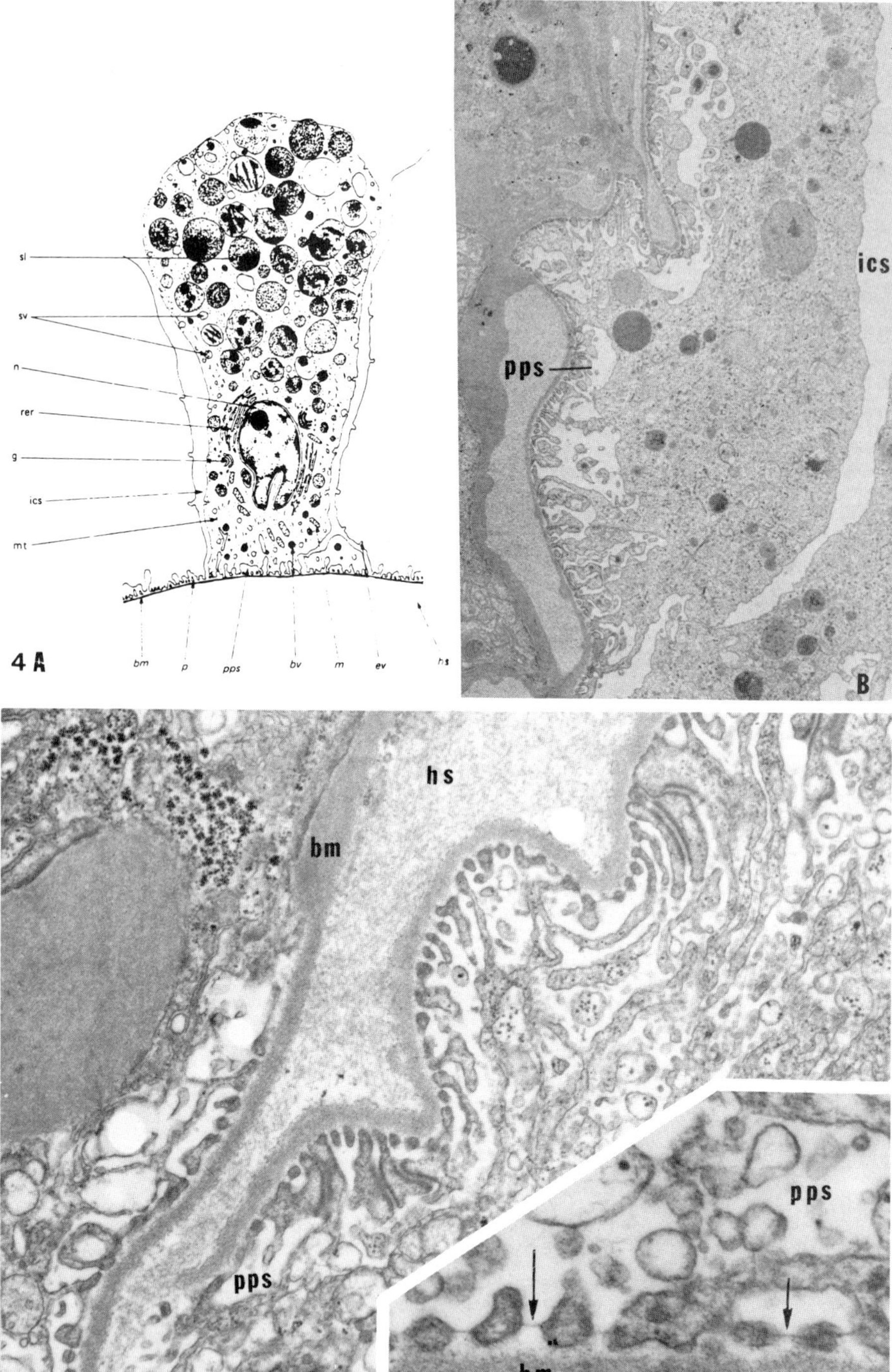
sl
sv
n
rer
g
ics
mt
4 A
bm
p
pps
bv
m
ev
hs
ics
pps
B
hs
bm
pps
C
pps
bm
D

TABLE 9.1.   Measurements of Crayfish Slit Diaphragm

		Number of measurements	Mean (nm) +/− SD
a.	Overall width	9	36.8 +/−0.3
b.	Diameter of helix	17	14.8 +/−0.2
c.	Pitch of helix	8	13.0 +/−0.3
d.	Width of helical filament	10	7.3 +/−0.2

SOURCE: From Schaffner and Rodewald (1978).

the coelomosac cells that extend to the basal lamina (Fig. 9.4D). Desmosome-like attachments are also often seen between foot processes in the basal regions of the cells.

Between each foot process is an epithelial slit diaphragm which, together with the basal lamina, serves as a two-layered filtration barrier (see Table 9.1). The diaphragm appears as two parallel, helical filaments which are joined to each other and the plasmalemma of adjacent podocytes (Fig. 9.5A–F). This slit diaphragm can act as a barrier to ferritin and presumably hemocyanin (Schaffner and Rodewald, 1978).

Numerous vesicles, vaculoes, and dense bodies occupy the cytoplasm of the apical ends of the cells. Microvilli are absent on cells in this region (Fig. 9.4A,B). The vesicles, which are of varying electron density, include lysosomes associated with the Golgi complex. Rough endoplasmic reticulum is also seen in this region.

FIGURE 9.4. Drawing of the fine structure of a cell in the end-sac in *Balanus balanoides*. (B) End-sac cell from the coelomosac of the hermit crab *Coenobita clypeatus* exhibiting several podocytes. ×4896. (C) Basal region of an end-sac cell with several podocytes adjacent to the basement membrane. ×38,400. (D) Enlargement of basal region of end-sac cell showing the presence of the epithelial slit diaphragm (arrows). ×36,000. *Key:* bm, basement membrane; bs, basal vesicles; ev, endocytoytic vesicles; g, Golgi body; hs, hemolymph space; ics, intercellular space; m, mitochondrion; mt, microtubule; n, nucleus; p, pedicel; pps, peripodocyte space; rer, rough endoplasmic reticulum; sl, secondary lysosome; sv, small vesicles. [(A) From White and Walker, 1981.]

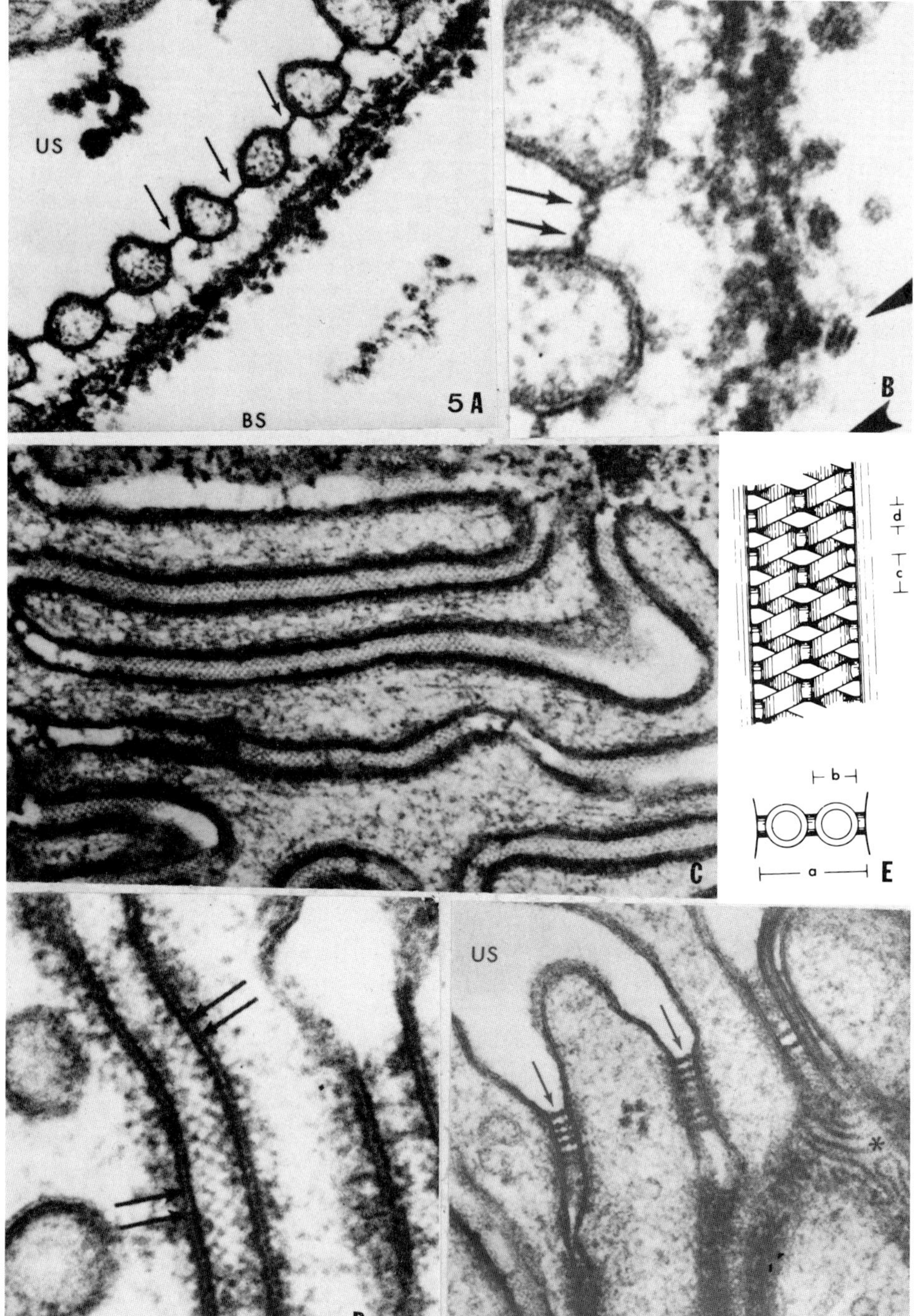
US
BS
5 A
B
C
d
c
b
a
E
US
D
F

9.4.   The Labyrinth (or Efferent Duct)

9.4.1.   *Light Microscopy*

The tubular labyrinth varies in length among the different species which have been studied. The cells of the epithelium vary in height in different regions of the tubule. There appear to be striations that run parallel to the long axis of the cells, and a brush border occupies the apical surface. The nuclei in general are apical in position (Figs. 9.3B,C). In the isopod *Mesidotea entomon*, alkaline phosphatase activity is limited to the cells of the excretory duct (Hryniewiecka-Szyfter and Tyczewska, 1984).

9.4.2.   *Ultrastructure*

The epithelial cells extend from the basal lamina to the lumen of the tubule. The apical surface is covered with microvilli (Figs. 9.6A and 9.7A,B). In *Artemia*, the microvilli may be branched and form apical blebs (Tyson, 1969). Small pinocytotic vesicles may be found between the microvilli. The nuclei (Figs. 9.6A and 9.7A) as well as aggregates of smooth endoplasmic reticulum and glycogen may be found in the apical cytoplasm (Fig. 9.7B).

Numerous infoldings of the basal cytoplasmic membranes are typical of the labyrinth or efferent duct cells. These infoldings extend for varying

FIGURE 9.5. (A) Coelomosac wall perfusion-fixed with tannic acid and 1% glutaraldehyde. Slit diaphragms are cut in cross section to reveal two side-by-side circular profiles within each slit (arrows). Filaments radiate from the foot processes to the basement membrane. ×95,060. (B) Circular profiles (arrows) of the slit diaphragm are clearly evident and appear to be joined to each other and adjacent plasma membranes by electron-dense material. Small rectangular structures with internal banding (arrowhead) appear in the hemolymph side of the basement membrane. ×155,200. (C) Tangential section of the basement membrane of the coelomosac sac wall showing the slit diaphragm as two parallel components between adjacent plasma membranes. ×93,120. (D) Enlargement of C showing how the parallel components of the slit diaphragm exhibit a zigzag pattern (arrows). ×135,800. (E) Diagrammatic representation of the slit diaphragm: (a) overall width; (b) diameter of helix; (c) pitch of helix; and (d) width of helical filament. Measurements found in Table 9.1. (F) Coelomosac wall showing multiple slit diaphragms (arrows) and in side view (asterisk). ×87,300. *Key:* US, urinary space; BS, blood sinus. (From Schaffner and Rodewald, 1978.)

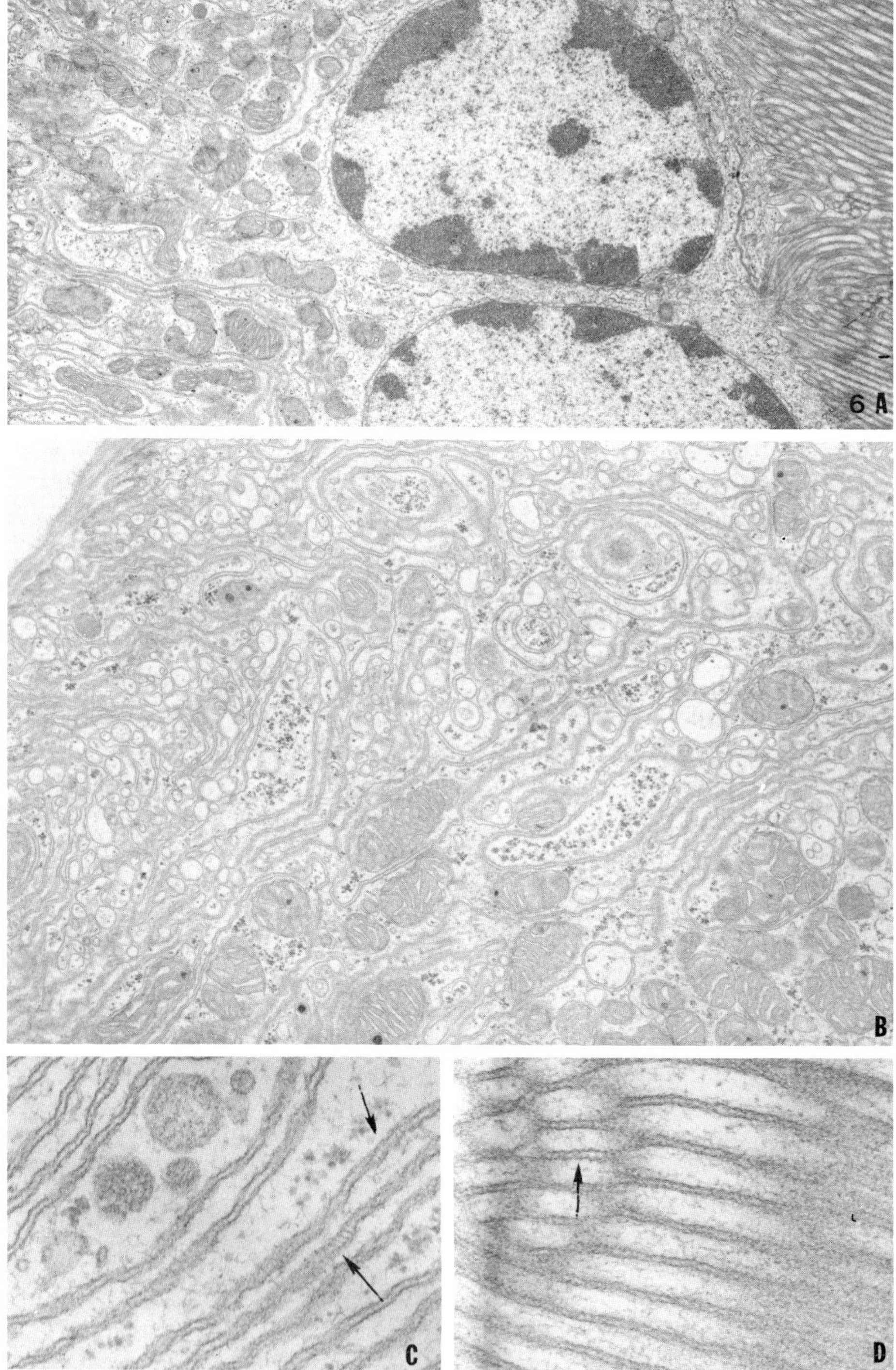

distances toward the apical regions of the cells (Figs. 9.6B and 9.8A). Mitochondria are prominently oriented along the long axis of these basal infoldings, and the intercellular spaces between adjacent infoldings are quite narrow. Septate desmosomes appear in these spaces (Fig. 9.6C), as well as in the spaces between adjacent microvilli (Fig. 9.6D).

Another type of cell is seen in the labyrinth of *Uca mordax* (Schmidt-Nielsen et al., 1968) and *Coenobita clypeatus* (G. W. Hinsch and B. F. Komm, personal observations). These cells tend to be flattened, without the numerous basal infoldings, have fewer mitochondria, and have fewer and shorter microvilli than the typical labyrinth cell (Figs. 9.8B and 9.9A,B).

Freeze-fracture studies of the labyrinth cells reveal two types of intercellular junctions. One type includes long septate junctions which consist of many parallel rows of 80- to 180-Å-diameter intramembrane particles on the PF membrane face (Shivers and Chauvin, 1977) (Figs. 9.10A–C). The other includes fine filaments which cross the extracellular space between plasma membranes of the infoldings of the basal regions of the labyrinth cells (Figs. 9.11A–C).

9.5.   The Valve System

In the antennal glands of the amphipod *Corophium volutator,* a valve system has been reported (Icely and Nott, 1979). Cells of two types replace the podocytes of the coelomosac as it opens into the efferent duct. Three type I cells are flattened and lack foot processes. They have few microvilli on their apical surfaces and have limited contact with the podocytes. However, they have extensive septate junctions with adjacent valve cells and are connected apically with zonula adherens.

The three type II valve cells surround the aperture between the coelomosac and efferent tubule. Each cell has an attachment region adjoining the type I cells and a bulbous region which protrudes into the lumen of the efferent tubule (Icely and Nott, 1979).

FIGURE 9.6. (A) Labyrinth cell from antennal gland of *Coenobita clypeatus* with microvillus bordering the lumen. The nuclei are located in the apical end of the cells, and the basal regions of the cells contain numerous invaginations of the plasma membrane. Many mitochondria are found in this region. ×7000. (B) Basal region of a labyrinth cells. ×16,500. (C) Septate junctions (arrows) appear between adjacent plasma membranes in the basal invaginations. ×50,500. (D) Similarly they appear to be present between the plasma membranes of the microvilli at the apical end of the cell. ×47,500.

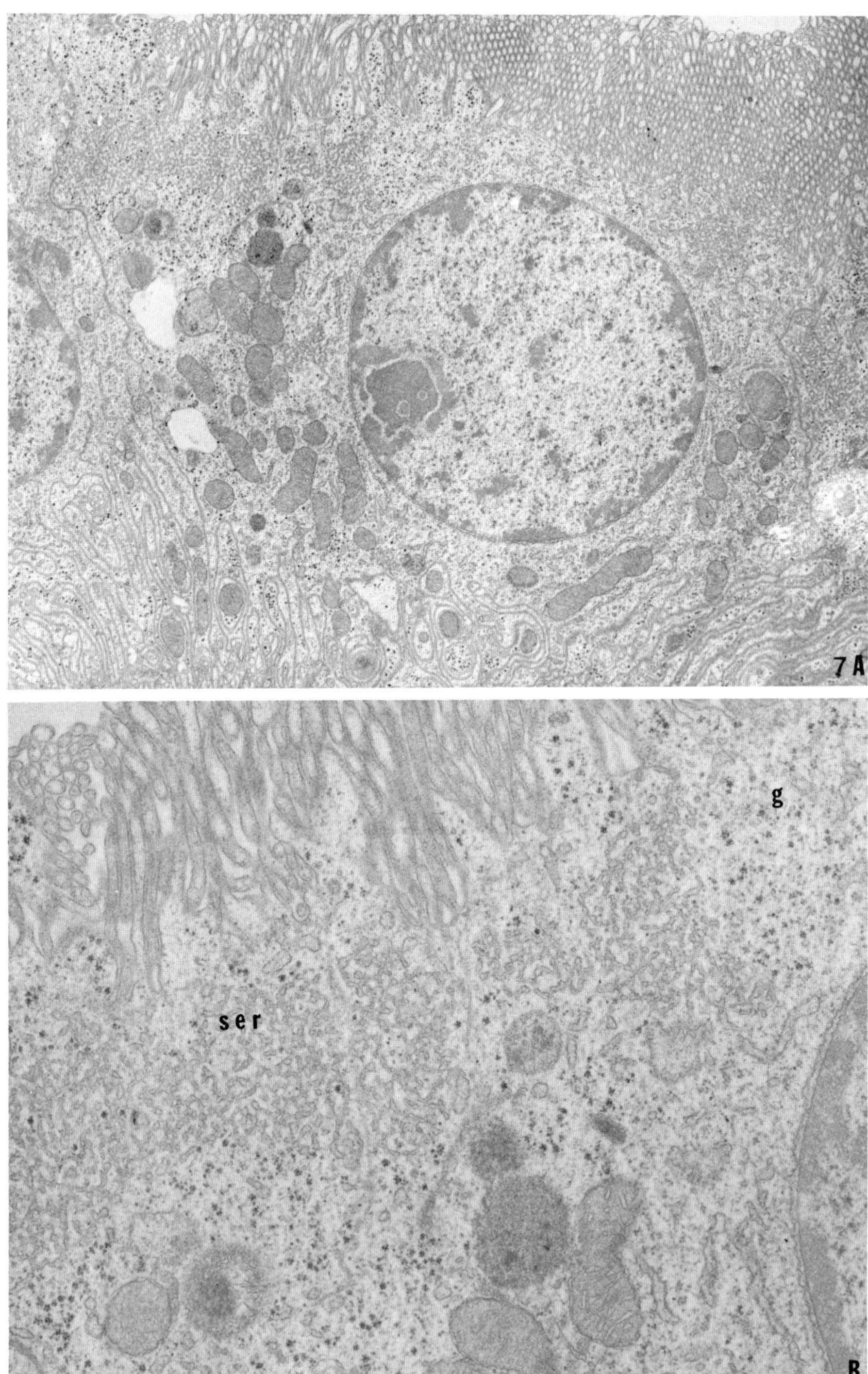

FIGURE 9.7. (A) Apical end of a labyrinth cell from *Coenobita clypeatus*. ×6790. (B) Enlargement of apical end of a labyrinth cell showing the arrays of smooth endoplasmic reticulum (ser) and clusters of glycogen (g). ×20,273.

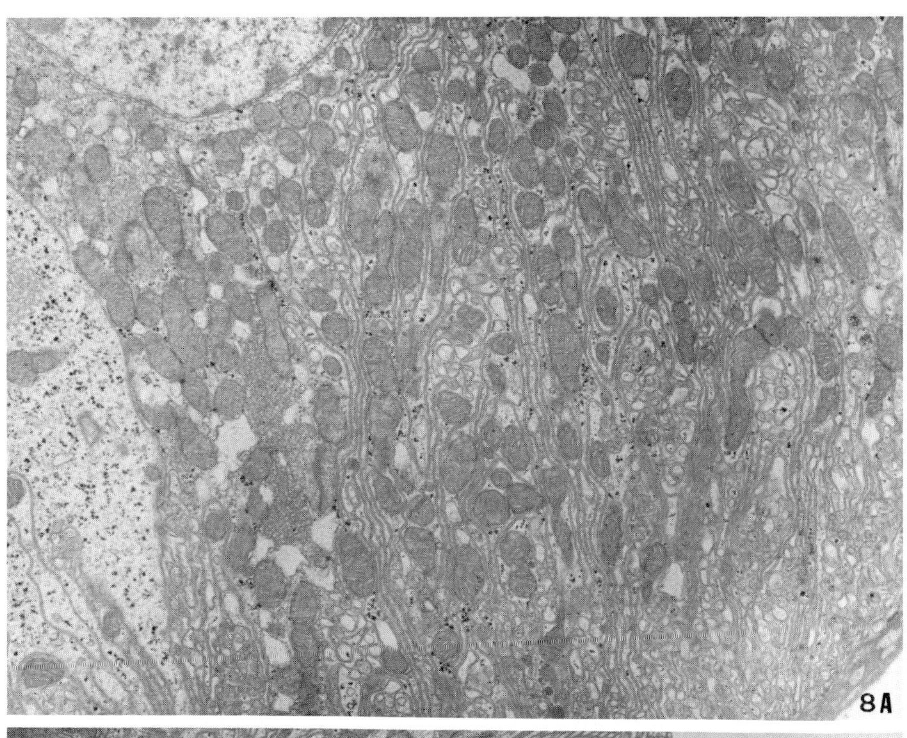
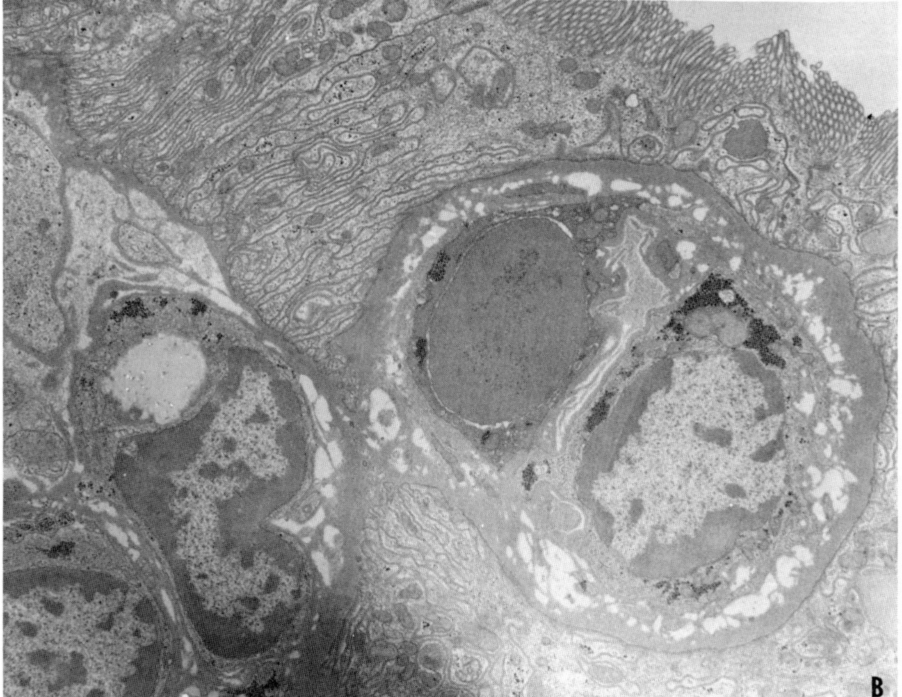

FIGURE 9.8. (A) Basal region of labyrinth cells showing the numerous invaginations of the basal plasma membrane. ×10,670. (B) Cells from the labyrinth which surround branches of one of the blood vessels which penetrate the antennal gland and contain hemocytes. ×5432.

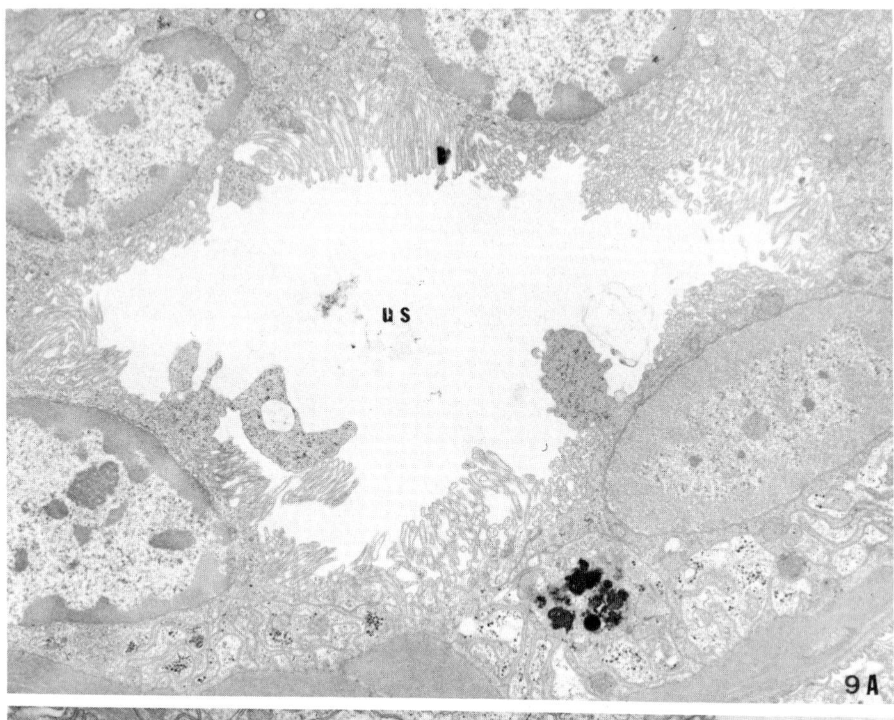
us
9A

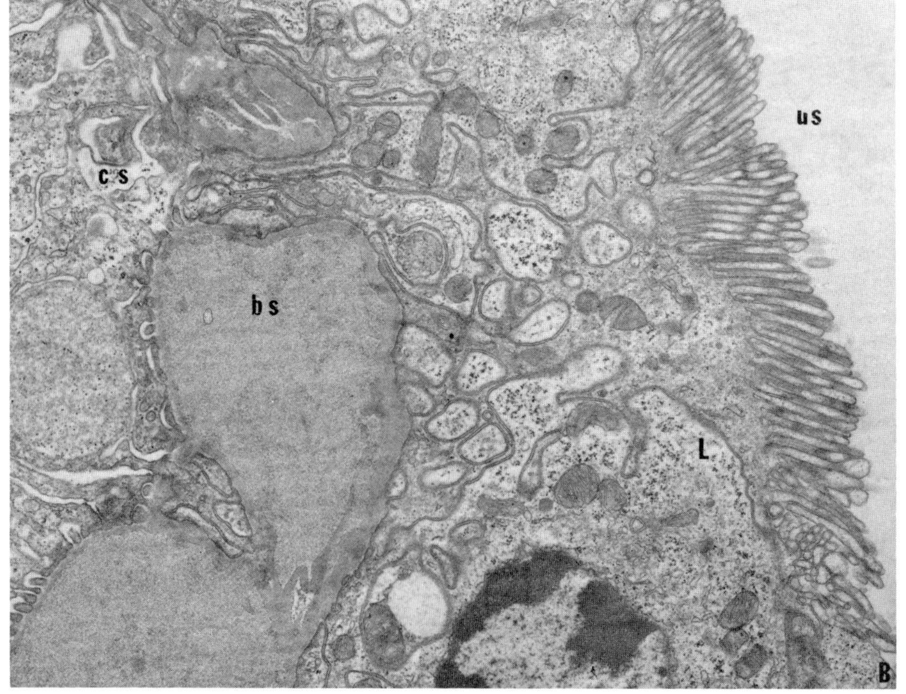
us
cs
bs
L
B

In the barnacle, a complex valve apparatus appears between the end-sac and efferent duct as well (White and Walker, 1981). It is composed of four cells which set upon a collar formed from both end-sac and efferent duct cells. They appear to be derived from podocytes, as indicated by modified cytoplasmic constituents and their possession of a reduced number of foot processes. They have an increased number of lysosomes.

9.6.   Function

Despite the large number of crustacean species, the excretory glands of relatively few have been studied. Ultrastructural studies have been primarily restricted to the antennal glands of the decapods. These have included studies on *Cambarus* sp. (Anderson and Beams, 1956; Beams et al., 1956), *Orconectes affinis* (Kummel, 1964), *Callinectes sapidus* (Johnson, 1980); *Procambarus* spp. (Peterson and Loizzi, 1974; Miyawaki and Ukeshima, 1967; Schaffner and Rodewald, 1978), *Uca mordax* (Schmidt-Nielsen et al., 1968), and *Coenobita clypeatus* (G. W. Hinsch and B. F. Komm, personal observations).

More recently, non-decapods including the brine shrimp *Artemia salina* (Tyson, 1968, 1969), the amphipod *Corophium volutator* (Icely and Nott, 1979), the isopod *Mesidotea entomon* (L). (Hryniewiecka-Szyfter et al., 1983; Hryniewiecka-Szyfter and Tyczewska, 1984), and the barnacle *Balanus* spp. (White and Walker, 1981) have been studied. These studies have all shown the basic similarity of structure among the varying species.

The physiological studies of the antennal gland in the crayfish (Kirschner and Wagner, 1965) indicate that the gland is analogous to a vertebrate nephron in filtering a primary urine from the blood; this urine is then secondarily modified by reabsorption and secreted in its definitive form. The end-sac or coelomosac epithelium possesses podocytes comparable to those of the Bowman's capsule of the vertebrate nephron and presumably is the site of ultrafiltration (Kummel, 1964, 1967; Schmidt-Nielson et al., 1968). The presence of pinocytotic pits along the cell membranes suggests that uptake of protein may occur, with its subsequent digestion within the lysosomes in the apical regions of the cell.

FIGURE 9.9. (A) Cells of a more distal area of the labyrinth of *Coenobita* surround the urinary sinus (US). These cells have few basal invaginations and much shorter microvilli than seen elsewhere in the labyrinth. ×6624. (B) Enlargement of one of the above cells adjacent to the blood sinus (BS). Podocytes of a coelomosac cell abut on the blood sinus as well. ×10,080.

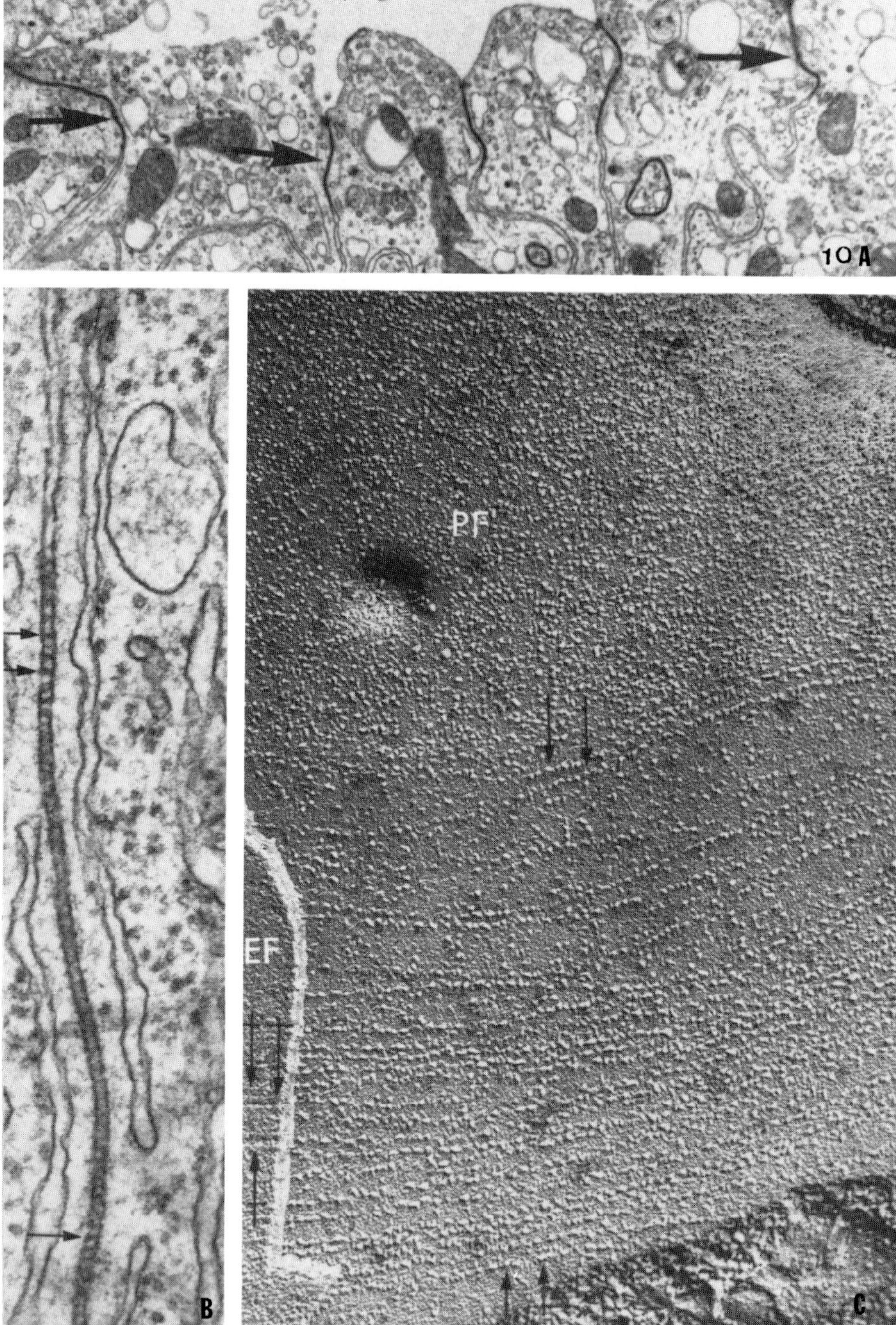
10 A
B
PF
EF
C

The cells of labyrinth apparently play an active role in the modification of the luminal contents absorbed by ultrafiltration in the end-sac. The structure of the labyrinth cells is that which is characteristic of epithelial cells involved in the movement of fluids. These cells may be involved in the reabsorption of sodium in *Uca mordax* (Schmidt-Nielsen et al., 1968).

In the rapidly expanding field of chemical communication between crustaceans, the role of the excretory organs has been recognized (Dunham, 1978; Ryan, 1966; Atema, 1985). Numerous ancedotal accounts have appeared in the literature involving many species. Ryan (1966) reported on an attractant produced by premolt females in portunids. Chemokinetic and chemotaxic reactions to pheromones were reported in *Homarus americanus* (Atema and Engstrom, 1971).

The precise role in and mode of transport of the chemical signals through the excretory organs is unknown. It is known that ecdysone is secreted by both adult and juvenile *Carcinus maenas* during the entire molting cycle (Seifert, 1982). Many authors have concluded that ecdysone acts as a sex pheromone, whereas others have been unable to demonstrate its role as such. An unknown pheromone is probably excreted by premolt females of many species via the antennal glands. In *Carcinus* it is secreted rhythmically by such females (Seifert, 1982).

9.7.   Summary

The paired segmental excretory organs, antennal or maxillary, of crustaceans show many similarities among species. The coelomosac, or end-sac, appears to be similar in structure to the podocytes of the vertebrate Bowman's capsule. It functions primarily in the ultrafiltration of urine from the hemolymph. Unique slit diaphragms are found between the pedicels of the podocytes. Cells of the labyrinth are characteristically those of epithelial cells involved in the modification and transport of fluid contents.

FIGURE 9.10. (A) Nephridial epithelial cells joined appically by septate junctions (arrows) which may extend basally one-third to one-half of cell height. ×8730. (B) Septate junctions with thin septa (arrows) which traverse extracellular space and insert into the plasmalemma of the apposed cell. ×58,200. (C) Freeze-fractured labyrinth cells showing parallel rows of particles (arrows) on outer surface of cytoplasmic leaflet (PF face) of plasma membrane. Parallel rows of pits (crossed arrows) on inner surface of outer leaflet of plasmalemma (EF face) complementary to rows of particles. ×72,750. (From Shivers and Chauvin, 1977.)

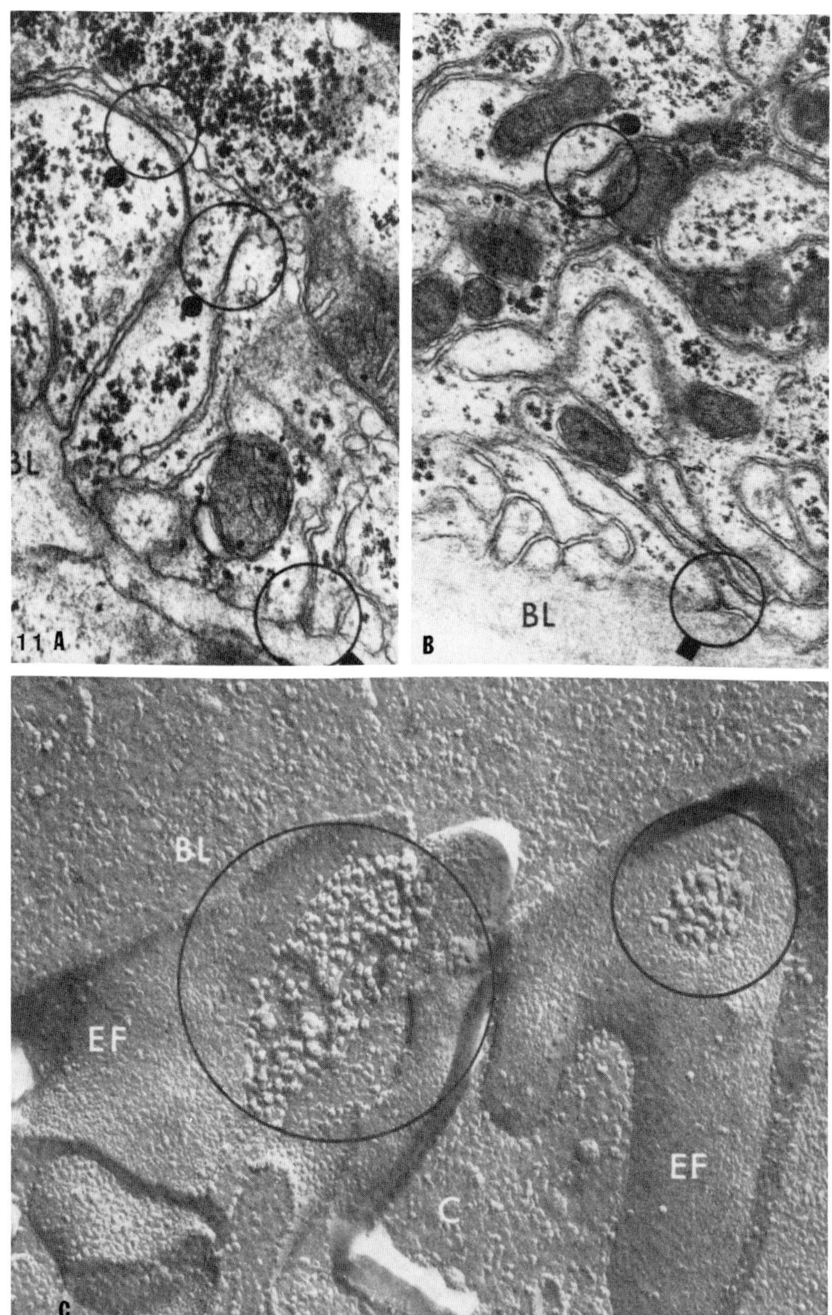

FIGURE 9.11. (A&B) Junctions between adjacent basal processes of labyrinth cells at base of cell (circle with square) next to basal lamina (BL). Fine filaments bridge extracellular space between adjoined plasma membranes. Other junctions at short distance from base of cell (circles with dot, A). (C) Replica of basal-most region of labyrinth cells showing many aggregates of junctional particles (circled). No evidence of filamentous bridgework as seen in A and B is apparent. A, ×31,104; B, ×21,049; C, ×58,200. (From Shivers and Chauvin, 1977.)

The excretory organs of crustaceans not only filter and excrete wastes but are involved in the excretion of ecdysone as well as sex pheromones. This antennal gland excretion/secretion provides a means of chemical communication among crustaceans.

References

Anderson, E. and H. W. Beams. 1956. Light and electron microscope studies on the cells of the labyrinth in the "green gland" of *Cambarus* sp. Proc. Iowa Acad. Sci. 63: 681–685.

Atema, J. 1985. Chemoreception in the sea: adaptations of chemoreceptors and behaviour to aquatic stimulus conditions. *In* M. S. Laverack (ed.), *Physiological Adaptations of Marine Animals.* Symp. Soc. Exp. Biol. 39: 387–423.

Atema, J. and D. G. Engstrom. 1971. Sex pheromone in the lobster *Homarus americanus.* Nature (Lond.) 232: 261–263.

Beams, H. W., E. Anderson, and N. Press. 1956. Light and electron microscope studies on the cells of the distal portion of the crayfish nephron tubule. Cytologia (Tokyo) 21: 50–57.

Dunham, P. J. 1978. Sex pheromones in crustacea. Biol. Rev. 53: 555–583.

Goodrich, E. S. 1945. The study of the nephridia and genital tracts since 1895. Q. J. Microsc. Sci. 86: 113–492.

Hryniewiecka-Szyfter, Z., W. Barbasiewicz, and J. Bielawski. 1983. Morphology, Histology, and morphometry of the excretory organ of *Mesidotea entomon* (L.) (Isopoda, Crustacea). Bull. Soc. Amis Sci. Lett. Poznan, Ser. D; Sci. Biol. 23: 141–146.

Hryniewiecka-Szyfter, Z. and J. Tyczewska. 1984. The excretory organ of *Mesidotea entomon* (L.) (Crustacea, Isopoda): The fine structure and localization of alkaline phosphatase. Bull. Soc. Amis Sci. Lett. Poznan, Ser D, Sci. Biol. 24: 55–60.

Icely, J. D. and J. A. Nott. 1979. The general morphology and fine structure of the antennary gland of *Corophium volutator* (Amphipoda: Crustacea). J. Mar. Biol. Assoc. U.K. 59: 745–755.

Johnson, P. T. 1980. Histology of the blue crab, *Callinectes sapidus:* a model for the Decapoda. Praeger, New York.

Kirschner, L. B. and S. J. Wagner. 1965. The site and permeability of the filtration locus in the crayfish antennal gland. J. Exp. Biol. 43: 385–395.

Kume, M. and K. Dan. 1968. *Invertebrate Embryology* (transl. by J. C. Dan). Prosveta, Belgrade.

Kummel, G. 1964. Das Colomosäckchen der Antennendrüse von *Cambarus affinis* Say. (Decapoda: Crustacea): eine elektronmikroskopische Untersuchung mit einer Diskussion über die Funktion. Zool. Beitr. 10: 227–252.

Kummel, G. 1967. Die Podocyten. Zool. Beitr. 13: 245–264.

Miyawaki, M. and A. Ukeshima. 1967. On the ultrastructure of the antennal gland of the epithelium of the crayfish, *Procambarus clarkii.* Kumamoto J. Sci. 8: 57–73.

Peterson, D. R. and R. F. Loizzi. 1974. Ultrastructure of the crayfish kidney: coelomosac, labyrinth, nephridial canal. J. Morphol. 142: 241–264.

Ryan, E. P. 1966. Pheromone: evidence in a decapod crustacean. Science (Wash., D.C.) 151: 340–341.

Schaffner, A. and R. Rodewald. 1978. Filtration barriers in the coelomic sac of the crayfish, *Procambarus clarkii.* J. Ultrastruct. Res. 65: 36–47.

Schmidt-Nielsen, B., K. H. Gertz, and L. E. Davis. 1968. Excretion and ultrastructure of the antennal gland of the fiddler crab *Uca mordax.* J. Morphol. 125: 473–496.

Seifert, P. 1982. Studies on the sex pheromone of the shore crab, *Carcinus maenas,* with special regard to ecdysone excretion. Ophelia 21: 147–158.

Shivers, R. R. and W. J. Chauvin. 1977. Intercellular junctions of antennal gland epithelial cells in the crayfish, *Orconectes virilis.* Cell Tiss. Res. 175: 425–438.

Tyson, G. 1968. The fine structure of the maxillary gland of the brine shrimp, *Artemia salina:* the end sac. Z. Zellforsch. Mikrosk. Anat. 86: 129–138.

Tyson, G. 1969. The fine structure of the maxillary gland of the brine shrimp, *Artemia salina:* the efferent duct. Z. Zellforsch. Mikrosk. Anat. 93: 151–163.

White, K. N. and G. Walker. 1981. The barnacle excretory organ. J. Mar. Biol. Assoc. U.K. 61: 529–547.

Morphology, Histology, Ultrastructure, and Functions of the Mandibular Glands in Crustaceans

10

GERTRUDE W. HINSCH

10.1. Introduction

Several different structures have variously been described as the Y-glands, molting glands, or mandibular glands of decapod crustaceans. Initially, there was much confusion in the literature, particularly because few anatomic or cytological details were published. The Y-glands of more than 100 species of malacostrancan crustaceans were described (Gabe, 1953, 1954, 1956; Echalier, 1959) and a function in molt control was suggested. Chaudonneret (1956) and Durand (1960) described glands in the crayfish, which Le Roux (1968) indicated were not the Y-glands of Gabe but rather the mandibular glands. Others (Connell, 1970; Iwakura, 1970; Miyawaki and Taketomi, 1970; Miyawaki et al., 1971; Hinsch and al Hajj, 1975) described the mandibular glands and identified them as the ecdysial gland. Sochasky et al. (1972) pointed out that studies of Y-organs may have involved two or more different organs. Further, not all studies of the physiology of the Y-organs have involved the molting organs. Rana (1975) and Rana et al. (1977) reported on the histology and histochemical nature of glands in *Paratelphusa* spp. that they referred to as the mandibular glands. These structures, however, appear to be associated with a digestive function and, as such, differ from the mandibular glands of Le Roux (1968).

This chapter will describe the morphology, histology, and ultrastructure of the mandibular glands as they appear in reptantian and natantian crustaceans and their probable reproductive functions.

10.2. Location and Morphology

The mandibular gland (MG) has been studied in many crustacean species. In the lobster, *Homarus*, the MGs are large, highly vascularized, multilobate structures, attached to the ventral hypodermis along the posterior face of the mandibles, immediately posterior to the bases of the posterior mandibular adductor muscles (Byard et al., 1975) (Figs. 10.1A,B,D).

In the brachyurans, it is described as applied against the posterior of the base of the apodeme of the posterior mandibular adductor muscle, above the muscles of the maxilla (Le Roux, 1968), or posterior to the mandibles in association with the muscles, and surrounding the chitinous tendons extending from the mandibles to the dorsal carapace (Hinsch and al Hajj, 1975; Yudin et al., 1980). They appear spherical and multilobate in shape (Figs. 10.1C).

The MGs of the anomurans spread transversely as a thick sheet (*Pisidia longicornis*) or as a single layer (*Anapagurus hyndmanni*) in the space between the circumesophageal connectives and the anterior apodemes of

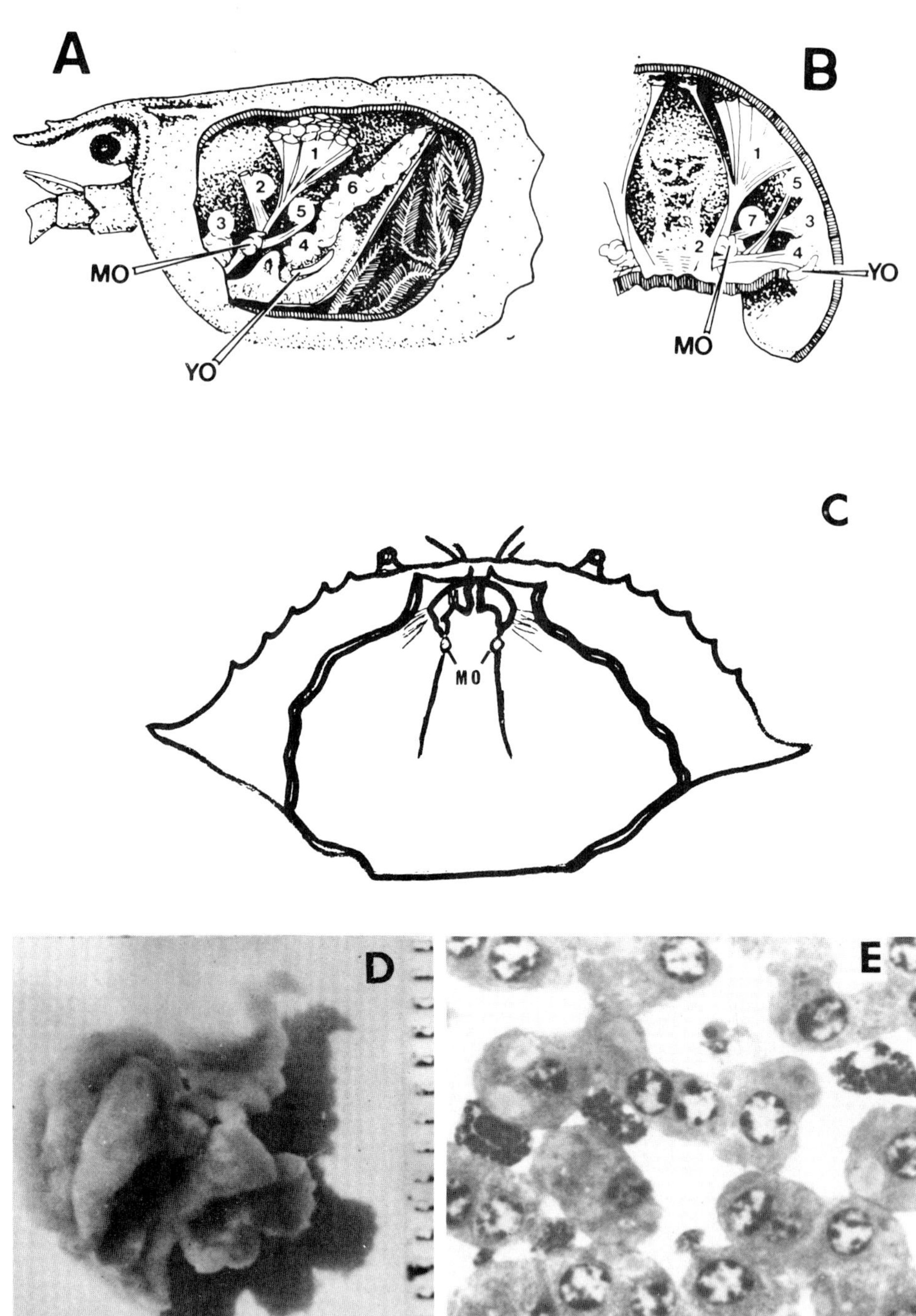

A
1
2
3
4
5
6
MO
YO
B
1
2
3
4
5
7
MO
YO
C
MO
D
E

the mandibles. In *Palaemon squilla*, they are found against the membrane between the mandible and the maxilla (Le Roux, 1968).

The MGs are present in both sexes. The size and shape of these organs varies with species and probably with different states of the life cycle. In many species of brachyurans, they are a pale cream or yellowish color (Le Roux, 1968; Hinsch and al Hajj, 1975; Yudin et al., 1980). In the lobster, they are pale- to dark-green structures (Byard et al., 1975).

Their embryonic origin was studied in *Inachus phalangium* (Le Roux, 1968). They are initially visible in the stage II zoea, where they develop on the internal face of the hypodermis of the mandibles. They move during the megalops stage and come to occupy their definitive place seen during the juvenile and adult stages. The MG of *Carcinus maenas* is vascularized by two arterioles, which arise as a bifurcation of the sternal artery. In *Inachus*, a single branch of the sternal artery enters the MGs (Le Roux, 1968). The glands of all species have considerable contact with the blood sinus.

Many nerves pass in the vicinity of the MGs of these crustaceans, but little information exists concerning possible innervation.

10.3.   Histology

10.3.1.   *Light Microscopy*

The MGs are connected to neighboring tissues by peripheral filaments. Intercellular hemolymph channels subdivide the glands into numerous cords of epithelial gland cells. The glands consist of lobules of epithelial cells enveloped by a layer of connective tissue. All epithelial cells are in contact with the hemolymph channels (Figs. 10.1E and 10.2A,B,C).

The epithelial cells of the MG vary in shape and tend to have eccentric nuclei with peripherally condensed heterochromatin. In *Callinectes*

FIGURE 10.1. Cephalothorzx of *Homarus:* (A) Lateral view. (B) posterior view showing the position of the mandibular organ (MO) and the Y-organ (YO). *Key:* 1 = posterior mandibular adductor; 2 = anterior dorsoventral; 3 = lateral mandibular adductor; 4 = posterior dorsoventral; 5 = maxillary abductor coxopoditis; 6 = epimeral muscles; 7 = major mandibular abductor. (C) Diagram of brachyuran crab showing mandibular organs (MO) surrounding the tendon from the mandibles. (D) Whole mount of the mandibular organ from the lobster, *Homarus.* (Scale to right = millimeters). (E) Light micrograph of the mandibular gland of *Homarus.* ×500. [(A&B) Redrawn from Sochasky et al., 1972, by D. E. Aiken. (D&E) From E. Couch, Texas Christian University.]

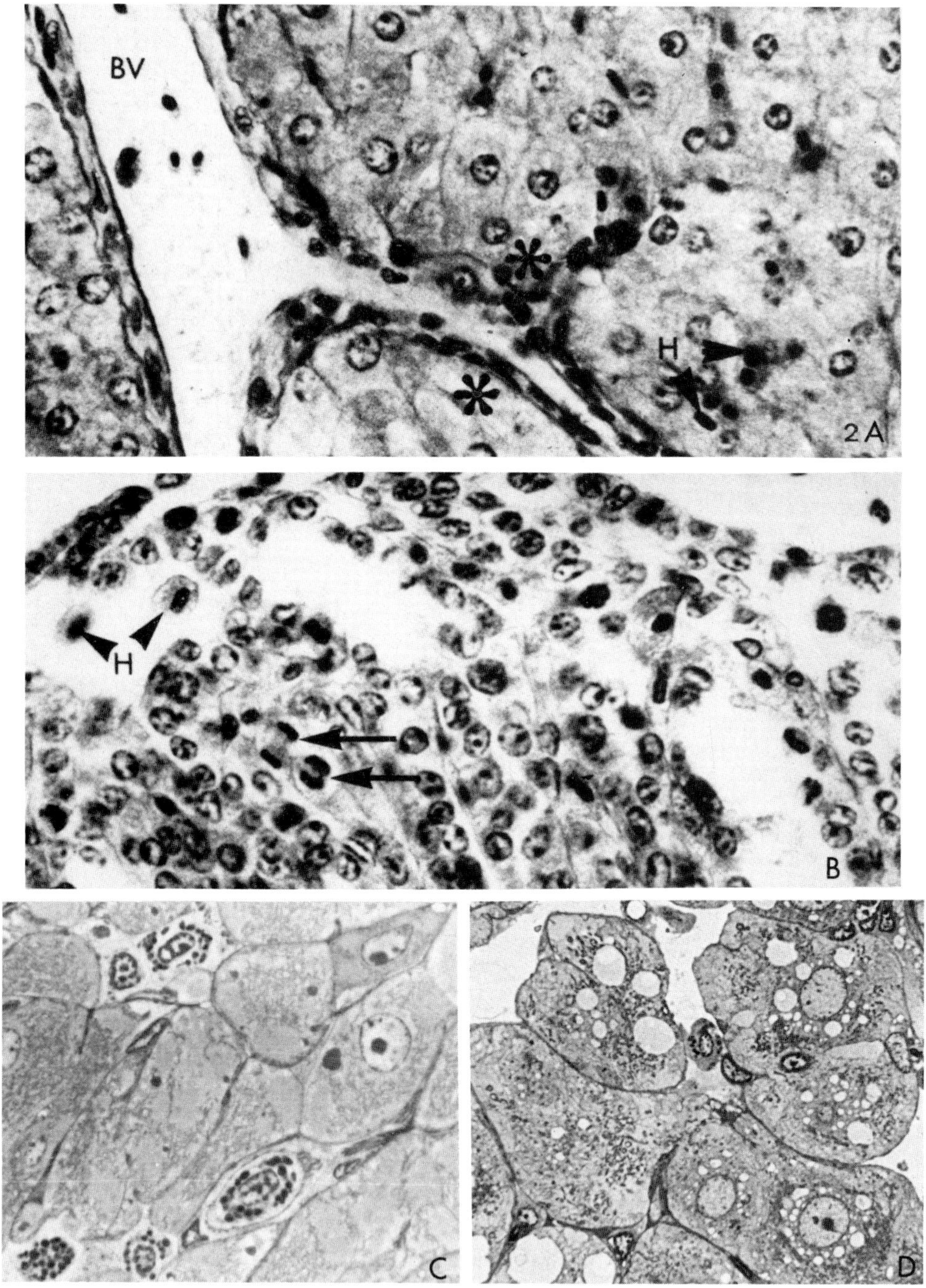

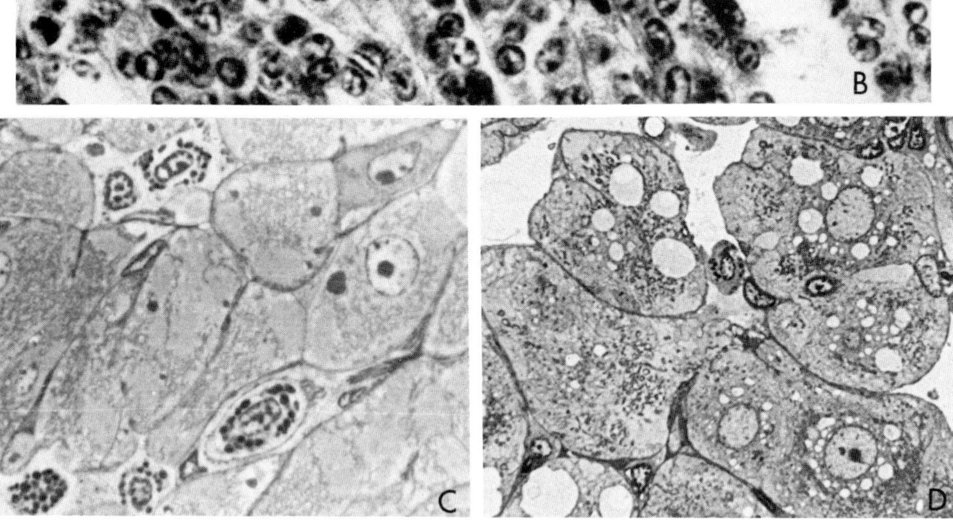

sapidus, the cells are sometimes binucleate (Johnson, 1980). One or more nucleoli are present in each nucleus. Aggregates of rod-shaped bodies and several light vacuolated circular areas are evident in the cytoplasm (Fig. 10.2C).

10.3.2. Ultrastructure

The cytoplasm varies considerably from species to species as well as at different stages in the life of the animals. The cytoplasm of the cells begins to vacuolize before the molt of puberty. The structure is relatively uniform in males but cyclically variable in females (Le Roux, 1968; Hinsch, 1981) (Figs. 10.3–10.6).

10.3.2.1. MALES AND NONOVIGEROUS FEMALES

Each lobule is composed of cords of epithelial cells, which rest on a thin basal lamina. The basal lamina lines the hemolymph channels that separate adjacent cords. Each epithelial cell is directly apposed to the hemolymph channel. The plasma membrane of these cells may be smooth, adjacent to the basal lamina, or folded up into numerous invaginations, and appears to be correlated with the types of organelles within the gland cells.

Occasional hemocytes are seen within the hemolymph channels (Figs. 10.3B and 10.4A). These channels are interconnected with the hemolymph sinuses.

The organelles of the MG cells appear relatively uniform within a single gland, but vary considerably from crustacean to crustacean, depending upon age, sex, and the reproductive state. The cytoplasm contains large numbers of mitochondria (Figs. 10.3A, 10.4A, 10.5A,B, and 10.6A,B). These are abundant and vary considerably in shape and size. Their cristae are tubular, are oriented perpendicularly to the long axis,

FIGURE 10.2. (A) Light micrograph of mandibular organ of male *Homarus americanus*, stage D_2. Note branching blood vessel (BV) with rapid reduction in diameter (asterisk). Small branches seem continuous with tissue spaces admitting entry of hemocytes (H) into parenchyma. ×450. (B) Cells from an area close to attachment to posterior mandibular adductor muscle in *Homarus*. ×625. (C) Mandibular organ of normal, intact male *Libinia emarginata*. Note the gray homogeneous areas of cytoplasm and the hemocytes with granular cytoplasm between cells of the mandibular organ. ×500. (D) Mandibular organ cells from an eyestalk-ablated male *Libinia*. ×550. (From Hinsch, 1977.)

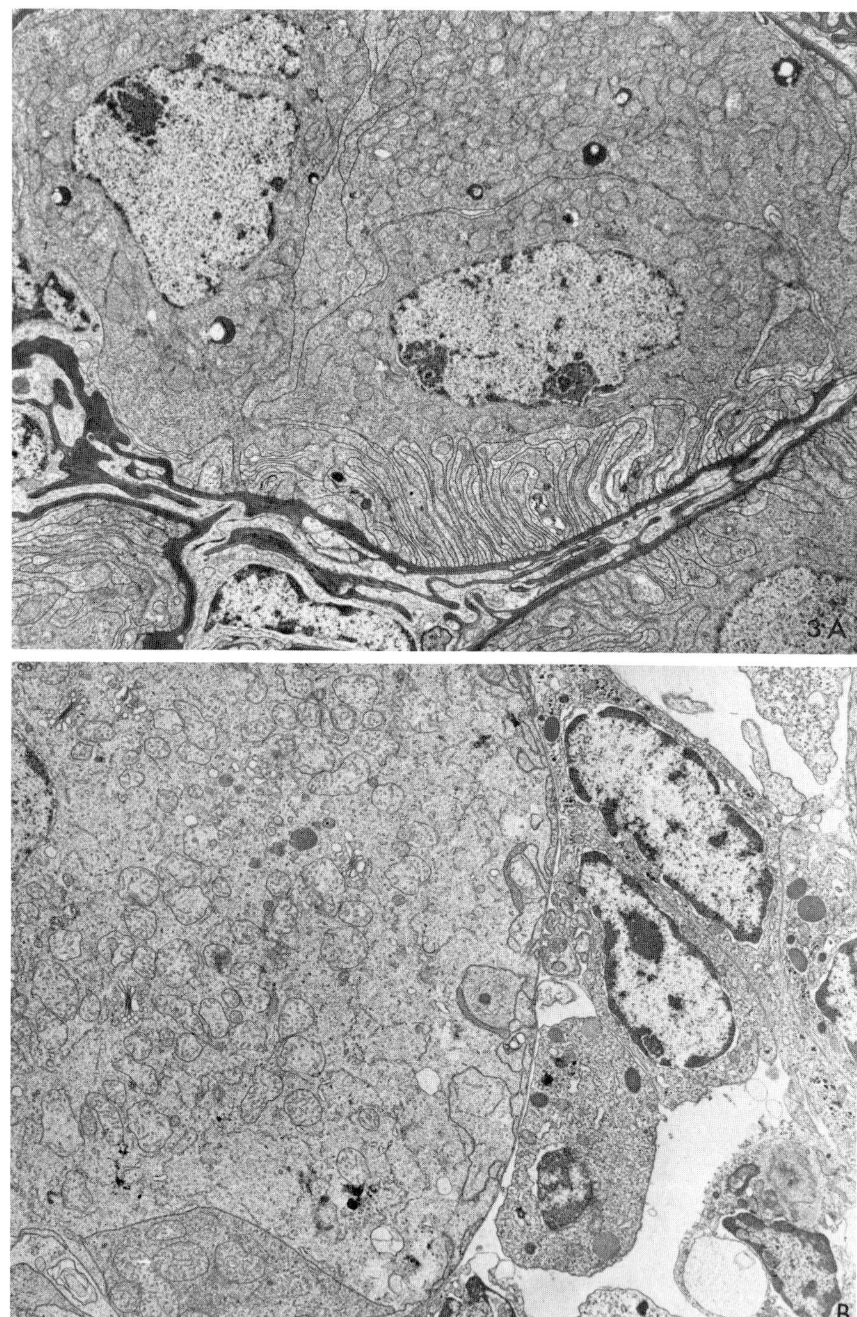

FIGURE 10.3. (A) Cell of mandibular gland of *Cancer irratus* showing typical structure of the mandibular gland in many species. The plasmalemma in contact with the densely staining basal lamina is highly convoluted. Numerous mitochondria are present in the cytoplasm of the cells. (B) Portion of a mandibular gland cell from *Callinectes sapidus* adjacent to a hemal space filled with hemocytes. A&B, ×3395. (G. W. Hinsch, unpublished work.)

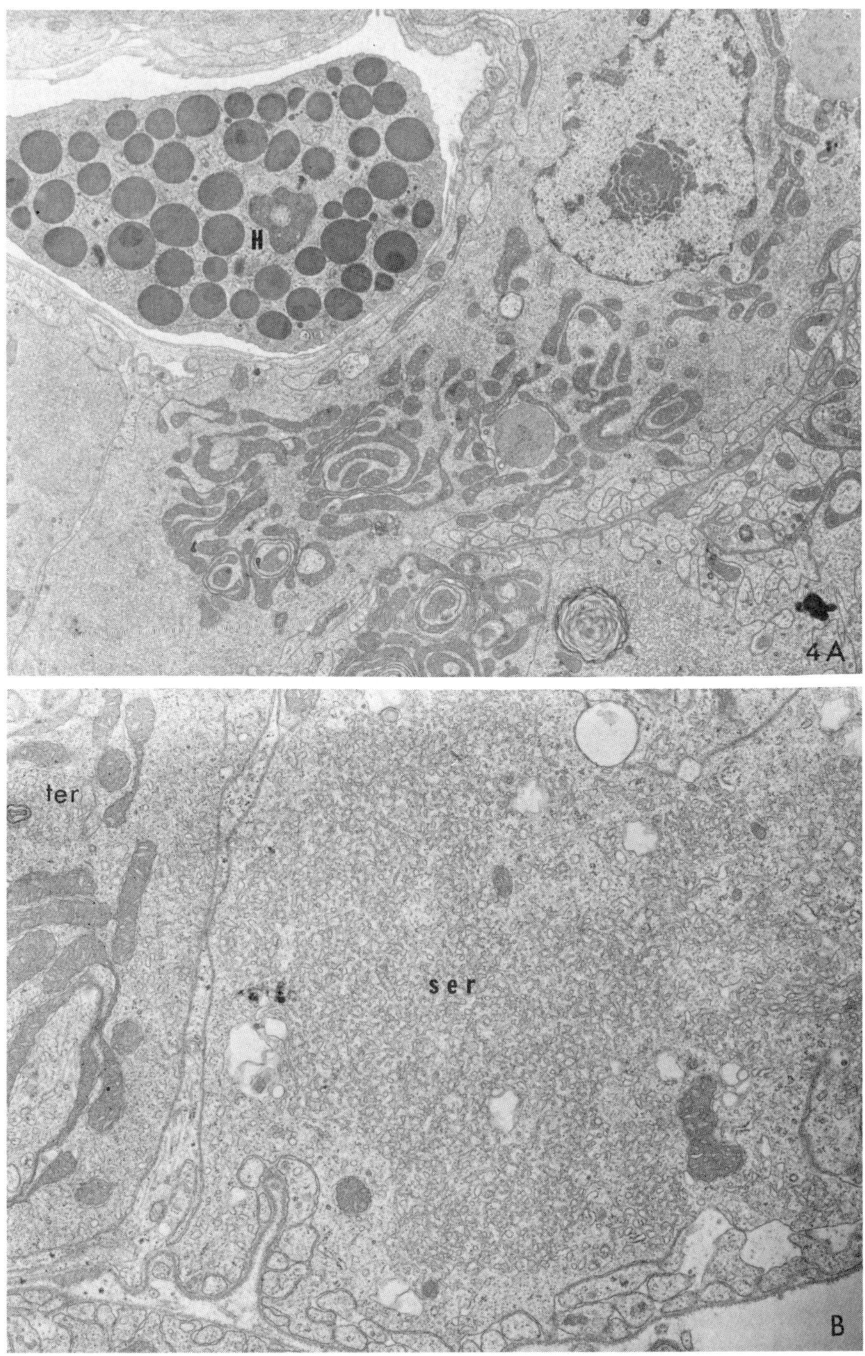

FIGURE 10.4. (A) Mandibular gland cell of *Carcinus maenas* demonstrating the various forms, which the mitochondria can assume. *Key:* H = hemocyte. (B) Large areas of the cytoplasm are filled with the tubular (ter) and vesicular forms (ser) of the smooth endoplasmic reticulum. A, ×4365; B, ×8730. (G. W. Hinsch, unpublished work.)

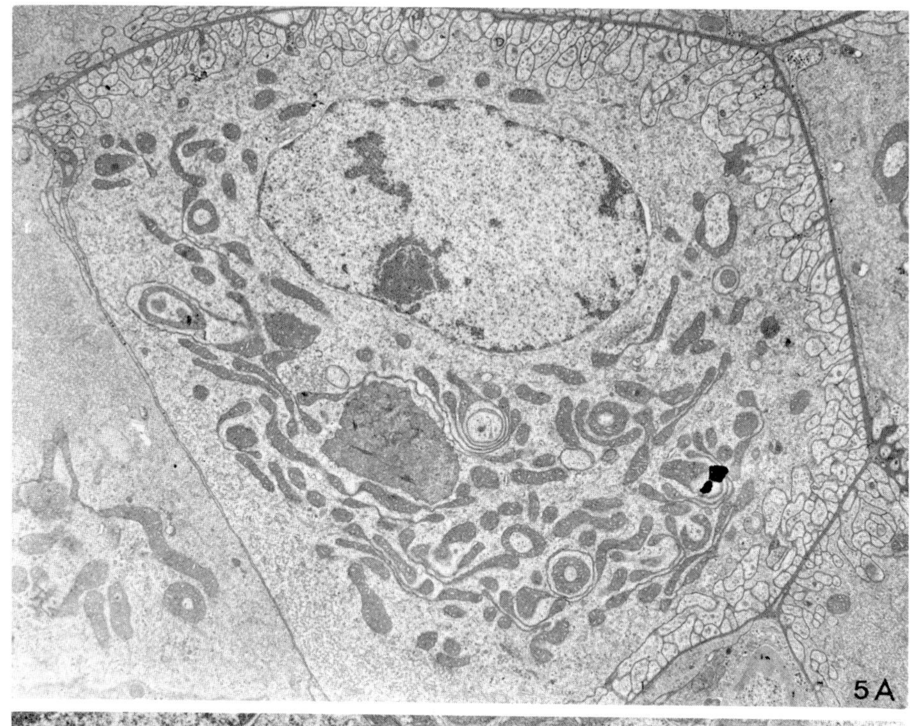

5 A

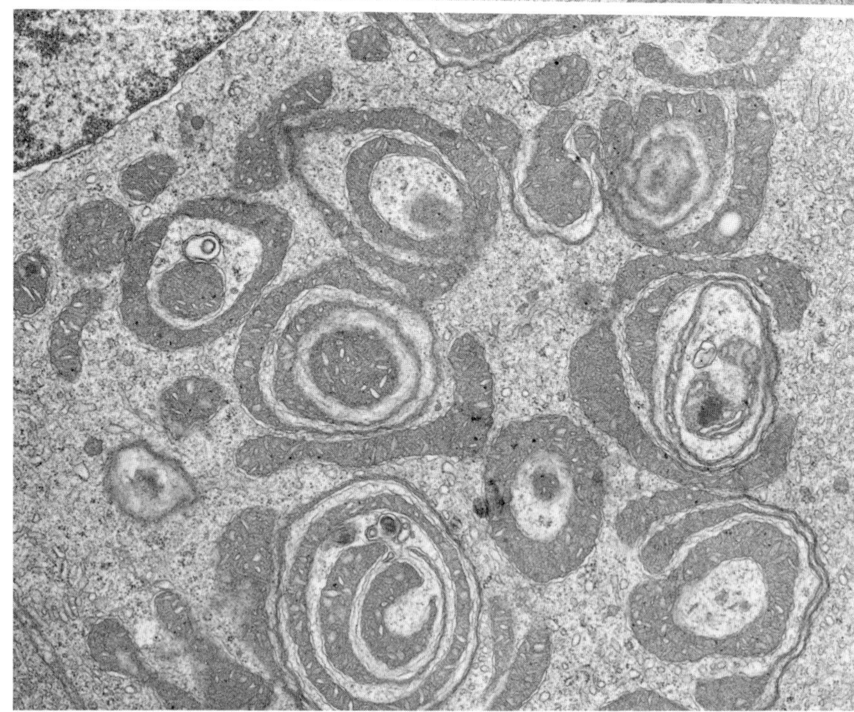

B

and are surrounded by a dense matrix. They frequently surround aggregates of tubules in the cytoplasm and encircle the nuclei. Large, concentric ring-shaped mitochondria are seen in some cells (Fig. 10.5B).

A prominent feature of the MG cells is the presence of dense cytoplasmic aggregates of smooth endoplasmic reticulum (Fig. 10.4B). The smooth reticulum is a system of highly anastomosed tubules (ter) or vesicles (ser). These aggregates are distributed throughout the cells, and the incidence of the two types of smooth reticulum varies considerably among glands and under different conditions. Small vesicles, which appear to originate from the smooth reticulum, appear to fuse with each other to form larger vesicles. These contain flocculent material and are located near the periphery of the cells (Fig. 10.4B).

The Golgi complexes consisting of saccules, vesicles, and flattened sacs are frequently observed in association with accumulations of smooth endoplasmic reticulum. The complexes may be associated with dense lipid-like inclusions, which are scattered throughout the cytoplasm or clumped together (Fig. 10.4B). Their origin and function is unknown. Few free ribosomes and little rough endoplasmic reticulum are present in the cytoplasm.

Maissiat (1980) described the ultrastructure of glands in the isopod *Ligia oceanica* which were called the antennal glands. The micrographs that accompany this paper suggest that the cells of these glands have a cytoarchitecture that is similar to that of the MGs of other crustaceans, as described above, and that they undergo similar changes associated with molting. The ultrastructure of these glands does not resemble that of the maxillary or antennal glands of other species of crustacean. It would appear, then, that the glands in this instance were misnamed.

10.3.2.2.   FEMALES

The structure of the MG cells is much more variable in female crustaceans. In the spider crab *Libinia emarginata*, the MGs of the female are much smaller than those of the male. Cells taken from females at different times in their life cycle differ histologically (Hinsch, 1981).

FIGURE 10.5. (A) Mandibular gland of *Carcinus maenas* in which the mitochondria have become lightly modified. (B) The mitochondria of the mandibular gland cells often become stacked inside of each other. In many cases, their middle regions become extremely narrow, often little more than the inner and outer mitochondrial membranes adjacent to each other. A, ×3395; B, ×10,476. (G. W. Hinsch, unpublished work.)

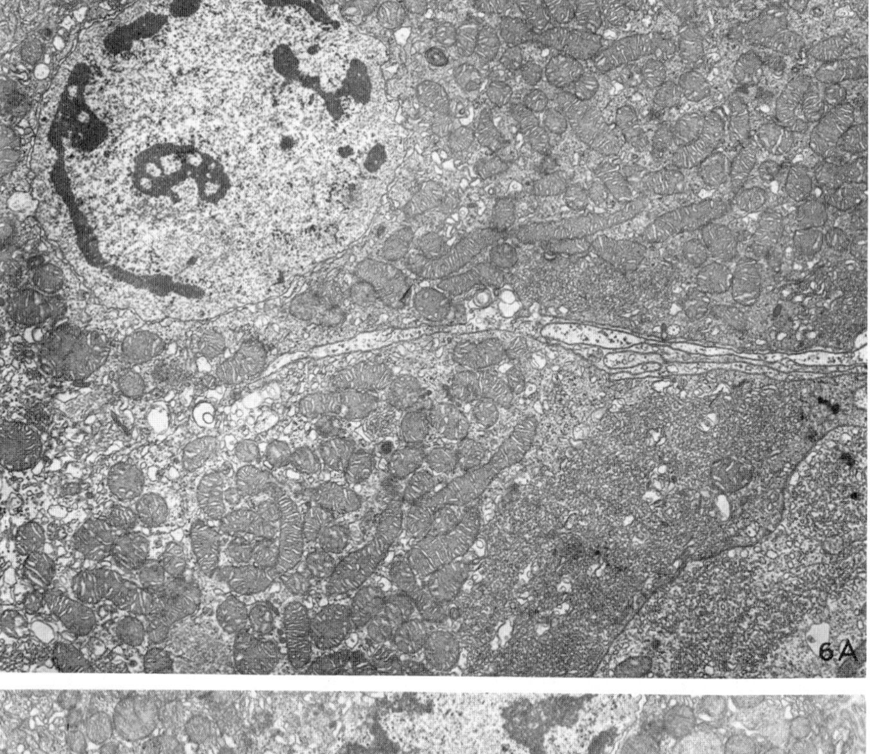

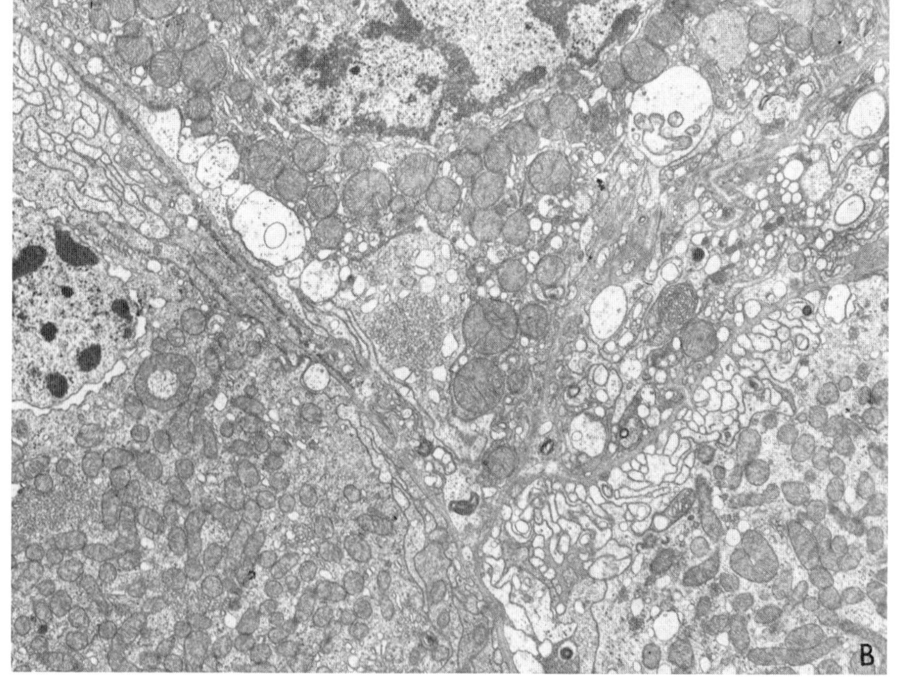

Immature Premolt Females

The MG cells contain numerous mitochondria, Golgi complexes, and numerous vesicles in association with small peripheral aggregates of smooth endoplasmic reticulum (SER) (Fig. 10.7B). Vesicles of rough endoplasmic reticulum (RER), and free ribosomes are visible in perinuclear areas.

Immature Postmolt Females

The MG of a female recently molted to an additional immature stage contains numerous vesicles exhibiting considerable size variation. Small vesicles usually are associated with the SER, while the largest lie adjacent to it. Considerable variation is shape of the mitochondria exists, while the matrix is very electron-dense. Free ribosomes appear in the cytoplasm, although profiles of RER are rarely seen. Golgi complexes are visible in many of the cells.

10.3.2.3. MATURE FEMALES

In the spider crab *L. emarginata*, variations in the organelles of the MG cells can be seen in females in different reproductive stages (Hinsch, 1981). The changes suggest that the MGs are involved in some way in the control of vitellogenesis (Fig. 10.7A,C,D). In *Carcinus*, Le Roux (1968) suggested that there is a parallelism between the function of the MGs and oogenesis, because of the variable appearance of the cells. When MGs were implanted into the hemocoel of immature female spider crab *L. emarginata*, the ovaries responded by showing active signs of vitellogenesis, including numerous pinocytotic vesicles in the oolemma (Hinsch, 1980). In these crabs, as is typical of majidid crabs, vitellogenesis normally commences only following the terminal molt to maturity (Hartnoll, 1963; Hinsch, 1968).

10.3.3. *Experimental Conditions*

The MG undergoes numerous changes in cytoarchitecture at different stages of the molt cycle. The mitochondria change in shape, and the smooth endoplasmic reticulum becomes associated with numerous vesicles (Aoto et al., 1974).

Eyestalk ablation produces profound changes in the cytoarchitecture

FIGURE 10.6. (A&B) Cells from different regions of the mandibular organs of recently molted male *Libinia emagrinata*. Few aberrant configurations of mitochondria are apparent. A, ×4320; B, ×3744. (G. W. Hinsch, unpublished work.)

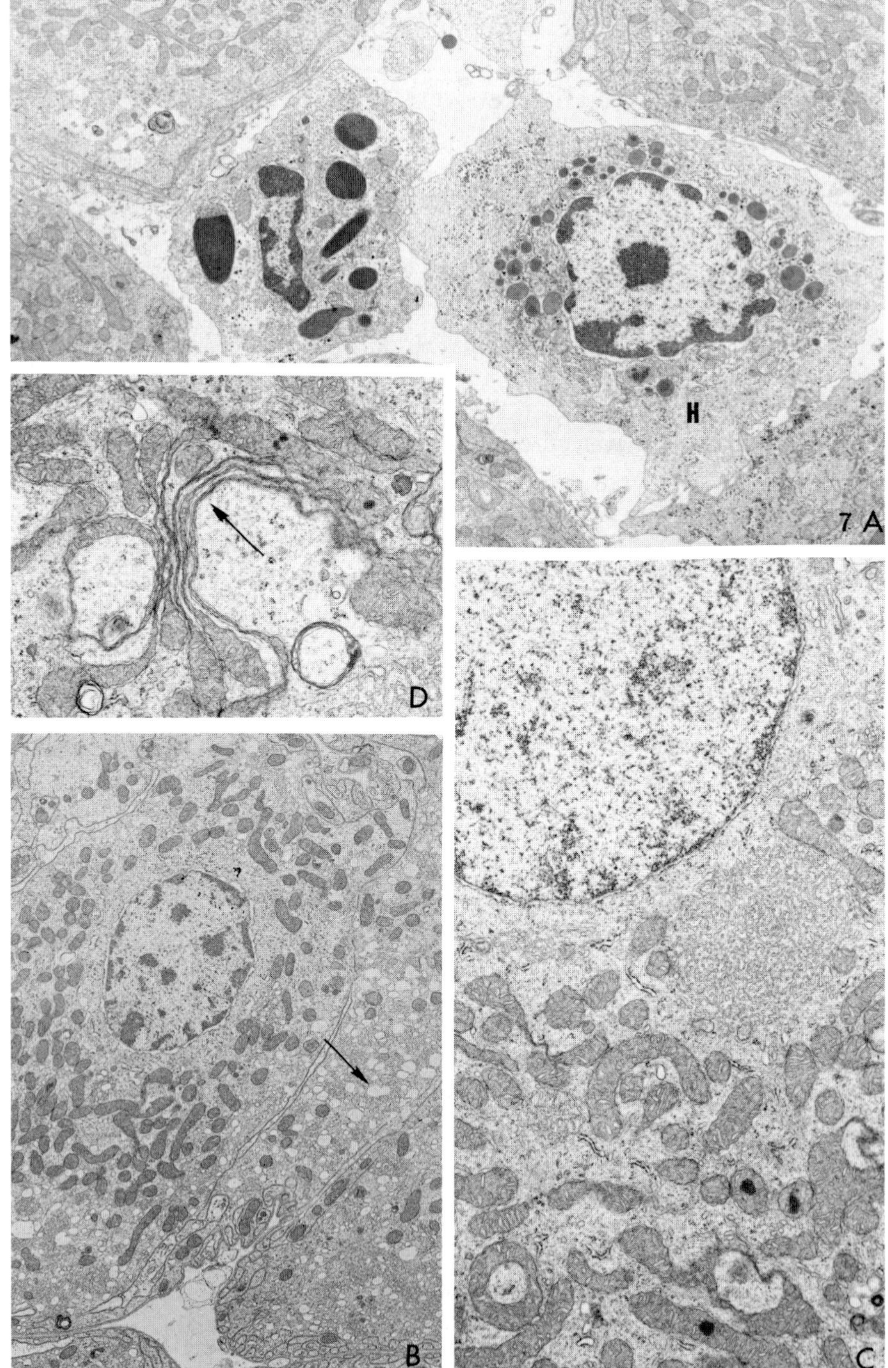

H
7 A
D
B
C

of the mandibular glands of *Carcinus maenas* (Bazin, 1976) and *Libinia emarginata* (Hinsch, 1977) (Fig. 10.2D). In *Libinia*, the glands hypertrophy and blebbing of the nuclear membrane is observed (Fig. 10.8A). Many free ribosomes appear in the cytoplasm, and the SER dissociates (Fig. 10.8B). Bizarre changes were noted in the structure of the mitochondria in *Libinia* (Hinsch, 1977) and in the shrimp *Palaemon paucidens* (Aoto et al., 1974). A banded structure of fibrils and granules is visible in the nucleus of MG cells of ablated *Libinia* (Hinsch, 1977) (Fig. 10.8C).

The MGs of eyestalk-ablated *Callinectes sapidus* contained a proliferating rhabdo-like virus (EGV-2) (Yudin and Clark, 1979). The viral proliferation involved the nuclear and plasma membranes and tubular smooth endoplasmic reticulum (Figs. 10.9A,B). The virus particles were typically bacilliform, with a length of 100–150 nm and a diameter of 25–30 nm.

10.4. Function

Several functions have been attributed to the MGs since the time of their first description. Le Roux (1968) considered them to be endocrine glands which resembled strongly the structure of the Y-glands, and suggested that a parallelism exists between their function and ovogenesis. Chaudonneret (1956) thought that the MG could be homologous to the corpora allata of insects. Sochasky et al. (1972) considered that they played no role in ecdysis. Byard et al. (1975) and Aoto et al. (1974) reported dramatic increases in activity of the cells just prior to molting, as did Le Roux earlier (1968) in *Carcinus.* Implants of the glands slow molting in shrimp (Connell, 1970).

In mammalian cells known to secrete steroids, the cellular organelles include extensive agranular endoplasmic reticulum, prominent Golgi

FIGURE 10.7. (A) Mandibular organ of a vitellogenic female showing hemocytes (H) in hemolymph between the cells. Mandibular organ cells contain numerous mitochondria and small aggregates of SER. (B) Mandibular organ of a premolt female spider crab. Cells contain numerous rod-shaped or oval mitochondria, peripherally located clumps of SER, and many small vesicles (arrows). (C) Cell of mandibular organ of nonreproductive female contains a variety of mitochondrial shapes, some with electron-dense inclusions. Little heterochromatin is apparent within the nucleus. Small cisternae of RER are in cytoplasm adjacent to the nucleus. (D) Stalked dumbbell-shaped mitochondria with much elongated portions (arrows). A, × 4656; B, ×4074; C, ×9215; D, ×12,222. (From Hinsch, 1981.)

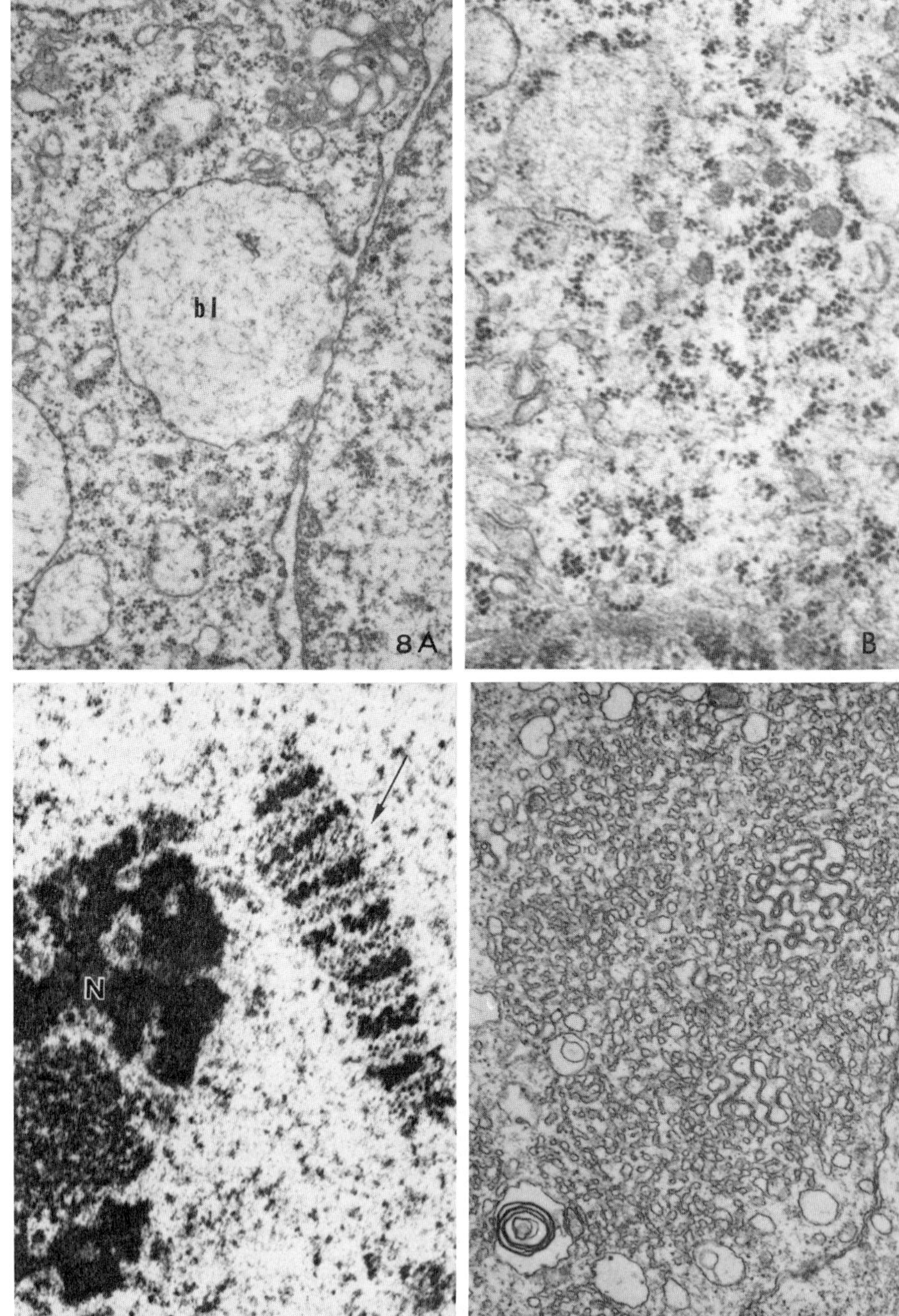

complexes, mitochondria of variable sizes and shapes, and lipid droplets (Christensen and Gillim, 1969). The first ultrastructural studies of the MGs demonstrated the presence of such characteristic cytoarchitecture (Aoto et al., 1974; Hinsch and al Hajj, 1975; Bazin, 1976). Thus it was suggested that the MGs play a role in steroid secretion.

Although the mandibular glands may play some role in the molting process and appear to have the ultrastructure of steroid-producing cells, they do not store or secrete ecdysteroids. The absence of ecdysone was indicated by radioimmunoassays (Keller and Schmid, 1979).

Implants of MGs into immature female spider crabs stimulates vitellogenesis (Hinsch, 1980). Sochasky et al. (1972) believed that the MGs of the lobster are involved in the female sexual cycle. Couch et al. (1978) identified progesterone in the hemolymph of the MG of *Homarus americanus*. Laufer et al. (1987) identified methyl farnesoate (MF) from isolated MGs of *L. emarginata*. MF has been isolated from the MGs of *C. sapidus* and *H. americanus* as well (Laufer et al., 1984). Thus, it has been suggested that the MGs are comparable to the corpora allata of insects and that they serve a regulatory role in reproduction of crustaceans.

10.5.   Summary

Epithelial glands are found in association with the mandibules of various crustaceans. These have been designated the mandibular glands. The cells of the MGs vary in shape and have eccentric nuclei. They exhibit a cytoarchitecture rich in mitochondria, Golgi complexes, and tubular and vesicular forms of the smooth endoplasmic reticulum. The cells are in contact with the hemolymph sinuses. The cytoplasm and its organelles may exhibit changes in vacuolarization and structure dependent on physiological state.

It is suggested that the MGs are homologous to the corpora allata of insects. They contain progesterone and methyl farnesoate, and play a regulatory role in reproduction of crustaceans.

FIGURE 10.8. (A) The outer nuclear membrane in destalked male *Libinia emarginata* often forms blebs (bl) filled with a flocculent material and covered with ribosomes. In areas more distant from the nucleus, vesicles filled with similar materials can be seen. (B) Clumps of ribosomes are common in the cytoplasm of the destalked male. (C) Banded structures frequently appear in the nucleus of the destalked male. (D) The smooth endoplasmic reticulum assumes modified forms. A, ×19,584; B, ×28,800; C, ×4416; D, ×19,200. (From Hinsch, 1977.)

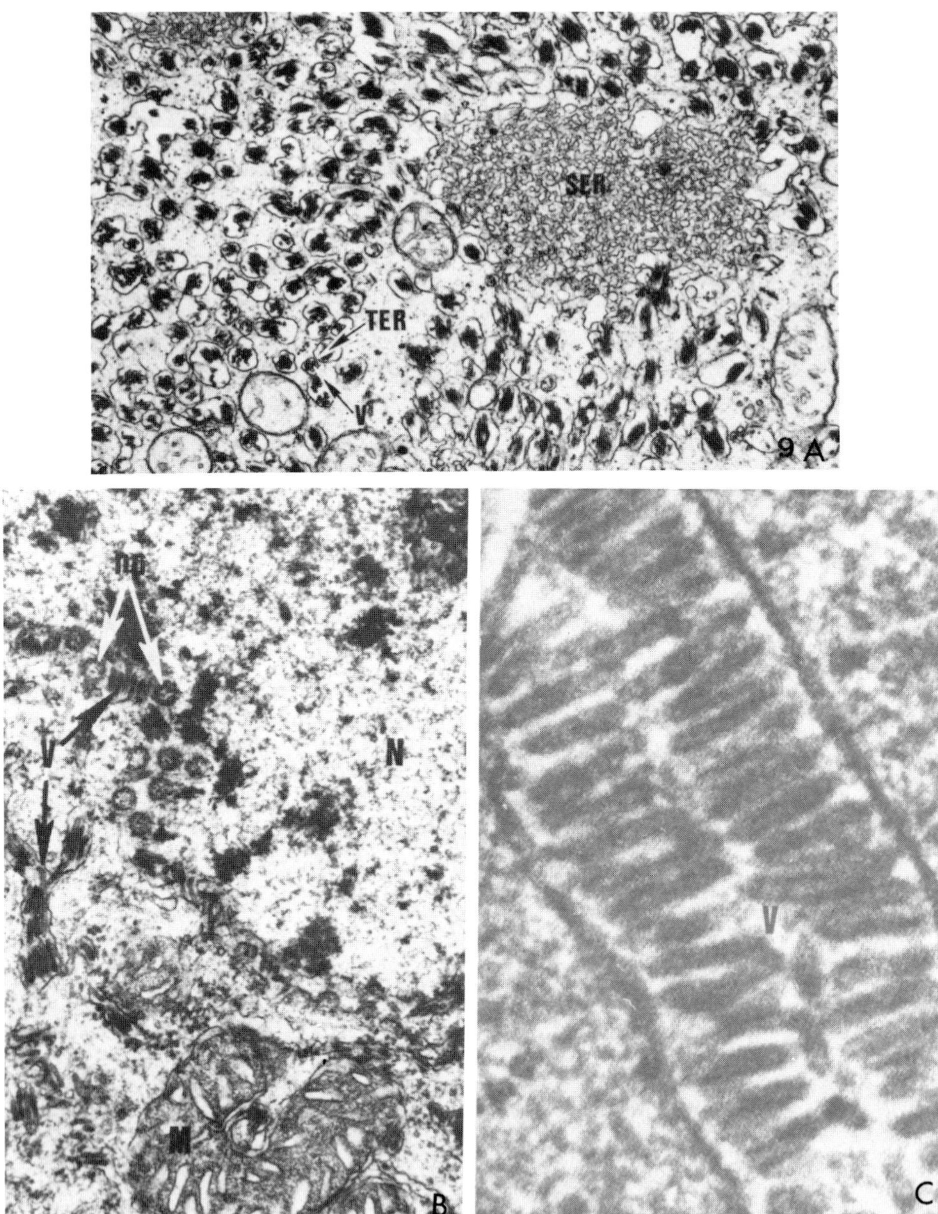

FIGURE 10.9. (A) Cells of mandibular organ of ablated *Callinectes* with virus particles (EGV-2) in all branches of the tubular smooth endoplasmic reticulum (TER) but none in the vesicular smooth endoplasmic reticulum (SER). (B) EGV-2 particles (V) in perpendicular alignment between the nuclear pores (np). *Key:* N = nucleus; M = mitochondria. (C) High magnification of virus particles (V) attached to the inner surface of the TER. A, ×28,900; B, ×32,300; C, ×119,000. (From Yudin and Clark, 1979.)

References

Aoto, T., Y. Kamiguchi, and S. Hisano. 1974. Histological and ultrastructural studies on the Y organ and the mandibular organ of the freshwater prawn, *Palaemon paucidens*, with special reference to their relation with the molting cycle. J. Fac. Sci. Hokkaido U. Ser. VI Zool. 19: 295–308.

Bazin, M. F. 1976. Mise en évidence des caractères cytologiques des glandes stéroïdogènes dans les glandes mandibulaires et les glandes Y du crab *Carcinus maenas* (L.) normal et épédonculé. C. R. Acad. Sci. Paris 282: 739–741.

Byard, E. H., R. R. Shivers, and D. E. Aiken. 1975. The mandibular organ of the lobster, *Homarus americanus*. Cell Tissue Res. 162: 13–22.

Chaudonneret, J. 1956. Le système nerveux de la gnathale de l'écrevisse *Cambarus affinis* (Say). Ann. Sci. Nat. Zool. Biol. Anim. [11] 18: 33–61.

Christensen, A. K. and S. W. Gillim. 1969. The correlation of fine structure and function in steroid secreting cell, with emphasis on those of the gonads. Pp. 415–488 *in* K. W. McKerns (ed.), *The Glands*. Appleton-Century-Crofts, New York.

Connell, P. A. M. 1970. The hormonal control of molting in the dwarf crayfish, *Cambarellus shufeldti*. Ph.D. thesis, Dept. of Biology, Tulane University, New Orleans.

Couch, E. F., C. A. Adejuwon, S. J. Segal, and S. S. Koide. 1978. Ultrastructural study and immunological evidence for progesterone production in the mandibular gland of the lobster (*Homarus americanus*). Biol. Bull. (Woods Hole) 155: 433.

Durand, J. B. 1960. Limb regeneration and endocrine activity in the crayfish. Biol. Bull. (Woods Hole) 118: 250–261.

Echalier, G. 1959. L'organe Y et le déterminisme de la croissance et de la mue chez *Carcinus maenas* (L.) Crustacé Décapode. Ann. Sci. Nat. Zool. [12] 1: 1–59.

Gabe, M. 1953. Sur l'existence, chez quelques Crustacés Malacostraces, d'un organe comparable a la glande de la mue des Insectes. C. R. Acad. Sci. Paris 237: 1111–1113.

Gabe, M. 1954. Particularités morphologiques de l'organe Y (glande de la mue) des Crustacés Malacostraces. Bull. Soc. Zool. Fr. 79: 166.

Gabe, M. 1956. Histologie comparée de la glande de mue (organe Y) des Crustacés Malacostraces. Ann. Sci. Nat. Zool. Biol. Anim. [11] 18: 145–152.

Hartnoll, R. G. 1963. The biology of Manx spider crabs. Proc. Zool. Soc. Lond. 141: 423–496.

Hinsch, G. W. 1968. Reproductive behavior in the spider crab, *Libinia emarginata* (L.). Biol. Bull. (Woods Hole) 135: 273–278.

Hinsch, G. W. 1977. Fine structural changes in the mandibular gland of
    the male spider crab, *Libinia emarginata* (L.) following eyestalk abla-
    tion. J. Morphol. 154: 307–316.
Hinsch, G. W. 1980. Effects of mandibular organ implants upon the spi-
    der crab ovary. Trans. Am. Microsc. Soc. 99: 317–322.
Hinsch, G. W., 1981. The mandibular organ of the female spider crab,
    Libinia emarginata, in immature, mature, and ovigerous crabs. J.
    Morphol. 168: 181–187.
Hinsch, G. W. and H. al Hajj. 1975. The ecdysial gland of the spider crab,
    Libinia emarginata (L.). 1. Ultrastructure of the gland in the male. J.
    Morphol. 145: 179–188.
Iwakura, C. 1970. Effects of Y organ hormone upon molting of the ter-
    restrial isopod, *Armadillidium vulgare*. Zool. Mag. (Tokyo) 79: 235–
    236.
Johnson, P. T. 1980. *Histology of the Blue Crab*, Callinectes sapidus: *a Model
    for the Decapoda*. Praeger, New York.
Keller, R. and E. Schmid. 1979. *In vitro* secretion of ecdysteroids by Y-
    organs and lack of secretion by mandibular organs of the crayfish
    following molt induction. J. Comp. Physiol. 130: 347–353.
Laufer, H., D. W. Borst, C. Carrasco, F. C. Baker, and D. A. Schooley.
    1984. The detection of juvenile hormone in Crustacea. Am. Zool. 24:
    33A.
Laufer, H., D. Borst, F. C. Baker, C. Carrasco, M. Sinkus, C. C. Reuter,
    L. W. Tsai, and D. A. Schooley. 1987. Identification of a juvenile
    hormone–like compound in a crustacean. Science (Wash., D.C.)
    235: 202–205.
Le Roux, A. 1968. Description d'organes mandibulaires nouveaux chez
    les Crustacés Décapodes. C. R. Acad. Sci. Paris 266D: 1414–1417.
Maissiat, R. 1980. Données préliminaires sur l'ultrastructure de la glande
    antennaire de *Ligia oceanica* (Crustacé Isopode Oniscoïde) et ses vari-
    ations au cours de cycle de mue. C. R. Acad. Sci. Paris 290: 1439–
    1442.
Miyawaki, M. and Y. Taketomi. 1970. Structural changes induced in the
    cells of Y gland of crayfish by an administration of ecdysterone.
    Zool. Mag. (Tokyo) 79: 150–155.
Miyawaki, M., Y. Taketomi, and E. F. Couch. 1971. The susceptibility of Y-
    gland cells of the crayfish, *Procambarus clarkii*, to exogenous steroids.
    Annot. Zool. Jpn. 44: 42–46.
Rana, S. V. S. 1975. Histological and histochemical studies on the man-
    dibular gland of fresh water crab, *Paratelphusa masoniana*. Biol.
    Zentralbl. 6: 693–701.
Rana, S. V. S., A. K. Sangal, and V. P. Agrawal. 1977. On the structure of
    the mandibular gland of fresh water crab *Paratelphusa lugubris* with

special reference to seasonal variation. J. Anim. Morphol. Physiol. 24: 251–254.

Sochasky, I., D. E. Aiken, and N. H. F. Watson. 1972. Y-organ, molting gland, and mandibular organ: a problem in decapod crustacea. Can. J. Zool. 50: 993–997.

Yudin, A. I. and W. H. Clark, Jr. 1979. A description of rhabdovirus-like particles in the mandibular gland of the blue crab, *Callinectes sapidus*. J. Invertebr. Pathol. 33: 133–147.

Yudin, A. I., R. A. Diener, W. H. Clark, Jr., and E. S. Chang. 1980. Mandibular gland of the blue crab, *Callinectes sapidus:* its structure and possible hormonal activity. Biol. Bull. (Woods Hole) 159: 760–772.

Part II Taxonomic Index

Part II Subject Index

Part 3
Roles in Histogenesis, Organogenesis, and Morphogenesis

Roles in Embryogenesis

Roles in Postembryonic Events

Contents